The European Discovery of the Indian Flora

The European Discovery of the Indian Flora

Ray Desmond

ROYAL BOTANIC GARDENS
OXFORD UNIVERSITY PRESS
1992

Oxford University Press, Walton Street, Oxford OX2 6DP
Oxford New York Toronto
Delhi Bombay Calcutta Madras Karachi
Petaling Jaya Singapore Hong Kong Tokyo
Nairobi Dar es Salaam Cape Town
Melbourne Auckland
and associated companies in
Berlin Ibadan

Oxford is a trade mark of Oxford University Press

Published in the United States
by Oxford University Press, New York

A catalogue record for this book is available from the British Library

Library of Congress Cataloging in Publication Data
Desmond, Ray.
The European discovery of the Indian flora / Ray Desmond.
p. cm.
1. Botany—India—History. 2. Gardens, English—Indic influences.
3. Gardens, European—Indic influences. I. Title.
QK358.D44 1992 581.954—dc20 91–44110

ISBN 0–19–854684–X

Set by Pure Tech Corporation, India
Printed in Hong Kong.

PREFACE

The underlying theme of this book is Europe's reaction to India's vegetation, its attempts to classify and name this exotic flora and its subsequent utilization. Although only European participation is examined in depth, passing reference is made to the not inconsiderable contributions of Indians themselves to the advancement of botanical research in their country. Their intimate knowledge of innumerable plants, especially those used in traditional medicine, was recognized by Van Reede, Royle, and other investigators; their artistic skills were frequently employed to make plant portraits in a western style indispensable for record and for study.

By 'India' is meant the whole subcontinent, including the neighbouring countries of Nepal, Sikkim, and Bhutan and the island of Sri Lanka (formerly Ceylon) in the south.

Although the profit motive guided the policies of the East India Company and other trading companies, it is not true, as some historians insist, that all the decisions and actions of the Court of Directors in London were influenced solely by commercial considerations. They founded and maintained, for instance, an unremunerative library and museum in Leadenhall Street in the City, were generous patrons of scholarly publications, and seldom opposed their officials pursuing oriental studies. India as well as British shareholders benefited from the intensification of crop production promoted by the Company. The introduction of European horticultural practices improved crop yields, new uses were found for native plants, and the resources of the country were diversified by the cultivation of crops from abroad, especially from the Americas. Sir Clements Markham sincerely believed that the cinchona plantations created by the British 'will be one of the most enduring monuments of our rule'. A brief resumé is given in Chapters 17 and 18 of the introduction of a few of the economically important crops such as cinchona, rubber, and tea.

Native herbal prescriptions attracted the attention of European physicians and naturalists. Garcia da Orta was one of the first to write about them; Van Reede enlisted the help of Indian doctors and scholars in recording indigenous medicines; Johann Koenig, the first professional botanist in India, diligently sought new drug plants; Sir William Jones encouraged the study of ancient Sanskrit manuscripts in order to learn more about plants of medicinal value; the botanical gardens at Calcutta and Saharanpur grew them experimentally; and nineteenth-century pharmacopoeias acknowledged their indebtedness to Indian remedies.

The collections and observations made by the amateur naturalist have always been a significant contribution to field studies and this is particularly true of India, where there were comparatively few professionals until well into the nineteenth century. The Asiatic Society, founded in Calcutta in 1784, offered direction and purpose to the amateur's research and provided an appreciative audience for the solitary field botanist. By the end of the eighteenth century, the casual and curious observer of natural history—traveller, trader, diplomat, or administrator—was supported and motivated by a corps of surgeons who had studied botany during their university days. Many of these surgeons in the Company's service had been trained at Edinburgh, Glasgow, and Aberdeen. It is therefore not surprising that nine of the superintendents of the Calcutta Botanic Garden (including those in an acting capacity) had been born and educated in Scotland. Botanists recruited from the Company's Medical Service in India included Roxburgh, Buchanan, Mitchell, Hyne, Shuter, Griffith, and Wight. Only a selection of all these botanists and plant collectors is considered, since a comprehensive survey would have impeded the narrative with a density of names.

Corporate bodies and institutions as well as individuals played a constructive role in floristic research. The Royal Society of Arts in London awarded its gold medal to William Roxburgh in 1805 and again in 1814 for his communications on vegetable products; other pioneers of India's economic crops received its silver medal. Briefly during the 1830s the Royal Asiatic Society, also in London, had a subcommittee steered by Royle to advise it on Indian products. In 1899 a Committee of the Royal Society for advising on scientific research in India was formed at the behest of the India Office, although little use appears to have been made of its services. The most influential institution was undoubtedly the Royal Botanic Gardens at Kew.

The earliest botanical gardens in Europe were physic gardens with representative collections of medicinal plants acquired and grown for drugs or for demonstrating to medical students. By the eighteenth century they had outgrown their physic garden origins to become custodians of novel and rare plants. One of the first tropical botanical gardens was established by the Dutch at Cape Town in the seventeenth century; the French acclimatized economic plants at Pamplemousses in Mauritius from 1735; 30 years later the botanical garden at St Vincent was created to provide British planters in the West Indies with a nucleus of crops for cultivation. All tropical botanical gardens shared a common goal: the introduction, cultivation, and distribution of plants primarily of commercial value. Such a role was proposed by Colonel Kyd for the garden at Calcutta, which became the first botanical garden in the subcontinent. It was always Sir Joseph Banks's aim to link Britain's colonial gardens by encouraging the transfer of plants between them. John Lindley's 1838 report on the moribund royal gardens at Kew sought to preserve this international traffic in plants by proposing a pivotal function for a reconstituted Kew 'round which all minor establishments of the same nature should be arranged . . . receiving their supplies and aiding the Mother country in everything that is useful in the vegetable kingdom'. The Royal Botanic Gardens at Kew under the direction of the Hooker dynasty energetically prosecuted that recommendation. Sir Joseph Hooker's son-in-law, Sir William Thiselton-Dyer, saw Kew as 'a sort of botanical clearing house or exchange for the empire'. Kew influenced the progress of Indian botany and horticulture through professional advice and help, the provision of trained personnel, and its active participation in plant introduction. The still unsuperseded *Flora of British India* was compiled at Kew, whose official initiative also resulted in the formation of the Botanical Survey of India.

Notwithstanding the evidence that during the first 50 years of the existence of the *Journal of the Asiatic Society of Bengal* only 80 botanical papers were published compared with 560 in zoology and 320 in geology, botanical investigation in India, especially its economic aspects, usually received preferential treatment. Calcutta's botanical garden preceded Calcutta's zoo by almost a century. At the beginning of the nineteenth century Lord Wellesley tried to give a boost to zoology by establishing a menagerie, unfortunately short-lived, at his country house at Barrackpore. During the early 1840s the geologist E. I. Murchison complained that geology and surveying in India had not received anything like the generous support accorded to botany. When the Government of India expressed reservations in 1889 about approving the formation of the Botanical Survey of India it was because, in their opinion, botany had already enjoyed greater patronage and expenditure than other equally deserving scientific disciplines. This point of view would have been disputed by Sir Joseph Hooker who regretted 'the present deplorably backward state of scientific Indian botany on the one hand, and the confusion of Indian economic botany on the other'. The truth lies somewhere between these two opposing interpretations of the situation.

The alarming prophecies of some Company officials regarding wholesale deforestation in India were referred by the Court of Directors to its newly formed Statistical Department. In July 1847 this department drew the attention of the Governor-General of India to 'the effect of trees upon the climate and the results of extensive clearances of timber', cogently making the point that 'the subject is one having a strong practical bearing on the welfare of mankind'. A few years later the Political Superintendent of Upper Sind began a replanting programme in an attempt to ameliorate local climate and rainfall. Hugh Francis C. Cleghorn, a surgeon in Mysore dedicated to the cause of forest conservation, became secretary of a committee appointed by the British Association for the Advancement of Science in 1850 'to consider the probable effects in an economic and physical point of view of the destruction of tropical forests'. It was also in 1850 that the Governor-General of India, Lord Dalhousie, expressed official fears of the consequences of deforestation to an audience of the Agri-Horticultural Society of India at Lahore.

> *During the last season . . . I have traversed the plains of the Punjab . . . there is one characteristic of the wide tract which could not fail to strike the least observant traveller . . . I refer to the almost total absence of forest trees and even of fruit trees and of bushes leaving the whole territory unadorned by the foliage which is its natural cover, nor stocked with timber requisite for a thousand purposes in the every-day life of the people who dwell in it . . . I feel strongly the urgent duty of endeavouring to give this country a clothing of forest trees from my knowledge of the well-ascertained and beneficial effect which trees produce on the health and fertility of the tracts in which they are found.*

India's officials and botanists were in the vanguard of a conservation movement that had its beginnings in the nineteenth century.

None of the research for this book could have been undertaken without access to the resources of the British Library and in particular its departmental library, the India Office Library and Records, and also the libraries of the Royal Botanic Gardens at Kew, the Natural History Museum, the Linnean Society of London, the Royal Horticultural

Society, the Royal Asiatic Society, the Royal Commonwealth Society, and the Royal Society of Arts. I am indebted to Alan Radcliffe-Smith, Professor W.T. Stearn, and Dr David Mabberley, who patiently read the manuscript, and to colleagues at Kew who checked the accuracy of botanical names. Most of the illustrations have come from the library at Kew; those which have been selected from the British Library, the Victoria and Albert Museum, and the Royal Horticultural Society are acknowledged in the relevant captions. I am also grateful to Pauline Rohatgi, John Falconer, and Norman Track for allowing me to copy illustrative material in their possession.

Kew
December 1990

R. D.

Abbreviations of libraries and collections mentioned in the footnotes

B.L.	The British Library (mainly the Department of Western Manuscripts)
B.M.(N.H.)	Natural History Museum, South Kensington
D.T.C.	Dawson Turner copies of the correspondence of Sir Joseph Banks in the Botany Library, Natural History Museum
I.O.L.R.	India Office Library and Records
L.S.	Linnean Society of London
S.R.O.	Scottish Record Office

CONTENTS

	List of plates	xi
1	The ancient world	1
2	The first European settlements	12
3	Arrival of the British	29
4	Botanical research begins	39
5	Sir William Jones	52
6	The Roxburgh era	61
7	Francis Buchanan	71
8	Calcutta Botanic Garden after Roxburgh	81
9	Other botanical gardens	106
10	William Griffith and Robert Wight	114
11	The Himalayas	123
12	Ceylon	154
13	Pursuit of the picturesque	173
14	A century of change	185
15	The search for useful plants	201
16	Green medicine	220
17	Tea and opium	231
18	Rubber, coffee, and forestry	245
19	Gardening in India	258
20	Indian flowers in European gardens	287
21	Transportation of plants	309
22	Epilogue	323
23	Bibliography	331
	Index	341

PLATES

The plate sections appears between pages 180 and 181

1 *Platanus orientalis*. Mughal miniature, *c*.1610.
2 *Luffa acutangula*. Hand-coloured engraving, *c*.1776.
3 *Coryphea taliera*. Hand-coloured engraving, *c*.1820.
4 *Caesalpinia sappan*. Hand-coloured engraving, *c*.1795.
5 *Erythrina stricta*. Watercolour.
6 *Gloriosa superba*. Stipple engraving, *c*.1802.
7 *Abutilon auritum*. Watercolour.
8 *Picea smithiana*. Watercolour, *c*.1832.
9 *Bergenia ciliata*. Watercolour.
10 *Hedychium gardnerianum*. Lithograph, *c*.1828.
11 *Vanda tessellata*. Hand-coloured engraving, *c*.1820.
12 *Arisaema speciosum*. Watercolour.
13 *Anemone vitifolia*. Hand-coloured engraving, *c*.1835.
14 *Clematis montana*. Hand-coloured lithograph, *c*.1831.
15 *Rosa clinophylla*. Stipple engraving, *c*.1817.
16 *Rosa macrophylla*. Watercolour.
17 *Hodgsonia heteroclita*. Lithograph, *c*.1855.
18 *Meconopsis simplicifolia*. Watercolour.
19 *Meconopsis simplicifolia*. Lithograph, *c*.1855.
20 *Rubus wardii*. Drawing.
21 *Cedrus deodara*. Lithograph, *c*.1884.
22 *Cinchona succirubra*. Lithograph, *c*.1862.
23 Bhairamghattee in the Lower Himalayas. Coloured aquatint, *c*.1820.
24 *Thespesia populnea*. Oil painting, *c*.1877.
25 *Primula sikkimensis*. Lithograph, *c*.1851.
26 *Primula denticulata*. Watercolour.
27 *Begonia rex*. Lithograph, 1859.
28 *Lilium nepalense*. Lithograph, *c*.1933.
29 (a) Judge's house and lawns at Dinajpur, Bengal. Watercolour, *c*.1800. (b) House in Calcutta showing water tank as a garden feature. Watercolour, *c*.1845.
30 (a) House and garden of Thomas Graham in Chowringhee, Calcutta. Coloured aquatint, 1809. (b) Garden of the Queen's Pavilion, Kandy, Ceylon. Watercolour.
31 (a) Garden designed by Sir Edwin Lutyens, adjoining the Viceroy's home in New

Delhi. Photograph. (b) Garden at the Ram Bagh Hotel, Jaipur. Photograph.
32 (a) Hindu temple at Sezincote, Gloucestershire. Photograph. (b) Charbagh-designed garden at Sezincote. Photograph.

THE ANCIENT WORLD

From the time of the Greek geographer, Hecateus of Miletus, about 500 BC, until the Portuguese discovered the sea route around the Cape of Good Hope and across the Indian Ocean towards the close of the fifteenth century, European scholars knew India only as a land of legend and fantasy. Writers of the classical world perpetuated such myths as the sun appearing larger there than elsewhere in the world, that it was home to gold-digging ants and to barbarous tribes contorted by bizarre deformities such as men with canine heads or spider legs or with fingers that bent backwards with supple ease. India was loosely synonymous with Asia in general, and as late as the seventeenth century a writer explained to his readers that 'by the name of India . . . we comprehend all the tract between India . . . on the west . . . unto China eastwards'. Exotic plants with an eastern provenance were often indiscriminately called 'Indian'.

The historical gossip about India emanated from commercial links, especially the spice trade, and from the stories of soldiers, missionaries, and solitary adventurers. About 515 BC, Darius, ruler of Persia, extended his empire by annexing the rich Indus valley. He despatched an expedition commanded by Scylax of Caryanda, a Greek officer, to explore the entire length of the River Indus. Scylax's fabulous tales of obscure tribes—of people, for example, endowed with enormous feet which they could conveniently adapt as sunshades—projected a landscape of India that was generally accepted by the credulous West. Herodotus (*c*.484–425 BC) never disputed such reports about a country he believed to be on the extremity of the inhabited world. His *Histories* which solemnly notes the existence of the gold-digging ants was the first classical work to speculate about the source of cinnamon. He also provides an early reference, possibly the first, to cotton: 'there are trees growing wild which produce a kind of wool better than sheep's wool in beauty and quality, which the Indians use for making their clothes'[1].

Ctesias of Cnidus attempted to correct Herodotus in *Indica*, written about 400 BC, the first separate treatise on India. As court physician in Persia for seventeen years he was in a privileged position to learn much about India from Persian officials who had visited the country. But his uncritical reportage simply added to the repertoire of Indian legends. Although his accounts border on the fanciful and absurd, it is nevertheless possible to hazard a guess at the identity of some of the plants he describes. 'Parebon, a plant about the size of the olive found only in the royal gardens, producing neither flower nor fruit, but having merely fifteen roots, which grow down into the earth, and are of considerable thickness, the very slenderest being about as thick as one's arm'[2], was identified by V. Ball[3] as *Ficus religiosa*, partly because its fruit are so small as to be inconspicuous. The 'purple flower which is used for dyeing purple, and is not inferior to the Greek sort'[4] could be a species of *Woodfordia*. Some descriptions defy identification like the trees 'as tall as the cedar or the cypress, having leaves like those of the date palm, only somewhat broader, but having no shoots sprouting from the stems'[5]. They yielded an oil with an exquisite perfume, 'the sweetest in the world'. Ball concluded that Ctesias had probably confused some species of *Laurus* with the screw-pine (*Pandanus tectorius*).

The Persian empire, whose hospitality Ctesias enjoyed, was conquered by Alexander of Macedon (356–323 BC) when he defeated Darius III in 330 BC. Alexander was not only well trained in military matters but from his tutor, Aristotle, had acquired a compelling curiosity about the natural world; consequently his officers were encouraged to report on the fauna and flora of the countries his army invaded. When Alexander marched east in 334 BC his knowledge of Asia was limited to Herodotus's writings and the advice of travellers. Darius's superior forces which included a small contingent of Indian troops and their elephants were annihilated by Alexander, and Darius himself murdered by defectors from his defeated army. With Persia absorbed into an expanding Greek empire, Alexander, with a diminished but still loyal army, crossed the Hindu Kush, crushed the tribes opposing him in Afghanistan, and reached the banks of the broad Indus river, beyond which lay India with its promise of untold wealth. In 326 BC he penetrated the north Indian plains, defeated Porus, King of the Jhelum region, and in less than two years subdued the Punjab as far as the Sutlej. When his exhausted troops refused to continue the march east, Alexander had a fleet of ships constructed from the trunks of Himalayan pine and deodar to sail down the Indus to the ocean, returning to Babylon via the Persian Gulf.

Contemporary accounts of Alexander's campaigns survive only in edited and fragmentary versions by later Greek historians. Onesicritus, who piloted Alexander's fleet down the Indus, provided a sceptical Strabo with a description of banyan trees of such girth that five men were needed to embrace their trunks, and a single tree alone could shelter 400 men. Strabo learnt from the same source that cinnamon and other spices grew in southern India. Strabo was also acquainted with the reports of Alexander's comrade in arms, Aristobulus, who recorded a tree 'not large, with pods, like the bean, ten fingers in length, full of honey . . . those who eat it cannot easily be saved from death'[6]. Alexander issued an order that his troops were to avoid it. Ball hazarded a guess that it was purging cassia, *Cassia fistula*, whose pods contain a sweet pulp traditionally used as a laxative[7].

The *Enquiry into plants* of the Greek philospher Theophrastus (370–*c*. 288 BC), sometimes exalted as 'the father of botany', incorporated accounts by Greek officers of plants beyond the eastern Mediterranean such as myrrh and frankincense, spices, and cotton. As these soldiers' tales were often his only source of information, his descriptions lack the clarity and precision of personal knowledge. A 'tree which is very large and has wonderfully sweet and large fruit'[8] could be the jack-fruit (*Artocarpus heterophyllus*) and perhaps the banana is meant by another 'whose leaf is oblong in shape, like the feathers of the ostrich'. In Bactria, the Greek colony in India, he is told there grows a 'sort of terebinth' with nuts as big as but sweeter than almonds—could this be the pistachio nut (*Pistacia vera*)? Cotton bushes 'have a leaf like the mulberry, but the whole tree resembles the wild rose'[9]. Bamboos also make their appearance.

The Indian reed is very distinct and as it were of a wholly different kind; the 'male' is solid [Dendrocalamus strictus] and the 'female' hollow [Bambusa bambos] . . . a number of this kind grow from one base and they do not form a bush, the leaf is not long, but resembles the willow leaf, these reeds are of great size and good substance, so that they are used for javelins[10].

The tree that attracted most attention and exaggerated comment was the banyan (*Ficus benghalensis*) (Fig. 1.1), so unlike any other the Greek invaders knew. Theophrastus duly described it.

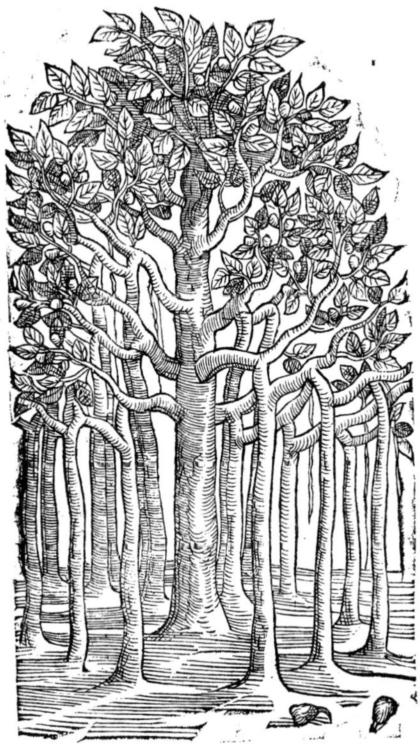

Figure 1.1 Banyan tree, *Ficus benghalensis*. (From J. Gerard (1597), *The herball or general historie of plantes*, London, p. 1331. Woodcut. 'This wonderous tree groweth in divers places of the east Indies, especially neere unto Goa, and also in Malacca, it is a stranger in the most parts of the world' (Gerard).

The Indian land has its so-called 'fig-tree' which drops its roots from its branches every year, . . . and it drops them, not from the new branches, but from those of last year or even from older ones; these take hold of the earth and make, as it were, a fence about the tree, so that it becomes like a tent in which men sometimes live. The roots as they grow are easily distinguished from the branches, being whiter, hairy, crooked, and leafless. The foliage above is also abundant, and the whole tree is round and exceedingly large. They say that it extends its shade for as much as two furlongs; and the thickness of the stem is in some instances more than sixty paces while many specimens are as much as forty paces through. The leaf is quite as large as a shield, but the fruit is very small, only as large as a chick-pea, and it resembles a fig. And this is why the Greeks named this tree a 'fig-tree'. The fruit is curiously scanty, not only relatively to the size of the tree, but absolutely[11].

This was a remarkably accurate description considering it was based on a soldier's report. Hindus planted this sacred tree along the verges of roads and in villages, providing shelter for people and a community of small mammals, birds, and insects. Of all India's trees it was the one that immediately attracted the attention of travellers. The hanging columns of its aerial roots reminiscent of the tenuous lines of Gothic tracery presented a dynamic composition for European artists like William Hodges and the Daniells. Bishop Heber experienced an aura of sanctity within its canopy: 'what a noble place of worship', he confided to his journal.

After Alexander's death in 323 BC, his generals fought among themselves for possession of his Asian conquests. Seleucus Nikator, having seized Babylon, resumed Alexander's Indian campaigns by rashly attacking Chandragupta Maurya's formidable empire which stretched from Afghanistan across the North Indian plains to Bengal in the east. The Indians proved too powerful, a treaty was concluded, and Seleucus sent Megasthenes as his ambassador to the Maurya court at Pataliputra near present-day Patna.

Megasthenes, the first Greek to reach the valley of the Ganges, remembered much of what he saw and heard. He was impressed by agricultural methods which could yield two grain harvests a year and admired the royal gardens with their tame pheasants and peacocks. He was one of the earliest travellers to mention Ceylon, 'a very large island, of which, as I hear, the name is Taprobane'. Gradually the geographical boundaries of India became more clearly defined, and its physical features in the north brought into sharper focus. India was introduced to a wider audience in the classical world through the chronicles of historians like Strabo, Pliny the Elder, and Arrian.

The *Geography* of Strabo (*c.*63 BC–AD 21), declared Humboldt, the distinguished German geographer and scientist, 'surpasses all the geographical writings of antiquity both in grandeur of plan, and in the abundance and variety of its materials'. Its author was born at Amasia in Pontus, enjoyed a generous patrimony, and passed much of his life in Rome and Egypt leisurely collecting material for his encyclopaedic account of the ancient world. Although Strabo was dependent on travellers' reports he assessed them with a healthy scepticism.

But it is necessary for us to hear accounts of this country [i.e. India] with indulgence, for not only is it farthest away from us, but not many of our people have seen it; and even those who have seen it, have seen only parts of it, and the greater part of what they say is from hearsay: and even what they saw they learned on a hasty passage with an army through the country. Wherefore they do not give out the same accounts of the same things, even though they have written these accounts as though their statements had been carefully confirmed. And some of them were both on the same expedition together and made their sojourns together, like those who helped Alexander to subdue Asia; yet they all frequently contradict one another. But if they differ about what was seen, what must we think of what they report from hearsay? Moreover, most of those who have written anything about this region in much later times, and those who sail there at the present time, do not present any accurate information either[12].

With frequent caveats, he proceeds to quote from these unreliable authorities; Eratosthenes, for instance, who, like Megasthenes, was impressed by the fertility of the soil which produced summer and winter crops and an abundance of fruit. He quotes Nearchos on reeds that yielded honey without bees, presumably sugar cane. Onesicritus is his witness for the banyan tree but he doubted his claims that it would shelter 400 horsemen, preferring Aristobulus's more modest figure of 50. 'But the accounts of all writers of the size of the trees have been surpassed by those who say that there has been seen beyond the Hyarotis a tree which casts a shade at noon of five stadia' (about 1100 metres)[13]. He mentions cotton, 'a self-grown grain, similar to wheat', and 'cinnamon, nard, and other products' of the south. India was, all writers concurred, a land 'abounding in herbs and roots both curative and poisonous, and likewise in plants of many colours'[14].

We learn from Strabo that after the death of Cleopatra in 30 BC and the annexation of Egypt by Rome, that many ships sailed to India from Myos Hormos, a major port in Egypt. The Red Sea had always been one of the principal maritime routes to eastern Asia. Indian merchants had occasionally visited Alexandria, an important entrepôt between East and West, as early as the second century BC. Some of the Indian plants and products mentioned in the Bible—bdellium, ebony, almug or algum, aloe-wood, aromatic grasses—came to Israel either through Egypt or overland by trade routes through Arabia, Mesopotamia, and Assyria. After Alexander's invasion of India, the volume of Indian merchandise reaching the Mediterranean region substantially increased, especially ivory, tortoiseshell, muslin, and spices. Theophrastus notes that 'as to all the other fragrant plants used for aromatic odours, they come partly from India whence they are sent over sea'[15].

For centuries, mariners had cautiously hugged the coast of Arabia, crossing the Arabian Sea to the Indian subcontinent. At some time towards the end of the millenium or in the early years of the first century AD the south-west monsoon was discovered, dramatically changing the pattern of traditional trade routes. Stoutly built ships of Greek merchants could now sail from Egypt in July, be blown by the winds to India by the end of September, and return in November on the north-east monsoon to the Red Sea, reaching Myos Hormos by the following February. The direction of the monsoon took ships farther south, to the Malabar ports of Barake, Nelkunde, Tyndis, and the capital Muziris (now Cranganore), called by Pliny the Elder 'the nearest market in India'.

The *Periplus of the Erythraean Sea* (Circuit of the Indian Ocean)[16] was a practical manual, written by a Greek merchant about the middle of the first century AD, on the navigation of the Indian Ocean with a gazetteer of the ports of the Red Sea, Somalia, southern Arabia, the Persian Gulf, and India's western coast. This comprehensive guide tells us that Africa imported Indian rice, ghee, oil of sesamum, sugar, and cotton; that Barygaza in the north of India exported logs of ebony, sugar, spikenard, and bdellium brought from the interior; and that Nelkunde was one of the principal pepper ports in the south.

During the Julio-Claudian and Flavian dynasties Roman trade with India flourished, the more direct monsoon route effecting a considerable fall in the price of Indian products. Roman gold and silver coins spanning issues of more than five centuries—the reigns of Augustus, Tiberius, Gaius, and Claudius predominating—have been found from Afghanistan to Ceylon. The greatest quantity have been excavated along the Malabar coast where highly prized black pepper was cultivated. A contempory Tamil poet wrote of 'the beautifully built ships of the Yavanas [the Greek traders] which come with gold and return with pepper'. In AD 92, Domitian built bigger pepper warehouses in the spice quarter of Rome. Pliny complained that the demand by the Roman aristocracy for Indian luxuries—spices, textiles, jewels, exotic animals, and birds—was proving a serious drain on the Roman exchequer. Indian merchants were paid mainly in gold and the coveted red coral of the Mediterranean. In Greece and Rome, spices were a familiar constituent of drugs, perfumes, and cosmetics. Hippocrates recommended the long pepper of northern India for its medicinal virtues. Cinnamon and spikenard perfumed fashionable Roman funerals; Nero used spices extravagantly in the embalming of Poppaea. Roman dishes were seasoned with spices, especially black pepper. M. Gavius Apicius, a notable Roman epicure, devised exotic recipes for spiced flamingo, ostrich, and parrot. Pliny was puzzled by this craving for pepper which he thought 'has nothing to recommend it in either fruit or berry. To think that its only pleasing quality is pungency and that we go all the way to India to get this. Who was the first person who was willing to try it on his food, or in his greed for an appetite was not content merely to be hungry?'[17].

Gaius Plinius Secundus (AD 23–79), Pliny the Elder, was a workaholic who regulated his life so that every precious moment was available for study. For his encyclopaedic *Natural history* he consulted no fewer than 2000 works by a hundred Greek and Roman authors—Herodotus, Ctesias, Theophrastus, and Strabo among them. Sixteen of its 37 books, covering every possible aspect of natural history, were devoted to plants. As it is a digest of information, not sifted or evaluated for accuracy, it consequently perpetuates many misconceptions and fallacies. Isaac Disraeli damned it as 'the awful repository of all the errors of antiquity'. But Pliny never claimed original thought; he was, moreover, meticulous in his citation of sources and his compilation is remarkably free from personal prejudice.

In considering the vegetation observed by Alexander's troops, Pliny takes the banyan tree first, selecting Theophratus's description. 'The leaf is quite as large as a shield,' wrote Theophrastus, 'but the fruit is very small, only as large as a chick-pea.' Pliny's version presents the 'broad leaves . . . the shape of an Amazon's shield; consequently they cover over the fruit and prevent it from growing, and it is scanty and not larger in size than a bean'[18]. As Professor W. T. Stearn has pointed out, the size of the leaf was exaggerated; banyan leaves are smaller than the smallest Greek shield[19]. Again with Theophrastus as his authority, Pliny states the banana to be the food of Indian sages and that one bunch could feed four people. Its leaf 'resembles birds' wings, and is a yard and a half long and a yard broad'[20]. There are also fleeting references in Pliny's text to ebony, the terebinth tree, the olive, sesame, rice, and indigo, but he expounds at length on spices and aromatics.

He distinguishes three kinds of pepper: long, white, and black. Long pepper (*Piper longum*) is a shrub found in northern India and the hills in the south; black pepper (*Piper nigrum*) is a climbing plant in the forests of Malabar and Travancore. Black pepper, declared Pliny, 'is more agreeable, but white pepper is of a milder flavour than either the black or the long pepper'[21]. He said there was an Indian grain resembling pepper 'but larger and more brittle, called the carophyllon which is reported to grow on the Indian lotus-tree'[22]. This is the clove, *Syzygium aromaticum*, probably brought to India from the Moluccas by Arab traders.

Bdellium (*Commiphora mukul*), often mentioned in ancient Sanskrit texts, is a spiny tree, Theophrastus's Indian thorn bush, exuding tears of a resin used medicinally and in incense and perfumes. 'It ought to be transparent like wax, to have a bitter taste, though without acidity', cautioned Pliny[23].

Pliny approved of the 'exquisite scent' of Costus (*Saussurea costus*), found high up in the Kashmir Valley and in the Punjab. Theophrastus likewise rated it highly as a perfume; it was also an ingredient of medicines, a condiment, and a fruit preservative.

Amomum (*Amomum subulatum*), 'a twisted shrub a hand high', reported Pliny, was purchased for the medicinal properties of its oil and as a cheap and inferior alternative to cardamom (*Elettaria cardamomum*), the true spice. (Fig. 1.2) Both Pliny and Dioscorides asserted that cardamom was an Arabian plant, whereas, in fact, it grows in West and South India and amomum, which all classical authorities from Aristotle to Galen associated with northern India, is more specifically located in Nepal and Sikkim. Another native of the Himalayas from Uttar Pradesh to South West China is *Nardostachys grandiflora*, the spikenard of the ancients. The plant which has a most distinctive odour, 'a musty smell like the gladiolus', wrote Pliny, was greatly esteemed as a perfume and drug.

Cinnamon oil was almost as highly rated as spikenard. Cinnamon and cassia were always grouped together by classical authors. Pliny ridiculed Herodotus's belief that they were obtained from birds' nests and dismissed another legend that cassia grew around marshes, protected by clawed bats and winged serpents, 'these tales having been invented by the natives to raise the price of their commodities'[24]. Ethiopia was believed to be the provenance of these plants but

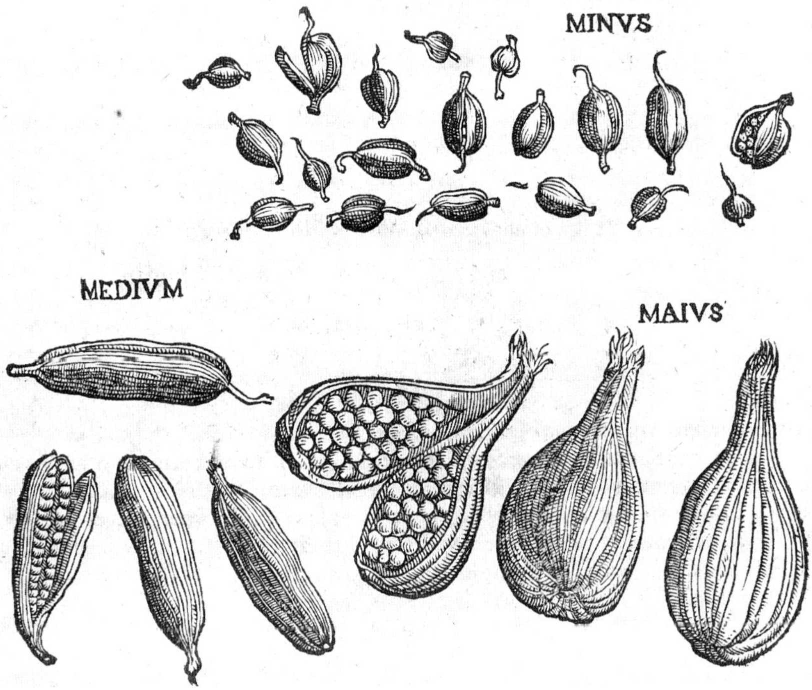

Figure 1.2 Cardamom, *Elettaria cardamomum.* (From P. Mattioli (1563), *New Kreuterbuch,* Prague, p. 220.) Woodcut.

Pliny interestingly reported that the inhabitants there obtained them from people whose ships were 'neither steered by rudders nor propelled by oars or drawn by sails, nor assisted by any device of art'[25]. The proposition has been advanced that these vessels could have been double outrigger canoes which made the long perilous voyage across the Indian Ocean to Madagascar[26]. *Cinnamomum macrophyllum* and *Cinnamomum aromaticum*, Pliny's cinnamon and cassia which came from South East Asia and southern China, respectively, indicate the geographical extent of the Roman spice trade. Small quantities of Indian cinnamon (Fig. 1.3) from the Himalayas and Ceylon may have been shipped from ports on the Malabar coast but the commercial exploitation of *Cinnamomum verum* in Ceylon did not really get under way until the arrival of the Europeans in the early sixteenth century.

When Rome was sacked by Alaric the Visigoth in AD 400, spices were still highly prized and his ransom included 3000 pounds of pepper. With the decline of Roman might the monopoly of the spice trade passed to Constantinople and the Byzantine Empire, and by the tenth century merchants of Venice and Genoa controlled this lucrative market.

After the fall of Rome, when sea communications with India declined and Muslim power dominated the Middle East, it was the legendary aspects of India that fired the imagination of writers in the Middle Ages. A few travellers like the Alexandrian monk, Cosmas Indicopleutes, reached the subcontinent. His *Universal Christian topography*, written about the middle of the sixth century, described several animals and a couple of plants he observed in Ceylon: black pepper twining itself around a tree, its fruit protected by a double leaf and 'a deep green colour like that of rue'[27]; and the coconut (*Cocos nucifera*). 'If the fruit is gathered ripe and kept, then the water gradually turns solid on the shell, while the water left in the middle remains fluid, until of it also there is nothing left over.'[28]

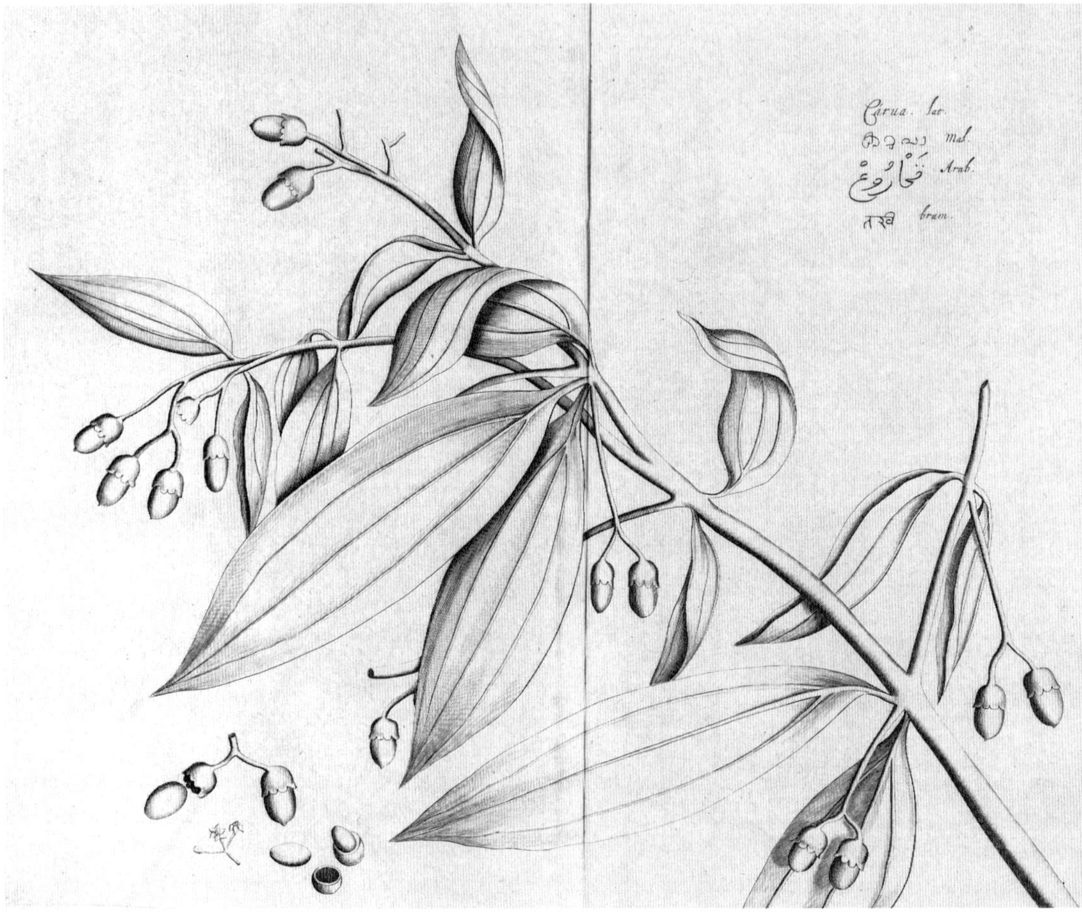

Figure 1.3 Cinnamon, *Cinnamomum verum*. (From Van Reede (1678), *Hortus Malabaricus*, Vol. 1, Amsterdam, fig. 57.) Engraving.

According to the *Anglo-Saxon Chronicle*, Sighelmus in AD 883 visited the shrine of St Thomas who, it was believed, had established Christian communities in India. Unfortunately, Sighelmus, who returned with 'costly spices, such as that country plentifully yieldeth', left no account of his travels.

In 1159 or 1160 the Spanish Jew, Benjamin of Tudela, began a journey which lasted some thirteen or fourteen years. During his travels he saw fields in India planted with pepper and noted that 'cinnamon, ginger, and many other kinds of spices also grow in this country'[29].

Marco Polo (1254–1325) together with his father and his uncle were the first Europeans to travel across Asia. They left Venice in 1271, arriving in China over three years later. After seventeen years in the court of Kublai Khan, Marco Polo returned home by South East Asia and India which he reached about 1292. He examined plants with the commercial acuity of a merchant. On the coast of Coromandel he notes the fine quality of 'brazil coilumin' (*Caesalpinia sappan*), approves the ginger, marvels at the abundance of pepper and succinctly describes the fermentation process in the manufacture of indigo. It seemed to him that Indians enjoyed an idyllic life, blessed with a cheap abundance of basic necessities; toddy was a capital drink and both men and women went naked 'save for a fine cloth worn about the middle'. Wistfully perhaps, he adds: 'They look not on any sin of the flesh as a sin.'[30] In Gujerat, where he left India, he was impressed by the size and age of cotton trees (presumably *Gossypium arboreum*): 'up to the age of 12 years indeed the trees give good spinning cotton, but from that age to 20 years the produce is inferior'[31]. His narrative was not taken seriously by his contemporaries, who regarded it as little more than an interesting collection of curious tales.

Friar John of Montecorvino was in South India making Christian converts about the same time as Marco Polo. His letters home contain terse insights on the people and the countryside. Indian nuts—the name mediaeval writers gave to the coconut—were as 'big as melons, and in colour green, like gourds. Their leaves and branches are like those of the date palm'[32]. Malabar grew the best ginger, which Friar John compared to a reed which could be dug up and transplanted. He was the first European to mention the export of cinnamon from Ceylon.

Jordanus of Séverac, a French Dominican friar, joined an Indian mission early in the fourteenth century. His journeys through the Mediterranean, the Levant, Armenia, Persia, and India (Gujerat, Malabar, and Travancore) recorded in *Mirabilia descripta* reflect a lively and spontaneous curiosity, an attractive naïveté, and an obsessive desire to be impressed. The fruit of the jack (*Artocarpus heterophyllus*) was, he announced, so large that it could easily feed five people, and the mango was 'so sweet and delicious as it is impossible to utter in words'. The coconut was truly a marvel: 'both flowers and fruit are produced at the same time, beginning with the first month and going up gradually to the twelfth; so that there are flowers and fruit in eleven stages of growth to be seen together'[33]. The trees of India overwhelmed him. There was the palmyra (*Borassus flabellifer*), for instance, which yielded toddy, 'a white liquor pleasant to drink', and of course the banyan, which provoked ecstatic astonishment: 'Tis marvellous! And truly this which I have seen with mine eyes, 'tis hard to utter with my tongue'. His enthusiasm is most engaging. 'I could tell very wonderful things of this India; but I am not able to detail them for lack of time.' Only the climate depressed him. 'The heat there is perfectly horrible and more intolerable to strangers than it is possible to say.'

After spending a quarter of a century travelling as far east as Java and Cochin-China, Nicolo de Conti returned to his native Italy in 1441. In South India he was intrigued to find leaves of palms used as writing material, one being the palmyra (*Borassus flabellifer*) whose leaves were so broad that three or four persons could shelter from the rain beneath one of them, the other the talipot (*Corypha umbraculifera*). He was possibly the first European to describe the durian (*Durio zibethinus*): 'a green fruit . . . the size of a cucumber. When opened five fruits are found within, resembling oblong oranges. The taste varies, like that of cheese'[34]. Surprisingly, he says nothing about its disgusting smell. He was the first westerner to attempt more than a superficial description of India's vegetation; that he speaks from personal observation is evident from his description of the cinnamon found in Ceylon (*Cinnamomum verum*).

> *It is a tree which very much resembles our thick willows, excepting that the branches do not grow upwards, but are spread out horizontally: the leaves very like those of the laurel, but are somewhat larger. The bark of the branches is the thinnest and best, that of the trunk of the tree is thicker and inferior in flavour. The fruit resembles the berries of the laurel*[35].

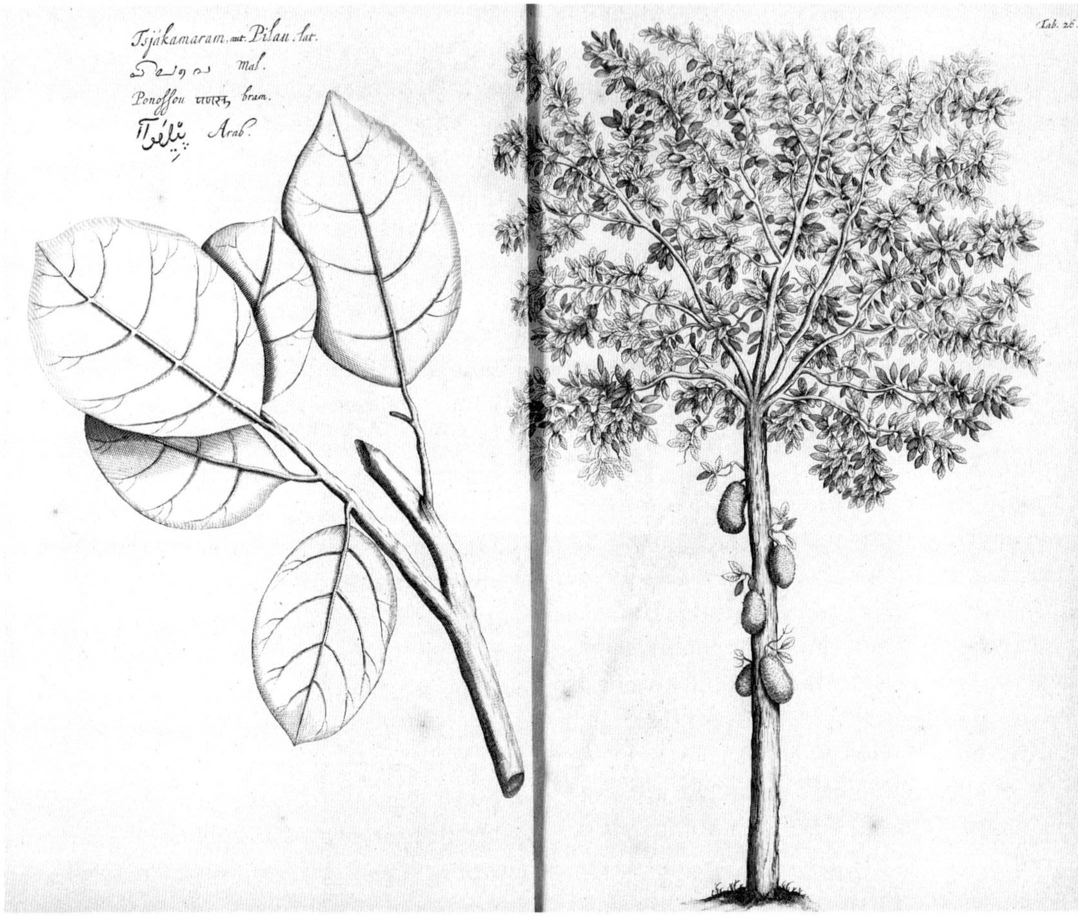

Figure 1.4 Jack, *Artocarpus heterophyllus*. (From Van Reede (1682), *Hortus Malabaricus*, Vol. 3, Amsterdam, fig. 26.) Engraving. Native to India, it is now grown in many tropical regions of the world.

He gives a striking account of the jack (*Artocarpus heterophyllus*) (Fig. 1.4).

> *There is a tree commonly found, the trunk of which bears a fruit resembling a pine-cone, but so big that a man can hardly lift it; the rind is green and hard, but still yields to the pressure of the finger. Inside there are some 250 or 300 pippins, as big as figs, very sweet in taste, and contained in separate membranes. These have each a kernal within, of a windy quality, of the consistence and taste of chestnuts, and which are roasted like chestnuts. And when cast among embers [to roast], unless you make a cut in them they will explode and jump out. The outer rind of the fruit is given to cattle. Sometimes the fruit is also found growing from the roots of the tree underground, and these fruits excel the others in flavour, wherefore they are sent as presents to kings and petty princes. These (moreover) have no kernels inside them. The tree itself resembles a large fig-tree, and the leaves are cut into fingers like the hand*[36].

A fragment of Conti's travels was printed in 1485–6 and a complete edition in *India recognita* (1492), the first work perhaps to discard the dead wood of mythology fostered by Greek and Roman writers. But as the century was drawing to a close the European vision of India was still for many scholars a confusion of dubious fact and fiction.

NOTES

1 Herodotus (1972), 248.
2 McCrindle (1882), 20.
3 Ball (1879–88), 338.
4 McCrindle (1882), 22.
5 Ibid., 29.
6 Strabo (1917), *Geography*, 15.1.21.
7 Ball (1879–88), 343.
8 Theophrastus (1916), IV.iv.5.
9 Ibid., IV.iv.8.
10 Ibid., IV.xi.13.
11 Ibid., IV.iv.4.
12 Strabo (1917), *Geography*, 15.1.2–3.
13 Ibid., 15.1.21.
14 Ibid., 15.1.22.
15 Theophrastus (1916), IX.vii.2.
16 McCrindle (1879).
17 Pliny (1960), *Natural history*, XII.xiv.29.
18 Ibid., XII.xi.23.
19 Stearn (1977), 15.
20 Pliny (1960), *Natural history*, XII.xii.24.
21 Ibid., XII.xiv.26.
22 Ibid., XII.xiv.30.
23 Ibid., XII.xix.35.
24 Ibid., XII.xlii.86.
25 Ibid., XII.xlii.87.
26 Miller (1969), 156–9.
27 McCrindle (1897), 362.
28 Ibid., 362.
29 Major (1857), xlvii-xlviii.
30 Marco Polo (1903), 376.
31 Ibid., 393.
32 Yule (1866), 213.
33 Jordanus (1863), 15.
34 Major (1857), Travels of Nicolo Conti, 9.
35 Ibid., 7–8.
36 Yule and Burnell (1903), 442.

THE FIRST EUROPEAN SETTLEMENTS

Upon the dissolution of the Roman Empire, the Indian Ocean was not penetrated again by Europeans until the Portuguese challenged the commercial supremacy of the Muslim world in Asia. During the fifteenth century Portuguese sailors cautiously extended their exploration of the Atlantic coastline of the African continent, reaching Sierra Leone by 1460. Under King John II Lisbon had become a centre for African commerce, laying the foundations of an administrative structure which would promote Portugal's future trade with the East. In 1487 Pedro de Covilhã left Portugal on a secret mission to investigate the spice trade, then a Muslim monopoly. Disguised as a Moorish merchant, Covilhã sailed from Aden to Cannanore on the Malabar coast, becoming in 1488 the first Portuguese to set foot on Indian soil. Also in 1487 Bartolomeu Diaz sailed out of Lisbon harbour with three small ships to continue the exploration of the African seaboard. They were the first European vessels to round the Cape of Good Hope and, although Diaz's apprehensive and mutinous crews forced him to turn back, the prospect of a route to the Indies had been established. Two years after the accession of Manuel I to the Portuguese throne in 1495, Vasco da Gama commanded four ships on a reconnaissance mission to India. At Mombasa on the east coast of Africa he engaged an experienced Arab pilot whose knowledge of the Indian Ocean brought the Portuguese fleet in May 1498 to Calicut, a prosperous Malabar port enjoying extensive trade links with the Arab world. When some curious Tunisian traders questioned the purpose of his visit, da Gama allegedly replied, 'Christians and spices'. He learnt that ginger, pepper, and cinnamon were abundantly available, that the cinnamon was inferior to that grown in an island called Cillan (Ceylon), and that cloves came from distant Melequa (Malacca). He departed from Calicut in late August with a modest but valuable cargo of spices. A delighted King Manuel informed Ferdinand of Spain that da Gama had returned with jewels and 'cinnamon, cloves, ginger, nutmeg, and pepper, as well as other kinds, together with boughs and leaves of the same'.

Encouraged by da Gama's success, a small armada of thirteen ships was assembled under the command of Pedro Alvares Cabral with orders to establish a trading post or factory at Calicut. The seven surviving ships of Cabral's fleet, having discovered Brazil *en route*, anchored off the Indian coast in September 1500. Cabral's tactless and aggressive behaviour provoked a confrontation with the Indians; da Gama returned in 1502 with a squadron of twenty-five ships to extort trade concessions from reluctant local rulers. Through their crusading spirit, their diplomatic skill in exploiting the rivalry between the little feudal states and, above all, through superior naval forces and fire power, the Portuguese were soon firmly entrenched on the Malabar coast. Goa was subdued in

1510, Malacca the following year, Ceylon was occupied in 1518 and three years later the Portuguese were a powerful presence in the Moluccas and in Sumatra. Venice, Europe's richest state, which had for many years dominated the spice trade channelled through the Red Sea, the Persian Gulf, and the Levant, was supplanted by Lisbon, where pepper could be purchased at a fifth of its price in Venetian markets. Neighbouring Spain, too, had good reason to celebrate when Juan Sebastian del Cano brought back the remnants of Magellan's fleet in 1522 laden with over 50 000 pounds of spices. To commemorate the event, del Cano requested the King of Spain to grant him a coat of arms bearing two sticks of cinnamon, three nutmegs, and two cloves.

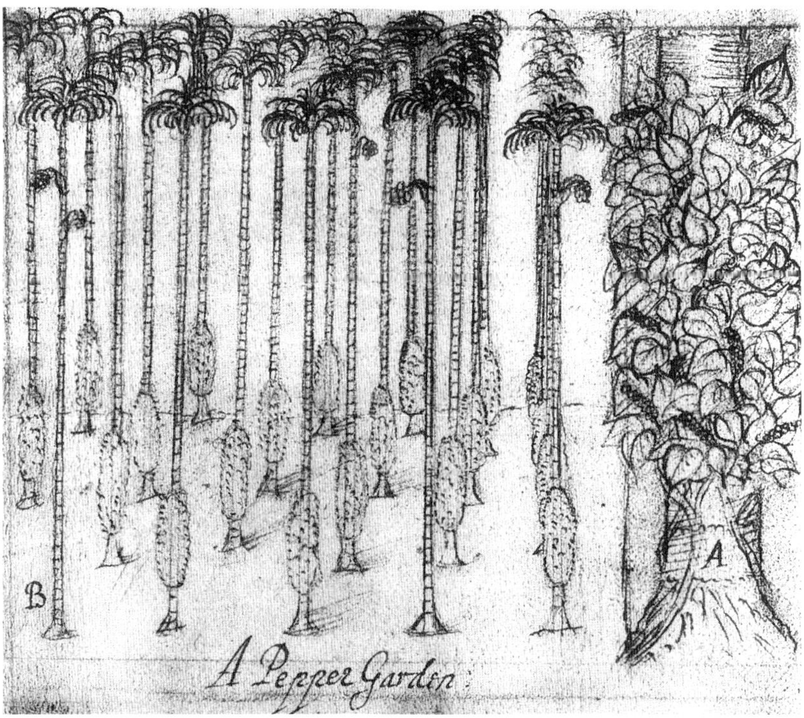

Figure 2.1 Sketch of a pepper plantation by Peter Mundy (From The *travels of Peter Mundy . . .*, Vol. 3, part 1: *Travels in England, western India, Macao and the Canton river, 1634–1637*, London, 1919.)

The lucrative spice trade was inevitably an obsessive theme in the letters and journals of Portuguese settlers and of European travellers to India (Fig. 2.1). Duarte Barbosa, who came to India with Cabral's fleet, meticulously listed 'the divers kinds of spices, where they grow, what they are worth at Calicut, and whither they are carried'[1]. During the sixteen or seventeen years that he lived in the subcontinent, at one time serving as Secretary of the Portuguese factory at Cannanore, his fluency in Malayalam (the language of Malabar) enabled him to investigate the customs and caste-system of the people of Malabar. The exhaustive utilization of the coconut palm (*Cocos nucifera*), a prolific coastal tree, commanded his attention. 'All the folk of Malabar have these palms, and by their means they are free from any dearth, even though other food be lacking, for they produce ten or twelve things all very needful for the service of man.[2] He enumerated these products which included a 'milk like that of almonds', 'a white fruit as large as an apple, which also is very sweet and dainty', copra from the dried kernel, charcoal from the shell, coir from the outer husk, toddy, leaves for thatching roofs, and timber for building and firewood.

Cesare de Fedrici, a Venetian merchant who travelled for eighteen years during the latter part of the sixteenth century similarly praised this bountiful tree. 'In the whole world there is not a tree more profitable and of more goodnesse than this tree is.'[3] He was less enthusiastic about the betel-nut palm (*Areca catechu*) and the habit of chewing the dried areca nut with the leaf of *Piper betle* which persists today (Fig. 2.2). Marco Polo

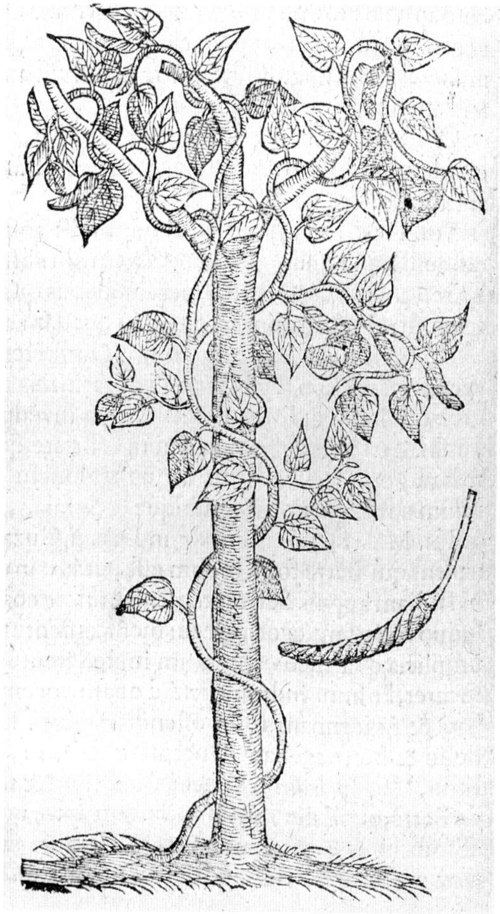

Figure 2.2 *Piper betle.* (From L'Escluse (1605), *Exoticorum libri decem,* Antwerp, p. 176.) Woodcut. The leaf of this creeper is chewed with dried areca nut and lime. Garcia da Orta observed that Indians kept their right thumb nail sharp and pointed to facilitate the removal of the leaf's midrib.

casually reported the custom without comment. Not so Fedrici. 'When this people eate and chawe this in their mouthes, it maketh their spittle to be red like unto blood, and they say, that it maketh a man to have a very good stomacke and a sweete breath, but sure in my judgement they eate it rather to fulfill their filthie lustes, and of a knaverie, for this Herbe is moyst and hote, & maketh a very strong expulsion.'[4]

The fullest description of the coconut palm is given in Ludovico di Varthema's *Itinerario* (1510) (Fig. 2.3). Little is known of the author of this popular travel book; he was born in Bologna and spent some five years ascertaining 'the situations of places, the qualities of peoples, the diversities of animals, the varieties of the fruit-bearing and odoriferous trees of Egypt, Syria, Arabia Deserta and Felix, Persia, and India'[5]. Varthema is never niggardly with his superlatives: the coconut palm is 'the most fruitful tree in the world' and the jack 'the most excellent fruit I ever ate'. The appearance of the jack reminded him of a ripe Persian quince and its flavour of musk melons.

It appears also, as though you were eating a preparation of honey, and it also has the taste of a sweet orange. Within the said fruit there are some pellicles like the pomegranate. And within the said pellicles there is another fruit which, if placed on the embers of the fire and then eaten, you would say that they were most excellent chestnuts[6].

He had a sharp eye for detail. The leaves of *Piper nigrum*, the indigenous pepper of Malabar, reminded him of those of 'a sour orange, but one more dry, and on the underneath part they are full of minute veins'. From the branches 'there grow five, six, and eight clusters, a little longer than a man's finger, and they are like small raisins, but more regularly arranged, and are as green as unripe grapes'[7].

Such exact observation would have met with the approval of Garcia da Orta who deplored the lack of curiosity displayed by most of his fellow Portuguese.

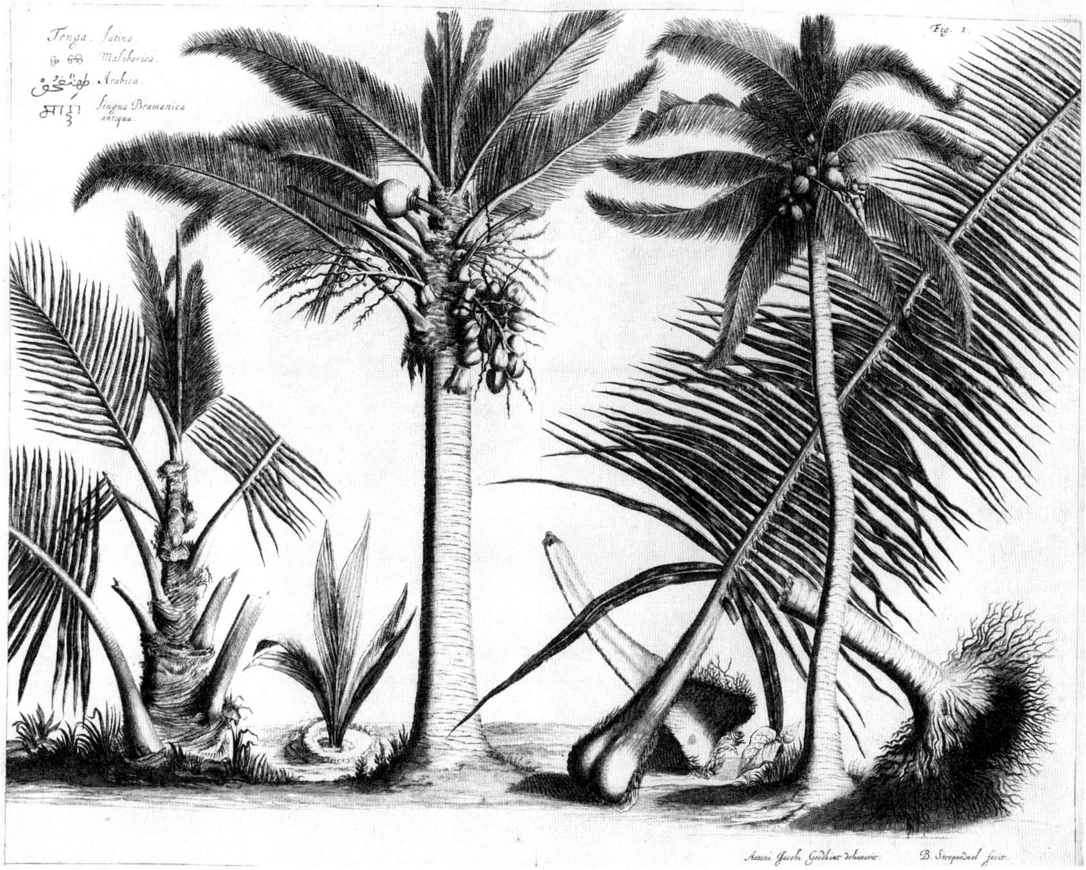

Figure 2.3 Coconut palm, *Cocos nucifera.* (From Van Reede (1678), *Hortus Malabaricus,* Vol. 1, Amsterdam, fig. 1.) Engraving. Drawn by Anton Jacob Goedkint, one of the author's artists. The original home of the coconut palm is unknown but some authorities favour South East Asia whence it spread to the Pacific and west to India.

They are not desirous of knowing anything about the things in the countries they visit. If they know a product, they do not seek to learn from what tree it comes, and if they see it they do not compare it with one of our trees, nor ask about the fruit, nor what it is like[8].

Garcia da Orta (*c.*1501–1568), the son of a Spanish Jew who had emigrated to Portugal and become a nominal Catholic Christian, had qualified as a physician before going to India in 1534. As personal physician to M. A. de Sousa, Captain-Major of the Indian Ocean, he travelled the length of the western coast, making a quick excursion to Ceylon before settling in Goa with occasional visits to the island of Bombay of which he had been granted a long lease. As a medical student he had read Dioscorides, Pliny, and Galen and although he would quote them with respect he was never reticent about challenging or questioning their pronouncements on tropical plants. 'Do not try to frighten me with Dioscorides or Galen, because I merely speak the truth and say what I know.'[9] Acutely observant and constantly inquisitive, he became the first European to investigate purposefully the flora of South Asia. He interrogated eastern merchants about the plants they traded in; he acquired and grew unusual specimens in his garden—he was especially proud of a tree which he said was called Jambos in Malacca (*Syzygium jambos*).

The flower is red and very sweet, with a taste like sorrel. The leaf is like the point of a lance; large and of a very pleasant green colour. The roots of this tree strike far into the ground to uphold the tree when it is loaded with fruit, which is the case many times in the year. Conserves are made both of the fruit and the flower[10].

In a rambling, discursive, and idiosyncratic narrative da Orta displays an intimate familiarity with Indian plants, especially those with commercial or medicinal values. *Coloquios dos simples e drogas he cousas mediçinais da India* was published in Goa in 1563, an early and now rare Indian incunable. The text is cast in the Renaissance literary convention of a dialogue, in this instance, between the author and a fictitious Spanish Doctor Ruano (from the Portuguese meaning 'man in the street'), a recent arrival in Goa, anxious to learn about drugs, spices, and fruit. Doctor Ruano defends classical authority whereas da Orta counters with the evidence of his own eyes. Altogether there are 59 colloquies mainly debating plants and drugs with digressions on diamonds, ivory, elephants, chessmen, a fight between a cobra and a mongoose, the caste system, the Elephanta Caves at Bombay, and much else. An examination of the structure, colour, or scent of a flower could be followed by a casual comment on a cure for snake bite or a Hindu legend. Omissions and errors, some typographical, in his book are serious flaws but these are counter-balanced by a critical perception and a readiness to dispute orthodox beliefs enshrined in classical texts. He was the first westerner to look carefully at Indian plants, in a work which is a landmark in the history of pharmacognosy.

The physician, Christobal Acosta (*c.*1525–*c.*1594)[11] met da Orta in Goa during his first tour of duty in the East (Fig. 2.4). Shortly after da Orta's death in 1568, Acosta returned to India, becoming physician at the royal hospital in Cochin. He found time to botanize before returning in 1572 to Europe where he obtained an appointment as town physician at Burgos in Spain. Inspired by da Orta's *Coloquios*, he wrote his own account of India's medicinal plants, *Tractado de las drogas, y medicinas de las Indias Orientalis* which was printed in Burgos in 1578. He leant heavily on da Orta's work without plagiarizing it. He rejected the dialogue presentation for a straightforward narrative embellished with rather crude woodblocks based on his own competent drawings.

Between them these two authors describe or discuss some hundred plant species yet their pioneer studies might never have been read beyond a privileged circle of scholars had it not been for the translation of their two books by Charles de l'Éscluse (1526–1609). L'Éscluse or Clusius, to use the latinized form of his name, was a native of Arras, then part of the Netherlands; he had studied law and medicine and as a botanist perceived the importance of da Orta's *Coloquios* when he saw a copy in either Lisbon or Coimbra in 1564; he translated it into

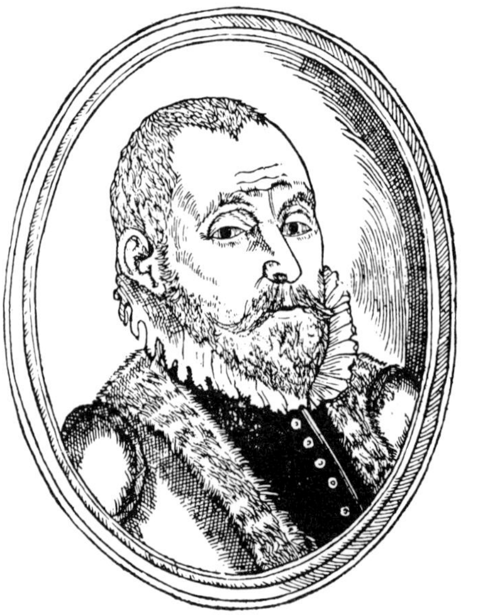

Figure 2.4 Christobal Acosta (*c.*1525–*c.*1594).

Latin and in an abridged form it was published by Christopher Plantin in 1567. *Aromatum et simplicium aliquot medicamentorum apud Indos nascentium historia* replaced da Orta's artificial dialogue with a conventional text and a commentary by Clusius; illustrations were based on a selection of vegetable products available in Antwerp's markets. A subsequent translation into French introduced da Orta to a wider audience. Clusius provided a similar service for Acosta, translating his *Tractado* in *Aromatum et medicamentorum in Orientali India nascentium liber* (1582). Both books were reissued in one volume in 1593 and were included in Clusius's collected works, *Exoticorum libri decem* in 1605, a publication that signalled a future Dutch commitment to a scientific and commercial investigation of the flora of the tropics. After a life of frustration and poverty, Clusius enjoyed a tranquil old age as Professor of Botany at Leiden University. This period of his life coincided with the importation of the exotic novelties of botanical exploration. Herman Boerhaave recalled that 'practically no captain, whether of a merchant ship or of a man-of-war, left our harbours without instructions to collect everywhere seeds, roots, cuttings, and shrubs and bring them back to Holland'[12]. Clusius patiently propagated these floristic offerings in the conservatory in Leiden's botanical garden. Intensification of interest in the world's flora, first of Europe, then of tropical countries, was undoubtedly a major factor in the sixteenth-century revival of botany, indeed of the origin of modern botany.

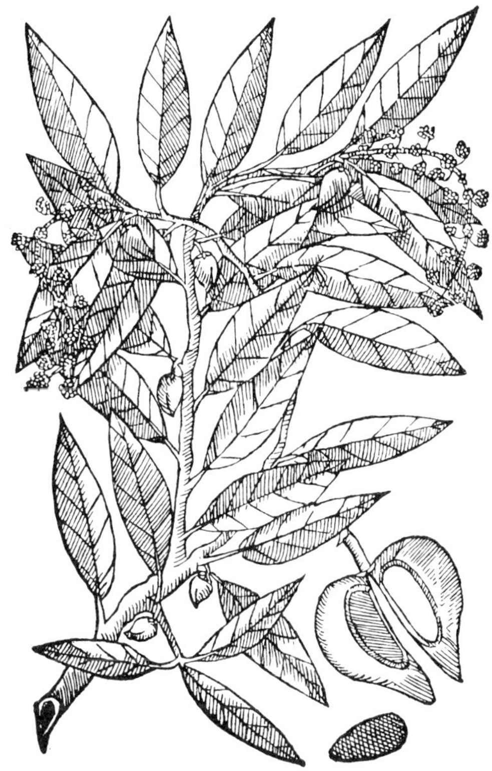

Figure 2.5 Mango, *Mangifera indica.* (From C. Acosta (1578), *Tractado de las drogas y medicinas de las Indias orientalis, Burgos,* p. 319.) Woodcut. This fruit, which was a particular favourite with the Mughals, has been cultivated in India for four millennia.

Portugal's supremacy in Asia was threatened when the Dutch appeared on the scene towards the end of the sixteenth century. The first Dutchman to set foot on Indian soil was Jan Huygen van Linschoten (*c.*1563–1611), a Catholic working in Spain when Philip II seized the Portuguese crown. To escape political unrest in the Iberian peninsula, he joined the retinue of the new Archbishop of Goa, reached India in September 1583, and remained there for five years. Although he appears seldom to have left Goa (possibly his movements were restricted by the suspicious Portuguese), he learnt a great deal about the trade routes to the East. It was the publication of this vital information on the navigation of the Indian Ocean that ensured the success of his *Itinerario*, issued in the Netherlands in 1595 and 1596. Within two years it was translated into German and English; in 1599 a Latin version was on sale in Amsterdam and Frankfurt and a French edition appeared in 1610.

It was not just the disclosure of maritime matters that the Portuguese had tried to keep secret that made the *Itinerario* a bestseller.

Linschoten was a perceptive commentator on life and landscape in Portuguese Goa. His descriptions of the vegetation were annotated by a Dutch physician, Bernard ten Broecke, latinized as Bernardus Paludanus. Paludanus, who had travelled in Syria and Egypt—returning, according to his epitaph, with 'chests filled with the wonders of nature'—had known Clusius when he was prefect of the royal gardens in Vienna. Paludanus complements Linschoten's botanical descriptions, enlarges on the culinary and medicinal properties of the plants, and frequently commends the reader to Clusius's version of da Orta and Acosta for further information. Here is Paludanus's contribution on the mango (*Mangifera indica*) (Fig. 2.5) , the fruit that some authorities believe Alexander banned to his troops.

This fruite is colde and moyst, although commonly they esteeme it to bee hotte, & say, it ingendereth a paine and gripping in the maw, of such as eate it, and more over the Heathen Physitians say, it is hotte, and reiecte it, because it ingendereth Saint Anthonies fire, Carbuncles, hotte burning Feavers, and swellings, with scurvines . . .[13]

Some of the plants mentioned by Linschoten, like the pineapple, cashew, and papaya, had been introduced from the Americas by the Portuguese and were already naturalized. Linschoten, too, was impressed by the multifarious uses of the coconut palm; he knew of no other fruit to compare with the exquisite flavour of the durion (*Durio zibethinus*); he lamented the scarcity of flowers during the brief greening of the fields in the monsoons. The behaviour of the wives of Portuguese settlers who had taken to chewing betel repelled him.

In the daytime wheresoever they doe sit, goe or stand, they are continually chawing thereof, like Oxen or Kine chawing their cud; for the whole exercise of many Portingale [Portuguese] women, is onely all the day long to wash themselves and then fal to the chawing of their Bettele[14].

Betel, added Paludanus, was also eaten by women before making love. It is, in fact, astonishing how many plants were credited with aphrodisiac qualities. Opium was used 'most for lecherie' and prostitutes resorted to bhang (*Cannabis sativa*) 'when they meane to have a man's companie'. The intoxicating properties of *Datura metel*—*Datura* comes from the Sanskrit meaning insane—were enlisted by both Indians and Portuguese women 'when they are disposed to be merie with their secret lovers'. The unsuspecting husband having been made insensible with this powerful drug, the lovers 'goe in his presence and performe their leacherie together . . . and when the time cometh that he reviveth out of his trance, he knoweth nothing what was done, but thinketh that hee had slept'[15].

In 1595 when the first part of Linschoten's *Itinerario* was published, a fleet of four Dutch ships anchored in Indian waters on their way to Batavia and spices. In 1602 the Dutch obtained permission to build a fort in Ceylon and presented three captured Portuguese ships to the Singhalese ruler. The Dutch wasted little time in systematically ejecting the Portuguese from their settlements in the Spice Islands and in 1641 the Portuguese lost Malacca, their last important stronghold in the East Indies. Their position in India was also being gradually eroded. In 1604 the Zamorin of Calicut, constantly harassed by the Portuguese, was a willing signatory to a treaty of friendship with the Dutch; five years later the Dutch built a fort at Palikat (Pulicat), just north of Madras, their first settlement in South India. Ceylon passed to their sphere of influence in 1658 and by 1663 the Portuguese had lost all their possessions in southern India. Only Goa, much further north, remained under their

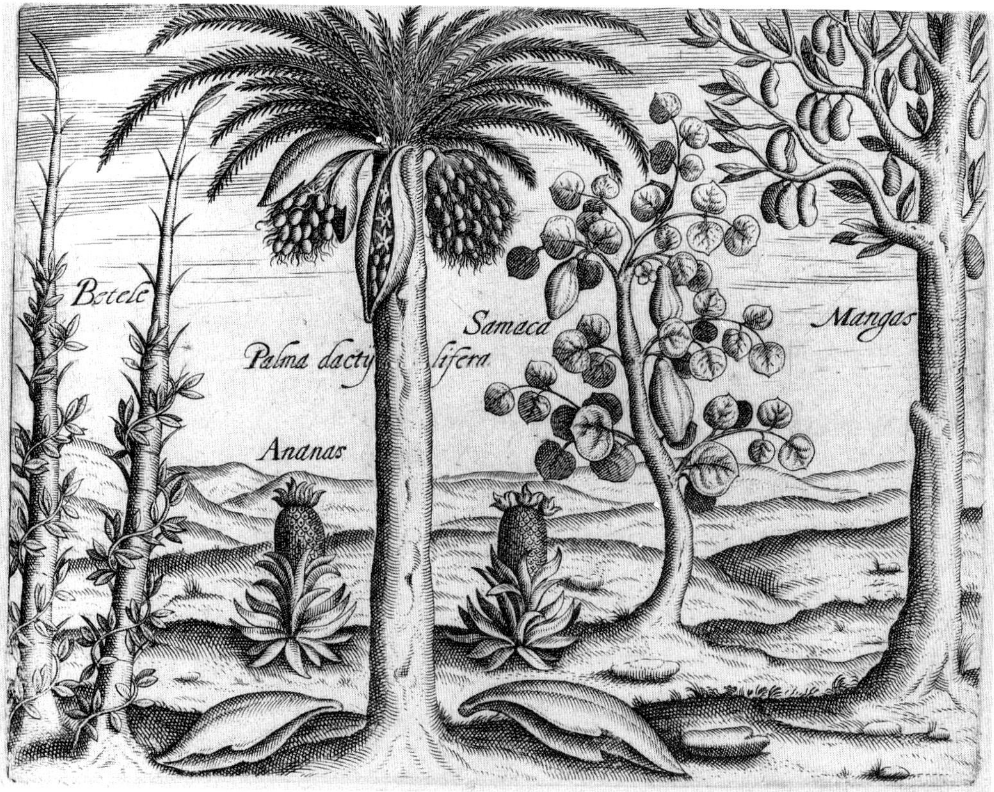

Figure 2.6 Some of the Indian plants that attracted the attention of early European travellers: 'Betele' *(Piper betle)*, 'Palma dactylifera' or date palm *(Phoenix dactylifera)*, 'Mangas' or mango *(Mangifera indica)*, and 'Ananas' or pineapple *(Ananas comosus)*, which was probably introduced from the New World to India by the Portuguese. (From Theodor de Bry (1601), *Collectiones peregrinationum in Indiam orientalem et occidentalem*, Part 4: *Indiae orientalis*, Frankfurt,) Engraving.

flag. The Dutch East India Company now had a monopoly of the cloves, nutmegs, and mace of the East Indies, the pepper of Malabar, and the cinnamon of Ceylon. In 1652 the Company established a station and garden at the Cape of Good Hope to provide fresh vegetables and fruit for the crews and passengers of their ships trading with India and the Malayan Archipelago. In the course of time plants in transit between the East and West were acclimatized in this Cape garden. In September 1656 a young Dutchman, Hendrik Adriaan Van Reede tot Drakenstein (1636–91)[16] enlisted as a soldier in the Dutch East India Company and joined a new expeditionary force destined for military action against the Portuguese in India. Van Reede fought in Malabar, was rapidly promoted, becoming State Officer to the throne of Cochin in 1663, First Captain and Sergeant-Major of Ceylon in 1667, and Commander of Malabar in 1669.

Malabar is part of the Western Ghats, that long chain of mountains rising from the western seaboard of the Indian subcontinent. Its tropical rain forests, drenched by the south-west monsoon, support a rich vegetation, some of it endemic. To the Portuguese and the Dutch it was a land of abundance: teak and sandalwood, groves of coconut palms and bananas, and the pepper, cardamom, and ginger that had attracted Arab and Greek traders centuries earlier. Van Reede's duties took him on long tours of inspection through the lush countryside and some years later, back in the Netherlands, he recalled 'the marvellous variety of the trees, which it was so great that it would be difficult to find two trees of the same kind in the same forest if one were to search for this'[17]. Their trunks and branches were festooned with climbing plants, and not only the fertile soils of the plains but even the seemingly desolate mountains were home to plant life. Vegetation grew in the most unlikely habitats—in fissures of rocks and on banks of rivers regularly covered by salt water at high tide. 'There was no piece, not even the smallest, even of the most barren soil which did not display some plants', wrote Van Reede. Confronted with this abundance, he was 'often seized by the desire to explore and examine the leaves, flowers, bark, and roots of these plants'[18]. He interrogated local Indians about their names and uses, particularly in medicine. From an initial curiosity emerged his desire to systematize and extend this random acquisition of botanical knowledge for the profit of the Company and the benefit of botanists back in Europe. On his own initiative he set up a laboratory in his Commander's residence in Cochin; he engaged a chemist to analyse the medicinal properties of the local flora in the hope that they might replace the vegetable drugs transported from Batavia at great expense.

In 1669 Batavia requested the Government of Ceylon to investigate the availability of medicinal plants in the island, hoping to reduce the cost of importing from the Netherlands drugs which often deteriorated during long voyages. Van Reede saw a floristic survey of Malabar as a useful contribution to self-sufficiency in medical supplies (Fig. 2.7). Lacking both botanical expertise and the leisure to undertake such a project himself, he sought competent assistance which he thought he had found in the person of an elderly Carmelite missionary visiting Cochin in 1673, Father Matthew of St Joseph (*c.*1617–1691), an Italian who had served his Order in the Lebanon and Iraq before coming to India in 1651. During his travels he always studied and sketched the local flora, eventually accumulating a vast collection of botanical records, known to his contemporaries as *Viridarium Orientale*, which he sent to Italy in 1663 for publication. Only a fragment ever appeared in print—in Giacomo Zanoni's *Istoria botanica* (1675) and Gaetano Monti's version of Zanoni, *Rariorum stirpium historia* (1742). Having seen some of Father Matthew's pen and ink flower drawings, Van Reede readily accepted his proposals for a flora of Malabar. The missionary began listing plants, adding such hasty sketches that Van Reede had reservations about his artistic skill.

Of course the excellent old man had not imagined that it would be necessary to take greater and more accurate pains in depicting and describing those plants, because he himself with the aid of his memory was indisputably acquainted thoroughly with the plants presenting themselves to him and their curative virtues. Moreover, these rough sketches had nearly always been made on his journeys, so that fairly often now the fruit of one plant, now the flowers of another plant were missing, and sometimes both'[19].

Discreetly, Van Reede did his best to supply the omissions but at the same time was exasperated that so little progress was being made. He needed little persuasion when Paul Hermann, medical officer and botanist in Ceylon, on a visit to Malabar in 1674 urged a fresh start, dismissing Father Matthew's contribution as inadequate for the purpose. Nevertheless, Van Reede later generously acknowledged the priest as the 'first founder and originator of this *Hortus*'[20]. Van Reede had engaged four native scholars and physicians to supervise the collection of plants and to supply information on their use in

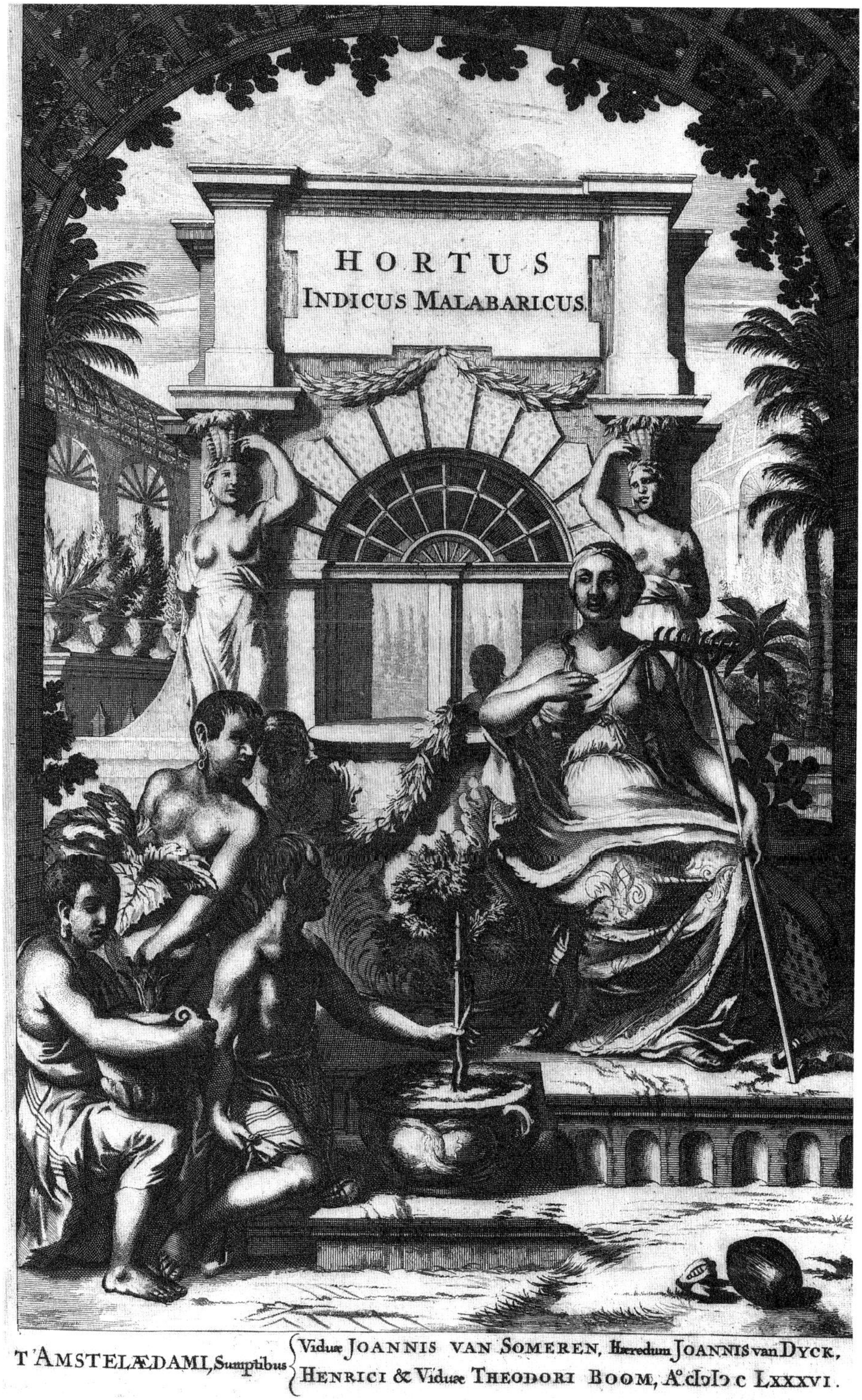

Figure 2.7 Engraved frontispiece of the first volume of *Hortus Indicus Malabaricus* (by Van Reede, Amsterdam, 1678.) The title varied during the publication of the 12 volumes: *Horti Indici Malabarici* and *Horti Malabarici.* It is now usually referred to as *Hortus Malabaricus.*

native medicine. The Company's translator diligently recorded their contributions in both Malayalam and Portuguese. As soon as plants were brought in by natives adept in climbing the tallest trees, they were drawn by three or four artists, at least two of whom had been recruited from the Cochin garrison: Marcelis Splinter and Anton Jacob Goedkint. Vegetation in the neighbourhood of Cochin received the closest scrutiny but whenever Van Reede had to visit other regions of Malabar, his officers and ship's crew would bring back armfuls of plants, like prized trophies, for the frantic artists to copy. These plant portraits were submitted to a committee of fifteen or sixteen scholars knowledgeable about the native flora (Fig.2.8).

When Van Reede left Malabar for Batavia in 1676 with Johannes Casearias (*c*.1640–1677), who had assisted Father Matthew with Latin translations, several hundred plants of Malabar had been drawn and described and the text of the first volume despatched to the Netherlands. In Batavia, probably waiting for a ship to take him home, Van Reede persuaded the physician, Wilhem ten Rhyne, to make a few contributions. When he arrived back in the Netherlands in 1678 the first volume of the *Hortus Malabaricus*, edited by Arnold Syen (1640–78), Professor of Medicine and Botany at Leiden University, had just been published. Van Reede was dismayed to discover that his original plan of allocating a volume each to trees, shrubs, and herbs had not been strictly observed; to balance the size of the volumes some shrubs had been included in the first volume on trees and some herbaceous plants

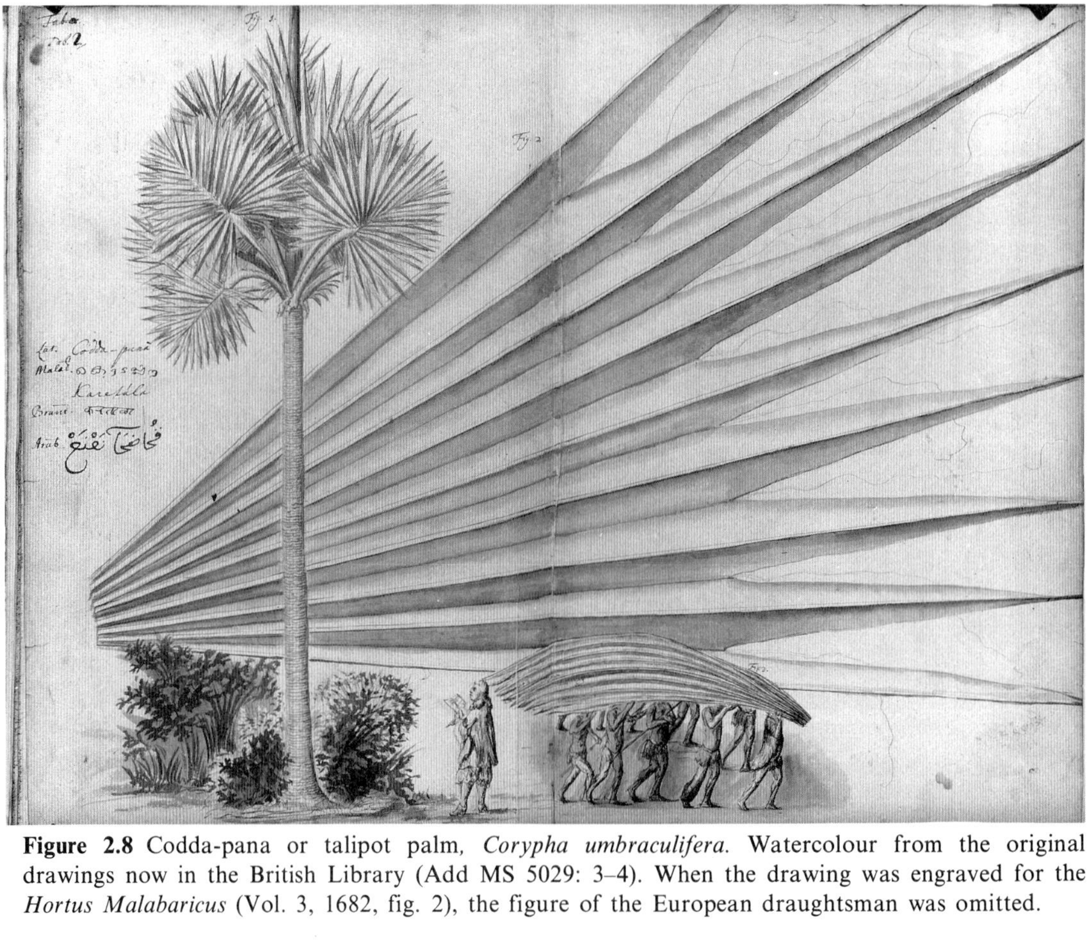

Figure 2.8 Codda-pana or talipot palm, *Corypha umbraculifera.* Watercolour from the original drawings now in the British Library (Add MS 5029: 3–4). When the drawing was engraved for the *Hortus Malabaricus* (Vol. 3, 1682, fig. 2), the figure of the European draughtsman was omitted.

were transferred to the impending second volume on shrubs. Syen had assumed there was nothing more to publish. On Syen's death, Jan Commelin (1629–92), a prosperous chemist in Amsterdam and a self-taught botanist, assumed the role of editor and commentator while Johannes Munnicks (1652–1711), Professor of Botany at Utrecht, attended to the descriptions and Latin translation. Commelin's position as a Commissioner of the Amsterdam Municipal Garden gave him access to new acquisitions of plants from the East and thereby the opportunity to check his commentaries before they were despatched to the printer. The twelfth and final volume of *Hortus Malabaricus* was published in 1693, the year following Commelin's death. Van Reede had returned to India in 1684 as Commander-General of the Western Quarters. He became ill while quelling a native rebellion in Malabar in 1691 and died on board ship off Bombay on 15 December. His body was taken to Surat for a solemn funeral; the coffin draped in black velvet was escorted by soldiers with their rifles reversed, mourning ribbons fluttering from halberds while drums rolled and field guns fired salvoes (Fig. 2.9). Representatives of the Indian, Armenian, and English communities paid their last respects. Van Reede's monumental mausoleum still stands at Surat; he is commemorated in the genus *Rheedia* but a more appropriate memorial is his *Hortus Malabaricus*.

It was the first regional floristic survey of the subcontinent. For about a century its only competitor was Rumphius's *Herbarium Amboinense* (1741–50) which described about a thousand plants found on the island of Amboina in Indonesia. The detailed descriptions and clear engravings of Van Reede's work made it an indispensable tool for any botanist in the eighteenth century studying the plants of South Asia. The morphology of some 690 species[21] is systematically presented: root, stem or trunk, leaf, flower, fruit, and seed. Even scent and flavour are recorded. Each description concludes with the plant's use in agriculture, commerce or medicine. Plant names are given in Portuguese, Malayalam, Arabic, several South Indian scripts, and in Konkani in devanagari script. This was to be a source of confusion for the great orientalist, Sir William Jones, who in some irritation complained to Sir Joseph Banks, 'Had Van Rheede exhibited the Sanskrit names with accuracy, we should not be puzzled in reading Indian poems & medical tracts; but in all his twelve volumes, I have not found above ten or twelve names correctly expressed in Sanskrit or Arabick'[22]. The reputation of the book relies as much on its illustrations as on its text. The 791 engravings, the majority of them double-folio, demonstrate not only the flexibility of copper engraving but also communicate the initial excitement generated by the discovery of the floral riches of India. A branch of a tree or a shrub is often portrayed diagonally across a double spread of pages with leaves sometimes abruptly cut off at the plate-line; floral dissections are frequently added. In austere outline and carefully controlled hatching the Dutch engravers have competently interpreted the original pen and ink and wash drawings[23], unfortunately omitting most of the incidental and decorative figures, animals, and views. Despite lapses in accuracy these plant portraits are recognizable today to anyone familiar with the flora of the state of Kerala.

The *Historia plantarum* (1686–1704) by the English botanist, John Ray, cited over eighty per cent of the plants featured in Van Reede's work. The Swedish botanist, Linnaeus, confidently based many of his new names on its plates and descriptions. 'I have not put my whole trust in any author,' he wrote, 'excepting the work *Hortus Elthamensis* by the very celebrated Dillenius and the work *Hortus Malabaricus* by the illustrious

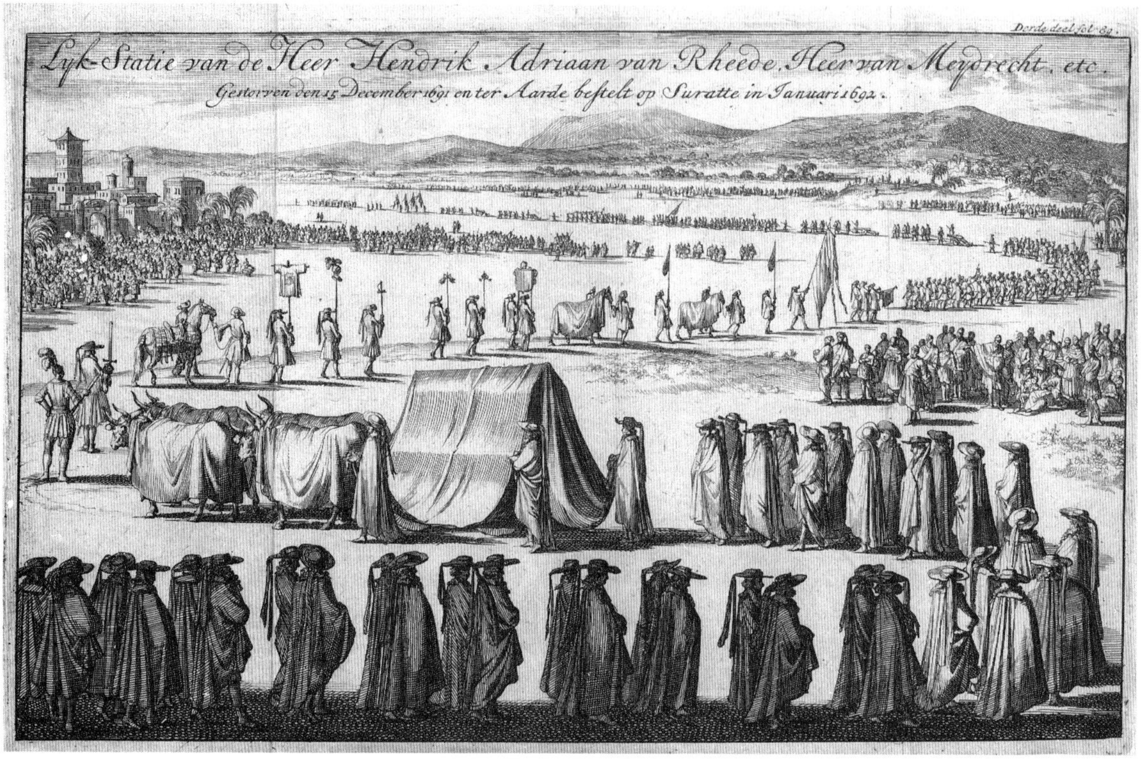

Figure 2.9 The funeral cortège of Van Reede at Surat, 3 January 1692. Engraving by Jan Luyken (in D. Havart (1693), *Op-en ondergang van Cormandel,* Amsterdam).

van Rheede, having my firm conviction in their accurate data.'[24] Distinguished botanists such as Burman, Lamarck, Jussieu, and de Candolle were likewise indebted to this pioneer Indian flora for a number of their new generic names. John Hill launched a London edition of the work in 1774 with reduced engravings and Linnaean names added, but with his death in 1775, only one volume was published.

It is often forgotten that the *Hortus Malabaricus* is a repository of ayurvedic medicine practised in Malabar. Since many of the palm-leaf manuscripts which recorded this traditional expertise have disappeared, this folk wisdom survives largely in the information communicated to Van Reede by his four principal Indian collaborators: Ranga and Appu Bhattan, Vinayaka Pandithan, and Itty Atchuthan. Van Reede wrote

They honour Antiquity and the first inventors of their sciences with the most pious reverence, and as regards medicine and botany, the knowledge of these sciences is preserved in verses, the first line of which begins with the proper name of the plant, whose species, properties, accidents, forms, parts, location, season, curative virtues, use, and the like they then describe highly accurately . . . The first invention of these disciplines (namely, medicine and botany) is considered to be so old that they show books by authors of whom all affirm by constant asseveration that they lived four thousand years ago[25].

Van Reede's Indian savants were the custodians of an ancient literature that had its origins in the culture of Aryan tribes, the descendants of the Indo-Europeans who had migrated from central Asia to Bactria and the Iranian plateau. The Aryans were nomadic pastoralists who settled in small rural communities after they had penetrated northern India about 1500 BC. They brought with them an oral tradition of prayers known as the Vedas which celebrated the sanctity of all life, the union of man and the natural world. The *Rig Veda*, the oldest of their sacred texts, comprises 1028 prayers or hymns addressed to their deities who personified cosmic and natural forces—Agni, the fire god, for instance, Surya, the sun god, or Soma a moon god and ruler of plants. Some 120 hymns are dedicated to Soma which was also the name given to a plant whose identity has provoked esoteric speculation among academics. The *Rig Veda* extolled it above all other plants in vague and rather fanciful descriptions. Scholars and botanists have proposed many candidates: a species of *Asclepias* or *Ephedra*, an alcoholic drink made from millet or barley, bhang or hemp (*Cannabis sativa*), rhubarb or—the latest submission—the bright red fly agaric mushroom (*Amanita muscaria*). Current knowledge of botany in ancient India derives from the study and interpretation of the *Rig Veda* and *Atharvaveda* and, in the post-Vedic classical period, from such works as the medical texts of Caraka and Susrata. These texts abound with references to roots, stems, branches, leaves, flowers, and fruits, and to the colour and texture of vegetation. A rudimentary classification divided the plant kingdom into trees, herbs, and creepers; another division was flowering and non-flowering, fruit-bearing and fruitless. Caraka distinguished between plants with purgative and astringent qualities and listed six categories of food crops.

The Vedas inspired and shaped Hindu religious thought which adopted the Aryan practice of linking plants with a pantheon of deities. Many trees are still sacred to devout Hindus. *Ficus religiosa* is associated with the triumvirate Brahma, Vishnu, and Siva; the three black spots on the husk of the coconut are symbolic of Siva's three eyes; the deodar is worshipped because it grows in the Himalayas, the abode of the gods. The lotus (*Nelumbo nucifera*) which emerges from the mud of a pool or river to flower splendidly above the surface of the water represents the transition from darkness to light, a symbol of purity and life. Lakshmi, the Goddess of Plenty, is depicted sitting on a lotus throne with a lotus footstool and holding a lotus flower. For Buddhists the lotus symbolizes divine birth. It is now the national flower of India. This veneration of nature and natural form inspired generations of Indian artists and craftsmen. The oldest plant decoration yet discovered in the subcontinent is the leaf of *Ficus religiosa* on a seal and pottery fragments found in the ruins of Harappa from the great Indus Valley civilization which was already in decline at the time of the Aryan invasion. The sculptors of the Buddhist stupas at Sanchi and Amaravati adroitly manipulated the rhythmic and decorative properties of the lotus flower and undulating stem, always, however, respecting its organic structure. A canopy of palm leaves and pendant fruit form a yellow sandstone capital to a pillar of the pre-Kushana period (third to first century BC). Nature's rich abundance has always been celebrated in Indian art.

There are Indians who would claim that their country's first biologist was the founder of the Mughal dynasty, the emperor Babur (1483–1530). The frequent references to the fauna and flora of Afghanistan and the plains of northern India in his memoirs, the *Babur-nama*, reveal the compulsive curiosity of a born naturalist. Firmly established in India after a decisive victory at Panipat in 1526, he proceeded to survey his kingdom. Everything was different from what he had known in Afghanistan: 'its mountains, rivers, jungles and deserts, its towns, its cultivated lands, its animals and plants, its peoples and their tongues, its rains and its winds'[26]. He marvelled at the rhinoceros and the peacock, pronounced the mango 'the best fruit of Hindustan', but found the jack-fruit excessively sweet. Any plant that particularly attracted his attention would be recorded in his memoirs. He approved of the scent of

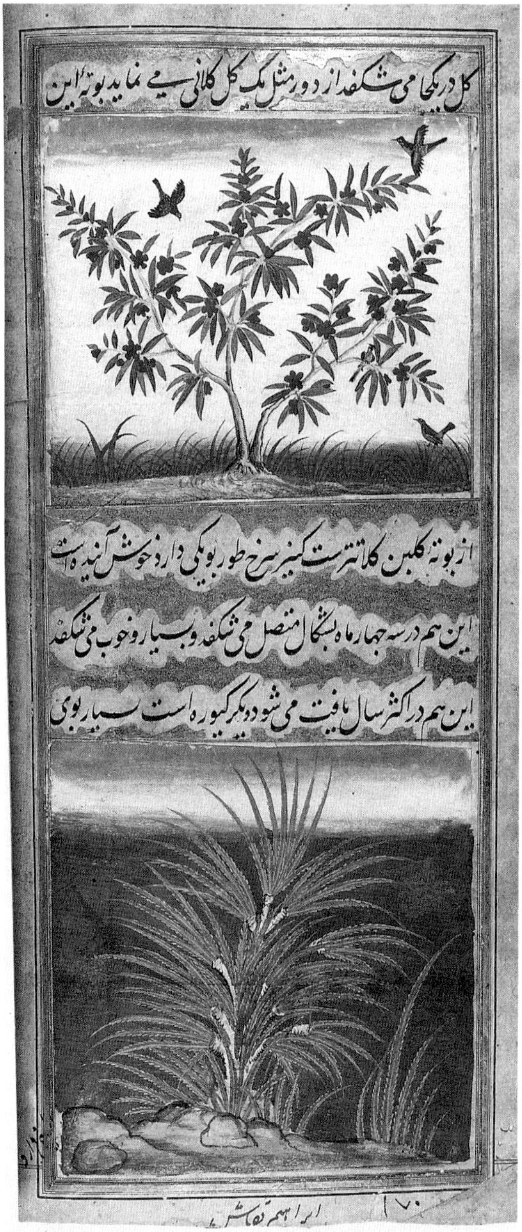

Figure 2.10 Oleander, *Nerium indicum* and (below) Screw-pine, *Pandanus tectorius*. Baburnama, Mughal, *c.*1590. Watercolour by Ibrahim Nakkash. (British Library (Or. 3714, folio 407v).)

the screw-pine (*Pandanus tectorius*) whose flowers he carefully noted were about thirteen to eighteen inches long (Fig. 2.10).

It has long leaves having the character of the reed gharau [Calamus scriptorius] and having spines. Of these leaves, while pressed together bud-like, the outer ones are the greener and more spiny; the inner ones are soft and white. In amongst these inner leaves grow things like what belongs to a middle of a flower, and from these things come the excellent perfume. When the tree first comes up not yet showing any trunk, it is like the bush of the male-reed, but with wider and more spiny leaves. What serves it for a trunk is very shapeless, its roots remaining shown[27].

He wrote at length about date and coconut palms and their uses, and the beautiful blossom of the jasun evoked a lyrical response. The jasun (*Hibiscus rosa-sinensis*) had reached India from China. Babur notes that the flower

is fuller in colour than that of the pomegranate, and may be of the size of the red rose, but, the red rose, when its bud has grown, opens simply, whereas, when the jasun bud opens, a stem on which other petals grow, is seen like a heart amongst its expanded petals. Though the two are parts of the one flower, yet the outcome of the lengthening and thinning of that stem-like heart of the first-opened petals gives the semblance of two flowers. It is not a common matter. The beautifully coloured flowers look very well on the tree, but they do not last long; they fade in just one day. The jasun blooms very well through the four months of the rains; it seems indeed to flower all through the year; with this profusion, however, it gives no perfume[28].

The miniatures in a version of the *Baburnama* of about 1590 are among the earliest surviving Mughal drawings of Indian plants and animals[29].

The year in which the first Jesuit missionary, Francisco Xavier, landed at Goa—1542—Babur's grandson, Akbar, was born. Abul Fazl, Akbar's friend and biographer, praised the emperor's military genius, his tolerance and hospitality. The *Ain-i Akbari*, his account of Akbar's court and administration, lists cultivated fruits, vegetables, and flowers. Abul Fazl is a master of vivid images: the trunk of the plantain or banana 'looks straight like a spear; the leaves come out of the trunk thick and soft, and resemble an unsewn plaited sleeve'[30].

Akbar's son, Jahangir, enthroned in 1605, shared Babur's enthusiasm for natural history but approached it with the deliberation of a scientist. After specimens of animals and birds had been precisely measured, his artists were commanded to draw them; abnormally large fruit—melons, pomegranates, quinces, and figs—were invariably weighed. He was infatuated by India's flora. 'From the excellencies of its sweet-scented flowers one may prefer the fragrances of India to those of the flowers of the whole world.'[31] Among his favourites were the Golden Champa (*Michelia champaca*) and the Flame of the Forest (*Butea monosperma*)—'it is so beautiful that one cannot take one's eyes off it'[32]. On one occasion his soldiers were ordered to decorate their heads with sprays of oleander blossom 'and whoever did not do so had his turban removed, a wonderful flower-bed was produced'[32]. Kashmir, the summer retreat of the Mughal court, was especially dear to him.

> *Its pleasant meads and enchanting cascades are beyond all description . . . The red rose, the violet, and the narcissus grow of themselves; in the fields, there are all kinds of flowers and all sorts of sweet-scented herbs more than can be calculated*[33].

Mansur, his chief painter, drew over a hundred Kashmiri plants. Jahangir's pleasure in the natural world was not just the childlike wonder of Babur; it was infused with the discipline of scientific enquiry, a combination also found in the approach of those Europeans who were soon to advance the development of botanical studies in India.

NOTES

1 Barbosa (1921), 227.
2 Ibid., 90–2.
3 R. Hakluyt (1904), *Principal navigations, voyages, traffiques & discoveries of the English nation*. Vol. 5, London, 379.
4 Ibid., 391.
5 Badger (1863), xxiii.
6 Ibid., 159.
7 Ibid., 157.
8 da Orta (1913), 86–7.
9 Ibid., 160.
10 Ibid., 237–8.
11 There appears to be some uncertainty regarding Acosta's dates of birth and death. Those given in Scribner's *Dictionary of scientific biography* have been accepted.
12 Stearn (1961), 20–1.
13 van Linschoten (1885), Vol. 2, 24–5.
14 Ibid., 64.
15 Ibid., 60.
16 There are variations in the orthography of van Reede's family name: Rheede and Reede, Draakestein, Drakestein, Draakenstein, Drakenstein, Draekestein. The form used in the authoritative biography by Heniger (1986) has been adopted.
17 Manilal (1980), 43.
18 Ibid., 44.
19 Ibid., 46.
20 Ibid., 47.
21 Nicholson *et al.* (1988), 14.
22 18 October 1791. Cannon (1970), 892.
23 The original drawings for the first ten volumes are now in the British Library.
24 C. Linnaeus, 1754, *Genera plantarum* (5th edn), Stockholm, xii.
25 Manilal (1980), 48.
26 Babur (1922), 484.
27 Ibid., 514.
28 Ibid., 513–14.
29 Now in the British Library.

[30] Fazl (1877), 74.
[31] Beveridge (1909), Vol. 1, 5.
[32] Ibid., 97.
[33] Beveridge (1914), Vol. 2, 145.

ARRIVAL OF THE BRITISH

In October 1579, Father Thomas Stevens (1549–1619), a Jesuit priest, stepped ashore at Goa from a Portuguese ship, the first Englishman to sail the sea route to India around the Cape of Good Hope. In a letter written in October 1583 to his brother Robert in Paris he praises the ubiquitous coconut palm and its refreshing milk: 'this is so abundant that after drinking the contents of one nut, you scarcely feel the need of another'[1]. His intervention helped to obtain the release of four of his countrymen from a Portuguese prison. John Newbery, Ralph Fitch, William Leedes, and James Storey, financed by the Levant and Muscovy Companies, had left London in February 1583 as the first English trade mission to India and the East, furnished with letters of introduction from Queen Elizabeth to Akbar and the Emperor of China. Suspected of spying, they were arrested on the island of Ormus in the Persian Gulf—a strategic Portuguese entrepôt for goods from the East—and sent to Goa where they were imprisoned. Through the intercession of Father Stevens and Linschoten they were released on parole after having assured the authorities that they were good Catholics. Fearful that their liberty would be short-lived, three of them fled to the safety of neighbouring Bijapur. Fitch, who eventually returned to London in 1591, was the party's only survivor. For eight years, much of the time alone, he wandered through northern India and the East, stayed briefly at Akbar's Mughal court at Agra, and was the first Englishman to reach Burma and the Portuguese settlement at Malacca. His brief, rather laconic account is probably the report he submitted to the London Companies that had sponsored the mission. The few plants he mentions are predictably spices: pepper ('like our ivy berry, but something longer like the wheat eare'), ginger (which 'groweth like unto our garlike'), clove (the 'tree is like to our bay tree')[2], and nutmeg, which reminded him of the walnut tree.

Fitch and his companions had travelled to India overland via Aleppo and Basra. Denied access by the Portuguese to the sea route around the Cape, British merchants had endeavoured to find a northwest passage across Arctic America to China and the spice islands of the Pacific. The defeat of the Spanish Armada in 1588 had given the nation a confidence in the skills of its mariners but all three ships comprising a private mercantile venture to the East in 1591 were lost. The Dutch who formed a Company of Far Lands in 1594 were more fortunate; the remnants of their fleet returned with a substantial cargo of spices. Within a few years the Dutch had established a profitable trade and exploited the demand for spices to preserve and flavour food by doubling the price to the British market. London merchants responded to this commercial opportunism by agreeing to finance another voyage to the East Indies in September 1599. On the last day of December 1600 Queen Elizabeth granted a charter for fifteen years to 'the Governor and Company of Merchants of London Trading into the East Indies'. The new Company, soon to be known as the Honourable East India Company, quickly assembled a fleet of five small ships which sailed from Woolwich in February 1601, returning in September 1603 with £1 000 000 of pepper and a handsome profit for the promoters of the venture.

At that time the British considered India primarily as a source of cheap cloth, especially Coromandel calicoes, to barter for spices in the East Indies where there was little demand for European goods. Consequently trade concessions from the Mughals were vital to develop the promising spice trade. The Mughal empire—the land of 'the Great Mogul'—now extended from the west coast across northern India to the Bay of Bengal, consolidated by excellent road and river communications. In August 1608 the first East India Company ship anchored off Surat on the west coast of India with William Hawkins bearing a letter from James I to Jahangir, requesting trading facilities. In 1612 the Emperor granted the Company permission to establish four trading posts or factories within his territories; one was at Surat, the principal outlet for goods from towns in the interior such as Agra, Delhi, and Ahmedabad. The massacre of the personnel of the Company's factory at Amboina in 1623 was a successful conclusion for the Dutch of their aggressive policy against the small British presence in the East Indies. Right up until the end of the eighteenth century the Dutch were to be the undisputed masters of commercial transactions there. This setback compelled the British East India Company to look to India for an expansion of their trade, a decision that was so vigorously pursued that by the 1640s the English had over twenty factories and settlements in India. When the island of Bombay with its magnificent harbour was occupied by the British in 1665 (it was part of the dowry of Charles II's Portuguese Queen), the Portuguese Viceroy at Goa was dismayed. 'I foresee that great troubles that . . . . . will result to the Portuguese and that India will be lost on the same day on which the English nation is settled in Bombay'. It proved a most perceptive prophecy. In 1674 the headquarters of the East India Company were transferred to Bombay, and Surat, 150 miles to the north, gradually lost its status as a major port.

William Finch was a member of William Hawkins's 1608 mission. During his four years in India he travelled extensively in the north on behalf of the Company, occasionally buying and selling indigo. He provides a first-hand assessment of indigo manufacture and a precise description of the *Indigofera* plant.

> *The herbe Nill [Indigofera] groweth in forme not much unlike cives [chives] or cich-pease [chickpea], having a small leafe like that of Sena but shorter and broader and set on a very short foot-stalke, the branches hard and of a woodie substance like unto broome. It usually groweth not above a yard high and with a stalke at the biggeste (which is at the third yeare) not much exceeding a man's thumbe. The seed is included in a small round codde about an inch long, resembling Foeni-graecum, save that it is more blunt at both ends, as if it had been cut off with a knife. It carryeth a small flower like that of Hearts-ease: the seed is ripe in November and then gathered . . .*[3].

Finch had a sharp eye for detail. Of the water chestunt (*Trapa bispinosa*) he wrote that 'the fruit is inclosed with a three cornered shell, of a hard woodie substance, having at each angle a sharpe picked prinking point, and is a little indented on both the flat sides like two posternes'[4].

The Reverend Edward Terry (1590–1660), chaplain to Sir William Roe during his embassy to Jahangir from 1615 to 1619, is less observant, more cursory in his reaction to plants. Among the 'great varietie of faire goodly trees', only the banyan prompted a moment's pause. The flowers, by and large, disappointed him. 'They rather delight the Eye, then affect the Sense, in colour admirable but few of them, unless Roses, and one or two kindes more, that are any whit fragrant.'[5] Cotton excited a more expansive response.

It blowes first into a yellow blossome, which falling off, there remaynes a cod about the bignesse of a man's thumbe, in which the substance is moyst and yellow, but as it ripene, it swels bigger till it breake the covering, and so in short time becomes white as Snow[6].

Such comments, interesting though they be, added little to the corpus of current botanical knowledge and it needed a trained mind and a disciplined eye to investigate and classify this bewildering flora. And these skills were to be provided by surgeons in the medical service in India during the eighteenth and nineteenth centuries. Their names constitute a roll-call of some of the most distinguished botanists in India: Koenig, Heyne, and Wallich (officials of the Danish settlements), Roxburgh, Griffith, Thomson, Anderson, Aitchison, King, and Prain. Some instruction in botany had always formed part of the curriculum of medical students and if they were fortunate enough to be at the University of Edinburgh under Professor John Hope they were assured of a thorough grounding in the Linnaean system. One of Hope's brightest pupils was William Roxburgh, who owned his botanical indoctrination to Edinburgh.

John Fryer (*c*.1650–1733), a medical graduate of Cambridge, was not in the same intellectual league but he had an enquiring and eclectic mind. He lightened the tedium of the long sea voyage to Bombay where he was to be surgeon by observing seabirds and marine life. As soon as he got to India he set about dissecting animals, examined birds' gizzards, and pondered the behaviour of the weaver bird. When he was transferred to Surat, he subjected its inhabitants, European as well as Indian, to his critical scrutiny. The vegetation in and around Surat inevitably came within his purview. He inspected the banyan tree worshipped by local natives and noted, but surprisingly without comment, that some banyan trees were reputed to be able to shelter 30 000 horsemen. That odd fact and all the other data he had diligently accumulated were conscientiously reported, as befitted a future Fellow of the Royal Society, in his *New account of East-India and Persia in eight letters* (1698) (Fig. 3.1). That he had consulted the *Hortus Malabaricus* is confirmed by his adaptation of some of its engravings for the botanical plates in his book (Fig. 3.2).

Another resident of Surat, John Ovington, enjoyed his two years there as chaplain at the English factory during the 1690s. The variety and volume of goods that passed through this 'most famed emporium of the Indian Empire' astounded him. The Dutch supplied the town with spices from their eastern plantations and Ovington was sceptical about their account of the propagation of nutmegs.

They relate a passage somewhat strange and surprizing concerning the nature of the nutmeg-tree, that it is never planted, and if it be it never thrives; but such of them as fructify and arrive at perfection, arise from a ripe nutmeg swallowed whole by a certain bird in those islands, which disgorges it again without digesting it, and this falling to the ground with that slimy matter it brought along with it, takes root and grows a useful tree: But this may be a subtle contrived story of the Dutch, to keep men from endeavouring to transplant them[7].

While the British were establishing bases, other nationalities were determined to par-

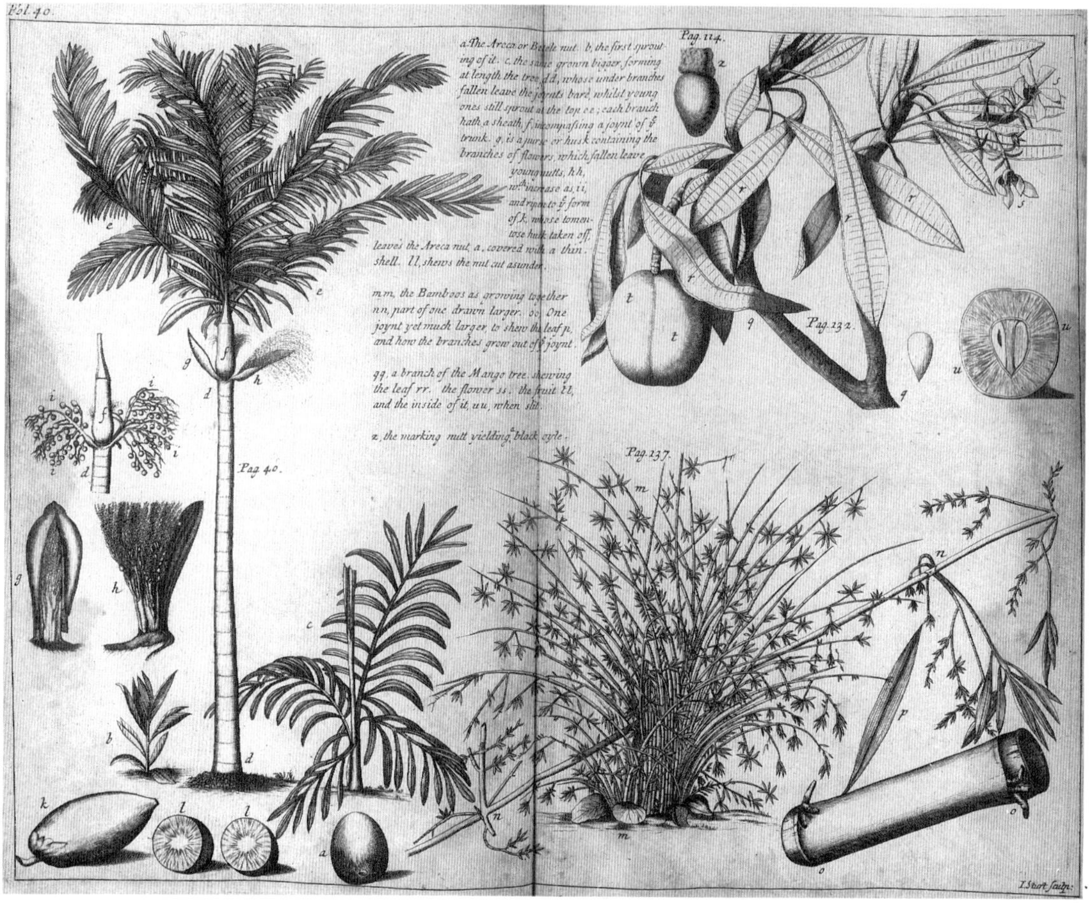

Figure 3.1 Betel-nut palm, *Areca catechu* (left); Mango, *Mangifera indica* (top right); Bamboo, *Bambusa bambos* (bottom right). (From J. Fryer (1698), *A new account of East-India and Persia; being nine years travels, 1672–1681,* London, plate 40.) Engraving. Fryer copied some of the plates in Van Reede's *Hortus Malabaricus* (see Fig. 3.2).

ticipate in the India trade. Emulating Dutch and British practice, they formed their own commercial companies. In 1620 the Danes obtained a lease of Tranquebar from the Governor of Tanjore. The French, who were to be serious contenders for supremacy in the subcontinent, formed six trading companies between 1604 and 1652 before launching the Compagnie des Indes in 1664; their main settlement at Pondicherry was purchased from Sher Khan, the Governor of Jinji ten years later. India was also a lure for individuals in search of adventure, employment or wealth including François Bernier, who was a physician to the Mughal nobility; Jean Tavernier, who practised his craft as a jeweller; and Niccolao Manucci, who served as an artilleryman in Dara Shikoh's army. Pietro della Valle was without doubt a romantic. After an unhappy love affair he left his native Italy in 1614, vowing never to return. He wandered through the Middle East to Persia and disembarked at Surat in 1623. He spent nearly two years in India, getting as far south as Calicut. He was the ideal traveller, filling long letters home with everything of interest he saw or heard. When he tasted 'many strange Indian fruits' he

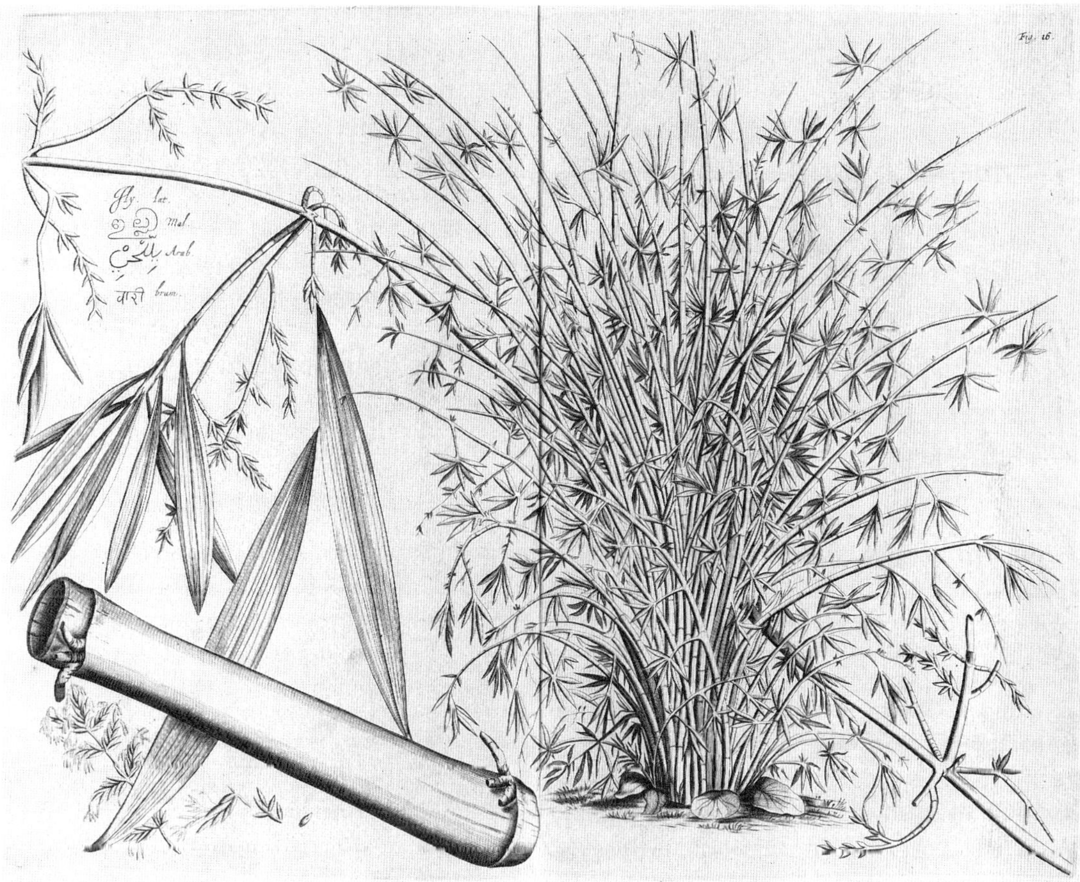

Figure 3.2 *Bambusa bambos.* (From Van Reede (1678), *Hortus Malabaricus,* Vol. 1, Amsterdam, fig. 16.)

remembered what Clusius had written about them in his editions of da Orta and Acosta. He was intrigued by Indian eating habits—the use of large banana leaves as disposable plates, for example, or the vegetable ingredients of curry.

Such random references to exotic trees and flowers in correspondence and books were tantalizing to botanists and gardeners back in Europe eager to see them for themselves. One of the most obsessive collectors in the natural history field was James Petiver (*c*.1664–1718), a London apothecary with premises at the White Cross in Aldersgate Street. He assiduously cultivated the friendship of ships' captains and surgeons, merchants, missionaries, and officials stationed abroad—anyone, in fact, who could be the means of obtaining specimens for him. He cajoled, he bullied, he persisted, he seldom gave up. He even printed a single sheet of 'Brief Directions for the Easie Making and Preserving Collections of all Natural Curiosities for James Petiver Fellow of the Royal Society London' which he distributed to his global network of correspondents. His friend, the naturalist John Ray, said 'he hath the greatest correspondence both in East and West Indies of any man in Europe'[8]. One of his victims was Samuel Browne (*d.* 1698), surgeon on the Madras Establishment at Fort St George. For some years he was a model correspondent, dutifully despatching dried plants and seeds to London. Petiver publicly acknowledged his indebtedness.

> *To the indefatigable Industry of this Worthy and Generous Person, I am beyond expression obliged, for his large and frequent Performances; is not only gathering himself, but constantly employed several Persons to make Collections for me, and many of them some days Journy up the Country, from such Places we never yet saw any Plants before. And to most of these he has added their Indian Names, and the Vertues to such as are known to have any ... He hath also procured me a Correspondence with divers Ingenious Persons residing in remoter Parts; and the like I could heartily wish my other Friends would do*[9].

But this cordial relationship had its moments of tension. Perhaps Petiver had been too importunate or, one suspects, Browne too easily provoked. His career in India had been marred by several unfortunate incidents. When he accidentally poisoned the Chief Justice, criminal negligence was imputed; a duel and an assault on a native brought him once again before the magistrates. After his post as surgeon was abolished he became Assay Master at Madras. No longer having any time to help Petiver, he readily approved Edward Bulkley's nomination as his successor as plant collector. Bulkley, he thought, 'has more leasure and also two Englishmen to help him but at present I have none, being discharg'd the surgery affair . . . I intend to make a clear house and to have no more to do with botany except as before I hinted, it be by the by'. His letter to Petiver ends on a conciliatory note: 'you had lesse cause to be angry then you suppos'd, let us talke more familiarly and tho you will not correspond wth me in plants pray let not our friendship dye'[10]. Presumably as a gesture of reconciliation, Petiver received in 1699 some twenty folio volumes of dried specimens collected in India, Ceylon, and distant China.

Two years earlier Browne had presented the East India Company with two volumes of Indian plants annotated with ethnobotanical notes. These were transferred to the Royal Society[11] and described in the Society's *Philosophical transactions*[12] in eight instalments by Petiver who identified many of them with the aid of the *Hortus Malabaricus*. It was Petiver's practice to publish additions of plants, animals, insects, and shells to his collections in a series of 'centuries' under the general title of *Musei Petiveriani*. He assumed his helpers would be flattered to see their names in print which also gave him

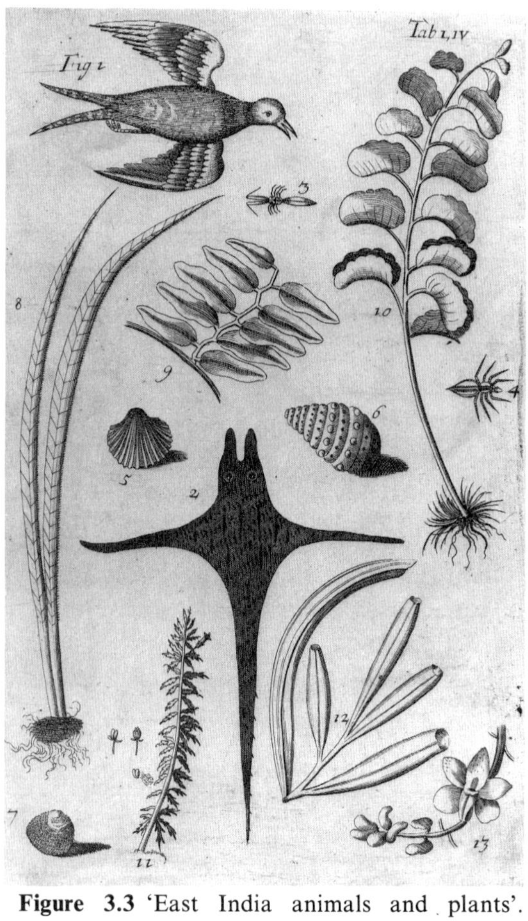

Figure 3.3 'East India animals and plants'. Mainly from Malabar. (From J. Petiver (1764), *Opera,* Vol. 1, London, plate 54.) Engraving.

an excuse to appeal for more volunteers to enrich his collections.

From these published expressions of thanks in the 'centuries' one can discover Petiver's other sources of Indian plants. There was Samuel Browne's successor as surgeon, Edward Bulkley (*c*.1651–1714), who promised annual deliveries including plants from other parts of the subcontinent; Captain John Conway, who not only safely brought back specimens but collected himself 'some quires of plants' from Fort St David, south west of Madras; and John Foxe, a surgeon, who forwarded material from Bengal. Other collaborators who were not so honoured in print were Richard Sambach and Benjamin Mewse.

Petiver permitted a fellow botanist, Leonard Plukenet (1642–1706), a physician at Westminster where he maintained a small botanic garden, to consult Browne's specimens. Notwithstanding his appointment as Queen Mary's Botanist and her Superintendent of the gardens at Hampton Court, his contemporaries had little regard for his botanical expertise. His *Phytographia* (1691–2) was at that time the largest anthology of plants, culled from many quarters, and inflated by careless duplication. Petiver complained of this habit in respect of one of Plukenet's Indian plants. 'That most celebrated Botanist, Dr Plukenet, may well boast of his innumerable species of Plants, if he multiplies others as he has done this, making it three different herbs.'[13] But Plukenet could give as good as he got; on a plant label he contemptuously dismissed Petiver as 'our Neophytous Centuriator'[14] and to add insult to injury he acknowledged Samuel Browne's plants to du Bois and not to Petiver.

Charles du Bois (1656–1740) was a London merchant and, like Petiver, a member of the Temple Coffee House Botany Club which met on Friday evenings to discuss plants, native and foreign. He took advantage of his position as Treasurer of the East India Company to acquire plants and seeds from India. Mr Randal, a gardener at Fort St George, Madras, was one of his suppliers; his stepbrother, Daniel, and his sister also collected for him in the neighbourhood of Madras. A scientific awareness, rather than a passing curiosity, was emerging and plants were now being collected for personal study and for botanists in Britain. Gatherings made by Mr Hynmers and Mr Gifford at Madras and Dr Daniel Waldo at Surat were presented to the Botanic Garden at Oxford; ship's surgeon, Alexander Brown, collected for Petiver and du Bois as well as for Oxford.

John Ray (1627–1705), the most distinguished botanist in Britain at that time, incorporated the work of these distant collectors in his monumental survey of the plant kingdom, *Historia plantarum* (1686–1704). Many of the 18 000 plants, arranged according to his own classification, were exotics, often passed to him by Petiver. '[I] return you many thanks', he told Petiver, 'for the great pains you have taken to enrich my History with such a multitude of rare & nondescript plants from China, India, Africa & America.'[15] But his greatest debt was to Van Reede from whose work he selected and succinctly described 615 plants[16]. The *Hortus Malabaricus* evokes for Ray a vision of a botanical paradise.

> *Who would believe that in the single province of Malabar, and that not so very large, there are found more than three hundred wild trees and shrubs, and probably many more? . . . This too is remarkable and peculiar to India that there are trees that shed their leaves when they flower and again burst into leaf when they set fruit . . . Who would believe that there is a botanical Europe in the middle of tropical India?—so that notable traveller Bernier assures us. In the land of the Mogul (Mogorum Imperator) he crossed high mountains; on their southern slopes Palms, Pepper, Sugar-canes, everything Indian and exotic; after six or seven hours travel, beyond the summits, a wholly new*

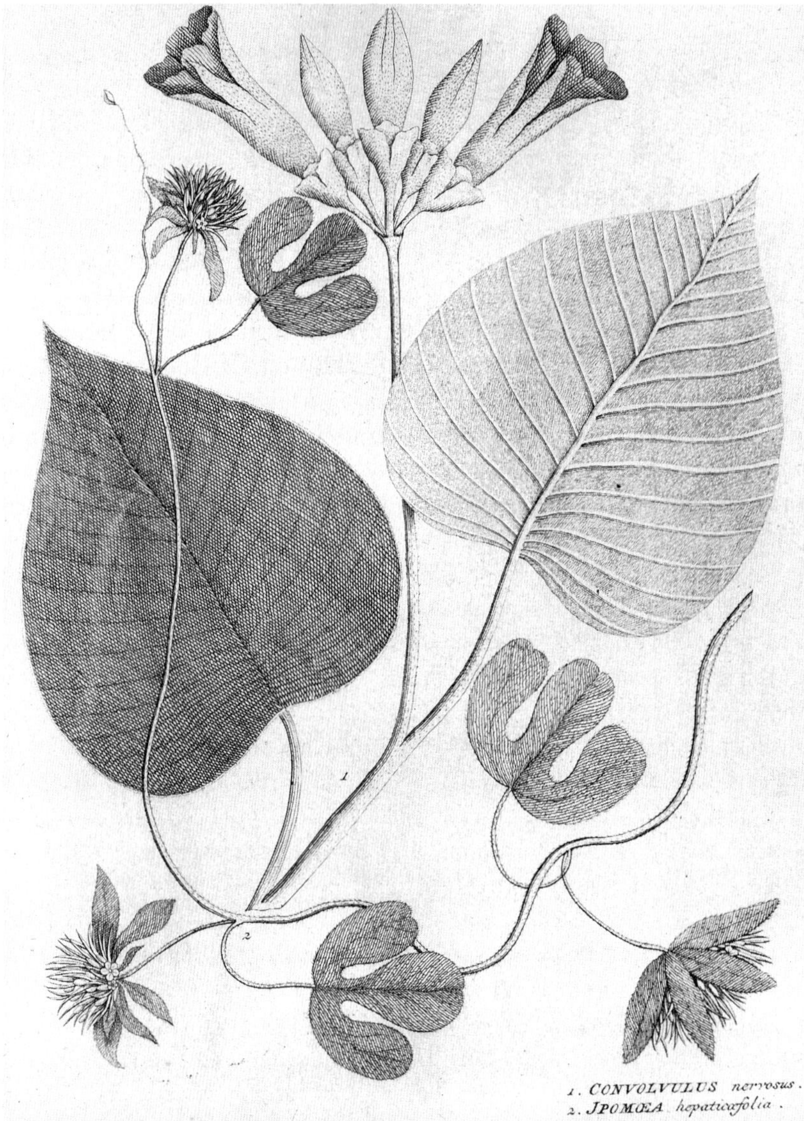

Figure 3.4 *Argyreia nervosa* and *Ipomoea pestigridis.* (From Nicolaus L. Burman (1768), *Flora Indica,* Leiden, plate 20.) Engraving.

scene, a swift passage into Europe, Oaks, Elms, Pines, Hyssop, Rosemary[17].

Both Ray and Petiver were recipients of plants from Georg Eberhard Rumphius (1627/8–1702), a German employed as a merchant by the Dutch East India Company on Amboina in the Moluccas. His devotion to botanical research on that small island is probably without parallel. His fortitude in the face of a succession of disasters which would have defeated a less resolute man is invested with tragic grandeur. In 1670 he became blind; four years later his wife, who had been his amanuensis, and his youngest daughter died during an earthquake; in 1687 a fire destroyed all his botanical drawings; in 1692 his manuscripts and replacement drawings, representing more than thirty years' labour, were lost when the ship taking them to the Netherlands for publication was sunk. Fortunately, a copy of the manuscripts had been retained and the old man set about preparing a fresh copy for publication which reached Amsterdam safely in 1696. Rumphius, however, died while the Dutch East India Company dithered. It was eventually rescued from the Company's archives by Johannes Burman (1707–79), Professor of Botany and Director of the Amsterdam Municipal Garden. He persuaded eight Dutch publishers to join him in printing *Herbarium Amboinense* in six volumes (1741–50). Many of the 1700 or so plants the work described were widely distributed throughout the Indian Archipelago. The descriptions are considered to be superior to those in the *Hortus Malabaricus* whereas the plates are inferior, being often reduced in size and inaccurate. But Rumphius cannot be held responsible for drawings which his blindness prevented him from correcting.

Johannes's son, Nicolaus Laurens Burman (1733–93) succeeded his father as Professor and Director at Amsterdam. He also inherited an incomparable herbarium of Asiatic plants which provided him with much of his data for a tropical flora, the *Flora Indica* (1768)—a misleading title since it embraces the New as well as the Old World, North and South America, the West as well as the East Indies and the Cape of Good Hope to China and Japan (Fig. 3.4). About one-third of the 1 300 species was collected in India, and occasionally precise localities are given. Burman also consulted contemporary botanical works and acknowledged some help from Linnaeus. It has 67 plates delineating 178 plants and Wight and Walker-Arnott preferred them to the text which they dismissed: '. . . the names adopted and the synonyms are frequently so erroneous, that it is scarcely entitled to rank among the scientific works of modern times.'[18] Harsh words and, indeed, it must be admitted that Burman not infrequently misinterpreted Linnaean species. Nevertheless, it still holds an honourable place in the history of Indian botany. It was a gallant attempt to compile a general flora of the tropics and a pioneer in the adoption of Linnaeus's binomials in Indian floristic studies.

In making little distinction between the tropical flora of the eastern and western hemispheres, Burman had before him the example of Linnaeus who was equally ignorant of the extent of regional diversity. Carl Linnaeus (1707–78) needs no introduction to anyone with any familiarity with botany. When he left Sweden in 1735 to obtain a medical degree at Harderwijk he saw for the first time flowers from Asia in well-stocked Dutch gardens; he had the opportunity to consult tropical plants in private herbaria and even found time to assist Johannes Burman with his *Thesaurus Zeylanicus*. The intellectual climate of the times was ripe for his intervention. Gradually, botany was being weaned from medical studies of which it had been an integral part for so long; it was at last becoming a legitimate study in its own right, not merely a source of new vegetable drugs. The quest of naturalists was for the key with which they could place all living things into God's grand design. Thus the classifying and naming of animals and plants became an honourable pursuit, an end in itself. Accumulating plant specimens, mounting them carefully on sheets of paper in books or folders was a routine matter but naming them in concise Latin phrases, descriptive of their essential features, and subsequently arranging them in some rational order, was problematical and controversial. There was little consistency, systems proliferated, and confusion reigned. John Ray produced a classification which had its adherents in Britain while the French were loyal to Tournefort's scheme. Linnaeus provided two sets of names for a plant: a concisely worded diagnostic name describing the characteristics of the species necessary for identification and, from 1753 onwards, a two-word designation comprising a generic and specific name. Linnaeus, being a pragmatist who doubted that a natural classification was ever attainable, devised an artificial scheme that was deceptively simple: twenty-three classes based upon the number of stamens in a flower and a final class for non-flowering plants. It offered an easily understood method for accommodating the deluge of plants from every known quarter of the globe. It enjoyed an uncontested popularity until it was superseded by natural classifications in the nineteenth century. What has endured is his binomial nomenclature. Linnaeus's students, or 'apostles' as he whimsically called them, loyally tested their master's precepts wherever their travels took them. His classification and nomenclatural rules were an incentive to botanical exploration and welcomed by botanists, especially in India and the East, bewildered and confused by the complexity of the vegetation around them.

NOTES

[1] Santapau (1958), 188–9.

[2] R. Hakluyt (1905), *Principal navigations, voyages, traffiques & discoveries of the English nation*. Vol. 4, London, 502–4.

[3] S. Purchas (1905), *Hakluytus posthumus or Purchas his pilgrimes*, Vol. 4, London, 45–6.

[4] Ibid., 43.

[5] Ibid., (1905), Vol. 9, 20.

[6] Ibid., 22.

[7] J. Ovington (1976), *A Voyage to Surratt in the year 1689* (ed. J. P. Gupta), New Delhi, 99.

[8] J. Ray (1928), *Further correspondence*, London, 279.

[9] Petiver (1699), 43.

[10] Browne to Petiver, 30 September 1698. B.L. Sloane MS 4062, folio 290.

[11] Now in B.M. (N.H.) together with Petiver's collections.

[12] 'An account of part of a collection of curious plants and drugs lately given to the Royal Society by the East India Company.' *Philosophical Transactions of Royal Society*, **22**' 1700–1, 579–94, 699–721, 843–58, 933–46, 1007–22; **23**' 1702–3, 1055–65, 1251–65, 1450–60.

[13] Ibid., **22**, 1700–1, 584.

[14] B.M. (N.H.) Sloane herbarium, Vol. 93, folio 119.

[15] 17 May 1704. B.L. Sloane MS 4063, folio 187.

[16] Heniger (1986), 172.

[17] J. Ray (1688), *Historia plantarum*, Vol. 2, London; translation of part of Preface given in C. E. Raven (1950), *John Ray: naturalist* (2nd edn), Cambridge, 231.

[18] Wight and Walker-Arnott (1834), x.

BOTANICAL RESEARCH BEGINS

In December 1771 Linnaeus was pleased to inform his friend, John Ellis, a London merchant, that his former pupil, Johann Koenig, had 'found a lot of new things in Tranquebar' in India. Tranquebar, a coastal town some 150 miles south of Madras, had been leased by the Danish East India Company from the Raja of Tanjore in 1620. Prosperity always eluded it and eventually in 1845 the Danes sold it without regret to the British. The natural history interests of some of the missionaries stationed there provided a focus for botanical activity among Europeans in the south. A Protestant Moravian mission, directed from Halle University in Saxony with the co-operation of the Danes, had been operating in Tranquebar since 1702. It would be reasonable to suppose that botanizing was only taken up seriously by some of the missionaries or 'United Brethren' as they called themselves, after Koenig's arrival there in 1768 as surgeon. He initiated them in Linnaean methodology; they exchanged specimens, formed a herbarium, named plants which until experience gave them confidence they did collectively, applying the word 'nobis' (named by all of us) to their plant labels. They sent plants to their European correspondents including Joseph Banks in London. In 1788 the Tranquebar Society, a learned society modelled on the recently formed Asiatic Society in Calcutta, attracted naturalists as well as orientalists, linguists, and historians. The missionaries were some of its most active members; one of their number, Christoph Samuel John, contributed several papers on the local flora.

Modern botany in India had its genesis in the researches of Johann Gerhard Koenig (1728–85) (Fig. 4.1). Of Baltic German origin, he became a student of Linnaeus in 1757 and, although he went to India as a physician in 1768, his real objective, according to his friend Patrick Russell, was to pursue his vocation as a naturalist and 'more covetous of fame than fortune', he quickly immersed himself in the study of the Madras flora. Finding it difficult to finance field trips on his modest income, he accepted in 1774 a post of naturalist to the Nawab of Arcot, a move which enabled him to make excursions to the nearby hills and even a voyage to Ceylon. While he was in the

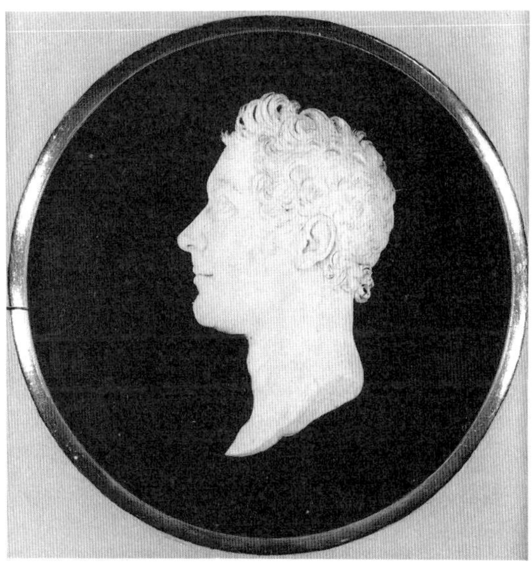

Figure 4.1 Johann Gerhard Koenig (1728–85). Small wax medallion at Kew.

Nawab's service he often stayed in Madras, sharing his botanical interests with fellow physicians in the British community, notably James Anderson, George Campbell, and William Roxburgh. His ultimate goal, a transfer to the British East India Company, was realized in 1778; the following year his new employers sent him to Siam and the Straits of Malacca to seek economic plants such as cardamom and gamboge for cultivation in India. His knowledge and botanical competence impressed all who knew him and William Roxburgh, sometimes his companion on field trips, admired his complete commitment to botany. Patrick Russell, who constantly urged him to submit to the Court of Directors of the East India Company a selection of his notes and drawings as evidence of his researches, was concerned that he left precise instructions for their dispersal in the event of his death, always imminent in India's unhealthy climate. He recommended Banks as a suitable custodian to whom Koenig, acting on his friend's advice, bequeathed all his manuscripts. Meanwhile Russell pressed him to employ an amanuensis to transcribe his illegible handwriting. Koenig's death in June 1785 from dysentery was an 'irreparable loss' to Russell, dismayed to find nothing among his papers fit for publication. A number of his plant descriptions were published in the *Mantissa plantarum altera* (1771) of Linnaeus, the *Supplementum plantarum* (1781) of the younger Linnaeus and in A. G. Retzius's *Observationes botanicae* (1779–91). Company officials at Fort William had 'little doubt, if he had lived, that his researches would have been crowned with extensive and useful discoveries'[1]. Joseph Banks, a beneficiary under his will, believed he had repaid the East India Company 'a thousand-fold over in matters of investment, by the discovery of drugs and dying materials fit for the European market'[2]. Sir William Jones in Calcutta sent his condolences to Russell. 'Dr Koenig's loss will be severely felt; he was a valuable man, with as much simplicity as nature itself, whose works he studied.' Then, seemingly, as a casual aside, he added: 'Do you know when his books are to be disposed of? I should wish to purchase his Linnaeus'[3]. Banks suggested that a suitable successor might be found in one of the universities of northern Europe 'where young men who shew a natural disposition for such studies, are educated principally with a view to natural history. They are taught not botany only but zoology, mineralogy and chemistry'[4]. Francis Masson, Kew Garden's plant collector in the Cape, was keen to fill the vacancy but Banks had his eye on another candidate who had studied medicine at Uppsala in Sweden, was a good all-round naturalist, and had botanized in the West Indies and in North America. Banks petitioned Lord Liverpool on behalf of his favourite, Olof Peter Swartz[5], and confided his choice to (Sir) James Edward Smith, but the salary was insufficient to tempt his protegé and it was Patrick Russell who filled Koenig's post.

John Peter Rottler (1749–1836), a graduate of Strasburg University, joined the Moravian Mission at Tranquebar in 1776. Within a year of his arrival he had married the widow of a Dutch sea captain and delivered his first sermon in Tamil. He botanized in the countryside around the Danish settlement and two excursions to Ceylon gave him an opportunity to investigate the island's flora. The intensity with which he conducted his botanical researches alarmed Father John who confided to William Roxburgh that 'my dear Rottler is lingering under several indispositions & I fear he has hurt his health by running about too much after plants & by sitting up night & day to bring his herbarium in a systematical order which is now finished & indeed in an elegant manner containing near 2000 plants'[6]. While he was in charge of missions in Vepery and Madras from 1803 he began compiling a *Dictionary of the Tamil and English languages* which included a list of vernacular names of South Indian plants. Both the University of Erlangen and Willdenow in Berlin were indebted to him for many specimens. He published descriptions of new species—'very imperfectly' according to R. Wight—in several European periodicals. When he died his herbarium had grown to almost 5000 specimens[7] and Wallich, the Superintendent of the Botanic Garden at Calcutta, honoured him as 'a great orientalist, and ardently attached to the study of plants'.

Rottler obligingly verified the names of the plants collected by two other members of the Moravian Mission—Klein and Heyne. Johann Gottfired Klein, born in 1766 in Tranquebar where his father was a missionary, went to school at his father's birthplace in Prussia, studied medicine at the Universities of Halle and Copenhagen, and returned to India about 1791 to practise medicine. When the post of surgeon became vacant at Tranquebar he was an obvious candidate. Professor Kratzenstein of the Mission College in Copenhagen whose opinion was sought as to his suitability was cautiously non-commital about the proposed appointment. 'I do not know Klein, but I think it would be advisable to grant the doctor to the missionaries who is proposed by themselves: for the situation of a medical man in Tranquebar is not an enviable one: anyone who wishes to obtain such a place must either be a zealous pietist desirous of gaining a higher place in Elysium, or an enthusiastic Linnéanist like Koenig.'[8] While the extent of Klein's piety is a matter for speculation, his enthusiasm for botany is beyond dispute. He searched for plants in the far south of the subcontinent from Quilon on the west coast to the mountain range of the Eastern Ghats. C. L. Willdenow in his edition of Linnaeus's *Species plantarum* (1797–1810) based some of his identifications on Klein's specimens.

Benjamin Heyne (1770–1819), also a physician, joined the missionaries at Tranquebar about the same time as Klein. Becoming quickly disenchanted with the tenor of life there he resolved to return to Europe. Father John, concerned that his departure would be a serious loss to Indian studies, asked his friend, William Roxburgh, in July 1793 whether he could employ Heyne on the pepper plantations he supervised at Samalkot. When Roxburgh was called to Calcutta the following month to be considered for the post of Superintendent at the Botanic Garden, he recommended Heyne to take charge of the plantations during his absence. Heyne, appointed as temporary botanist, became the second resident of Tranquebar to enter the service of the British East India Company. His duties were to supervise the introduction of useful trees and plants to the territory above the Coromandel coast known as the Northern Circars, in particular, those that would offer alternative sources of food in time of famine such as breadfruit from the Nicobar Islands and potatoes (Fig. 4.2). In 1799 he was drafted as botanist on Colonel Colin Mackenzie's survey of Mysore. When the Madras Presidency decided to close the Samalkot plantation in 1800, the Governor, Lord Clive, instructed Heyne to seek a suitable site in Mysore for a new botanical garden. Heyne chose a former royal garden belonging to the late Tipu Sultan at Bangalore. When ill-health necessitated Heyne's withdrawal from the Mysore survey he obtained the post of the Company's Naturalist and Botanist in the Madras Presidency in 1802. The Company's expectations of their Naturalist were excessively optimistic. He was presumed to be a linguist, demographer, antiquarian, meteorologist, mineralogist, and zoologist. What was required of him as a botanist alone was more than enough to keep him fully occupied. He was to report on agricultural methods and the crops cultivated by Indians:

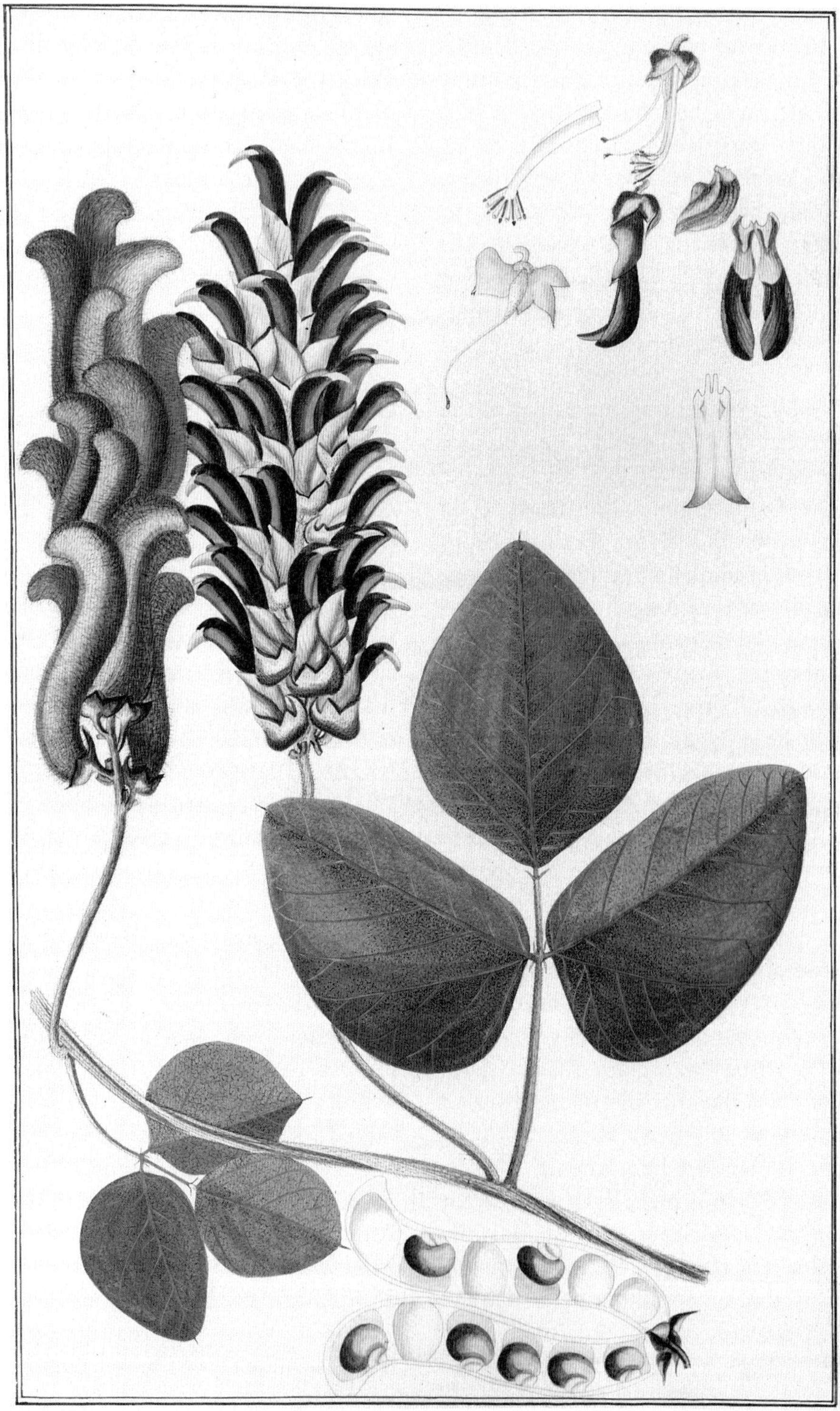

Figure 4.2 *Mucuna pruriens*. One of a series of watercolours of the vegetation of the Coromandel coast executed for Benjamin Heyne, probably by an Indian artist.

Trees and shrubs which are natives of the soil & which introduced; hints respecting the improvement [with] prevailing kinds of woods & shrubs, which are used in husbandry as medicines & for what disorders or food by artificers and for fuel [and] hedges; drawings of these as well as other rare plants with scientific descriptions & specimens of the same[9].

Responding to a request from Heyne, Lord Clive made available a couple of rooms at the Government Garden House for the reception of a 'cabinet of curiosities' or small museum of minerals, insects, birds, and plants. 'In fact without such a repository supported by the Government', Heyne assured the Governor, 'nothing but drawings will perpetuate the labours of a natural philosopher in India and it is well known that these, let them be ever so correctly attended to, still require in the eyes of the learned a positive proof only obtained by specimens.'[10] Heyne's submission in September 1802 for an establishment of a clerk, an Indian artist and two plant collectors to accompany him on his tours was also granted. Within a year he sought a further favour from Clive—a temporary posting to an outstation where he could recuperate and, at the same time, organize his vast collections of minerals, plants, and drawings 'which if longer put off may become unintelligible even to myself'. Lord Bentinck who succeeded Clive as Governor of the Madras Presidency in the summer of 1803, in due course questioned the benefits derived from Heyne's tours. He considered his time would be better employed in supervising the Bangalore garden and extending its cultivation of useful plants. Three years later, pleading ill health, Heyne obtained a post as surgeon at Cuddapah; in 1812 he left on three years' European leave, resuming his position as Naturalist and Botanist on his return. On his death at Vappera in February 1819 he was succeeded by W. S. Mitchell who also died within a year; assistant surgeon George Hyne became acting Naturalist ('and Botanist' was dropped from the designation of the post in 1825) until James Shuter, newly arrived from England, was appointed in 1822. Shuter returned home in 1826, a very sick man, and Robert Wight became Naturalist until 1828 when the post was abolished. All the occupants of this post had been surgeons—Koenig, Russell, Roxburgh, Heyne, Mitchell, Hyne, Shuter, and Wight. When Heyne had visited London during the course of his leave he had consulted Banks's herbarium to verify the identification of his plants before depositing them in the India Museum belonging to the East India Company. He had intended presenting duplicate specimens to Willendow but, learning of his death, he offered them to his fellow countryman, Albrecht Wilhelm Roth (1757–1834) who selected some 200 species for description in *Novae plantarum species praesertim Indiae Orientalis* (1821). Heyne's surviving collections in India, deemed in too bad a state to ship to England, were transferred to the custody of the Calcutta Botanic Garden. Robert Wight thought Heyne to be 'a diligent collector [rather] than a practical botanist'[11].

Heyne's progress and success as Naturalist and Botanist to the Madras Government were scrutinized and assessed on behalf of the Company by Sir Joseph Banks (1743–1820) to whom Heyne had been instructed to submit annual reports. Banks, as President of the Royal Society (an office he held for over forty years) and adviser to the King on the management of the royal gardens at Kew, and as a skilful entrepreneur, was a desirable confidant and ally (Fig. 4.3). Although eclectic in his interests, botany always remained his favourite relaxation. Having sailed round the world on Captain Cook's *Endeavour* with his own entourage of artists, a naturalist, and servants, the discovery of new lands with the revelation of their flora and fauna, never failed to excite

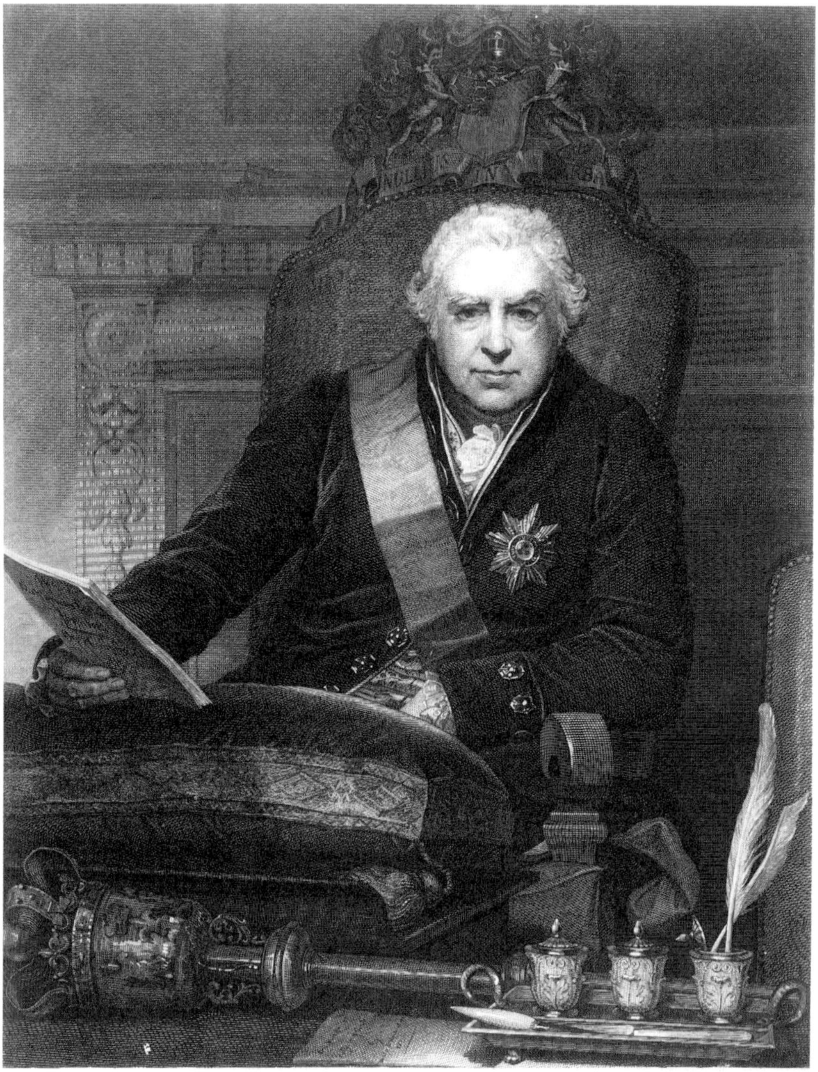

Figure 4.3 Sir Joseph Banks (1743–1820) in the presidential chair at the Royal Society. Engraving by Nicholas Schiavenilli of a portrait by Thomas Phillips.

him. India, its history and culture, stirred the romantic in him, the potential wealth of its natural resources aroused his business instincts. Despite the demands of office and the burden of a vast correspondence, conducted almost entirely in his own hand, he was always ready to advise the Company's Court of Directors. He undertook to find a suitable successor to Koenig; he was consulted about the formation of a botanical garden at Calcutta and until his death concerned himself with its progress; he urged the Company to establish tea plantations in India and to develop other economic products such as cochineal, indigo, and cotton. In return for his services he was content to receive plants for Kew and his own modest garden at Spring Grove in Heston, and for specimens to add to his personal herbarium.

In November 1787 the Company consulted him about Patrick Russell's proposals for a book on Indian natural history. Two years earlier Patrick Russell (1727–1805) had been selected by the President and Council at Fort St George at Madras as Koenig's replacement as Naturalist. Russell, a medical graduate of Edinburgh, had in 1753 followed his half-brother, Alexander, as physician at the

English factory at Aleppo in Syria. In 1781 he joined his brother, Claud, who was administrator at Vizagapatam in South India. There he agreeably filled his time looking at plants and fishes, and observing snakes at a safe distance. His contentment permeated the last letter he wrote to his friend Daniel Solander, Banks's naturalist companion during the *Endeavour* voyage and, subsequently, assistant librarian at the British Museum, not knowing that Solander had died six months earlier. Half in fun, he tried to entice his London friends to India.

> *If you could prevail on our friends at the Shakespeare to make a trip with you, I promise to entertain you all most sumptuously in a grove of Mango, Coconut, Cassia, Palmyra and Tamarind trees. There is not a Gentoo in the country who can make a haggis; but you shall have black puddings, curries of all kinds, non-descript fish & sauce, antelope, wild hog, tongues of the Phanocopterus, roasted floyerchins, flying foxes in ragout, grilled shark and hotch-potch composed of all the culinary vegetables in the Hortus Malabaricus, or mentioned by the learned Don Garzia Dall 'horto*[12].

The same letter *en passant* discloses some of the botanical works he had taken with him to India: *Hortus Malabaricus*, *Herbarium Amboinense*, and most of Linnaeus's publications. He was eagerly awaiting Plukenet's *Phytographia* and Burman's *Thesauras Zeylanicus* from England. He had formed a congenial relationship with Koenig; they botanized together and spent evenings correcting and annotating their copies of Van Reede and Rumphius. His competence and enthusiasm convinced the Madras Government that he was well-qualified to be their Naturalist. He established a cordial correspondence with Sir William Jones, President of the newly formed Asiatic Society in Calcutta, who regretted that Koenig had not bequeathed his papers to Russell as he feared that Banks would be far too busy to do anything about them. Evidently he was not aware that Russell himself had suggested sending them to Banks.

Russell solicited Banks's support in December 1784[13] to persuade the Company to publish and embellish with perhaps a few engravings, Koenig's descriptions of useful and new plants. His friend's unexpected death had made him even more determined to pursue this project to fruition, seeing it as a fitting culmination to Koenig's labours. Almost exactly a year after his appointment as Naturalist, he petitioned the Governor of Madras to sanction the publication of a book on the plants of Coromandel for the use of the Company's personnel[14]. The absence of any botanical guide had hindered European residents in India who were dependent upon descriptions, often inadequate or inaccurate, scattered throughout many books and periodicals, some expensive and bulky like the *Hortus Malabaricus*, and few easily accessible in India. Koenig's papers would yield many new descriptions which Russell was prepared to supplement—he was confident, too, of the collaboration of other botanists in India. He anticipated the first number describing 30 or 40 plants and calculated that four numbers would complete this publication devoted to plants of 'established utility'. He wondered whether the East India Company, a notable patron of other works, could be encouraged to finance the venture with guidance from Banks. His submission was sympathetically received; the Medical Board envisaged it as a compendium of medical plants which would help the young practitioner in India. Pending the approval of the Court of Directors in London, a list of the plants to be included in the first number was circulated to surgeons in the Presidency for their comments and contributions. In a personal letter to Banks, Russell stressed that 'without your countenance we shall cut no figure, and an opportunity may be lost of propagating in this country a spirit of useful research'[15]. Believing he could be more useful by remaining in India, collecting specimens and answering queries, Russell left the choice of editor to Banks but diffidently proposed Jonas Dryander (1748–1810), Banks's librarian. He also favoured William Curtis's elegant and costly *Flora Londinensis* as a model.

The Court of Directors cautiously committed themselves to Russell's project, but conscious of yet another financial crisis threatening the Company, stipulated a close supervision of Russell's progress. 'And you are to require from Dr Russell so long as he shall continue to receive the Company's pay, an annual communication of all his discoveries and improvements, which is to be regularly transmitted to us that we may lay the same before the Royal Society. But whenever Dr Russell shall fail in making such an annual communication as you shall deem worthy [of] the perusal of that learned body his allowance from the Company must cease.'[16] A draft despatch of 31 July 1787[17] instructed Madras to consult Anton Hove, then in India collecting plants for Kew, about Russell's book, but this directive got deleted in the final version. Instead the Court mildly censured Madras for not providing some indication of the cost of the project.

While this dialogue between London and Madras was continuing, Russell was busy in the field. By September 1787 he had a chest of dried plants ready for despatch to Banks; in October the Madras Hospital Board, having received about 900 herbarium specimens from Russell in the Circars, tried unsuccessfully for the services of a Mr Butler, a former pupil of Koenig at Tranquebar, to take charge of the collection. Russell hoped this collection would be the start of a 'Herbarium Coramandeliense', to assist in the identification of the local flora and as a stimulus to botanical study among new recruits to the Medical Service. Ignorant of the commercial aspects of publishing, Russell was unable to cost the book. 'It is my intention therefore in the interim', he told the Governor, 'to persist in the prosecution of my part of the task in expectation of further advice.'[18] The frustration engendered by the protracted official correspondence introduced a jarring note of irritation. Impetuous Russell apologized to Banks 'for using your name with such freedom without your permission'[19]. In November 1788[20] Banks magisterially gave his blessing to the project, suggested some modifications to Russell's outline plan and, with typical thoroughness, provided a breakdown of the production costs. He recommended that each part should comprise 20 rather than 50 plants. He agreed that Curtis's *Flora Londinensis*, currently and very slowly being issued in parts, was an admirable model, combining 'elegance with perspicuity'; '500 copies may be provided elegantly for about £23 a plant (decently for about £18 a plant); a considerable part of which money would be received back by the sale, which should not be less that £1.5 each, if elegant, or £1.1 if decent'. He advised that 'commercial utility' should not be the sole criterion for a plant's inclusion. Russell was to supply dried specimens, coloured drawings, and descriptive notes. Banks offered to find a suitable person to edit the work which he himself was prepared to supervise. He concluded with estimates for an 'elegant' and 'common' edition of 500 copies. James Sowerby, botanical artist and engraver, who supplied the calculations for hand-colouring the plates, had expressed a willingness to undertake the commission. Banks reckoned that a retail price of one guinea for each part of the 'common' edition would yield a modest profit if the edition were completely sold. Impressed by his confidence and assured of his participation, the Court of Directors informed the Madras Government that they were willing to sponsor the book.

Russell, meanwhile, had resigned his post in February 1789 and was on his way home before this despatch reached India. His last acts were to present a cabinet of dried plants to the Madras Presidency and to nominate as his successor 'Mr Roxburgh of Samalcotah, of whose skill and practice in Indian botany, I have unequivocal proofs, and this, conjoined with his local knowledge acquired in the course of a long residence in the country, induces me to think him singularly qualified to be of essential service to the Company in the way the Hon'ble Court of Directors seem to wish, that is in the conducting of researches into Natural History, more with a view to immediate utility, than to the extension of pure botany'[21]. Can a hint of disillusionment or of rancour even be detected in that last comment? Russell, a pioneer herpetologist who was the first European in India to distinguish harmless from poisonous snakes, spent his remaining years checking Roxburgh's descriptions and preparing for publication his *Account of Indian serpents, collected on the Coast of Coromandel* (1796–1807), a folio that in the quality of its engravings matches Roxburgh's *Plants of the Coast of Coromandel.*

William Roxburgh, born in Ayrshire in 1751, studied medicine at Edinburgh University where he had the benefit of being initiated in the rudiments of botany and Linnaean taxonomy by Professor John Hope. Roxburgh was later to name *Hopea odorata*, a species of Asian timber tree, after his old tutor. He entered the service of the East India Company, first as a surgeon's mate on the Company's ships and in 1776 accepted a post of assistant surgeon in the Madras General Hospital. He met Koenig whose knowledge of the India flora stimulated his own studies. Roxburgh, who was at his side when he was dying, recalled years later that one of the last plants Koenig saw was *Roxburghia gloriosoides* (now *Stemona tuberosa*). 'He did attempt to examine it but was unable; for the cold hand of death hung over him. He desired that I would describe it particularly, for he thought it was new, and uncommonly curious and beautiful.'[22]

Promoted to full surgeon in 1780, Roxburgh spent most of his early service in the Northern Cicars at the garrison station at Samalkot some 200 miles north of Madras. He perceived that pepper and coffee might profitably be cultivated here and obtained permission to establish experimental gardens. Had this not been allowed he had resolved to return to Scotland with his wife and two small children in the hope of better prospects. Such intentions were fortunately dispelled by the success of his plantations of over 50 acres. He grew coffee, pepper, and breadfruit, carefully cossetted some cinnamon and indigo specimens, experimented with *Opuntia*, the host plant of the cochineal insect, and planned a pilot scheme for silk production from Bengal silk-worms. Madras, pleased with his horticultural expertise, allowed him to extend his plantations to a capacity of 100 000 plants. Roxburgh was now sufficiently experienced and confident to float the idea of a botanical garden in the Circars. He still found time to botanize and in October 1789 could boast that two native artists he employed had produced between 400 and 500 botanical drawings[23]. The reputation of this promising botanist eventually reached Joseph Banks who in March 1778 invited the young man to send him plants for identification. He was clearly a strong candidate for Russell's vacant post and in 1790 the Company in London readily endorsed the appointment of the Madras Government's nominee. Sir William Jones was confident of the valuable contributions he would make to the botany of India's east coast provided he could enjoy some remission from severe migraine. Fortunately, neither pain nor the loss of his entire library of manuscripts in a flood could deflect Roxburgh from a whole-hearted dedication to

Russell's project. In September 1790 he sent London the first of many consignments of drawings and botanical descriptions. While informing Banks that he had already accumulated about 700 drawings, one third of which were of grasses, he requested his frank opinion of their quality and the confirmation of his continued support. Banks evidently made some quibble and provoked Roxburgh's aggrieved response. 'It would have given me much satisfaction if you had mentioned what the defects were that my drawings and descriptions had. I would then, probably, have been able to rectify them in those that are still to finish, but you have left me in the dark.'[24] This criticism, although invited, had irked Roxburgh who turned to Sir James Edward Smith, President of the Linnean Society of London, for his opinion. 'Some two years ago Sir J. Banks wrote me that my drawings labourd under the same defects Dr Russell's did, but to this hour I know not what those defects are so mine must continue to be wanting.'[25] Sir Joseph's silence was not, however, an expression of disappoval for in July 1794, having now examined some 500 drawings from Roxburgh, he unreservedly commended them to the Court of Directors[26]. He praised the skill of the Indian artists, particularly the precision of their diagnostic details, confident that European botanists would find them superior to the engravings in both the *Hortus Malabaricus* and the *Herbarium Amboinense*. He submitted for the Directors' consideration specimens of coloured and plain engravings of some of the drawings. In order, no doubt, to influence their decision, he pointedly reminded them that Roxburgh had himself paid for all the drawings to be made. Banks now recommended a publishing schedule of two fascicules a year, each one containing 25 plates. With an edition of 500 copies he calculated the unit cost would be fifteen shillings a fascicule. The public, he was sure, would find a retail price of one guinea very reasonable. Knowing his clientele, he belaboured the profit motive—the book would be invaluable to those officials in India interested in exploiting the commercial potential of the country's vegetation. The Company's response was prompt and positive. They accepted Banks's recommendations and advanced £300 for the first fascicule which appeared in May 1795. Of the hundred copies which were at the disposal of the Company, 60 were hand-coloured for presentation to George III, to past and present Directors (this varied between 30 and 33 copies of each fascicule), to Banks and Russell for their collaboration, and to Roxburgh as a 'testimony of the Court's approbation of his services'; the remainder were distributed to distinguished personages and to officials of East India House. Forty plain copies were despatched to India for sale at 12 rupees each—20 to Calcutta, 12 to Madras and 8 to Bombay.

Roxburgh never got his coloured copy of this first number which was lost at sea; he had to be content with a plain copy but before he received it he had written to Banks apologizing 'for a degree of stiffness [that] runs through the whole of my drawings, particularly the first'[27] but he was confident that his artists were improving. Banks concurred. 'You are right in saying that your draughtsmen improve. Many of the last drawings are charming, the grasses in particular are valuable in the highest degree to botanists as the fructification is delineated with accuracy.'[28] The first two fascicules had been well received but after the publication of the fourth number Banks admitted that sales were disappointing; there was a recession in book selling due, so Banks said, to taxation. The Company's margin of profit was further eroded by their profligate presentation of so many coloured copies. Having paid the publisher, George Nicol, over £2000 up to 1799, the Court of Directors demanded a statement of all sales. The news from India was equally depressing. Not a single copy had been purchased in the Madras and Bombay Presidencies despite advertisements in local newspapers. In the summer of 1801 with six fascicules published and total receipts a disappointing £401, the Committee of Warehouses which was handling the book, thought it time to take stock of the venture. It expressed a hope that on the return of peace to war-torn Europe, the book might find a market there. But even should it fail to cover production costs, the Committee still recommended proceeding with it because 'its tendency to promote the study of natural history, and its affording an inducement to the Company's servants to devote a portion of their time and attention to scientific researches were objects well meriting the Company's countenance and support'[29].

Robert Thornton, author of another ambitious work, the *Temple of Flora* (1799–1807), also blamed the Napoleonic wars for its disastrous sales which eventually ruined him. Extravagantly conceived flower books such as W. Curtis's *Flora Londinensis* (1777–98), G. Brookshaw's *Pomona Britannica* (1804 12), J. Sibthorp and Smith's *Flora Graeca* (1806–40), and J. Bateman's *Orchidaceae of Mexico and Guatemala* (1837–43) were the product of national pride (Thornton saw his book as 'a national undertaking'), the skill of superior printers like Thomas Bensley, William Bulmer, Richard Taylor, and Charles Whittingham who made a conscious effort to improve the physical appearance of books, and, of course, the patronage of a coterie of bibliophiles. In 1815, Goethe expressed astonishment at 'the mad luxury carried on in England with regard to books', mentioning in particular 'a botanical work exclusively treating of pine trees, costing 80 guineas' (A. B. Lambert's *Description of the Genus Pinus* (1803–24)). But size proved often to be a decisive factor with the public as the bookseller, Henry Bohn, warned Audubon, who ignored his advice: 'productions of taste are purchased with delight by persons who receive company, particularly, and that to have your book be laid on the table as a pastime piece of entertainment is the principal use made of it, and that, if in compass it needs so much room as to bring shame on other works or encumber the table, it will not be purchased by the set of people who now are the very life of the trade'. Perhaps their elephant folio format contributed to the failure of both Roxburgh's and Russell's books. There was another cogent factor as a Commitee of the House of Commons learnt from a printer in 1818: 'books are a luxury . . . [and] those who would be disposed to purchase books, have not the means of so doing, and are obliged to be frugal'. George Nicol presented his own analysis of the book's failure; lacking regular subscribers it depended on casual sales; surprisingly, no copies had been sold in Germany, 'the great mart for books of natural history'; frequent delays in its publication consequently raised doubts about its completion. When the twelfth and final fascicule was published in 1820, he informed the Court of Directors that he held the letterpress of about 150 to 160 copies of each fascicule but not the plates since they were printed only when required. Despite this depressing report he remained amazingly optimistic about its future, declaring it way ahead 'in beauty and science' of the *Hortus Malabaricus* which now was appreciating in value in the book world.

Coromandel, the name given to the eastern coast of the Madras Presidency, defined the geographical limits of Roxburgh's book. *Plants of the coast of Coromandel* owed its realization to many contributors: Patrick Russell who first conceived the project; Sir Joseph Banks who enthusiastically promoted it; William Roxburgh who diligently collected the plants and provided all the data; his team of anonymous Indian artists who illustrated it; Jonas Dryander who edited the text; Daniel Mackenzie who engraved most of the drawings[30]; William Bulmer who printed it; and the East India Company who continued to finance it even after it became evident they would never recoup their investment.

Many of the plants illustrated were selected on the basis of their relevance to commerce and medicine. Each of the 300 plates is accompanied by a taxonomic description and, where known, the plant's uses by Indians in food, medicine, and agriculture. The drawings sometimes depict the entire plant but, more frequently, just a branch and inflorescence with dissections and nearly always life-size. The English engravers took no liberties in copying the drawings whose stiffness and formality are faithfully reproduced. There is a minimal amount of hatching on the plates and colour was added in simple washes, seldom any attempt being made to suggest solidity and texture. These plant portraits lie flat on the page, more suitable as motifs for textile designs than as specimens of dynamic botanical art.

The original 2542 drawings[31] are bound in albums in the Indian Botanic Garden in Calcutta; the copies Roxburgh had made for Banks's perusal are now in the library of the Royal Botanic Gardens at Kew[32]. Further copies of selected drawings were made from time to time for Roxburgh's friends and colleagues—Lord Wellesley, Patrick Russell, John Fleming, and Thomas Henry Colebrooke. James Hare had nearly 1200 copied before he was stopped. India owes Hare a debt of gratitude for retrieving the series of Roxburgh drawings after Colebrooke had given them to John Roxburgh in 1815 for transmission to his father's executor. Hare took this step believing them to be the Company's property, unaware, evidently, that the earlier drawings had been paid for by Roxburgh himself and lawfully belonged to the family. The Bengal Government rejected a belated claim by Bruce Roxburgh in 1833, maintaining 'it would be difficult at this distance of time to state what portion was [Roxburgh's] private property or to distinguish the drawings executed at his private cost from the majority which are the property of the Government'[33].

Astonishingly, 194 of the original copper plates from which the engravings in *Plants of the coast of Coromandel* had been printed survived until 1881 when they were offered by the India Office to Kew Gardens which politely declined them. 'Looking at the extreme imperfection of the series and the antiquated and cumbrous character of the work which has been practically superseded by the more recent publications of Wight and Beddome, Sir Joseph Hooker, while thanking you for bearing the establishment in mind, is of the opinion that the only value of the plates at the present day is for disposal as old copper.'[34]

NOTES

1 Public letter to London, 1 August 1785. *Fort William India House correspondence*, **9**, 1959, 576.

2 Banks to E.I.C., 22 February 1787. I.O.L.R. E/1/80 folio 169.

3 Jones to Russell, 8 September 1785. Cannon (1970), Vol. 2, 181.

4 Banks to E.I.C., 22 February 1787. I.O.L.R. E/1/80 folio 169.

5 Banks to Lord Liverpool, 30 March 1787. B.M. (N.H.), D.T.C., Vol. 5, folios 173–4.

6 John to Roxburgh, 17 August 1792. I.O.L.R. MSS Eur. D.809.

7 Rottler's herbarium is now divided between Kew and Liverpool.

8 *Liverpool Libraries, Museum and Arts Committee Bulletin*, **6**, 1857, 34.

9 Extract of Fort St George Public Consultations, 18 June 1802. I.O.L.R. F/4/152, item 2601.

10 Ibid., 2 April 1802.

11 Wight and Walker-Arnott (1834), xiii.

12 Russell to Solander, 21 December 1782. B.M. (N.H.), D.T.C., Vol. 2, folios 228–31.

13 Russell to Banks, 26 December 1784. B.M. (N.H.), D.T.C., Vol. 4, folios 94–6.

[14] Russell to Governor of Madras, 3 November 1786. I.O.L.R. Range 240, Vol. 65, folios 2046–57.
[15] Russell to Banks, 10 January 1787. I.O.L.R. MSS Eur. F18/1, item 217.
[16] Despatch to Madras, 16 February 1787. I.O.L.R. E/4/873, folios 157–9.
[17] Despatch to Madras, 31 July 1787. I.O.L.R. E/4/873, folios 730–1.
[18] Russell to Governor of Madras, 5 March 1788. I.O.L.R. Range 241, Vol. 5, folios 872–4.
[19] Russell to Banks, 10 October 1788. B.L. Add MSS 33978, folios 206–7.
[20] Banks to E.I.C., 25 November 1788. B.M. (N.H.), D.T.C., Vol. 6, folios 91–3.
[21] Russell to Madras Government, 23 February 1789. I.O.L.R. Range 241, Vol. 11, folios 711–24.
[22] Roxburgh (1832), *Flora Indica*, Vol. 2, 236.
[23] Dalrymple (1793), *Oriental repository*, Vol. 1, 22.
[24] Roxburgh to Banks, 17 August 1792. B.L. Add MSS 33979, folios 1761–73.
[25] Roxburgh to Smith, 14 January 1794. L.S. Smith letters, Vol. 8, folio 186.
[26] Banks to E.I.C., 4 July 1794. B.M. (N.H.), D.T.C., Vol. 9, 52–6.
[27] Roxburgh to Banks, 19 December 1795. B.L. Add MSS 33980, folios 41–2.
[28] Banks to Roxburgh, 29 May 1796. B.L. Add MSS 33980, folios 65–6.
[29] E.I.C. Committee of Warehouses, 10 June 1801. I.O.L.R. MSS Eur. F 25, folios 75–97.
[30] Girton, Peake, and Weddell engraved the plates of the third volume.
[31] This is the figure quoted by C. B. Clarke in his 1874 reprint of W. Roxburgh's *Flora Indica*.
[32] 2512 drawings with another 150 which may have the same provenance.
[33] Bengal Government to E.I.C., 19 October 1833. I.O.L.R. E/4/144.
[34] W. Thiselton-Dyer to L. Mallet, 29 August 1881. Kew Archives. Herbarium donations to 1900, Vol. 2, 654.

SIR WILLIAM JONES

In the summer of 1793 William Roxburgh was summoned to Calcutta to be considered for the post of the first salaried Superintendent of its botanical garden. The English flag had been hoisted over a small settlement on the banks of the River Hooghly in Bengal in 1690 and within a century Calcutta's population had swollen to half a million. Its trade had outstripped the ports of the Coromandel coast and when Surat, once the dominant commercial centre of all India, was weakened by internal wars, its position became unassailable. The East India Company created a Georgian city on a former malarial swamp. The Victorians, admiring its broad roads fronting large mansions with classical facades, stuccoed and whitewashed, and set in smooth lawns with shrubberies and shade-giving trees, called it the City of Palaces. But it was also a city of contrasts: European elegance was surrounded by open drains, mud huts, and the stench and filth of narrow streets in crowded bazaars. Parliament's Regulating Act of 1773 elevated the Governor of Bengal, Warren Hastings, to Governor-General with authority over the Presidencies of Madras and Bombay. Calcutta became the seat of the Government of British India and the location of the new Supreme Court of Judicature. It boasted theatres, an orchestra and choir that performed Handel's Messiah to home-sick Britons, a race-course, and a learned society presided over by (Sir) William Jones (1746–94)[1].

Jones, manifestly a polymath, had an amazing facility for languages (his fluency in Persian had earned him the affectionate soubriquet 'Persian Jones'); he wrote poetry, knew enough about musical theory to attempt a serious study of Hindu music, and gave as much time as he could to natural history. He had switched to law when he realized that oriental studies alone would never guarantee a remunerative career. Newly wed, he left England in 1783 for the lucrative post of Puisne Judge in the Supreme Court of Calcutta. The challenge of Indian culture, still largely unknown in the West, appealed to his inquisitive intellect. On the voyage out he prepared an intensive programme of the research he wished to pursue in the arts and sciences; 'it is my ambition to know *India* better that any other European ever knew it'. On 15 January 1784, just a few months after his arrival in Calcutta, he persuaded 29 residents to gather in the Grand Jury Room of the Supreme Court to discuss his proposition 'On the institution of a Society for inquiring into the history, civil and natural, the antiquities, art, sciences, and literature of Asia'. This society should have as its grand objective the study of '*Man* and *Nature*; whatever is performed by the one, or produced by the other'[2]. His audience, won over by his infectious enthusiasm, unanimously voted the formation of an Asiatick Society. Warren Hastings declined the presidency of the new society in favour of

Jones whom, he modestly declared, was a much more deserving candidate for that honour. As a recently elected Fellow of the Royal Society, Jones adopted the distinguished London institution as an examplar for the Asiatick Society which adhered to the same high standards of scholarship, adopted the same administrative structure and modelled its journal, *Asiatick Researches*, on the *Philosophical Transactions of the Royal Society*. Without his leadership and vision and his constant encouragement, the Society might well have foundered after a few years or degenerated into a congenial club for the local intelligentsia. Members' papers and reports were selected by Jones for inclusion in *Asiatick Researches*. This journal, first published in 1788, was well received in London where an octavo edition of the early volumes was subsequently published; some volumes were even translated into French and German. In 1814 the botanist Nathaniel Wallich proposed that the Society should form a museum, offered his services as curator and generously donated his duplicate specimens. In 1868 the museum's collections, apart from its library, pictures and coins, were transferred to the new Indian Museum, financed by the Government. The Society which had been housed in its own building from 1808, changed its name to the Asiatic Society of Bengal in 1851 and a royal charter commemorated its 150th anniversary in 1934.

Jones shared the conviction of Patrick Russell and William Roxburgh in South India and of Joseph Banks and the East India Company in London that an investigation of Indian vegetation could yield rich rewards in new or improved crops and medicine. In his second presidential address in 1785 he directed the attention of members to Indian practices in agriculture, to the use of vegetable drugs and 'who can foretel what glorious discoveries of other oils, roots, and salutory juices, may be made by your Society? If it be doubtful whether the *Peruvian* bark be *always* efficacious in this country, its place may, perhaps, be supplied by some indigenous vegetable, equally antiseptick and more congenial to the climate'[3].

Botany, 'the loveliest and most copious division in the history of nature', always took precedence over zoology for Jones. 'I never could learn by what right, nor conceive with what feeling, a naturalist can occasion the misery of an innocent bird, and leave its young, perhaps, to perish in a cold nest, because it has gay plumage, and has never been delineated, or deprive even a butterfly of its nocturnal enjoyment, because it has the misfortune to be rare or beautiful.'[4] His colleague, Henry Colebrooke, Chief Judge of the High Court of Appeal, expressed a similar sensitivity when he renounced animals for flowers. 'My chief disgust to ornithology arises from feeling the cruelty of keeping birds pining for several days, while a draftsman is making a drawing of them.'[5] Perhaps their compassion had been subtly moulded by or infused with the eastern reverence for all life. Botany afforded Jones a relaxation he shared with his wife who painted some of the plants he collected for study, and it was a comfort when he was not well. 'In a state of convalescence from my severe illness three years ago, my physician [John Fleming] forbad all intense application, but, as he was a botanist he permitted me to examine *flowers*, and lent me *Linnaeus* with whose system I was delighted.'[6] His admiration for Linnaeus never blinded him to what he perceived were his shortcomings. He judged the *Philosophia botanica* (1751) 'harshly, not to say barbarously written. I grieve to see botany imperfect in its two most important articles, the *natural orders* and the *virtues* of the plants, between which I suspect a strong affinity'[7]. In his 'Design of a Treatise on the Plants of India'[8] and 'Botanical Observations on Select Indian Plants'[9] he argued the superiority of romanized Sanskrit names over Linnaean. 'I am very solicitous to give Indian plants their true Indian appellations, because I am fully persuaded that Linnaeus himself would have adopted them had he known the learned and ancient language of this country.'[10] To illustrate this thesis he listed 419 plants under their Sanskrit names[11]. He encouraged fellow botanists to learn Sanskrit in order to discover prescriptions for drugs in ancient medical manuscripts and allusions to plants in epic poetry. He drew attention to the 300 medicinal plants listed in the *Amarakosa*, a Buddhist encyclopaedia, concerned that this untapped source of native botanical expertise should not be disregarded in any attempt to compile an Indian flora.

The realization of such a flora was now made feasible by the easily applied sexual classification of Linnaeus. Jones, however, disliked the names the Swedish naturalist had chosen for his 24 classes and his immodest metaphors to illustrate the number of stamens and styles in each class. The class *Diandra*, for instance, was reserved for plants with two stamens likened to 'two husbands in the same marriage' and these comparisons became progressively more outrageous with *Polyandra* denoting 'twenty males or more in the same bed with the female'. This colourful imagery offended Jones. 'The allegory of *sexes* and *nuptials*, even if it were complete, might, I think, have to be discarded, as unbecoming the gravity of men, who, while they search for truth, have no business to influence their imaginations.'[12] He was not alone in his moral indignation; even Sir James Edward Smith, President of the Linnean Society of London, gravely agreed that 'a literal translation of the first principles of Linnaean botany is enough to shock female modesty'.

Jones proposed a procedure to be followed in all floristic surveys. He recommended using large note-books, devoting a page to each specimen collected, its description and identification, its vernacular name, and any facts gleaned from ancient manuscripts and Hindu doctors on its medicinal, dietary, or commercial uses. Whenever possible the validity of such data should be checked by experiments. Engravings could be added although Jones had reservations about their clarity and accuracy. Unfortunately his premature death forestalled this first attempt to plan a national flora.

His qualified remarks about Linnaean nomenclature apart, Jones had, as he assured his friend Banks, the highest regard for Linnaeus. Banks and Jones had been to school together at Harrow; they were both Fellows of the Royal Society and must occasionally have met in London's social and cultural circles. Once settled in India, Jones offered to send his friend plants and animals. Banks asked him to report on Dacca cotton, and solicited his help in getting Asian plants to the botanical garden at St Vincent in the West Indies. It is more than likely that Banks sounded out Jones on spikenard which had been the subject of a paper by Gilbert Blane to a meeting of the Royal Society in March 1790[13].

Hebrew and Roman citizens usually preserved their precious store of spikenard in sealed alabaster boxes for special occasions. Mary annointed the feet of Christ with this costly ointment, Roman matrons regarded it as one of their choicest perfumes, and it was often used in burial rituals. Theophrastus mentioned it in his *History of plants*; Dioscorides distinguished two kinds of spikenard, 'the one Indian, the other Syrian; not that it is found in Syria, but because one part of the mountain where it grows turns towards Syria, and the other towards India'[14]. Assyrian nard was noted by Horace; according to the anonymous author of the *Periplus*, it was shipped down the Ganges and also produced in the region of the Upper Indus; Ptolemy identified Randamarta in Nepal as its source. Pliny who stated it came from Syria, Gaul and Crete, described it as a shrub with oily, brittle roots and shoots which 'sprout into ears, and

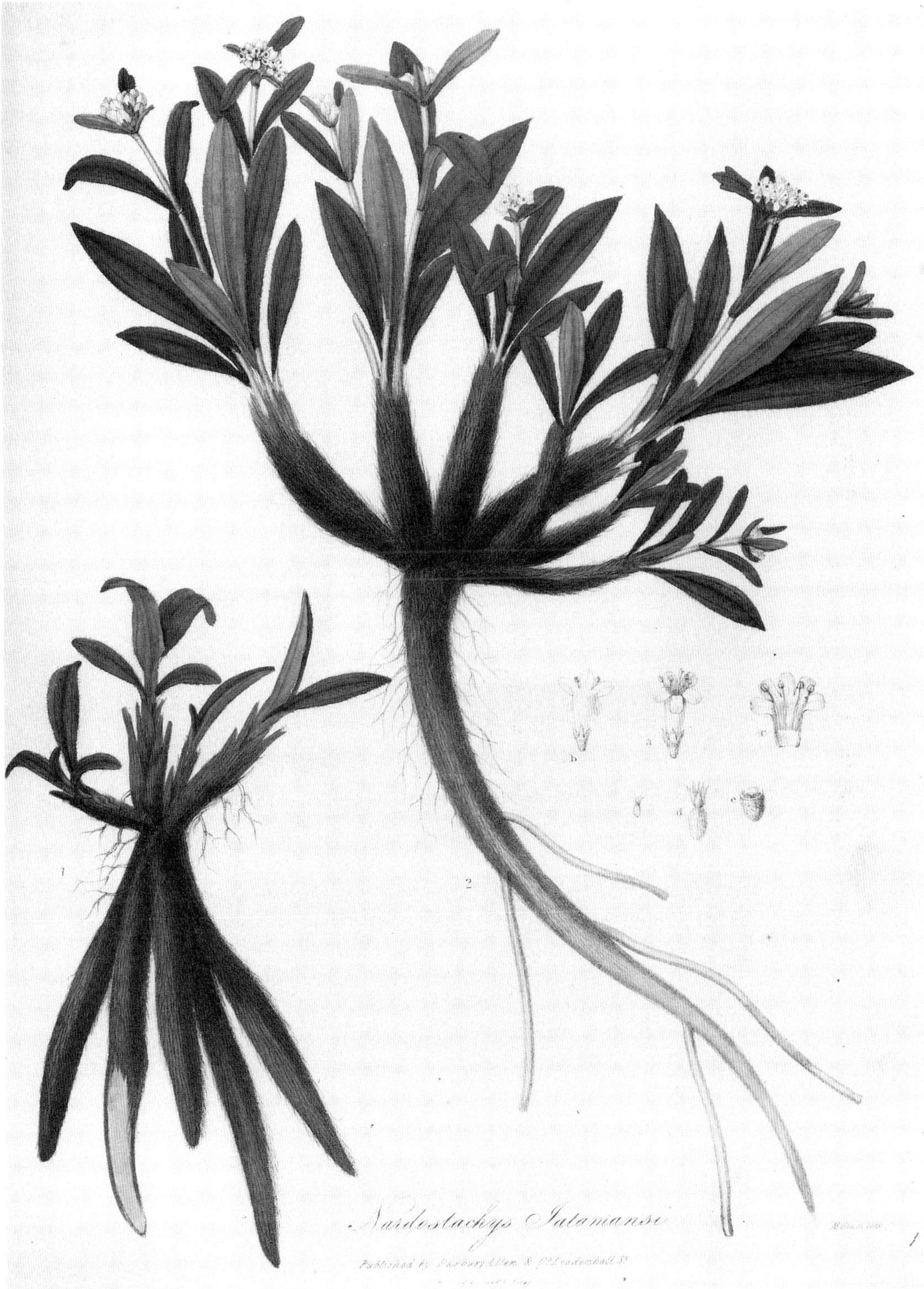

Figure 5.1 Spikenard, *Nardostachys grandiflora.* (From J. F. Royle (1839), *Illustrations of the botany . . . of the Himalayan mountains and of the flora of Cashmere,* London, plate 54.) Hand-coloured lithograph.

consequently both the spikes and the leaves of the nard are famous'[15]. He dismissed Indian nard 'growing by the Ganges' because of its 'poisonous smell' (Fig. 5.1).

It was this confusion among classical authorities regarding the plant's identity and location that Gilbert Blane undertook to resolve when his brother William sent him a dried specimen, a grass he had collected while travelling in the company of the Vizier of Oudh. His attention had been drawn to it by the strong scent from grasses trodden on by elephants and horses in his host's retinue. Banks identified the grass as a species of *Andropogon*[16]. Gilbert Blane recalled Arrian's account of Alexander's army in India trampling nard beneath their feet, perfuming the air about them. Garcia da Orta had written that there was only one nard in India and, convinced that the Portugese physician's description of it resembled his own specimen, Blane confidently proposed that the Indian nard was none other than this scented grass. Even Linnaeus recognized *Andropogon* as the true nard.

Quite independently, Jones was also investigating this botanical enigma. 'What is spikenard', he asked Patrick Russell. 'I mean botanically, what is the natural order, class, genus, etc. of the plant? What was spikenard in the alabaster box of the Gospel? What was nardi parvus onyx? What did Ptolemy mean by the excellent nard of Rhangamutty in Bengal? I have been in vain endeavouring for above two years to procure an answer to these questions.'[17] He doubted whether it could be a grass. Nard was a Persian word, probably meaning a stem; the equivalent word in Arabic was 'sumbal', an ear or spike, and 'there can be no doubt, that by the *sumbul* of India the Musselmans understand the same plant with the *nard* of Ptolemy and the *Nardostachys* or *Spikenard* of Galen'[18]. Hearing from Muslims that Indian 'sumbul' was also called 'Jatamansi', he obtained a specimen of the latter from a native physician who informed him that it came from a mountainous country north-east of Bengal. Could that country be Bhutan? He enquired of officials who had penetrated that little-known land whether jatamansi grew there. He needed to examine fresh specimens. A few jatamansi plants were eventually brought back by some pilgrims to Nepal, cultivated at Gaya in Bihar where a Mr Burt drew and described them for Jones. On the basis of Burt's information Jones named this historic plant of the ancient world, *Valeriana jatamansi*. Some years later Wallich and Royle discovered the true jatamansi which David Don included in his *Prodromus florae Nepalensis* (1825) as *Patrinia jatamansi*; De Candolle transferred this alpine plant in 1830 to the genus to which it now belongs—*Nardostachys*—but also described it as *N. grandiflora* where it is now botanically correct. Jones's analysis of evidence demonstrates the range of his scholarship and the flexibility of his approach—his patient pursuit of clues in historical records, in philology, and anthropology.

The prospects of oriental scholarship may have lured Jones to India but so did the possibility of acquiring a modest fortune to enable him to enter Parliament or to retire to his study. As his generous salary allowed him to save so did his ambition for public office diminish; all he now wanted was a country house with a garden which he and his wife could enjoy together. 'The last twenty years of my life I shall spend, I trust, in a studious retreat', he told Banks, '& if you know of a pleasant country house to be disposed of in your part of Middlesex with pasture-ground for my cattle and garden-ground enough for my amusement have the goodness to inform me of it. I shall be happy in being your neighbour.'[19] Prolonged illness compelled his wife to return to England with a promise from her husband to follow within two years at the most. But he died suddenly of inflammation of the liver in April 1794, just 47 years of age. He was buried in South Park Street Cemetery where his slender obelisk towers above all other monuments, just as in life he towered above his contemporaries (Fig. 5.2).

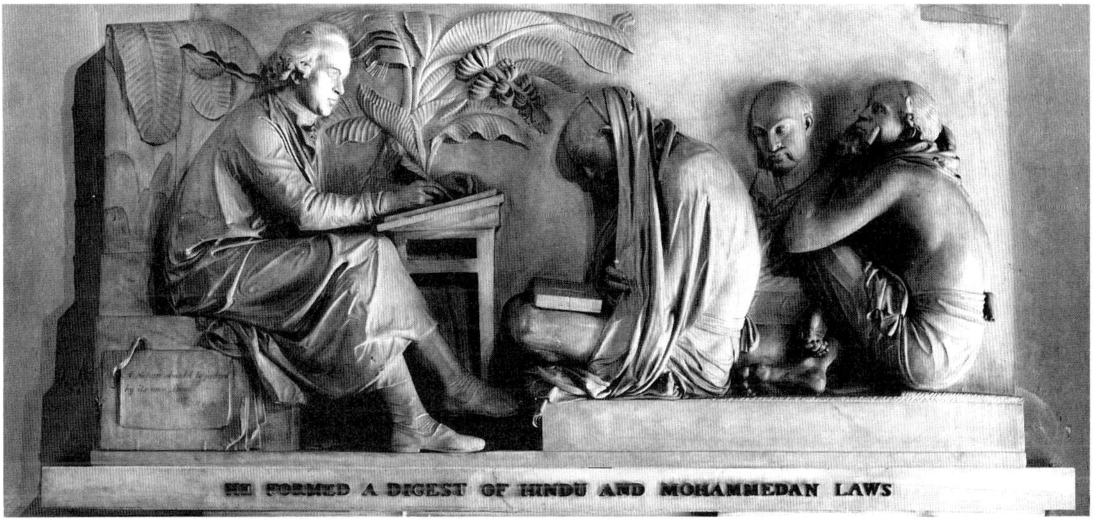

Figure 5.2 Detail of John Flaxman's memorial to Sir William Jones (1746–94) in University College, Oxford.

Botanical research, so diligently promoted by Sir William Jones, received an impetus with the establishment of a botanical garden in Calcutta. The proposal for its formation came from a most unlikely source, the Military Secretary to the Government, Colonel Robert Kyd. In 1786 he conveyed to the Bengal Government his concern that alternative sources of food should be available to the native population in times of famine. He recommended two palms, the sago from Malaya and the Persian date, both nutritious and easily cultivated. A stock could be imported, cultivated in a Calcutta nursery and distributed throughout British India[20]. At this stage he made no specific mention of a botanical garden but, having received a sympathetic response, he felt emboldened to submit a much more ambitious scheme. His bait was the profitable cinnamon which he claimed he had discovered growing on the Assam borders; it was also, he said, a native of Bhutan. He reminded the Government that some cinnamon trees procured from Ceylon some five years earlier were now flourishing in Warren Hastings's former garden, proof that Bengal's climate and soil were suitable for their cultivation[21]. The Company, should it so wish, could cultivate cinnamon on a commercial scale. He concluded his submission with a piece of persuasive rhetoric.

> *But I take this opportunity of suggesting to the Board the propriety of establishing a Botanical Garden not for the purpose of collecting rare plants (although they also have their use) as things of mere curiosity or furnishing articles for the gratification of luxury, but for establishing a stock for disseminating such articles as may prove beneficial to the inhabitants, as well as to the natives of Great Britain, and which ultimately may tend to the extension of the national commerce and riches*[22].

He appended an ambitious list of commercially viable crops for this garden: Dacca cotton, indigo, sarsaparilla, tobacco, coffee, sandalwood, lac, teak, pepper, cardamom, ipecacuanha, copal, asafoetida, camphor, nutmeg, clove, and tea.

The acting Governor-General, John MacPherson, who accepted Kyd's proposal, notified London of his intention to implement it. The Court of Directors, assured that the monthly running costs of the garden would not exceed a modest 200 rupees, readily gave their approval. In the summer of 1786 a site was selected on the west bank of the River Hooghly at Sibpur a few miles from Calcutta. It would not appear to have been particularly suitable—derelict, overgrown with brushwood and frequently flooded—but it was convenient for Kyd whose country house adjoined it. In March 1787 the Government of Bengal despatched letters requesting plants to stock the garden—to the Governor of Bombay for Malabar black pepper, cardamom, and sandalwood together with some natives familiar with their cultivation; to Madras for teak; to the British Resident at Mocha in Arabia for coffee; to the Resident at Basra for the Persian date palm; to the Prince of Wales Island (Penang) for sago and breadfruit from Malaya, and camphor from Borneo; to the Moluccas for nutmegs and cloves; to Siam for 'the tree affording the famous gum with which the Chinese lacker ware is prepared'; and to China for tea and some experienced gardeners.

The new garden was off to a good start and in Colonel Kyd (Fig. 5.3) it had a resolute manager who enjoyed the support of the Governor-General. He ditched and embanked his plot of ground, erected a bamboo fence to keep out wild buffalo, and waited for the plants to arrive. He was pleased to inform John White, Assistant Secretary to the Government, in May 1787 that 'the first donation was yesterday made to the botanical institution by Colonel Pearse of a mangoe stem and nanna tree, thro' the means of Lieutenant Cox, lately returned from the eastwards'[23]. Other donations in due course followed from officers returning from eastern stations; Lieutenant Gillespie presented a flower, possibly an iris, he had found growing among some ruined buildings; cinnamon seeds came from Warren Hastings's old garden at Alipore; and Hastings despite the distress of growing hostility towards him at home sent Kyd a medicinal plant from Abyssinia called 'Woogginoos'. Kyd's optimistic prediction that within a few years he would propagate peaches, plums, apricots, cherries, apples, pears, grapes, and Chinese fruit in sufficient numbers to stock the gardens of British India was never realized. He blamed the damage caused by a hurricane for his lack of success; his failure to cultivate tea he attributed to Bengal's climate and the absence of skilled Chinese labour; above all, he was disheartened by the poor response from British officials in Bombay, Bencoolen, and China. The plants of commercial value he desperately wanted were frequently ignored by these officials who, Kyd believed, thought 'that such articles were not so much the object of a Botanical Institution as flowers and fruit only'[24]. There was, however, one person who had no misconceptions, no doubts about the role and destiny of the botanical garden at Calcutta, and Sir Joseph Banks submitted his positive views—which had been invited—to the Court of Directors of the East India Company.

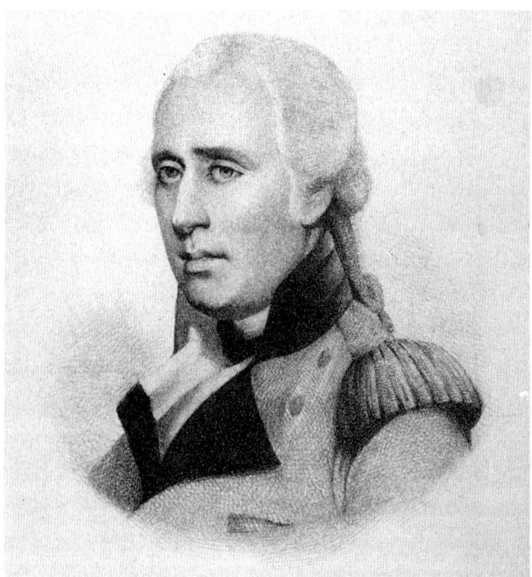

Figure 5.3 Robert Kyd (1746–93). Lithograph.

In a letter of 15 June 1787[25] Banks had endorsed Kyd's desire to reduce the severity of famine by the extensive cultivation of the sago and date palms. He saw the Calcutta garden as a nursery for useful plants; the availability of cheap Indian labour in the production of crops and drugs for export would give the Company a competitive advantage in world trade. He cited French achievements in cultivating nutmegs and cloves. But in another letter some fifteen months later[26], Banks questioned the need for a site of 50 acres for the botanical garden. It would require a large labour force, not inconsiderable maintenance expenses, and would encourage the likelihood of malpractice; 'it appears therefore to me that if the garden is established on this extensive plan the seeds of its certain dissolution will be sown at the very moment of its institution'. Banks reminded the Company that the royal garden at Kew grew more than 5000 trees and plants in no more than two acres and that even the large London nurseries seldom exceeded ten acres. He was confident that Kyd's management would be more effective if the garden were confined to a couple of acres, a calculation that Kyd vigorously disputed. The garden, he argued, needed its acreage not only for a nursery but also for an 'embryo plantation' where a range of vegetables new to India could be experimentally grown[27]. Kyd required more than Banks's few acres for the many consignments of plants and seeds being received by Calcutta: ash, oak, chestnut, birch, willow, poplar, and plum from England; peaches, lychees, pears, oranges, mangosteens, bamboos, and flowers from China. An inventory of over 4000 plants growing in the Botanic Garden in January 1790[28] represented an impressive achievement but no mention was made of their condition. The high failure rate, particularly with European fruit trees, persisted. Nevertheless, Kyd seemed satisfied with the progress he was making. He invited Sir William Jones's wife to visit the garden where she would see 'a mulberry tree more beautifully loaded with fruit than anything in her *dear, dear, England*'.

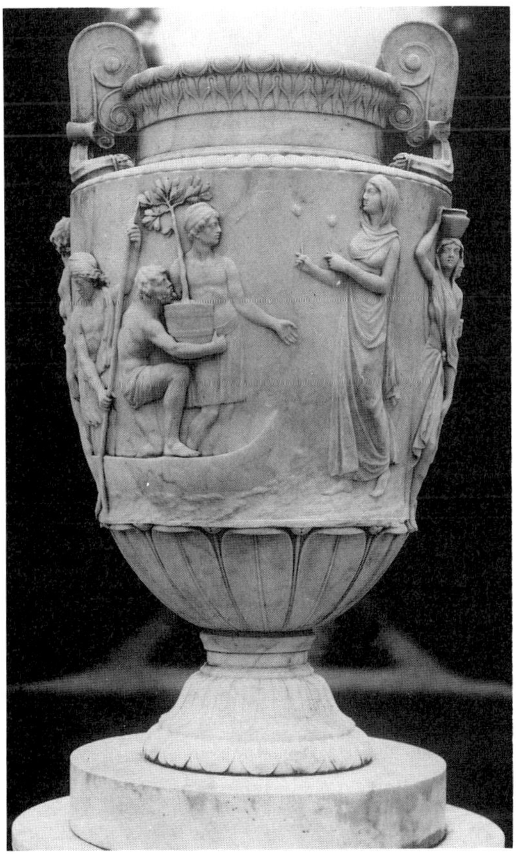

Figure 5.4 Marble urn by John Bacon surmounting the memorial to Robert Kyd. It was erected in the Calcutta Botanic Garden on a site chosen by his successor there, William Roxburgh. Europeans in India at that time had a predilection for urns.

His optimism was not entirely shared by Banks who usually saw all despatches from India regarding the Botanic Garden. He observed that no action had been taken on his November 1788 letter querying the size of the garden. He presumed that the primary function of the garden still remained 'the promotion of public utility and science'[29]. He clearly disapproved of Kyd's comment that it would be 'a not inelegant retreat for such of the members of administrations whose indisposition may preclude [them] from the duties of their station'. Somewhat testily he informed the Court of Directors that if they wanted his advice in future he would welcome regular statements on the plants in cultivation similar to the lists he received from the botanical garden at St Vincent whose progress he monitored on behalf of the War Office. If Kyd, who had no botanical expertise, would send him dried specimens and also fruits of any unknown plants preserved in spirits, he undertook to have them identified and named.

Colonel Kyd died on 26 May 1793 and the surgeon, John Fleming, was appointed temporary Superintendent. Kyd had requested that he be buried without any religious ceremony near a favourite avocado pear tree in his own garden, but his wishes were ignored and his remains were interred in fashionable South Park Street Cemetery (Fig. 5.4).

NOTES

[1] William Jones was knighted in 1793.
[2] *Asiatick Researches*, **1**, 1788, xii–xiii.
[3] Ibid., 409.
[4] Ibid., **4**, 1795, 12.
[5] T. E. Colebrooke (1873), *Life of H. T. Colebrooke*, London, 136.
[6] Jones to Lord Spencer, 18 August 1787. Cannon (1970), 752.
[7] Jones to Russell, 22 September 1787. Cannon (1970), 776.
[8] *Asiatick Researches*, **2**, 1790, 345–52.
[9] Ibid., **4**, 1795, 237–312.
[10] Ibid, 237.
[11] 'A Catalogue of Indian plants, comprehending their Sanskrit and as many of their Linnaean Generic names as could with any degree of precision be ascertained.' *Asiatick Researches*, **4**, 1795, 229–36.
[12] Ibid., **2**, 1790, 348.
[13] 'Account of the Nardus Indica or Spikenard.' *Philosophical Transactions*, **80**, 1791, 284–92.
[14] Miller (1969), 89. He speculates that the mountain range in question was the Hindu Kush or the Pamirs.
[15] Pliny (1960), *Natural history*, XII.xxvi.42.
[16] The aromatic grasses have since been transferred to the genus *Cymbopogon*.
[17] Jones to Russell, 22 September 1787. Cannon (1790), 777.
[18] 'On the Spikenards of the Ancients.' *Asiatick Researches*, **2**, 1790, 406–7.
[19] Jones to Banks, 18 October 1791. Cannon (1970), 895.
[20] Kyd to Bengal Government, 15 April 1786. B.M. (N.H.) D.T.C., Vol. 4, folios 29–32.
[21] Kyd was unaware that he was referring to two distinct species: *Cinnamomum tamala* of the Himalayas and *C. verum* of Ceylon.
[22] Kyd to Bengal Government, 1 June 1786. B.M. (N.H.) D.T.C., Vol. 7, folios 57–67.
[23] Kyd to White, 16 May 1787. I.O.L.R. Home Miscellaneous 799, folio 93.
[24] Kyd to Bengal Government, 6 March 1788. I.O.L.R. Home Miscellaneous 799, folio 185.
[25] Banks to E.I.C., 15 June 1787. B.M. (N.H.) D.T.C., Vol. 5, folios 184–91.
[26] Banks to E.I.C., 25 November 1788. I.O.L.R. E/4/634, folio 41–3.
[27] Kyd to Bengal Government, 23 September 1789. B.M. (N.H.) D.T.C., Vol. 7, folio 68.
[28] B.M. (N.H.) D.T.C., Vol. 7, folios 65–7.
[29] Banks to E.I.C., 17 January 1791. I.O.L.R. E/1/86, folios 57a–c.

THE ROXBURGH ERA

William Roxburgh's successful management of the horticultural station at Samalkot made him an obvious candidate to succeed Kyd. Invited to Calcutta by the Governor-General, Lord Cornwallis, he was persuaded, albeit reluctantly, to accept the appointment as Superintendent of the Botanic Garden which, he told Banks, was 'a good Indian orchard & nursery but of too great extent'[1]. He collected his family from the South including, presumably, his young son John, boarded out with Father Christoph Samuel John (1747–1813) at the Tranquebar Mission. One gets a very distinct impression from the correspondence between Father John and Roxburgh that the missionary took leave of his charge with little regret: 'he causes me more trouble [than] all the others together. I would certainly not keep him for any sum in a year & would have sent him back long since if my heart was not so much attached to you & if I was not determined to give you all satisfaction possible'. With prophetic insight he saw a promising future as a plant collector for this high-spirited and unacademic youngster; 'his greatest skill & pleasure [is] to climb on the highest coconut trees like a squirrel. Then he may *collect* for you all plants; & seeds you wish from foreign countries being very fond of running about, but to describe them, will never become his business'[2].

A few months after Roxburgh (Fig. 6.1) had been installed as Superintendent in November 1793, he was joined by a new nurseryman for the garden, Christopher Smith (*d.* 1807). When his predecessor, Thomas Hughes, died in 1792 the Bengal Government had requested the Court of Directors to send a replacement. 'He should be practised in the methods observed in the gardens in the environs of London, and if he be a man of some science, and has had the advantage of seeing well laid out grounds, and possesses a mild disposition, his services will be the more acceptable'[3]. Banks was consulted and Smith was his choice. His creden-

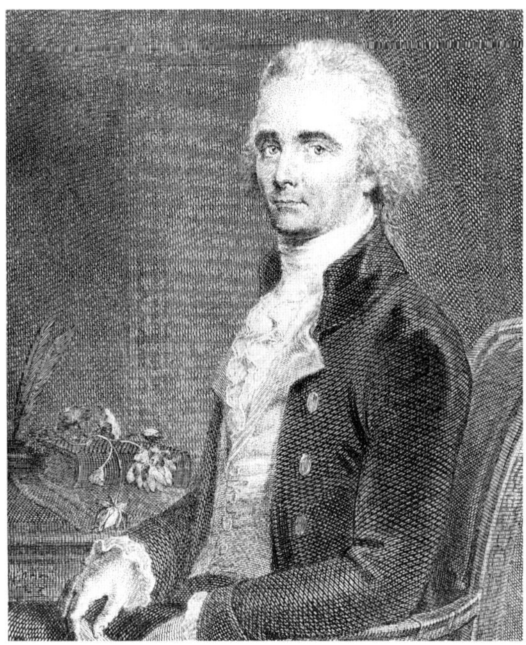

Figure 6.1 William Roxburgh (1751–1815). Engraving by C. Warren.

tials were impressive; he had served an apprenticeship in William Pitcairn's botanical garden in Islington, had successfully tended the breadfruit which HMS *Providence* had brought from the Pacific to the West Indies; furthermore, the ship's captain, William Bligh, had vouched for his 'sobriety & quietness of disposition'. His appointment as botanical gardener or nurseryman at Calcutta was confirmed by the Court of Directors in January 1794. In September he sailed for India with Peter Good, a Kew gardener, as his assistant to tend a large consignment of plants from the royal gardens, many of them new to Bengal. Good returned on the same ship in 1795 with a presentation of plants from Calcutta, the largest single collection, after the HMS *Providence* cargo in 1793, to be delivered to the royal gardens.

Roxburgh took up the challenge of his new post with characteristic zeal, propagating and distributing seedlings of teak, Bengal hemp, Virginia tobacco, Arabian coffee, and indigo to different parts of the subcontinent for trial cultivation. By January 1794 over 2000 plants had been despatched throughout India, to England, the West Indies, and St Helena. Amongst the new introductions he was trying to grow were French olives and Jamaican allspice.

Measures to mitigate the worse consequences of endemic famine had been proposed by Roxburgh who had witnessed the appalling effects of starvation in the Northern Circars—he had favoured the large-scale cultivation of coconuts as an insurance against the failure of traditional cereals. He now deployed all the authority of his present post to achieve his objective of reducing the dependency of impoverished peasants on a few grain crops. The Madras Presidency were asked to grow breadfruit trees imported from the Nicobar Islands for consumption during any famine.

Roxburgh's status and large family required the spacious three-storied house which was being built for him in the Botanic Garden; it still stands, rather shabby, its stucco peeling. Never a robust man, his health suffered in Bengal's notorious climate; indeed doubts were raised about his being fit enough to carry on. Christopher Smith, alert to the prospect of early promotion, prepared the way with an ingratiating letter to Sir Joseph Banks. 'So far as I can learn the Dr. will not remain long here—I know there are gentlemen here who wish to get his place, tho I know of none that is fit for it; I shall therefore push for it myself, and if the Doctor favours me with his interest I flatter myself with success here. . . . May I beg', he concluded, 'a continuance of your friendly offices which will lay me under additional obligations to you, never to be forgotten.'[4] A month later he tells Banks that the deterioration in Roxburgh's health had necessitated his withdrawal to the coast and possibly convalescence at the Cape of Good Hope. 'I don't suppose he will return here again. He told me yesterday that I was to take charge of everything in his absence, and to conduct business as I should think most advisable.'[5] Smith's confident expectations suffered a setback in the autumn of 1795 when Roxburgh sent him to the Moluccas and the Spice Islands on a collecting expedition with nutmegs and cloves as his primary objective.

Great Britain became custodian of the Dutch settlement at the Cape when Napoleon's forces occupied the Netherlands. Roxburgh, who saw this settlement providing a strategic staging post for plants in transit between India, Europe, and the West Indies, approached Banks with the proposal of forming a small garden there with himself in charge where, at the same time, he could recover his health. He was confident he could 'make it one of the first gardens in the world, & if the Company is allowed to defray the expense I can engage to reduce that of the garden here so much as to defray all that would be necessary for the Cape garden'[6]. He would commute between the two gardens with Smith deputizing for him in Calcutta and another gardener in a similar role at the Cape. Such a scheme, patently a product of opportunism and self-interest, had little chance of ever being considered by the East India Company. In December 1797 Roxburgh was given permission to proceed to the Cape and thence to Scotland to recuperate. During his absence John Fleming once again assumed charge of the Botanic Garden.

When Roxburgh returned to Calcutta in the autumn of 1799 he was able to report to the Bengal Government the success of his teak trials with several thousand young trees ready for distribution. He was disappointed with Christopher Smith who, although he had collected nutmegs and cloves, had ignored specific instructions to harvest also the ripe seeds of other plants. A later despatch from Smith which followed an expedition to the Prince of Wales Island (Penang) extended his selection by the inclusion of sago, mango, pepper, and epidendrums. His nutmeg seedlings were now apparently well established except on the Coromandel coast but the few cloves that had survived transportation were sickly specimens. Smith was directed to arrange for a Dutch gardener or an intelligent native at Amboina to collect seeds and tubers for Calcutta. Before the Moluccas were returned to Dutch sovereignty under the terms of the Peace of Amiens in 1802, Roxburgh obtained permission to send his son William there on a final plundering expedition to transfer cloves, nutmegs, and sago palms to the newly-created plantation at Penang. 'Mr Smith who has been between five & six years employd, at a very great expense, amongst these islands, has scarce sent anything but nutmeg plants. He has lost us a most valuable opportunity.'[7] Smith defended his performance in a letter to the Governor-General, Lord Wellesley[8]. He claimed that on his last mission to the Moluccas he had despatched over 71 000 nutmegs, 55 000 cloves, and 29 000 other plants, the majority destined for Penang. He took the opportunity to indicate his disappointment that the surgeon, William Hunter, had been placed in charge of the plantation there, a post which he had hoped to fill. Smith was notified that he was to succeed Hunter as Superintendent of the spice plantations at Penang in 1804. One hopes he was at last placated but as he died in 1807 he enjoyed his promotion for but a few years.

Lord Valentia on a botanical tour of India paused at the Calcutta Botanic Garden in February 1803 and bestowed his accolade of approval: it was well laid out though he regretted that no area had been reserved for a scientific presentation. He understood that commercial considerations dictated the garden's policy; the flourishing multitudes of nutmeg, cinnamon, loquat, grafted mango, and teak were a visible testimony of its success—mangosteen was the one notable failure. New discoveries from Nepal and Chittagong were neighbours of introductions from the West Indies. His lordship especially admired the banyan tree, spectacularly clothed in epidendrums and ferns[9].

So prolific were Roxburgh's nurseries that he propagated more plants than he could possibly distribute in a season and so, in December 1804, obtained an enlargement of the Garden. Once more his work was disrupted by illness; in 1805 he was again granted sick leave but on this occasion it was his son William and not Dr Fleming who became acting Superintendent. When he was in London Roxburgh lost no time in consulting Sir Joseph Banks about the future management of the Botanic Garden should ill-health prevent his return to India. Adopting a magisterial manner, Sir Joseph addressed the Court of Directors, extolling Roxburgh's achievements, reminding them of his remarkable botanical skill which he had acquired from Koenig, a pupil of the great Linnaeus. His son William, now supervising the garden during his father's absence, though an amicable young man lacked those qualities essential for a competent naturalist. If Roxburgh senior chose not to resume his post at Calcutta then, in Sir Joseph's opinion, Dr Buchanan would make an admirable successor. 'I am bold in pleading this cause, because it is in favour of a science to which I have devoted the best part of my life, & for which I have voluntarily exchanged my prospects of Parliamentary consequence, & possibly of high office.'[10] The Superintendent should be supported by a well-qualified assistant who would be a potential successor, and if he were not to be found in India, then he should be recruited in Europe. It comes as no surprise when Sir Joseph, a practised manipulator, disclosed that he had the ideal candidate for such an appointment. Having a wide circle of friends and acquaintances, Banks was never at a loss to recommend suitable candidates for any post. On this occasion he offered the Company Charles Koenig, an assistant in his own library for the past five years, '& I cannot now part with him without regret'. Sir Joseph's magnanimous gesture, however, depended on a reciprocal generosity towards William Roxburgh's son. Perhaps feeling he had betrayed Roxburgh senior in sponsoring Buchanan, Sir Joseph insisted that the Court of Directors found young William a post on its Indian establishment and 'unless the honest claims of his father & himself on the Company can be properly adjusted, I must beg leave to withdraw all concern in this plan for the advancement of the public interest in the due administration of the funds provided by the Company for the advancement of natural knowledge'. When Roxburgh left for India in 1807 a despatch from London notified the Bengal Government of his imminent return[11]. It faithfully reflected the main thrust of Banks's letter—that the objectives of the Botanic Garden were twofold: namely, the introduction of useful plants together with the improvement of horticultural practices and the encouragement of botanical research through the collection and despatch of specimens to European botanists. This was the first time the Company had acknowledged a scientific role for its garden. The attainment of these objectives required a scientist as Superintendent and when the post next became vacant it was to be offered to Francis Buchanan of their Medical Establishment. As for William Roxburgh junior, London directed that he was to be found other employment for which his abilities equipped him.

William Roxburgh's three marriages produced about fifteen children—the exact number seems uncertain—and, botany apart, his constant preoccupation was careers for his sons and husbands for his daughters. The future of his eldest, William, caused him the most concern and effort. As early as 1790 he had obtained Banks's promise to use his influence with the East India Company to get a writership for young William. With a view to his son eventually succeeding him, Roxburgh had made sure he was taught natural history and especially botany at his school in Edinburgh[12]. In 1796 Banks assured an agitated father that the time was not yet right to press for his son's preferment. In April 1798 Roxburgh, returning to England on sick leave, reminded Banks of his anxiety about his son's career. 'My disease, which is obstinate & nervous, gives me reason to fear my life can not be a long one', he told Banks. 'It would therefore give me the more pleasure to see him soon provided for.'[13] At the same time he solicited the help of Sir James Edward Smith, President of the Linnean Society of London, to get his son appointed a writer or, failing that, a botanical assistant at the Calcutta Botanic Garden. Banks, having eventually met young William and being favourably impressed by 'his open

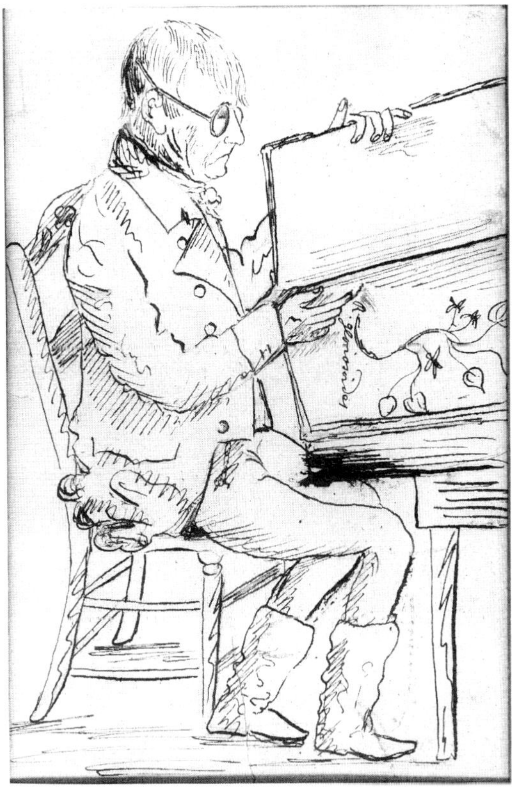

Figure 6.2 Pen and ink sketch of Roxburgh in old age by (Sir) William Jackson Hooker. (Kew Archives.)

and honest countenance', promised Roxburgh senior to press his case once a new and more co-operative Chairman had been appointed by the Court of Directors. Success came early in 1800 when the Company engaged him to assist his father in his botanical duties. The draft despatch to Calcutta announcing his appointment included an instruction that on the death or resignation of his father, his son was to succeed him, only to be deleted in the final version. Shortly after William arrived in Calcutta, his father sent him collecting in Bihar and Chittagong, but he failed to get him attached to Captain Knox's embassy to Nepal. While William junior was plant collecting in the Moluccas and Sumatra in 1803, a despatch from London authorized his appointment to the post of Superintendent when it next became vacant[14]. And indeed, as we have already seen, when Roxburgh senior was granted sick leave in 1805 his son automatically assumed temporary control of the garden. When Doctor Roxburgh returned to India he was greeted with good news concerning another member of his family—his eldest daughter was going to marry Brigade Major Gough, 'a very good looking young man'. A solution to the vexed problem of young William's future arose when the superintendency of the spice plantations in Sumatra became vacant in 1807, a post he was well qualified to fill. Before his father's retirement, William left the Company's service to become a planter, thus relieving the Court of Directors of any further embarrassment concerning his future.

On a cool November morning in 1810, before breakfasting with the Roxburgh family, Maria Graham (sister of Robert Graham, Professor of Botany at the University of Edinburgh) toured the Botanic Garden. The collection of trees impressed her: the huge banyan splashed with the colours of flowering epiphytes, the largest mimosa she had ever seen, a squat African baobab, tall talipot palms, and a lone Norfolk Island pine. The garden had now become the country's principal centre for acclimatizing newly introduced plants. Whenever the Mughal Palace at Delhi wanted rare and unusual plants for the royal gardens it would approach Calcutta with a list of its requirements: flowers, apple trees, grape varieties from Europe, fruits from China, or avocados from Mexico.

Roxburgh's health was again causing concern as Robert Brown (Sir Joseph Banks's Librarian) learnt from Colonel Hardwicke. 'The poor man I fear is yet not to be pronounced out of danger from the progress of his complaint—still his energy of mind supports him, and flatters him with the hope of getting back to what he emphatically terms his 'Paradisus' on the banks of the Hooghly.'[15]

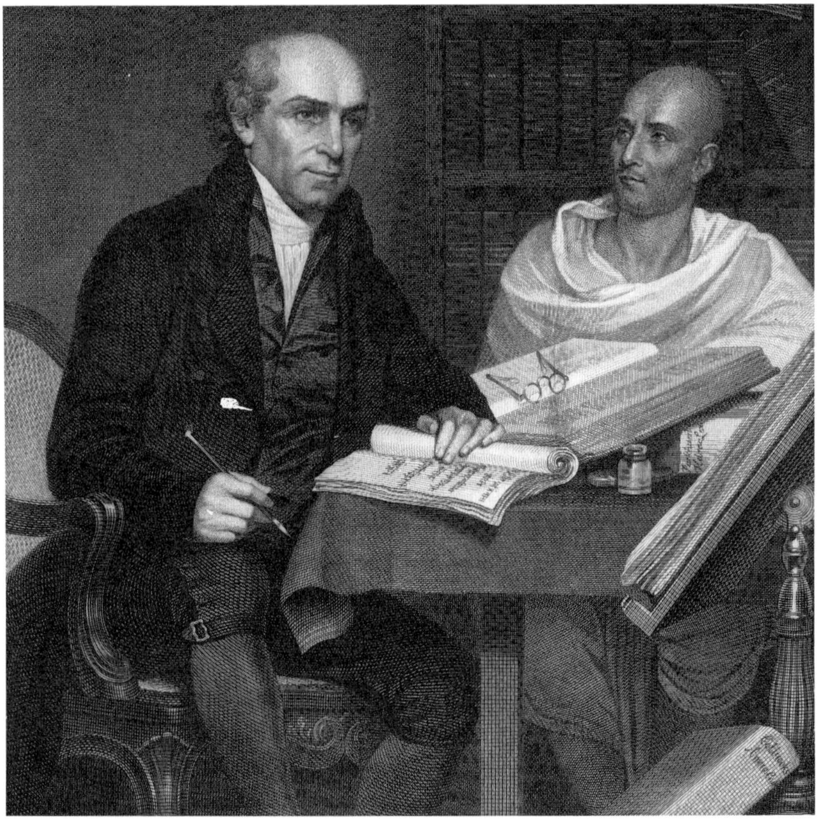

Figure 6.3 William Carey (1761–1834) with his Indian teacher. Engraving by W. Worthing of a portrait by Robert Home.

In 1813 Roxburgh left the country for good and was replaced temporarily by Henry Thomas Colebrooke (1765–1837), Chief Judge of the Supreme Court in Calcutta, an amateur naturalist, and an active member of the Asiatic Society. He engaged Thomas Swinden, a seaman on one of the Company's ships as nurseryman in the Botanic Garden, accepting his claim to have been a professional gardener; it is likely he was related to Nathaniel Swinden, a nurseryman in Brentford in Middlesex. As his performance did not match his alleged horticultural experience he was banished to less onerous duties at Barrackpore Park.

Barrackpore, north of Calcutta, was the country residence of the Governor-General. The broad river Hooghly separated it from the Danish settlement at Serampore on the opposite bank where William Carey (1761–1834) was a Baptist missionary. Modesty and humility characterized Carey, which perhaps were not ideal qualities for militant evangelism. The son of a Northamptonshire weaver and parish clerk, he began his working life as a journeyman shoemaker before entering the Baptist ministry. The success of the Moravian missions overseas prompted Carey and several other brethren to form a Particular Baptist Society for the Propagation of the Gospel amongst the Heathen. Carey, its first missionary, arrived in Calcutta in 1793 when Roxburgh was embarking on his new career at the Botanic Garden. During his early years in India, adjusting to his new life, he found employment as an indigo planter while preparing himself for his mission by studying Bengali, Urdu, and Sanskrit (Fig. 6.3). Prevented by the East India Company's prohibition on missionaries from preaching in British India, he moved to the more permissive administration of Danish Serampore in 1800. There Carey and his missionary colleague, Marshman, began printing their

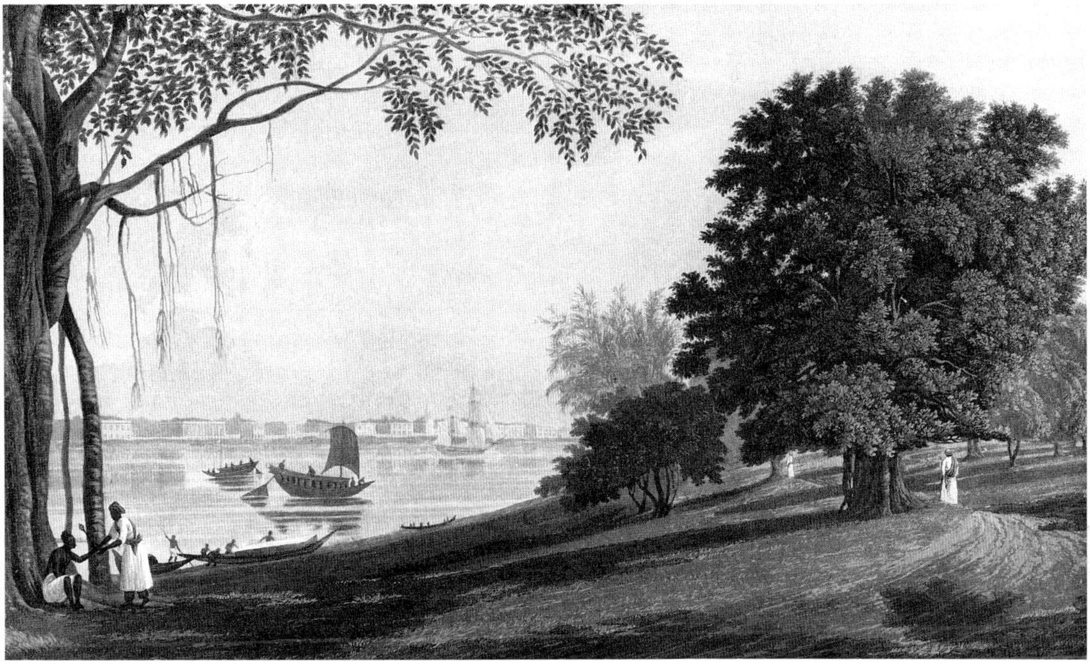

Figure 6.4 A view of the Danish settlement at Serampore on the far bank of the River Hooghly, as seen from Barrackpore, the residence of the British Governor-General. William Carey maintained a small but well-stocked garden at Serampore. (From J. B. Fraser (1824–6), *Views of Calcutta and its environs*, London, plate 21.) Coloured aquatint.

translations of the Bible on their own printing press with the assistance of an Indian craftsman who knew how to punch and cast type. From 1801 Carey taught Bengali at the newly formed College of Fort William in Calcutta where Indian languages, law, and history were mandatory subjects for new entrants to the Company's service. Some plants which Carey had collected on a trip near the Bhutanese border and presented to the Botanic Garden established a friendship with Roxburgh who reciprocated with choice items for Carey's own garden at Serampore (Fig. 6.4).

In 1813 Carey printed at his Mission Press *A Catalogue of plants described by Dr Roxburgh in his MSS Flora Indica, but not yet introduced into the Botanical Garden*. This was distributed in 1814, bound with Roxburgh's *Hortus Bengalensis*, a catalogue of 1510 species cultivated in the Botanic Garden. In his Introduction Carey praised Roxburgh's achievements—the enlargement of the collections from some 300 species in 1794 to about 3500 in 1813 and the description with drawings of nearly 2000 species (there were, in fact, over 2500 drawings). The catalogue gives the donors and provenance of the living collection. Most of the acquisitions had come from European residents in India, confirmation of at least some public support for the garden. Over 150 species from Mysore, Nepal, Chittagong, and Bengal were credited to Francis Buchanan; other major donors included Carey himself, Colonel Hardwicke, Benjamin Heyne, and Andrew Berry of the Madras Medical Service. Nearly 200 species came from M. R. Smith, Superintendent of a teak plantation at Sylhet in the Khasia Hills although it would appear that he received a Government allowance to employ native collectors. William Kerr, a former Kew gardener located at the Company's factories at Canton and Macao,

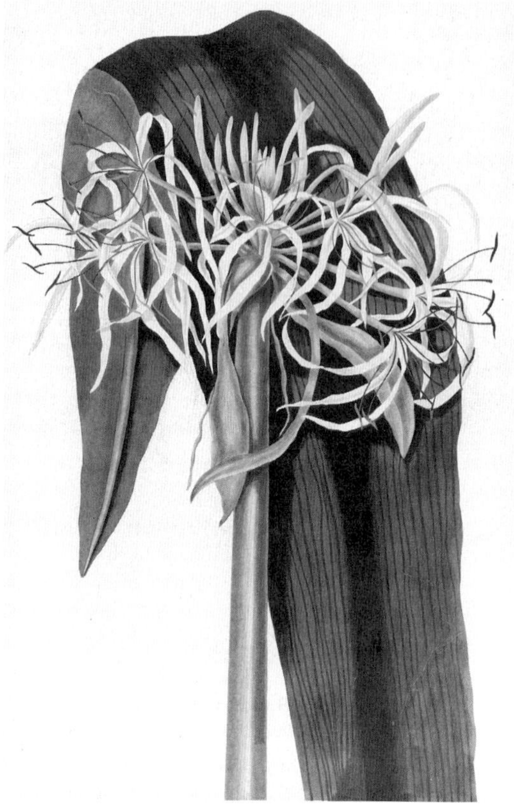

Figure 6.5 Crinum asiaticum. (From Mrs E. Bury (1831–4), *A selection of Hexandrian plants belonging to the natural orders Amaryllidae and Liliacae,* London, plate 43.) Hand-coloured lithograph. One of the many plants sent by William Carey from Serampore to the Liverpool Botanic Garden.

enriched Calcutta with some 50 Chinese species. William Hamilton of Philadelphia, noted for his collections of native and foreign plants, presented over 130 species from North and South America and the West Indies. Both Sir Alexander Hume and A. B. Lambert in England reciprocated the many generous gifts of seeds or dried specimens from Roxburgh with items from their own collections.

The flow of plants to the Botanic Garden, many of them new genera, the facilities to observe them at all stages of their growth, and with native artists at hand to produce pictorial records, encouraged Roxburgh to set about compiling an Indian flora. In November 1802 he informed Sir Joseph Banks that he had written half of it[16]. According to Colebrooke who had promised to contribute vernacular plant names, there was no intention of illustrating the work since many of the species were being depicted in *Plants of the coast of Coromandel*[17]. By December 1803 Roxburgh had completed the arrangement of all his taxonomic notes. In March 1805 the Bengal Government invited the Company to sponsor its publication, a work that had involved Roxburgh in nearly thirty years' research[18]. No reply to this request has been discovered but the Court of Directors were hardly likely to be enthusiastic at a time when financial prudence was operative and the sales of *Plants of the coast of Coromandel* failed to cover costs. Roxburgh himself was reluctant to publish his flora until he had consulted botanists in Britain. In September 1814 he left his manuscript with Robert Brown, possibly the most eminent botanist in the country. Then he decided he would publish it at his own expense, hoping that the Company would underwrite the cost of engraving, in a modest manner, those species not represented in *Plants of the coast of Coromandel.* His health becoming progressively worse, he despaired of ever seeing his book in print, and was resigned to depositing his manuscript and drawings in India House in London. He died in Edinburgh on 18 February 1815 (Fig. 6.6) while Robert Brown was slowly revising his manuscript but neither encouragement nor exhortation by colleagues stirred Brown, well known for procrastination, into finishing it. Colebrooke, anxious to honour a request in Roxburgh's will that he should decide what manuscripts should be published, wrote to Brown offering to supervise the printing of the flora. John Fleming, on leave in England, had indicated his willingness to finance it when the first volume appeared in 1820, printed at the Mission Press at Serampore.

William Carey, aware that its publication was at a standstill in England, had resolved

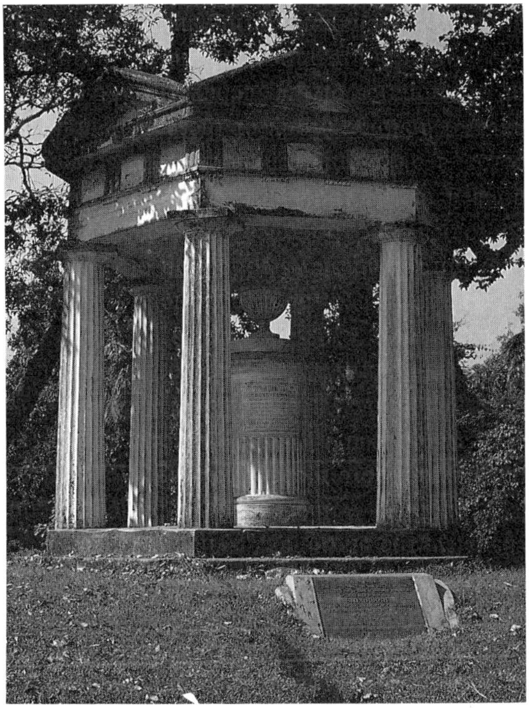

Figure 6.6 Memorial urn commemorating William Roxburgh, erected in 1823 in the Calcutta Botanic Garden. His son, Captain James Roxburgh, had the canopy over it built in 1837.

in 1817 to begin printing it at Serampore. He owned a copy of Roxburgh's manuscript and had the promise of Nathaniel Wallich, the new Superintendent of the Calcutta Botanic Garden, to contribute additional material from his own collections. Francis Buchanan's warm commendation of this first volume of the *Flora Indica* was qualified by his discovery of erroneous descriptions of species. 'I suspect indeed that our friend Roxburgh sometimes mistook his own plant and took his specific characters from one different from what he had described.'[19] Roxburgh's Latin also displeased him and he found fault with his synonymy. Four years elapsed before the second volume was published in 1824. The delay was due to Wallich's Nepal expedition which lasted 18 months, followed by illness which required convalescence in Penang and Singapore. A further volume was optimistically promised for the following year; it never materialized and with only one-third of Roxburgh's manuscript published it remained incomplete. William Griffith complained that it was marred by typographical errors attributable to Carey's 'debility of great age'. But age never deterred Carey; when he was nearly seventy he undertook to complete the *Flora Indica* at the request of Roxburgh's two sons, Bruce and James. It was unanimously agreed to publish the original manuscript without Wallich's numerous additions culled largely from the Nepalese and Burmese floras. The ostensible excuse was that since Wallich was in England, obtaining permission to include his contributions would delay its publication. When the three volumes were published in 1832 a reviewer in the Journal of the Asiatic Society of Bengal[20] announced a fourth volume. Devoted to cryptogams and edited by William Griffith, it eventually appeared not as a separate volume but as an article in the Calcutta Journal of Natural History for 1844[21]. It was still unchallenged as a national flora—only two parts of J. D. Hooker's *Flora of British India* had yet been published—when the 1832 edition together with the article on cryptogams were reprinted in 1874. The editor, C. B. Clarke, saw a continuing role for Roxburgh's *Flora Indica:* his single-volume reprint would be more portable as a handbook than Hooker's multi-volume compilation ranging far beyond the geographical limits of the subcontinent; economic botany, one of the excellent features of the *Flora Indica,* was being excluded from the larger work. But Clarke's fundamental justification for reprinting a work that had been ready for the press at the beginning of the century was that 'Roxburgh's work is so excellent, and his species so well conceived, that they form a solid framework, which being once put together, all the other species are easily fitted into their due places'[22]. It was a model for all those botanists who subsequently attempted less ambitious regional floras.

NOTES

[1] Roxburgh to Banks, 1 December 1793. B.L. Add. MSS 33979, folio 224.

[2] John to Roxburgh, 25 July 1793. B.M. (N.H.). Roxburgh letters.

[3] I. B. Banerjee (1974), *Fort William-India House correspondence*, Vol. 2: *1789–92*, Delhi, 546.

[4] Smith to Banks, 24 August 1795. B.M. (N.H.). Copies of Banks's correspondence.

[5] Smith to Banks, 17 September 1795. Ibid.

[6] Smith to Banks, 10 December 1796. Ibid.

[7] Roxburgh to Banks, 18 July 1802. B.L. Add. MSS 33981, folio 42.

[8] Smith to Private Secretary of Governor-General, 30 September 1802. I.O.L.R. F/4/142, item 2485, folio 85.

[9] Viscount Valentia (1809), *Voyages and travels to India*, Vol. 1, 64.

[10] Banks to E.I.C., 2 August 1806. I.O.L.R. E/1/114, folio 36.

[11] E.I.C. to Bengal Government, 7 January 1807. I.O.L.R. E/4/661, folio 5.

[12] Roxburgh to Banks, 28 December 1794. B.L. Add. MSS 33979, folios 290–1.

[13] Roxburgh to Banks, 24 April 1798. B.L. Add. MSS 33980, folios 143–5.

[14] E.I.C. to Bengal, 17 August 1803. I.O.L.R. E/4/654, folios 591–2.

[15] Mabberley (1985), 196.

[16] Roxburgh to Banks, 28 November 1802. B.L. Add. MSS 33981, folios 67–70.

[17] Sir T. E. Colebrooke (1873), *Life of H.T. Colebrooke*, London, 212.

[18] Bengal to E.I.C., 22 March 1805. I.O.L.R. E/4/67, paras 62–7.

[19] Buchanan to Wallich, 16 October 1821. B.M. (N.H.) Buchanan letters.

[20] No. 2, 1832, 131, 138.

[21] **4**, 1844, 463–520.

[22] *Flora Indica*, 1874, v.

FRANCIS BUCHANAN

When William Roxburgh left Calcutta to regain his health early in 1813, very few of his friends expected him to return. It was generally known that Francis Buchanan was the Company's next nominee for the post and if he refused there would be no shortage of contenders. Among them was Thomas Hardwicke (1755–1835), an all-round amateur naturalist with a predilection for zoology. Early in his military career he had botanized on a tour to Srinagar in Kashmir and described his gatherings in *Asiatick Researches*. He discovered *Rhododendron arboreum* in the Siwalik Hills and sent seeds of it to England, the first Indian rhododendron to be introduced to European gardens. Sir James Edward Smith was pleased to copy Hardwicke's drawing of this rhododendron in his *Exotic botany* (1804). The opportunity, however, for Hardwicke to change career never arose and he eventually retired from the Bengal Artillery in 1823 with the rank of Major-General.

For many years Francis Buchanan (1762–1829) had set his sights on the superintendency of the Botanic Garden, seeing it as a fitting culmination to his chosen vocation in India. He was born in Stirlingshire; his education at Glasgow University was followed by medical training at Edinburgh where he made friends with James Edward Smith, the future President of the Linnean Society of London. His career began with the post of ship's surgeon on an East Indiaman, with four voyages to the East between 1785 and 1794, reaching China and the Philippines on at least one occasion. He abandoned seafaring when he entered the Bengal Establishment in September 1794 as an Assistant Surgeon. Within a few months he found himself attached to Captain Michael Symes's embassy to the Court of Ava, the ancient capital of Burma. He owed his selection to William Roxburgh whom he may have met during one of his eastern voyages. A similarity in background and interests helped the initial acquaintance to progress rapidly to friendship: they were both Scots, had studied medicine and botany under John Hope at Edinburgh University, and were dedicated naturalists. Roxburgh used his influence to engineer opportunities for the younger man to botanize by recommending him for postings to places where the vegetation invited exploration and study. The relationship appears to have withstood for a number of years the strains of an obvious incompatability—Buchanan forceful, arrogant, and combative; Roxburgh modest and, one suspects, more than a little intimidated by his protégé's superior scholarship.

Buchanan regarded the Burmese expedition as 'a new field for botanical researches'[1]. A diligent collector, he conscientiously sent seeds to Roxburgh, and sought out a tree called 'launzan', rumoured to be the source of a commercially viable oil[2]. On his return to India, his specimens and descriptions were despatched to the Court of Directors in London who expressed their approval of 53 coloured drawings before passing them to Sir Joseph Banks[3]. Eight of the drawings were engraved by Daniel Mackenzie in Symes's *Account of an embassy to the Kingdom of Ava sent by the Governor-General of India in the year 1795* (1800). They had been chosen by Banks whose help Symes acknowledged, probably quite unintentionally failing to make it clear that Buchanan and not Banks was the author of the published descriptions, an omission which understandably offended Buchanan. A note attached to this Burmese herbarium reveals the extent of Buchanan's indebtedness to Roxburgh on his first botanical foray. 'I must acknowledge my obligations to my friend Roxburgh, who with a liberality inspired by the true regard for science, has communicated to me his own drawings and descriptions of plants, and diligently compared with them such plants I brought with me; and I cannot but admire his infinitely superior knowledge of the plants of India.'[4] Buchanan was posted to an isolated station in south-eastern Bengal where dampness and insects hindered the formation of a personal herbarium. He engaged a young Indian artist to draw fishes, his main preoccupation at the station. It was a comfortable existence lacking only two requisites—books and a wife—to make him perfectly content but he had not been forgotten by Roxburgh who tried in 1797 to get him included in an expedition to Manila. In 1798 he enjoyed a three-month interlude in Chittagong seeking sites suitable for the cultivation of spices, thanks to Roxburgh's intervention with the Board of Trade. His specimens collected during this brief excursion in eastern Bengal eventually reached Banks. In October 1798 Roxburgh's influence obtained him a posting to Baruipur, only a day's journey from Calcutta whose libraries would be accessible to book-starved Buchanan. While he was at Baruipur and Roxburgh was on leave, Buchanan pursued his researches on Gangetic fishes, botany having temporarily taken second place. A visit to Nepal early in 1800 was terminated at the frontier when he was recalled by the Governor-General, Lord Wellesley, to undertake a survey of Mysore in the south.

Richard Colley Wellesley (1760–1842), brother of the future Duke of Wellington who was serving in India at the same time, was, as Lord Mornington, appointed Governor-General in 1798 and immediately faced with a serious threat to British rule by the ruler of Mysore, Tipu Sultan, and his French advisers. Tipu was killed in the decisive battle of Seringapatam in 1799; a Hindu acceptable to the British replaced the former Muslim dynasty and Lord Mornington, expecting to be suitably honoured by a grateful British nation, was dismayed to be created Marquis Wellesley in the inferior Irish peerage. It should not have been unexpected since the Younger Pitt's India Act of 1784 had resoundingly declared that the territorial enlargement of British India was 'repugnant to the wish, the honour and policy of this nation'. It was an inadmissible anomaly that a trading company whose shareholders were interested only in profit should find itself responsible for an expanding empire it was reluctant to govern. Through the abilities and ambitions of its officials—the military genius of Clive and the administrative skills of Hastings—the British became the most powerful European presence in India. The French were subdued, the Dutch withdrew to the East Indies, and the Portuguese had long since ceased to be a threat. After the 1784 Act the Company's administration of India was regulated by a government-nominated Board of Control. The 24 members who comprised the Court of Directors of the East India Company debated policy behind the imposing classical façade of its headquarters in Leadenhall Street in the City (Fig. 7.1), constrained not only by the Board of

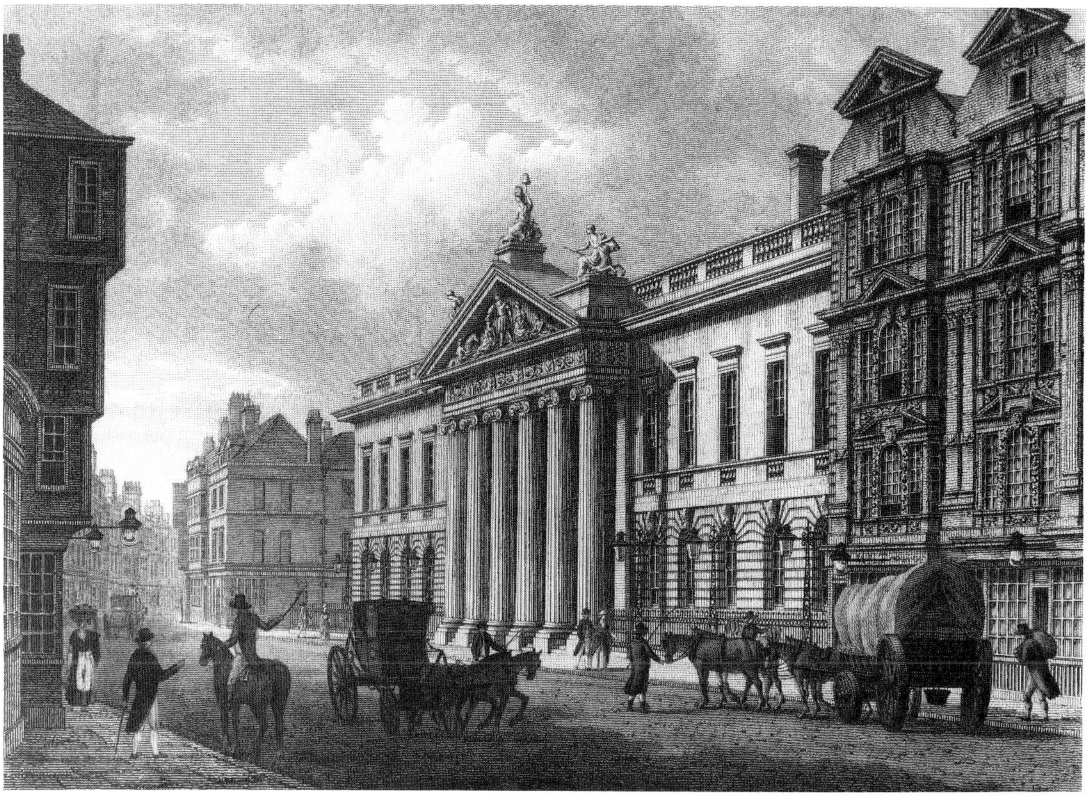

Figure 7.1 East India House, headquarters of the East India Company, in Leadenhall Street, City of London. Engraving, *c.*1800.

Control but also by an intolerable burden of staggering debts which exceeded 14 million pounds in 1800 and had nearly doubled seven years later. Against its inclinations it financed an army and a corps of civil servants, and was reluctant to support a Governor-General whose vision of imperial grandeur could only hasten the reality of bankruptcy. Unfortunately for the Company, Wellesley saw himself as an instrument of enlightened rule in India. He created a college at Fort William in Calcutta where the Company's servants could be taught oriental languages to equip them as effective administrators. Always conscious of the dignity of his office, he devised an elaborate ceremonial and rebuilt Government House in Calcutta after the design for Kedleston Hall in Derbyshire: a domed Ionic façade heralded greater splendours within—a Throne Room, a Marble Hall, and a Ballroom, all of monumental proportions. He was sublimely indifferent to enquiry and instruction from the London Directors, alarmed by rumours of his extravagance. When in 1799 Lord Clive, Governor of Madras, recommended a comprehensive survey of Mysore, Wellesley approved, confident that the report would confirm the economic benefits that would follow the extension of the British sphere of influence to that state. Colin Mackenzie (*c.*1753–1821) of the Madras Engineers was detailed in 1799 to undertake this survey of the culture, geography, and natural resources of Mysore. As it was unlikely to be finished within a few years (Mackenzie did not in fact complete it until 1807), Wellesley wanted an interim report to submit to the Court of Directors. Roxburgh considered Buchanan to have the energy, stamina, and breadth of knowledge to tackle the shorter survey. He had never entertained any doubts about Buchanan's capabilities but when he extolled him to Banks as 'the best botanist in India'[5], the President of the Royal Society mildly disputed the claim.

You mention Dr Buchanan, I admire his talents both as a botanist & a geographer & hope sometime to become acquainted with him. I cannot however subscribe to his being the best botanist in India, tho his descriptions & drawings of Birmah plants are in my hands, because I think you are a better one[6].

When Lord Clive asked Roxburgh to nominate an officer to carry out the survey, Buchanan remained his first choice.

He is the best botanist I know of in India and in every other respect the best qualified to furnish you with an account of the vegetable productions of these countries. He is also a good zoologist; in short every way well qualified for the research. I have had a good deal of conversation with him on the subject and find him willing to be employed by your Lordship in the above manner. Could Mr Heyne accompany him as mineralogist and chemist, I have no doubt but their joint endeavours would soon furnish a more complete account of the natural history, state of agriculture, etc., of these provinces than we at present have of any of the British possessions in Asia[7].

It is a testimony to the respect accorded to Roxburgh's judgement that Buchanan found himself in late April 1800 in Madras to begin a survey which would take him through Mysore to Malabar on the west coast recording history, customs, commerce, and natural products. The Governor-General's objectives had economic implications: varieties of vegetables and their methods of cultivation, breeds of native cattle, the cultivation of cotton, pepper, sandalwood, and cardamoms, and also mining, manufacturing trades, and climate. In addition, Buchanan was expected to send plants and seeds with informative comments to the Calcutta Botanic Garden. Such a demanding brief was beyond any one person's resources, no matter how talented. Consulting his manuscript copy of Roxburgh's *Flora Indica*, Buchanan soon found that most of the plants he examined had already been described leaving his painter with little to record. The extracts from the *Hortus Malabaricus* which Roxburgh thoughtfully sent him were of limited use since he had arrived in Malabar in winter when little of the vegetation was in fruit. He was further frustrated by natives 'who could or rather would not inform me of the names of plants. The obstinacy of the people of Malabar is astonishing and every man you meet suspects you have an evil design in every question you ask—without therefore making some stay in the province and breeding up a man or two to collect plants and procure their names nothing can be further done than to collect specimens and describe such plants as you find in a proper state'[8]. Obtaining the names of plants orally rather than from script introduced variations in orthography, but more disturbing was the discovery that 'a native seldom hesitates in giving some name or other to every plant which you meet, although it is probable that with a great many he is quite unacquainted'[9]. He recognized trees introduced by the Dutch from the East Indies and by the Portuguese from the West Indies. Buchanan's brief but encouraging introduction to the flora of Malabar stimulated an interest in Van Reede's great work which consoled him in his old age.

Buchanan's report pleased both the Governor-General and the Court of Directors who authorized its publication in three volumes in 1807[10]. Buchanan had intended an abridgement of his sheaves of notes but when he arrived in England in 1805 the printing of the book was already under way.

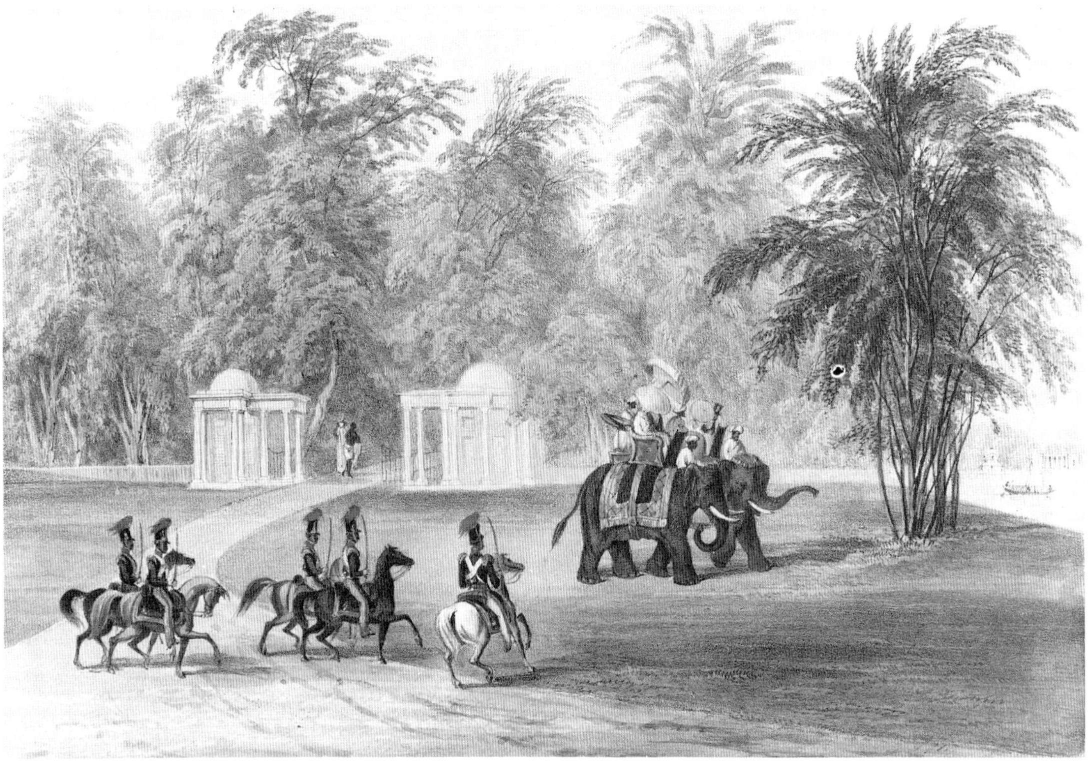

Figure 7.2 Entrance to Barrackpore Park, which was first enlarged and landscaped when Lord Wellesley was Governor-General. (From Sir C. D'Oyly (1848), *Views of Calcutta and its environs*, London, plate 16.) Hand-coloured lithograph.

Obedient to his brief, Buchanan concentrated on economic botany although he had promised James Edward Smith that there would be an appendix of plant descriptions. In the event, the Court of Directors excluded it on the grounds of expense. When the publishers declined to accept the additional cost of illustrating it, Buchanan passed his drawings to Smith, hoping that some of them might be published in his *Exotic botany*.

In March 1802 Buchanan was back in Bengal, having determined that when he had saved £6000 he would return to Britain. 'I intend', he informed Smith, 'to make out a Flora Indica with short characters of the species as much as possible resembling those Jussieu gives of genera as I am able.'[11] In April Buchanan was stationed in Khatmandu as medical officer of the small contingent under Captain Knox, the first British Resident in Nepal. The innate restlessness that permeates even Buchanan's correspondence caused him to weary of Nepal, far from the intellectual stimulus of Calcutta. It was indirectly his request for a transfer and the resistance of the suspicious Nepalese to a replacement for him that precipitated Knox's abrupt departure from Nepal in March 1803. Back in Baruipur, Buchanan resumed his former position as Assistant Surgeon, dejected by the lack of progress in his career and his uncertain future. Then came an unexpected turn in his fortunes. In 1803 he was appointed personal physician to Lord Wellesley who also put him in charge of his small menagerie at Barrackpore.

Barrackpore, some fourteen miles from Calcutta, had been a military cantonment before Wellesley chose it as a suitable site for his country residence in 1801. The Commander-in-Chief's bungalow, patently inadequate for its new occupant, was demolished and work began on building a dwelling to match the splendour of the new Government House in Calcutta; at the same time the adjacent park was enlarged and landscaped (Fig. 7.2). Wellesley even contemplated joining the two houses by an imperial tree-lined avenue. A small collection of animals and birds brought together in 1800 at Garden Reach in Calcutta for incorporation in a proposed natural history department at the College of Fort William (subsequently vetoed by London) was transferred to Barrackpore as the nucleus of a national menagerie announced in the Governor-General's minute of 26 July 1804[12].

Wellesley, conscious that zoological research in India was lagging behind botany, proposed official support and encouragement by creating an 'establishment for promoting the natural history of India'. Without having consulted the Court of Directors in London, he proclaimed that the 'illustration and improvement of that important branch of the natural history of India, which embraces an object so extensive as the description of the principal parts of the animal kingdom, is worthy of the munificence and liberality of the English East India Company, and must necessarily prove an acceptable service to the world'. Buchanan was to manage the project which aimed to describe and draw animals and birds sent by medical personnel in the three Presidencies in India, Ceylon, and Penang. Regular despatches of reports and illustrations were to be forwarded to London for eventual publication. Buchanan's studies on fishes[13] had taken him into the zoological field but Thomas Hardwicke who might well have been a better candidate with his greater expertise as an amateur zoologist was absent on home leave. Buchanan, however, would still probably have been the Governor-General's choice because of the special rapport between them, a bond strengthened by mutual interests and similar character traits—arrogance and intolerance predominating.

Wellesley's profligacy eventually proved too much for the patience of the Court of Directors. Aghast at the unauthorized expenditure of £63000 on Government House, they recalled him in 1805 and ordered the cessation of all work on the residence at Barrackpore. The new Governor-General, an old Indian campaigner, Lord Cornwallis, arrived in July 1805 with a mandate from 'the most loathesome den of the India House' (Wellesley's words) to impose stringent economies. Wellesley's colourful ceremonial was the first casualty, the status and syllabus of the College of Fort William were contracted, and the Barrackpore menagerie (Fig. 7.3) had its budget reduced by sixty per cent. The dispersal of its beasts and birds began and only a few exotic creatures occupied the charming Gothic and Classical cages when Lord Lytton abolished the menagerie in the seventies. Lord Wellesley's collection of over 2600 zoological and botanical watercolours, now housed in the India Office Library and Records, are the only reminder of his service to natural history. Many of these drawings are of the inmates of the menagerie; the botanical portraits are mostly copies of drawings done for Buchanan during his Mysore and Nepal tours or work executed by Indian artists working for Hardwicke and other amateur naturalists.

With the departure of his patron and the prospect of draconian financial constraints which would doubtless hinder his advancement in India, Buchanan accompanied Wellesley to England in August 1805. While he was on furlough he renewed friendships, made useful contacts and petitioned influential people. He impulsively presented his Mysore and Nepalese collections with his notes and drawings[14] to James Edward Smith,

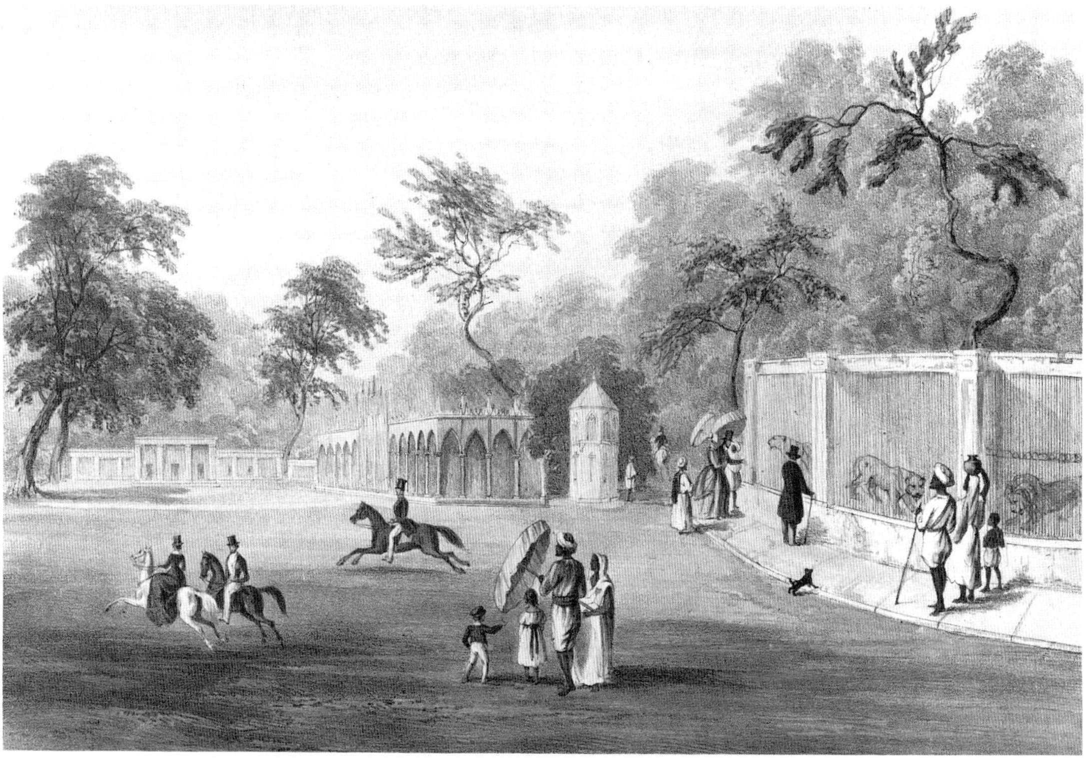

Figure 7.3 Menagerie formed at Barrackpore by Lord Wellesley. (From Sir C. D'Oyly (1848), *Views of Calcutta and its environs*, London, plate 16.) Hand coloured lithograph.

an act of generosity he was later to regret. Cornwallis's tenure as Governor-General was terminated by his death in October 1805. His eventual successor was Lord Minto who had been appointed President of the Board of Control in 1806, the year in which Buchanan joined the distinguished membership of the Royal Society of which Minto was already a Fellow. Minto evidently liked or was impressed by Buchanan who was invited to return to India with his lordship.

In the meanwhile William Roxburgh, taking advantage of his sick leave in Britain, was trying to secure the succession of the Botanic Garden in Calcutta for his son William. Conveniently forgetting the many favours he had received from Roxburgh, Buchanan campaigned energetically on his own behalf. He asked Smith to enlist the help of influential shareholders in the Company such as James Bosanquet and Sir Abraham Hume. Buchanan could be very single-minded when his own interests were at stake. 'What I wish therefore is to be appointed successor to old Roxburgh whenever he chooses to retire. He of course wishes to get his son appointed and says that science is not necessary; that a cultivator alone is required.'[15] Self-effacing Roxburgh was no match for Buchanan who obtained the Court of Directors' nomination[16]. The Court had already decided in February 1806 that when Buchanan eventually returned to India they preferred to employ him as an itinerant surveyor and naturalist rather than a deskbound Superintendent at Barrackpore. Their recommendation that Buchanan should undertake a statistical survey of Bengal was consistent with this decision; 'nor do we consider that this appointment can interfere with his other duties, in the event of his succeeding Doctor Roxburgh as Botanical Superintendent, for his journies will give him a much better opportunity than he could otherwise have of procuring specimens and descriptions of Indian plants for the learned in Europe'. Both Buchanan and Roxburgh returned to India in 1807, their former mutual regard marred by the undignified lobbying in which both of them had indulged in London.

Shortly after Buchanan was promoted Surgeon in June 1807, he received instructions from the Bengal Government regarding a survey of the Presidency. The survey was motivated by a need for an accurate and detailed knowledge of the territory in order to improve its administration and commerce. Colin Mackenzie's exhaustive survey of Mysore, now in its final stages, set exacting standards for Buchanan to emulate. A heavy schedule of duties indicated that this was going to be a protracted tour—indeed when Buchanan was recalled to Calcutta in 1814 the survey was still far from completion. It was encyclopaedic in scope: topography embracing physical features, climate, archaeological remains, demography, religions and customs, animal, vegetable, and mineral products, agriculture, horticulture, animal husbandry, arts and crafts including architecture, and commerce in its broadest connotation. Whenever possible Buchanan was to send plants and seeds to Calcutta. The work force allocated to him comprised a native officer, a 'learned Brahmin', two clerks, two artists, a few men specifically to collect 'natural curiosities', a military escort of twelve sepoys, labourers and two elephants to carry the tents and equipment. He set out from Dinajpur in the north of the Bengal, proceeding westwards and systematically investigating every district, all the while his notebooks, statistical tables, and maps inexorably growing[17]. Often the botanical yield was disappointing especially in regions under intensive cultivation. Buchanan published only a fragment of the plethora of data he had accumulated and it was unfortunately left to Montgomery Martin to produce a slipshod, uncritical edition of the papers, omitting any acknowledgement to Buchanan on its title-page[18].

The unremitting strain of such intensive surveying took its toll of Buchanan's health. At the same time his desire for Roxburgh's post progressively declined, so much so that in November 1811 his friend, John Fleming, asked him outright whether he was still interested in it. Buchanan made no secret of the fact that he was preparing to leave India for good—probably in 1813, he told Smith. An active and productive retirement devoted to sorting his collections and writing papers was beginning to have a greater appeal than the management of the Botanic Garden. When Colebrooke, following the death of his wife, requested in November 1814 to be relieved of the temporary charge of the garden, Buchanan was his automatic replacement. However, Buchanan had now resolved to return to Britain as soon as possible, no longer wanting the post he had coveted for years. His last administrative act—on 22 February 1815—recommended the revision of the regulations governing the Botanic Garden in order to prevent the wanton and thoughtless destruction or theft of plants by the public. Two days later he happily handed over his superintendency to Nathaniel Wallich.

His departure, which should have been an occasion for pleasant farewells and agreeable memories, was spoilt by a disagreement with the Governor-General, Lord Moira, who refused to allow him to take his drawings with him even though they were destined to be presented to the Court of Directors in London. Lord Moira argued that since the drawings were already the property of the Company, Buchanan could not donate what was not his to give. The crux of his objection, however, was the separation of the drawings from the plant descriptions they illustrated. Lord Moira acted in good faith following the advice given him that the natural history sections in Buchanan's survey were 'so vague and indistinct as to be absolutely useless' without access to the accompanying drawings[19]. Buchanan protested that his desire to deposit the drawings in India House was simply to make them more widely available. He added—and here one suspects is the nub of the matter—that if he were personally deprived of them he could not proceed with the taxonomic description of his plant collections after he reached England. The Governor-General was not persuaded and Buchanan boarded the ship with his boxes of herbarium specimens, minerals, and manuscripts but the disputed manuscripts remained in Calcutta. He took the quarrel to the Court of Directors and, having convinced them of the impropriety of Lord Moira's prohibition, Calcutta was ordered to surrender the drawings. The Government of Bengal were reluctant to relinquish their claim to the drawings which had not reached London by October 1815 when Buchanan informed Smith that 'the Directors, it is true have ordered the drawings to be sent home immediately but what that animal [i.e. Lord Moira] may do with them in the interim it is difficult to say. At present I intend publishing at an early period the letter press descriptions and shall afterwards when I receive the whole publish the figures'[20]. Once more London insisted on Calcutta surrendering the drawings now in the custody of James Hare, the Botanic Garden's new Superintendent. Hare told the Chief Secretary to the Government of Bengal that the drawings which had accompanied Buchanan's survey papers were of buildings and agricultural implements. The disputed volume of natural history drawings included work executed at the menagerie at Barrackpore while the flower paintings appeared to be a continuation of Roxburgh's icones. It was Hare's contention that not to have the drawings at hand for consultation would 'put an end to natural history for the present in India'[21]. He recalled Colebrooke's indiscretion in releasing Roxburgh's drawings to his son for transmission to the executor of Roxburgh's will. 'A recent error had nearly deprived this establishment of the thirty years' labour of Doctor Roxburgh and had the accomplishment of Doctor Buchanan's plan been added, every trace of natural history in India would have been blotted out, and if ever the thirst of knowledge in England should depute another Buchanan to India, he must have robbed the very repository he is anxious to enrich.' London's command, however, could not be ignored but before parting with Buchanan's drawings, Hare obtained authorization to copy them for the Calcutta archive.

Buchanan had withdrawn to his native Scotland, disenchanted with his treatment by the East India Company, resolved never to ask them for drawings, manuscripts or specimens for his research. 'The Court of Directors has indeed received my collection with such contempt and arrogance that I would neither ask nor receive any favour from so scoundrelly a body.'[22] He retired to Leny in Perthshire, married, had children, and changed his name to Hamilton to comply with the terms of a legacy. He declined an invitation from Wellesley, now Lord Lieutenant of Ireland, to join his staff. In 1819 he published a limited edition of 50 copies of *Genealogies of the Hindus extracted from their sacred writings* and Constable issued his *Account of the Kingdom of Nepal and of the territories annexed to this dominion by the House of Gorkha*. The work that undoubtedly gave him the most satisfaction was an *Account of the fishes found in the River Ganges and its branches* (1822). Two years earlier he had decided to write a commentary on the *Hortus Malabaricus* and

Herbarium Amboinense. 'I would give my opinion concerning each plant in these works, describe more fully such as I have seen, reconsider the synonames and uses mentioned in them and introduce descriptions of such analogous plants as I have discovered.'[23] He asked Smith whether such a commentary would be suitable for the *Transactions of the Linnean Society*, using the Mysore drawings which he had earlier deposited with him to illustrate the articles. For the next five years he busied himself interpreting the species described in these two pioneer floras. He inflicted his conclusions on the long-suffering members of the Linnean Society at 31 meetings between 1821 and 1852[24]. Buchanan offered his commentary on the *Herbarium Amboinense* to the Wernerian Natural History Society in Edinburgh but, he told Smith, 'whether or not it will appear is doubtful as [Robert] Jameson [President of the Society] I suppose has not forgotten a lesson of honesty which I was under the necessity of giving to him'[25].

One would like to think that Buchanan, who contributed so much to Indian natural history, mellowed with his advancing years, but a contentious temperament evidently soured a well-merited enjoyment of his considerable achievements. Disappointment, disillusionment even, pervades his counsel to Wallich. 'My advice to you as an old man of a good deal of experience, both in India and Europe, is along with your research after science, to collect money as fast as possible and whenever you have a competence to return to your native land.'[26]

NOTES

1 Buchanan to Smith, 14 February 1795. L.S. Smith letters, Vol. 2, folio 199.

2 It was the subject of his first botanical paper: 'Description of the tree, called by the Burmese Launzan'. *Asiatick Researches*, **5**, 1798, 123–5.

3 Buchanan's Burmese specimens, manuscripts and drawings are now in the B.M. (N.H.)

4 *Journal of Botany*, 1902, 280.

5 Roxburgh to Banks, 13 July 1797. B.L. Add. MSS 33980, folio 102v.

6 Banks to Roxburgh, 9 August 1798. B.L. Add. MSS 33980, folio 159.

7 Prain (1905), xii.

8 Buchanan to Roxburgh, 31 January 1801. B.M.(N.H.) Buchanan letters, folio 47.

9 Buchanan to Wallich, 1 November 1828. B.M.(N.H.) Buchanan letters, folio 165.

10 *A journey from Madras through the countries of Mysore, Canara and Malabar*. London.

11 Buchanan to Smith, 3 March 1802. L.S. Smith letters, Vol. 2, folio 206.

12 Bengal Public Council, 1804. I.O.L.R. Range 6, Vol. 1.

13 *An account of the fishes found in the River Ganges and its branches*, 1822, Edinburgh.

14 1500 plants collected in Mysore and Nepal and 200 drawings.

15 Buchanan to Smith, 22 November 1806. L.S. Smith letters, Vol. 2, folio 219.

16 London to Bengal. Public letter, 7 January 1807. I.O.L.R. E/4/661, folios 5–14.

17 Now in I.O.L.R.

18 R. M. Martin (1838), *The history, antiquities, topography and statistics of eastern India*. London, 3 vols.

19 Lord Moira to Vice-President in Council, 5 January 1815. I.O.L.R. F/4/513, item 12340.

20 Buchanan to Smith, 27 October 1815. L.S. Smith letters, Vol. 2, folio 228.

21 Hare to Government of Bengal, 27 July 1816. I.O.L.R. F/4/559, item 13709.

22 Buchanan to Wallich, 4 February 1817. B.M.(N.H.) Buchanan letters.

23 Buchanan to Smith, 23 July 1820. L.S. Smith letters, Vol. 2, folio 243.

24 Only one-third of Buchanan's commentary on the *Hortus Indicus Malabaricus* was accepted by the editor of the *Transactions*.

25 Buchanan to Smith, 25 May 1824. L.S. Smith letters, Vol. 2, folio 250. 'Commentary on the Herbarium Amboinense', in *Memoirs of the Wernerian Society of Edinburgh*, **5**, 1823–4, 307–83; **6**, 1827–31, 268–333.

26 Buchanan to Wallich, 16 October 1821. B.M.(N.H.) Buchanan letters.

Calcutta Botanic Garden after Roxburgh

Francis Buchanan's temporary successor at the Botanic Garden, Nathaniel Wallich (*né* Nathanael Wulff) (1786–1854), was born in Copenhagen, the son of a Jewish merchant. He had been a pupil of Professor Martin Vahl, the author of *Symbolae botanicae* (1790–4), at the University of Copenhagen, and, on qualifying as a surgeon in 1806, left for the Danish settlement at Serampore on the River Hooghly opposite Barrackpore with an additional commission to collect plants and seeds for the Botanical Garden in Copenhagen. When he took up his appointment as the settlement's surgeon in 1807, the British fleet had bombarded Copenhagen and following the Company's annexation of Serampore in 1808, Wallich and his fellow countrymen became prisoners of war. He was released from captivity in June 1809 at the behest of William Roxburgh with the authorization of the Bengal Government to employ him in the Botanic Garden. Roxburgh proposed sending him to the eastern frontier of the Company's territory to collect plants, an assignment that appealed to Wallich who had professed that knowledge rather than wealth was to be his motivation in life. It comes as a surprise, therefore, that he should soon relinquish this potentially attractive mission to return to Serampore. It would appear he declined the opportunity to be a field botanist because he considered his travelling allowance to be inadequate.

In 1811 he petitioned the East India Company for a post of Assistant Surgeon on their Bengal Establishment. London approved his application in February 1814 and within months he was ordered on active service with the British forces invading Nepal, but the posting was cancelled upon his nomination as acting Superintendent at the Botanic Garden. The day after he took up this temporary appointment he confided his anxiety about his future in a letter to Buchanan. 'What I dread is this eternal instability of employment which renders a man's endeavours, let his exertions be however strong, vain and fruitless. I wish to be employed henceforth solely in Natural History and thus in time be enabled to lay something not unworthy of their approbation before my employers.'[1] For the time being he had the management of four nurseries, a kitchen garden, an orchard, a farm, and a garden laid out in Linnaean order, while a considerable portion of the central part of the Botanic Garden lay fallow. At his disposal was a work force of about a hundred backed up by about 70 convicts chiefly occupied in flood prevention schemes. In March 1815 he was allocated another 50 labourers to restore the neglected teak plantations.

In July 1815 he summarily dismissed his head nurseryman, John Roxburgh, the illegitimate son of William Roxburgh, who had been entrusted to Father John of the Tranquebar Mission for his education. He had been his father's companion to the Cape of Good Hope in 1799 where he remained until 1804 collecting native plants and European cultivars for the Calcutta Botanic Garden. On his return to India, his father, impressed by his proficiency in flower painting, made him head painter in charge of the Indian artists. When Wallich objected to botanizing in the eastern region in 1809, John went in his place, collecting in the neighbourhood of Chittagong. Roxburgh's son continued his combined duties as nurseryman and head painter until his dismissal by Wallich. Reinstated by James Hare who had been appointed Superintendent in April 1816, his reprieve, however, was short. With Wallich's reappearance as Superintendent he was again dismissed in March 1819, his post being given to George Potter, Overseer of the labourers (Fig. 8.1). John Roxburgh protested his innocence, speculating that the provocation for his dismissal must have been his occasional lapses in punctuality occasioned by his courtship and subsequent marriage. He appealed in vain to the Court of Directors for justice and compassion 'for the sake of the revered father of your Lordship's memorialist whose name is identified with the science of botany in this country & the Company's Garden at this place'[2]. As not all the full facts of the case have survived, it is impossible to judge the propriety of Wallich's action, but it would appear he behaved harshly. Hare had rejected Wallich's original charges against John Roxburgh as trivial. One can only speculate whether Wallich's departure from the Botanic Garden in 1809 had been precipitated by a clash with William Roxburgh which subsequently marred relations with his son. As we shall see in the incident regarding the confiscation of Hare's drawings, Wallich could be vindictive.

Nevertheless Wallich had many friends who admired his botanical competence—William Carey was convinced he 'would do more for the science of botany than any other man in India', at the same time recognizing his volatility and impetuosity. Carey saw him as 'a man of great warmth [but] perhaps too lively in his resentment'[3]. Thomas Hardwicke cautioned Wallich not to

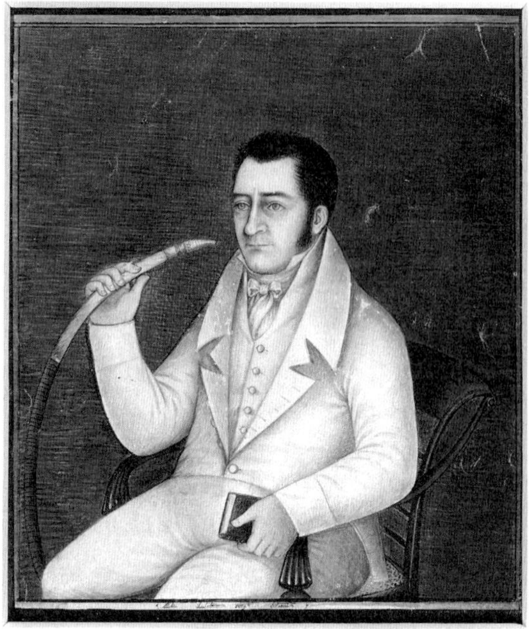

Figure 8.1 George Potter (*d.* 1833) smoking a hookah. Gouache by Lachman Singh, *c.*1828, a botanical artist at the Calcutta Botanic Garden. Potter replaced John Roxburgh as head nurseryman in the Botanic Garden in 1819. (India Office Library and Records.)

allow 'the irritability of his temper to outreach his reason'[4]. With such an unstable personality, it was inevitable that Wallich would sooner or later clash with Hare after the latter's appointment as Superintendent. Wallich had been retained on the staff and allowed to remain in the official residence with his family while Hare chose to live at Barrackpore. Wallich resented Hare's treatment of him, complained about debts he had incurred in the preparation of an illustrated account of some of the rarities in the Garden, using the Roxburgh icones, and in pique resigned. When illness compelled Hare to retire, the Governor-General, Lord Moira, offended by Wallich's behaviour, gave the superintendency in December 1816 to Thomas Casey: 'a man who was never known to have the smallest taste for the science'[5] complained Carey, who feared that his ignorance and incompetence would do irreparable harm to the garden. Colebrooke and John Fleming, President of the Bengal Medical Service, brought this unsatisfactory state of affairs to the attention of Sir Joseph Banks. Sir Joseph, always a vigilant guardian of the garden's interests, persuasively advocated Wallich's cause. The Court of Directors, heeding his advice, on 13 February 1817 directed the Bengal Government to install Wallich as Superintendent.

In August 1817 Wallich was once more in command with a timely opportunity to settle old scores with James Hare. He was confronted by a team of Indian artists who had been engaged by Hare to copy the Botanic Garden's drawings. He learnt that Hare had already taken 573 copies with him to England and another 619 had been executed since his departure. Wallich immediately took charge of these painters to prevent further damage to the original drawings which had been copied 'by the aid of glass pains [panes]' and forthwith submitted a formal complaint to the Government of Bengal. He contended that the 20 albums housing some 2600 drawings, done either for Roxburgh or copied from other collections such as those of Colebrooke and Buchanan, were official property not to be copied without authority. His superiors concurred, instructing him to transfer the 619 drawings to the East India Company's library in London. His insistence that Hare should be required to surrender those already in his possession posed a delicate problem. The Principal Secretary of the Government, knowing Hare's 'correctness and particularity on all occasions', had taken the precaution of consulting the Governor-General who, although he could not remember with complete confidence having given Hare permission to copy the drawings, confessed that he 'certainly did so by clear and unavoidable implication'[6]. Hare was consequently allowed to keep the disputed drawings but forbidden from publishing them and Wallich had the pleasure of dismissing his artists in Calcutta.

Wallich, appreciative of Sir Joseph Banks's support in his appointment, showered him with gifts of plants from Sylhet and Nepal; they included seeds of red and white species of rhododendrons. Banks was delighted.

> *When I took upon myself to recommend to the Directors of the East India Company to reinstate you in the charge of the Botanic Garden at Calcutta, I was conscious I was doing a good & a proper thing, from the activity, intelligence & zeal you had so amply evinced while the Garden was under your administeration [sic], but I confess I was not aware [of] the increased energies which have been this year exercised in favour of my favorite establishment, the Royal Gardens of Kew'*[7].

He did not live to receive Wallich's most exotic gift, a musk-deer acquired from the British Resident in Nepal. It was accustomed, wrote Wallich in his last letter to Sir Joseph, 'to live upon boiled rice and bread or biscuits'[8]. Between August 1817 and August 1818 a succession of chests and baskets of plants and parcels of seeds from Calcutta were directed to many notable gardens in Britain. Other parts of the world, too, were indebted to Wallich's generosity—France, the Netherlands, USA, South Africa, New South Wales, Mauritius, St Helena and, of course, his native Denmark which awarded him the Golden Cross, Fourth Class, of the Order of Dannebrog. His request to the Court of Directors to be allowed to assume the equivalent rank and style of an English knight was politely refused.

The reduction of the 300 acres of the Botanic Garden by 40 acres for a site for Bishops College in 1820 did not lessen the flow of ornamental and useful plants to the public and private gardens of Europe. With the support of the Medical Board a plot of land was reserved for experimental trials with medicinal plants, both native and foreign. This followed a recommendation from the Apothecary to the East India Company, Dr Clarke Abel, who was concerned that imported European drugs lost much of their potency during long sea voyages. It was additionally proposed that Calcutta's efforts should be supplemented by small physic gardens established in the foothills of the Himalayas. Titty-ghur, near Barrackpore, became a nursery for the introduction of useful plants into Bengal.

Their inability to refer to standard texts constantly frustrated the researches of botanists in India. William Roxburgh had told Banks in 1790 that he thought the Company should supply its medical staff with relevant books. When Roxburgh suffered the disastrous loss of his library in the floods of 1793, London considerately shipped some replacement volumes to him. At that time standard works included Van Reede's *Hortus Malabaricus*, Rumphius's *Herbarium Amboinense*, Burman's *Thesaurus Zeylanicus* and Linnaeus's many publications. In December 1819 Wallich advocated the formation of a 'public library' at the Botanic Garden[9] since the libraries at the College of Fort William and the Asiatic Society had few botanical books. He offered to sell over 500 volumes from his personal collection to form the nucleus of this library. It was a heterogeneous selection: among them Van Reede and Rumphius, Philip Miller's *Gardeners dictionary* and *Figures of plants*, Mrs Blackwell's *Curious herbal*, Joseph Plenck's *Icones plantarum medicinalium*, A.B. Lambert's *Genus Pinus*, several works by Linnaeus, Redouté's *Les Liliacées*, Rees's *Cyclopaedia*, Lamarck's *Encyclopédie methodique: botanique*, nine volumes of G.C. Oeder's *Flora Danica*, J. Sowerby's *English botany*, and 44 volumes of *Curtis's Botanical Magazine*. His valuation of 8500 rupees (about £1063 at that time) was accepted by the appointed assessors, Thomas Hardwicke and William Carey, who only rejected a few items. They queried, for example, Thornton and Redouté and objected to the *Philosophical Transactions of the Royal Society* on the grounds that the periodical was light in its botanical content. While the Court of Directors declined to authorize the purchase, it recognized the need for a botanical library by sanctioning an annual vote of £200 for the next ten years to acquire books and periodicals. The embryonic library attracted donations: Ludwig Blume presented parts of his *Flora Javae* and von Martius his *Nova genera et species plantarum* and *Reise in Brasilien*. In 1828 government officials at Fort St George paid a thousand rupees for a copy of Redouté's *Les Liliacées* belonging to the late George Hyne, briefly Naturalist in the Madras Presidency.

The creation of a library incurring ongoing expenditure required London's authorization but the trifling sum needed to buy a bell for the Botanic Garden received swift local approval. Wallich was having difficulty in summoning his labourers to work in the morning and after the midday break. Large Chinese gongs specially obtained from Canton could not be heard throughout the length and breadth of the grounds and, moreover, failed to withstand the vigorous hammering they got. Even blasts on a trumpet were ineffective. A large well-cast bell—a difficult item to find in Calcutta—might have greater resonance. And when, at last, such a rarity, moderately priced, appeared on the market, Wallich was its triumphant purchaser (Fig. 8.2).

Wallich's submissions to Government and letters to friends testify to the pride he took in the Botanic Garden, locally known as 'Wallich's pet', which was, he ceaselessly preached, the most informative exposition of India's bewildering flora. He encouraged his fellow surgeons to make use of all its facilities including the new library. He urged the attendance of pupils from the school for native doctors. Victor Jacquemont, the French botanist, who saw the garden in 1829 when Wallich was on leave in England, was

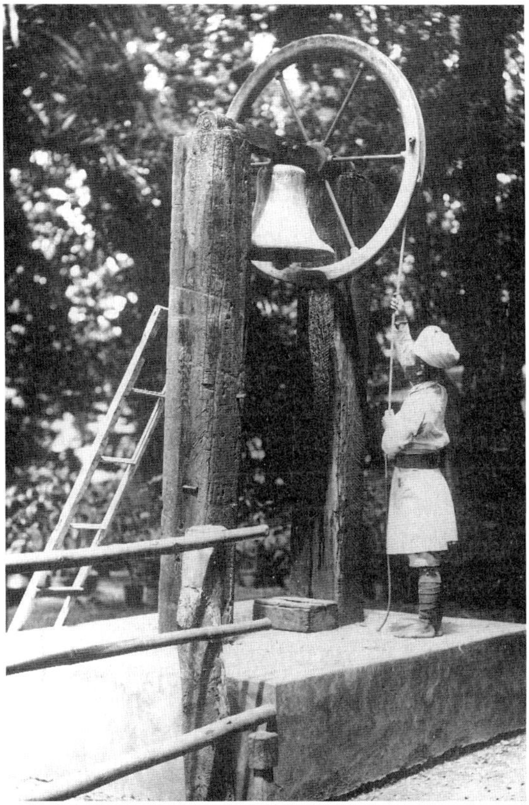

Figure 8.2 Gardeners being summoned to work at the Calcutta Botanic Garden by the bell probably purchased by N. Wallich. Photograph, 1908–12

impressed by its collections and library but his comments on the absent Superintendent displayed ungracious envy. 'A Danish botanist, of ordinary talents, but who passes here for the first in the world, is the director of this establishment; and he certainly has the best income of any *savant* in existence.'[10] Wallich's standard of living bedazzled him as much as the establishment itself—'72 000 francs a year and a superb house'.

Visitors to the garden were invariably awed by the vast, sprawling tangle of branches and aerial roots of the banyan tree. One November morning in 1823 the Reverend Reginald Heber, Bishop of Calcutta, accompanied the Governor-General's wife, Lady Amherst, preceded by her escort of attendants with gilt-sticks, maces, swords, and ceremonial spears, on a perambulation of the Botanic Garden. An avenue of sago palms nostalgically recalled for the Bishop slender Gothic columns. Milton's concept of Paradise came to mind as he admired nature's luxuriant vegetation from the South Seas, Australia, America, Africa, India itself, and neighbouring countries (the elephantine bulk of the *Adansonia* (baobab) was the only jarring feature). The attractions of the garden, its order and colourful displays, owed as much to Wallich's skilled labour force as to his own management. In 1820 he requested salary increases for those members of his staff he particularly relied upon: the head gardener, Kurrim Khan, who had been an employee since Colonel Kyd's days and Shaik Mooty, the head man in the general nursery. He gave a warm endorsement of the latter's abilities. 'The knowledge of the treatment and cultivation of plants, his acquaintance with the Linnaean system and what is more surprising, with the natural classification are such, as would do honor to any European gardener.'[11]

Wallich's indisputable devotion to the garden did not inhibit his external activities and interests. He contributed new plant descriptions to Carey's printing of Roxburgh's *Flora Indica*. He oversaw Government teak plantations at Sylhet and elsewhere. And then in December 1820 the British Resident at Khatmandu, Edward Gardner, a frequent donor of plants to the Botanic Garden, enticed him to Nepal. When he returned a year later, illness contracted in the forests of the Terai made him a refugee from Bengal's hostile climate. He was granted six months' sick leave to proceed to China but got no further than the East Indies. On 10 August 1822 he sailed from Calcutta leaving the judge, William Leycester, in charge of the garden; with him went his head Overseer, George Potter and two apprentices to assist in the collecting of plants. Wallich seldom wasted a moment or an opportunity—leave in England in 1828 was similarly profitably employed. During the eight days Wallich was in Penang, he obtained permission to create a new garden or nursery. The original one, founded in 1800 and looked after first by William Hunter and then Christopher Smith, had been disbanded and its contents sold. A site of 130 acres was chosen and George Potter[12] with one of the apprentices remained to develop it. The next stop in the voyage was Singapore when Wallich renewed his acquaintance with Sir Stamford Raffles, the founder of the British settlement there. Raffles recruited him to a committee considering the erection of warehouses and presumably had little difficulty in persuading Wallich to request formally the establishment of both a botanical and an experimental garden. Wallich deployed all the arguments in favour of the proposal[13]. Singapore was part of an archipelago without rival in respect of climate, fertility, and vegetation. Raw materials for the construction of ships were abundant; the commercial prospects of Malayan fruits, vegetables, and grains should be investigated; trials already made indicated that clove and nutmeg grew well, coffee was a promising candidate, 'even tea grows freely and seems to lose nothing in luxuriance of flower and fruit by the change from its natural climates'. Raffles invited Wallich to select a suitable site, promising financial help and it is clear from all the extant correspondence that Raffles considered the gardens at both Penang and Singapore would be subordinate to the Calcutta Botanic Garden. London qualified their approval of both establishments with an insistence on financial prudence. William Montgomerie, an Assistant Surgeon with the garrison troops at Singapore, was placed in charge of the garden and Wallich left, promising the secondment of one of his experienced gardeners. When Montgomerie failed to receive this assistance or any communication from Wallich, he concentrated on the successful cultivation of spices. In 1829, just two years after he was posted to India, the Singapore botanic garden was closed down.

1824 was a gratifying year for Wallich: it saw the publication of the second volume of his edition of Roxburgh's *Flora Indica* and the first fascicule of his *Tentamen florae Napalensis illustratae*. In 1825 the Bengal Government despatched their peripatetic botanist to the Kingdom of Oudh and the province of Rohilkhand to report on submontane forests.

The treaty of Yandabo in 1826 terminated two years of jungle fighting with the Burmese, ceding to the British not only Assam, but also the coastal regions of Arakan and Tenasserim, and part of the province of Martaban east of the Salween river. Wallich, once more on the move, investigated the rich timber resources of Martaban and Tenasserim. A passenger on the first steamboat to navigate the Irrawaddy river, he botanized with satisfactory results whenever the ship moored. It took a month for the party to reach Ava, the capital of Upper Burma, where Wallich explored the neighbouring hills, discovering many new species and marvelling at the abundance of teak and oak, 'the two greatest glories of the forests of Europe and Asia'. Before he returned to Calcutta he found the 'Thoka' tree close to an abandoned monastery on a hill by the Salween river. Two solitary trees had been discovered by John Crawfurd a few months earlier. Wallich was not disappointed by Crawfurd's enthusiastic description of its spectacular flowers. 'They were both covered with pendant panicles of large vermilion red blossoms and formed a superb object, hardly equalled in the flora of the East Indies and I presume not surpassed in magnificence and elegance in any part of the world.' Wallich named it *Amherstia nobilis*, in honour of the Governor-General's wife, a notable patron of natural history in India (Fig. 8.3). Successful in raising it in the Botanic Garden in

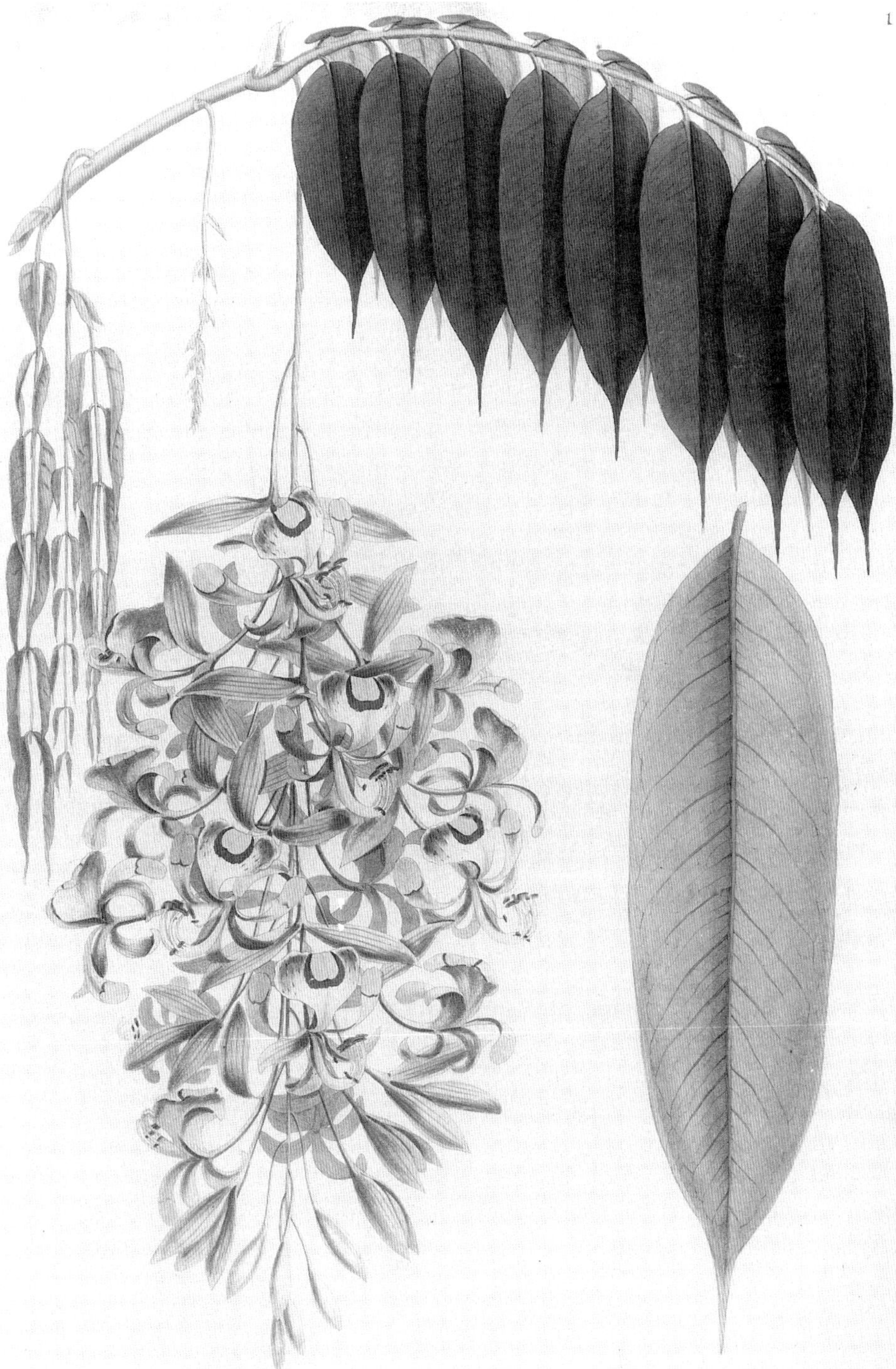

Figure 8.3 Amherstia nobilis. (From N. Wallich (1829), *Plantae Asiaticae rariores*, Vol. 1, London, plate 1.) Hand-coloured lithograph of drawing by Vishnupersaud, one of the Calcutta Botanic Garden artists. Wallich named this spectacular Burmese flower after Countess Amherst and her daughter, Lady Sarah, 'who, after a residence of nearly five years in India . . . have returned to England with a large and very interesting collection of preserved specimens of plants, gathered and excellently preserved by their own skill and industry'.

Calcutta, he failed, however, to introduce it into cultivation in England.

Since his appointment as Superintendent in 1817, Wallich had been absent from his garden for a total of more than three years. Twenty years' endurance of India's climate, the demands of administration, and the privations of botanical tours had taken their toll. An application for home leave was supported by a recital of his achievements[14]. Species in cultivation in the Botanic Garden had doubled since the publication of the catalogue in 1814; some 6000 were stored in the herbarium which also housed a collection of plants preserved in spirits and a reference set of wood specimens; pictorial records of the Nepalese and Burmese flora had been added to the drawings collection. He claimed a record discovery of many new species: 'my harvest far exceeds that reaped by any travelling naturalist or any party of naturalists during a corresponding period of time'. The management of the garden denied him time to exploit the herbarium he had amassed or to publish the results of his researches while the primitive state of graphic art in India discouraged him from attempting to reproduce his drawings locally. He consequently requested leave in England, not only on legitimate grounds of ill-health, but also to be able to employ skilled craftsmen to lithograph his drawings illustrating an account of his collections which would be published at no expense to the Company. His application granted, Wallich arrived in London in July 1828 with his entire herbarium packed in thirty crates together with chests of living plants and packets of seeds for friends and horticultural institutions. Accompanying him was one of his plant collectors, William Gomez, to unpack the cases and sort the contents and James Watson, a garden apprentice, deputed to assist Gomez and also to expand his horticultural experience whilst in England. This vast herbarium exceeding 8000 species represented the combined efforts of Wallich and his collectors—Robert Blinkworth (Kumaon), Kamrup (Sylhet), William Gomez (Tavoy and Tenasserim), George Potter (Penang), Henry Bruce (Chittagong), and Govan and Royle, superintendents of the Botanic Garden's satellite establishment at Saharanpur. Wallich also fostered a network of unofficial collectors who, during the course of their duties, contributed living and dried specimens: Edward Gardner in remote Khatmandu, Major-General Thomas Hardwicke on active service, the French botanist Louis Leschenault de la Tour in the Nilgiri Hills and South India, and Raffles in Malaysia. Lady Amherst brought flowers from Simla and, nearer to Calcutta, William Carey at Serampore shared his acquisitions with Wallich[15].

The East India Company rented and furnished two floors of a small house in Frith Street in London for this herbarium and sanctioned the distribution of its duplicates amongst institutions and botanists in Europe and America. Wallich found colleagues willing to share the labour and rewards of this distribution: John Lindley, one of the country's ablest botanists, George Bentham who rejoiced in the company of 'joyous Wallich' and his oyster lunches, and Robert Graham from Edinburgh University; in due course continental botanists made their appearance, offered their services and claimed a share of the spoils—von Martius, Kunth, Lehmann, Nees von Esenbeck, and Alphonse de Candolle among them. Specimens were sorted, lithographed hand-written labels with identifying numbers were attached and these were accumulated in the *Numerical list of dried specimens of plants in the East India Company's Museum, collected under the superintendence of Dr Wallich of the Company's Botanic Garden at Calcutta* which appeared in parts from 1828 to 1849. Wallich, 'generous, liberal, expansive'[16], relishing every moment of his philanthropic mission, had sought the Court of Directors' permission in November 1829 to dispose of the duplicates of other herbaria languishing in the Company's India Museum. These comprised plants collected in South India by Patrick Russell, Gottfried Klein, Benjamin Heyne, Johan Rottler, the considerable collections of Francis Buchanan and Robert Wight, a small gathering by William Roxburgh, and George Finlayson's plants from Siam and Indo-China. The Company which had always adopted an ambivalent attitude towards its museum readily agreed, subject to the retention of one complete set. The wide distribution of this material provided an invaluable opportunity for the study of the Indian flora by competent botanists in Europe.

As the India Museum was always starved of space, funds, and staff, one suspects that the Court of Directors secretly welcomed Wallich's offer to diminish the size of its herbaria by the dispersal of duplicates. Some British botanists, in particular Robert Brown, disapproved of his generosity towards continental institutions but the Linnean Society of London received the most favoured treatment. In 1832 the type set of specimens was presented to the Linnean Society with a grant of £200 to purchase paper on which to mount them[17]. The formal acknowledgement of the gift by the Council of the Linnean Society praised the Company's patronage of 'those distinguished naturalists who have cultivated science in Asia, so much to their own honour and to the credit of the service to which they belonged'[18]. William J. Hooker added to the public acclamation by dedicating the first volume of his *Botanical Miscellany* (1830) to the East India Company as 'a testimony of the important services rendered to science in their princely patronage of botany, by which the plants of the vast possessions under their sway, especially the more interesting and useful kinds, have been made known and munificently distributed throughout the whole civilized world'.

Without the participation of the team of British and continental botanists who sorted and described this vast accumulation of plants, Wallich could not have accomplished so much in so brief a period of time. Self-interest undoubtedly motivated his collaborators who subsequently published the results of their labours in taxonomic monographs. Some of them also contributed their expertise and knowledge to Wallich's selective account of the India flora, *Plantae Asiaticae rariores*. The first number appeared in September 1829 and the second had just been published when in the following April Wallich applied for an extension of his leave to finish it. The work was programmed to continue at quarterly intervals, 'the cheapest of its kind published in any branch of natural history, each plate costing only two shillings, inclusive of every other expense of letter-press'[19]. Wallich insisted that his personal supervision was required to ensure that the publication did not suffer the inertia that had delayed the completion of Roxburgh's *Plants of the Coast of Coromandel*. He was granted a further year's leave, subsequently extended until September 1832 by which time the twelfth and final number had been published. It comprised 295 plates selected mainly from 1200 drawings by Indian artists. All but one drawing were printed by the relatively new process of lithography which represented a radical departure from the basic principles of graphic reproduction which had endured for some three hundred years. Lithography delivered the artist from the bondage of the engraver, enabling him to draw with a greasy chalk directly on the surface of a polished stone. Francis Bauer, the resident botanical artist at Kew Gardens, unreservedly welcomed the process. 'I certainly prefer the lithographs, as for colored plates they are much better than line engravings.'[20] There were those defenders of traditional methods like, for instance, Benjamin Maund, the proprietor of *The Botanist* (1836–41) who chose steel engravings for his periodical because 'the sandy and indefinite lines of lithography make it unsuitable for portraying the niceties of botanical dissections'[21]. This broken or dotted line was preferred by the artist and designer, Noel Humphreys, who applauded the softer image produced by the chalk line in Wallich's book, contrasting it with 'the cold effect of aquatinta shades . . . and the harshness of line-shading on copper or steel'[22]. The plates in *Plantae Asiaticae rariores* are bold and emphatic, hand-coloured in simple washes by John Clark. The three volumes represent a continuation of the Roxburgh tradition in their dependence on the artists employed at the Calcutta Botanic Garden: 146 drawings were by Gorachand, 109 by Vishnupersaud with a solitary example of the work of Runjia. The remaining plates executed by European artists include three by the young William Griffith to whom Wallich had been introduced by John Lindley, and who was later to be the cause of much distress for the author.

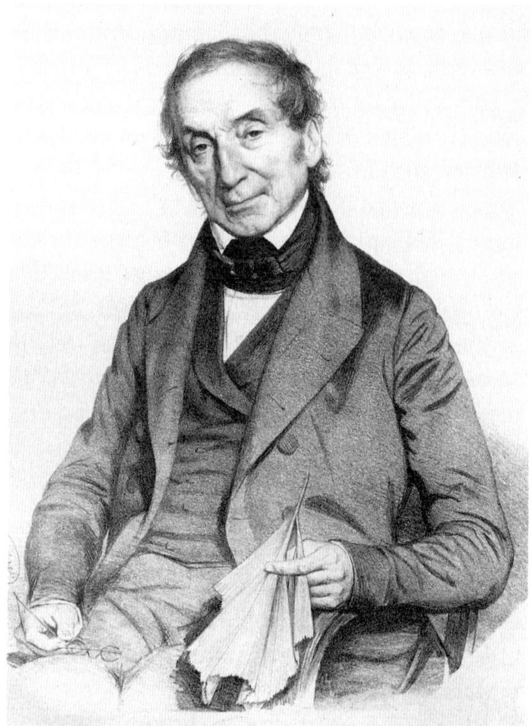

Figure 8.4 Nathaniel Wallich (1786–1854). Lithograph by T. H. Maguire of a portrait by Sir Daniel Macnee.

The work celebrates some of the spectacular finds by Wallich and his collectors. The very first plate, *Amherstia nobilis*, is a theatrical gesture, a splendid curtain-raiser with its pendulous racemes of vermilion flowers. Hard-pressed, Wallich readily accepted offers of help from dependable George Bentham who produced a synopsis of Indian Labiatae, from A. H. Haworth who described a few *Kalanchöe* species and from Robert Brown who contributed a token offering of a few descriptions. William Hooker had no doubt whatever that this was Wallich's *magnum opus*, 'a work, which, whether for the beauty or rarity and interest of the subjects, the execution of the plates, or the accuracy of the descriptions, is surpassed by no publication of this, or any other period'[23]. Two hundred and fifty copies[24] were printed of which the East India Company purchased 40, their usual quantity for any work they patronized. It was the last of the ambitiously conceived, extravagantly produced Indian floras; its successors were more modest in conception, more restrained in presentation. The illustrious quartet—Van Reede, Rumphius, Roxburgh, and Wallich[25]—between them laid firm foundations for the more intensive and systematic investigation of the vegetation of the subcontinent which was to be pursued later in the nineteenth century.

In between writing *Plantae Asiaticae rariores* and distributing specimens, Wallich spared time to sit for his portrait by John Lucas. It portrays a man content with his achievements, relaxed yet alert, his generous mouth, high forehead, and prominent nose proclaiming forceful character, resolution, and determination, qualities he was to need when he reached Calcutta in 1833 to resume control of the Botanic Garden from Sir Charles Metcalfe. He returned to find his institution a casualty of the retrenchment imposed by the East India Company on its Indian administration. As the finances of the Company continued to worsen, exacerbated by the Burmese war, draconian measures became inevitable—in 1830 professional posts at the College of Fort William and that of the Superintendent of Hindu Researches were abolished. An annual saving of nearly 7000 rupees at the Botanic Garden was effected by staff reductions; more disturbing, however, was the resolution that upon Wallich's retirement, the Superintendent's salary was to be cut by 1000 rupees a month—such a severe drop in salary would necessarily entail a loss of status. Other economies included the cancellation of the library grant. Expenditure on the Botanic Garden had declined from 58 000 rupees in 1826/7 to just under 40 000 rupees in 1828/9. A terse request from the Court of Directors for a statement on the practical benefits of the garden hinted at further rigorous measures to come. But the Court was circumspect and tactful in its approach:

> *the labors of Dr Wallich and his predecessors have been eminently successful in extending botanical knowledge and if they have not tended in any great degree to the general introduction of the culture of rare and valuable articles of produce, or otherwise materially promoted the agricultural and commercial interests of the country the circumstance is rather attributable to the condition of society and the state of the country, than to any inherent defect or inefficiency. We are induced to think that the chief cause of the ill success of the institution in the points last noticed is, that Europe has nothing to give to India in the form of a new article of culture which is at once suited to its climate and worthy of its reception*[26].

Wallich deliberately delayed submitting this crucial statement which demanded a carefully prepared presentation of facts and figures; instead he turned his attention to less momentous matters. A few weeks before his arrival in Calcutta George Potter, his head gardener, had died and Metcalfe had replaced him by James Kane, formerly employed at the Saharanpur Garden. When he was recalled to his Regiment of Artillery in 1834, Kane opted to purchase his discharge and became gardener for the Agricultural and Horticultural Society at Lucknow. Since qualified gardeners were a rarity in India, Wallich was delighted in the appointment of John White Masters as head gardener—'a better choice could not possibly have been made', he told the Government of Bengal in 1836. Yet two years later he dismissed Masters for allegedly supporting attacks on the management of the Botanic Garden. The combination of ill-health, age, and the frustrations of financial constraints had made volatile Wallich, irritable and suspicious. 'I am perfectly well—and yet I am sick—that is *I am sick of India*', he tells Nathaniel Ward[27]. He could be a fractious companion, unpredictable in his moods, as William Griffith discovered when he accompanied him on a mission to Assam in 1835.

Assam, one of the spoils of the recent Burmese war, held out the promise of being a most profitable acquisition after the discovery of the tea plant there in 1834. With the termination of the Company's monopoly of the Chinese tea trade imminent, the urgent need to establish tea plantations in India stirred officials in London and India into action. A Tea Committee despatched Wallich, William Griffith, and John McClelland who was essentially a geologist, to confirm a report of tea growing in Assam. Wallich was allowed to take three native artists and two gardeners with him. At first relations were harmonious between Wallich, who was now nearly 50, and his two much younger colleagues. The first hints of friction arose, according to McClelland, when Wallich accused Griffith of concealing some of the plants he had collected. We are told he was jealous of the size of Griffith's collections and, in insisting that there should be only one official collection, removed a number of Griffith's specimens. Griffith complained that he demanded unquestioning obedience. 'I repeat what I have often said, that it is utterly impossible to pull well with such a man, and that he is a compound of weakness, prejudice and vanity.'[28] This intense dislike of Wallich was to have unfortunate consequences; McClelland hounded him in the pages of the *Calcutta Journal of Natural History* and Griffith destroyed, but not with malicious intent, much of his work in the Botanic Garden.

Doubts about the role and administration of colonial botanical gardens like the one at Calcutta were raised in an official report to Treasury in 1838 on the state of all the royal gardens in England, and of the one at Kew in particular. The review committee consisted of John Lindley, Professor of Botany at University College, London and two gardeners, Joseph Paxton and John Wilson. In presenting a plausible case for converting Kew Gardens into a national botanical garden with an imperial role, the performance of colonial botanical gardens was perhaps deliberately understated.

There are many gardens in the British colonies and dependencies as Calcutta, Bombay, Saharanpore, in Mauritius, at Sydney, and Trinidad, costing many thousands a year. Their utility is much diminished by the want of some system under which they can be regulated and controlled. There is no unity of purpose among them; their objects are unsettled, their powers wasted, from not receiving a proper direction; they afford no aid to each other, and it is to be feared but little to the countries where they are established; and yet they are capable of conferring very important benefits upon commerce and of conducing essentially to colonial prosperity. A national botanic garden would be the centre around which all these lesser establishments should be arranged.

If such a committee, which one would suppose to be sympathetic, could be critical of Calcutta, it is not suprising that the Court of Directors, on behalf of all its shareholders, should insist on a review of its investment in the Botanic Garden. In 1834 and again in 1835 they reminded the Bengal Government they still awaited Wallich's report. Wallich, who could procrastinate no longer, pleaded pressure of official duties and his work with the Tea Committee in mitigation when at last he delivered his report to the Government of Bengal on 1 October 1836. The document paraded his accomplishments but his stance was defensive, always emphasizing the lack of the equipment and services which botanical gardens in Europe enjoyed. Calcutta had to depend on its own resources and 'unite the functions of botanist, gardener, nurseryman, and seedsman'[29]. He alluded to the difficulties in propagating plants from overseas in the plains of North India. He looked upon his garden as a kind of convalescent home until such plants could be moved to the more congenial climate of the Himalayan foothills. The primary objectives of the garden, conceived at the time of its formation were still apposite, 'namely to bring to light as many as possible of the vegetable stores which nature has so bountifully provided for the country, to cultivate and multiply all that has the least tendency or holds out any prospects of becoming of general utility, and to disseminate all such plants as widely as possible'. To this end some 16 000 plants had been supplied to European residents and Indians between June 1835 and June 1836 alone; 42 000 tea shrubs raised from Chinese seeds had been distributed. London, apparently mollified, blandly approved all the positive initiatives of their Superintendent. 'The Report of Dr Wallich has made a very favorable impression upon us with regard to the practical benefits enjoyed by the community through the establishment.'[30] It seemed that the threat of reductions other than those announced in 1830 had receded for the time being.

Wallich's next report, prefaced by apologies for its late delivery, covers the period from 1836 to 1840. He chronicles his successes, frankly admits his failures, but is clearly pessimistic about the future. Thousands of coffee seedlings raised from imported Bourbon seeds had been despatched abroad; tea plants, propagated from Assamese and Chinese stock, had been widely distributed; the mangosteen declined to flower; no seedlings had been obtained from the few ripe nutmegs; and the clove persisted in dying in Bengal's inhospitable climate. He also blamed the climate for his failures with apples, pears, oranges, and other fruit trees, breadfruit, and foreign grasses.

> *Many plants which have been imported over and over again, have resisted all endeavours to keep them alive, although at times arriving in the finest condition. The beautiful, but so far very rebellious species of Fuchsia, are most particularly of this description; nor will it excite wonder that the whole class of Orchideae, which justly constitutes the pride of gardeners at home, should be with us, in the Plains of Bengal, of such difficult management, especially the foreign sorts*[31].

It was a recital of lamentations. Robert Blinkworth, the only survivor of the Botanic Garden's team of collectors, now passed much of his time in the tea nurseries in Kumaon. Wallich pleaded for the reinstatement of his plant collectors whose services were vital to maintain the flow of plants; without them the garden was dependent on donations. Overseas donors whom Wallich singled out for their exceptional generosity included the royal gardens at Kew, the botanical garden at Liverpool, Jardin des Plantes in Paris, the botanical garden on the island of Bourbon, and the English nurseries of Loddiges of Hackney, Rollisson of Tooting, and Knight of Chelsea. Calcutta reciprocated with the despatch of some 190 000 plants to nearly 2000 gardens. This impressive traffic in plants was later gratefully acknowledged by Joseph Hooker at Kew; during 'the first twenty years especially of Dr Wallich's superintendence', he wrote, 'it had contributed more useful and ornamental tropical plants to the public and private gardens of the world than any other establishment before or since'[32].

Wallich was now a very sick man. 'I am so old in India—my powers of mind as well as of body are so feeble after 34 years of Indian career,' he confided to Sir William Hooker[33]. One instance of the economic stringency which impeded his management was the official reaction to his request for an additional European overseer and more native watchmen to combat the theft of plants from the garden. In refusing his application, the Government blandly advised him either to concentrate the valuable and rare plants near his residence or to place them in a secure enclosure! He left Calcutta in April 1842 to convalesce at the Cape, regretting the need to return in order to qualify for certain financial benefits applicable to the medical branch, but vowed that 'not more than a month shall elapse from the day of my return to Bengal before I am off for good on my final return home to Europe'[34].

William Griffith was nominated to manage the Botanic Garden during Wallich's absence and while he was proceeding to Calcutta from Malacca where he was stationed, another Dane, the surgeon Joachim Otto Voigt (1795–1843), took temporary charge. Since William Carey's death in 1834, Voigt had been looking after his five-acre garden at Serampore which, according to Jacquemont, was so intensively cultivated that it required a labour force of some 50 malis or native

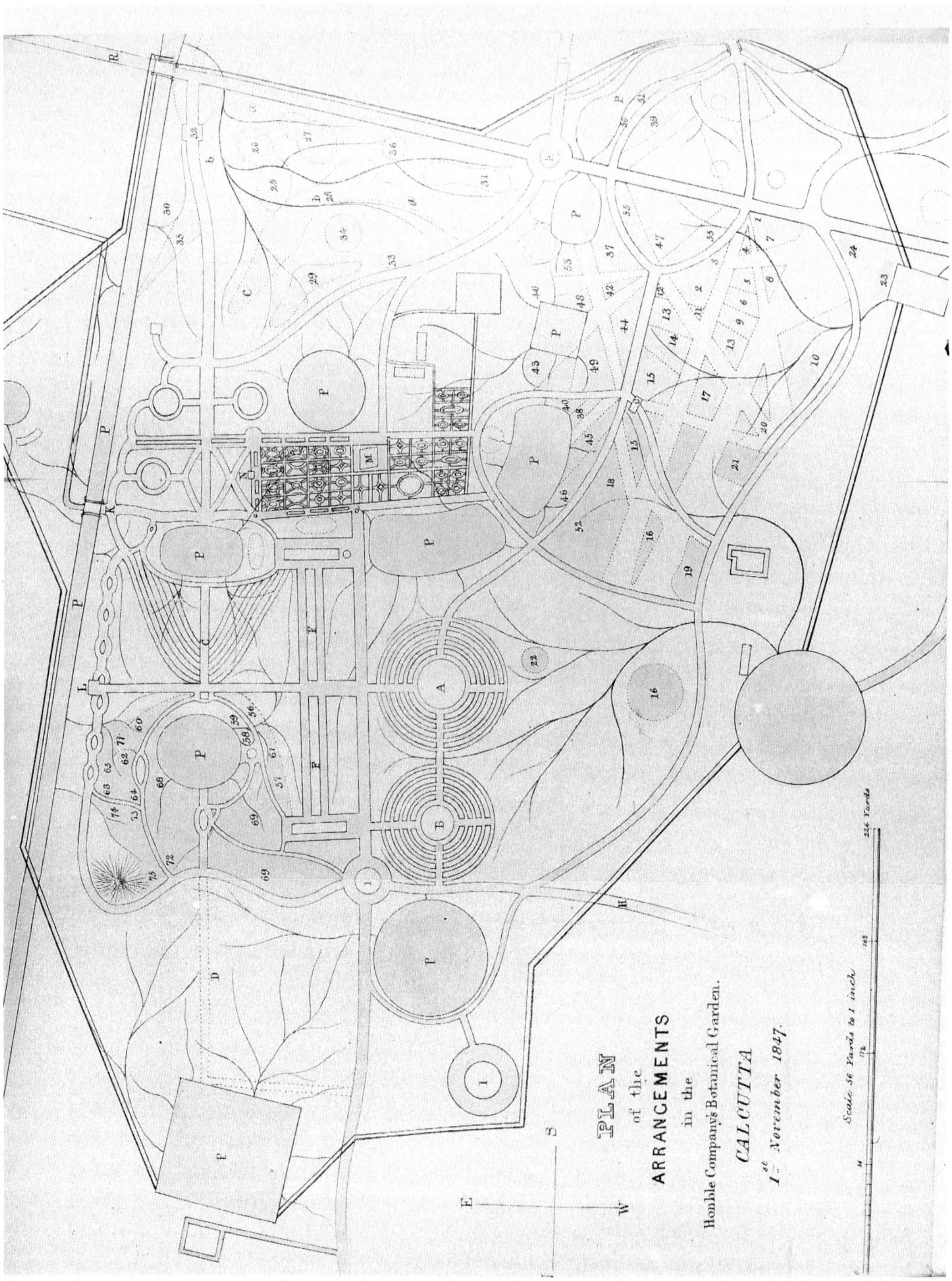

Figure 8.5 Plan of the Calcutta Botanic Garden in 1847 incorporating some alterations initiated by William Griffith including a natural garden (A); and a garden of medicinal plants (B).

gardeners. Carey had created a garden at Mudnapatty, filling it with local Bengal plants, specimens from Bhutan and English bulbs. With the compulsion of a born gardener, he began another garden as soon as he got to Serampore in 1800. Plants brought from Mudnapatty provided a nucleus of stock; others were acquired from friends, among them Roxburgh at the Botanic Garden. It soon became one of the best private gardens in the whole of Bengal and Voigt, now its dedicated custodian, conceived the idea of compiling a list of its treasures combined with the contents of the Botanic Garden in neighbouring Calcutta. When the Agricultural and Horticultural Society agreed to publish it, Voigt took advantage of the two months he was in charge of the Botanic Garden to incorporate its holdings. Unfortunately, illness enforced his departure from India early in 1843, his manuscript catalogue going with him. After his death in June 1843, the manuscripts came back to India where William Griffith, who examined them, was impressed by the clarity of their arrangement, the lucidity of the writing and the density of information. 'It would be a mortal sin not to publish them. The work will do Voigt very good credit . . . it will perpetuate his name as an Indian Botanist.'[35] He undertook to supervise its printing which was completed only a few days before Griffith returned to Malacca. Before he left he selected a monument to commemorate Voigt; Griffith died unexpectedly from hepatitis soon afterwards and the memorial intended for Voigt was purchased by his friends to mark his grave in Malacca. Griffith, a botanist of exacting standards, rated Voigt's *Hortus Suburbanus Calcuttensis* (1845) as a fitting successor to Roxburgh's *Hortus Bengalensis*, approving the arrangement of its 5515 species according to Jussieu's natural classification. With professional succinctness it provides the reader with details of habit, flowering and fruiting periods, properties and uses, both English and Bengali names, and bibliographical sources. It provided, said Griffith, the basis of a flora of lower Bengal.

William Griffith, who saw himself as Wallich's successor, solicited the support of Robert Brown in London. 'Wallich nearly succeeded in putting Lord Auckland's mind against me', he complained, 'and I can prove that he spread injurious reports about me in Calcutta, there is no reason why he should not resort to similar *manly* ends in England.'[36] He was consequently gratified to be made acting Superintendent during Wallich's sick leave and shortly after his arrival in December 1842, recommended a radical reorganization of the Botanic Garden with all the impetuosity and brashness of youth. Wallich, he asserted, had neglected it: it was overgrown with trees, some of them in a poor state; many of the plants were not labelled; both the garden herbarium and the library were in a mess. The Bengal Government accepted his proposals (Fig. 8.5) which would inevitably destroy a great deal of the existing lay-out but when the Court of Directors in London received 120 copies of his report they complained that it 'contained strictures on Dr Wallich which ought not to have been circulated and which have therefore been struck out of the copies we have distributed'[37]. Griffith envisaged a circular natural garden illustrating Jussieu's and de Candolle's classification of plants with a monument to William Jack, the distinguished Malayan naturalist, at its centre; complementing this would be a display of medicinal and useful plants of lower Bengal. A separate flower garden would allow the removal of the floral borders flanking the paths. The casual, informal layout that had evolved over the years was anathema to Griffith's orderly, scientific mind: there were to be distinct compartments demonstrating systematic botany, medicinal and decorative plants, and elsewhere nurseries and orchards. Trees that had protected plants from the full force of the sun were thoughtlessly removed and ferns died in the absence of shade. 'The avenue of Cycas trees (*Cycas circinalis*) once the admiration of all visitors . . . had been swept away by the same unsparing hand which had destroyed the teak, mahogony, clove, nutmeg, and cinnamon groves.'[38] When Joseph Hooker who made that comment saw the Botanic Garden in 1848 he was appalled by 'the indiscriminate destruction of the useful and ornamental which had attended the well-meant but ill-judged attempt to render a garden a botanical class-book'[39]. But Griffith was oblivious to the havoc he was wreaking and completely insensitive to any aesthetic considerations. The garden for him was no more than an open-air laboratory; he had ladders propped up against certain trees to examine more closely progressive changes in inflorescence and fructification. His energy was phenomenal, his aspirations boundless—in addition to implementing his plans for the garden, he had in mind the publication of a popular 'Flora Benghalensis.' But he fell foul of the Bengal Government when he offered his services as Secretary to the local Agricultural and Horticultural Society. They disapproved of their employees being officers in any society and London supported this view. 'The duties of the Botanic Garden are sufficient to occupy the Superintendent's time and you did right to prohibit him from taking a private appointment although no salary was attached to it.'[40] The Government was now less receptive to such requests from its officials, particularly those appointments which attracted honoraria. When Wallich had tried to claim an allowance for teaching botany at the Medical College, the curt refusal reminded him of his generous salary and of the commodious residence he occupied rent-free.

Wallich must surely have enjoyed his two years at the Cape where he met the prominent local botanists, C. W. L. Pappe and Baron C. F. H. von Ludwig. He botanized on several tours, corresponded with David Livingstone in neighbouring Bechuanaland and promised to supply Indian species for a proposed Cape botanical garden. On his return to Calcutta in August 1844 he was confronted with Griffith's transformation of the Botanic Garden, denuded of many of its fine trees and his favourite *Amherstia* almost dead through the careless exposure of its roots. 'Remaining is a source of much pain—more than I can possibly express to you,' he wrote somewhat incoherently to Sir William Hooker, 'for where is the stately, matchless garden that I left in 1842? Is this the same as that? Can it be? No–no–no! Day is not more different from night than the state of the garden as it was from its present utterly ruined condition.'[41] In fairness to Griffith, it must be remembered that the garden had suffered from a severe cyclone in 1842 which may have been responsible for the destruction of some familiar features.

Wallich believed he was the victim of a vendetta instigated by his former companions in Assam, Griffith and McClelland. He was anonymously critized in the *Calcutta Journal of Natural History* which McClelland had founded and edited. An unsigned review of G. T. F. Speede's *Indian handbook of gardening* interpreted the author's dedication to the Superintendent of the Calcutta Botanic Garden as an ironical comment on Wallich's condescending attitude to horticulture and accused him of neglecting his post. Seven pages were devoted to a hostile analysis of Wallich's report for 1836 to 1840[42]. A published extract from a letter by Robert Wight, the leading botanist in the Madras Presidency, especially wounded Wallich who saw himself innocent of the charge of despatching Roxburgh's herbarium to London. In a subsequent letter to McClelland, Wight attempted to assuage the hurt he had unintentionally inflicted on Wallich.

I was unconscious that I was doing Wallich an injury, and had not the most remote idea of reflecting on his conduct in the arrangement and distribution of the Indian Herbaria. On the contrary, I consider Botany, and especially Indian Botany, as owing him a large debt of gratitude for what he did, on that occasion, towards making the measures of Flora known to the scientific world[43].

Nevertheless, even charitable Wight still persisted in his allegation that a grave error of judgement had been committed in not retaining at Calcutta a complete set of specimens among which Roxburgh's herbarium would have had a distinguished place. An editorial comment added its condemnation by arguing that this massive disposal of specimens now hindered the researches of the Indian botanist, 'rendering him incapable of describing his own collections until he can first visit Europe'. Wallich protested to Wight and complained to Hooker.

Are you aware that I have neither taken home nor even sent home Dr Roxburgh's collections of dried plants They were sent home by himself long before I came to the garden . . . How is it then after so many years have lapsed I am now held up to the public in India, as having done something grievously wrong? None of the Superintendents of this garden had left a single specimen of their botanical collections there. The fact is it was thought that botanical specimens could only with great difficulty be preserved from decay in Bengal[44].

The preservation of dried specimens was indeed a problem. Wallich tried to deter foraging insects by placing the legs of his herbarium cabinets in trays of water, frequently replenished throughout the day. In 1819 he had declined an offer of duplicate specimens from William Swainson in London simply because of insect infestation and the effects of damp and mould during the monsoons[45]. Although Wallich may be exonerated from blame in respect of Roxburgh's collections he had still neglected to ensure that one set of all the specimens he distributed in London was ear-marked for Calcutta. It is true that he had requested the Council of the Linnean Society in 1832 to assemble 'the best set obtainable' for India from the specimens donated to them but there is no evidence of his reminding them after his return to Calcutta. It was only when Kew acquired the India Museum's herbaria in 1858 that Calcutta at last received its legitimate share of the labours of Indian botanists. On resuming his post of Superintendent in 1833 Wallich formed an herbarium of plants grown in the Botanic Garden but, according to Griffith, it had degenerated into complete disorder by the time he took charge nearly a decade later. One of Griffith's reforms was the creation of a 'public herbarium' which he intended to house on the well- ventilated top floor of the Super intendent's house. Wallich had unwisely stored his herbarium in the cellar.

Despite the reassurance of visitors like Dr Hoffmeister who, when he inspected the Botanic Garden in January 1845, enthused about the 'perfect Paradise', Wallich was now a spent force. Wight was not alone in perceiving that his 'botanical day is over'. A letter Wallich wrote to Hooker in February 1845 reveals the tragic extent of his paranoia.

If I have ever deserved from them, the smallest portion of their unceasing & fierce persecution—if on any occasion whatsoever, I have given cause for any feeling except that of obligation and thankfulness—aye gratitude in the fullest sense of the word—then let me be visited, if possible, with ten-fold torture & agony than I have sustained from them these many years[46].

In November he informed Hooker that he would be resigning early in 1846 and speculated on the succession. He had unofficially approached Wight who had declined, anticipating better prospects by remaining in the medical branch. His next choice, Hugh Falconer, Superintendent at Saharanpur Botanic Garden, was on sick leave in England. Then there was William Jameson, acting Superintendent at Saharanpur, Captain William Munro, 'an ardent and excellent botanist', Captain Nathaniel Vicary, 'a most excellent botanist', and Dr Thomas Thomson, 'your friend and pupil'[47]. A candidate had not been selected when Wallich left India for good in April 1846 and the garden passed to the temporary charge of his arch enemy, McClelland, who proceeded with the reorganization begun by the late William Griffith. Much of the centre of the western half of the garden still remained undeveloped and a Sumac plantation covered the eastern part. McClelland tackled the restructuring of the Botanic Garden in the full knowledge that he had the support of the Court of Directors who had endorsed Griffith's measures[48].

Meanwhile at India House in London, John Forbes Royle, formerly Superintendent at Saharanpur Botanic Garden, pondered a list of eligible candidates. It was known that R. Wight and also M. P. Edgeworth of the Bengal Civil Service were unlikely to accept the appointment. The Bombay Presidency could offer A. Gibson, Superintendent of the Dapoori Garden and Dr Gilder of the Bombay Medical Service; H. F. C. Cleghorn was prominent in botanical research in Madras; T. Thomson in Bengal had demonstrated his taxonomic skills in naming the plants he had sent home. Royle chose Hugh Falconer whom he had earlier recommended as his successor at Saharanpur but first there was the crucial question of the Superintendent's salary which he firmly believed should not be less than that paid to the Apothecary-General. In 1830 the Court of Directors had ruled that upon Wallich's retirement, the Superintendent's monthly salary should be reduced from 1500 rupees to 500 rupees. The injustice and folly of this drastic cut became manifest when a comparison was made with the 1000 rupees a month received by a regimental medical officer with no responsibilities beyond routine duties. When both Royle and the Bengal Government pointed out that the revised salary was unlikely to attract candidates of the right calibre, the Court of Directors relented, agreeing to maintain the existing salary provided the incumbent continued to fill the post of Professor of Botany at the Medical College[49].

A memorandum on the Calcutta Botanic Garden—author unknown but Royle is suspected—stipulated a triple role for it: as a scientific institution, as a centre for the cultivation of useful plants, and as an educational institution assisting the Medical College. It regretted that the abolition of plant collectors during Lord Bentinck's retrenchment had hindered the investigation of the Indian flora. Named specimens of every species ought to be housed in the herbarium. If European criteria for assessing the competence of a botanical garden were applied to Calcutta it would be found wanting. The existing vacancy must be filled by 'a qualified person who had at least pretensions to being a botanist, and not one who has assumed the pretensions of a botanist'[50].

There was never any doubt about the botanical qualifications of Hugh Falconer (1808–65) who relieved McClelland of his temporary command of the Botanic Garden in 1848. Born in Scotland, educated at Edinburgh University, one of Wallich's helpers in the London distribution of Indian herbaria, appointed Assistant Surgeon on the Bengal Establishment, he was in 1832, while still in his early twenties, given the Superintendent's post at Saharanpur. As a palaeontologist he played a major role in the discovery and cataloguing of the fossil fauna in the neighbouring Siwalik Hills. He also had charge of

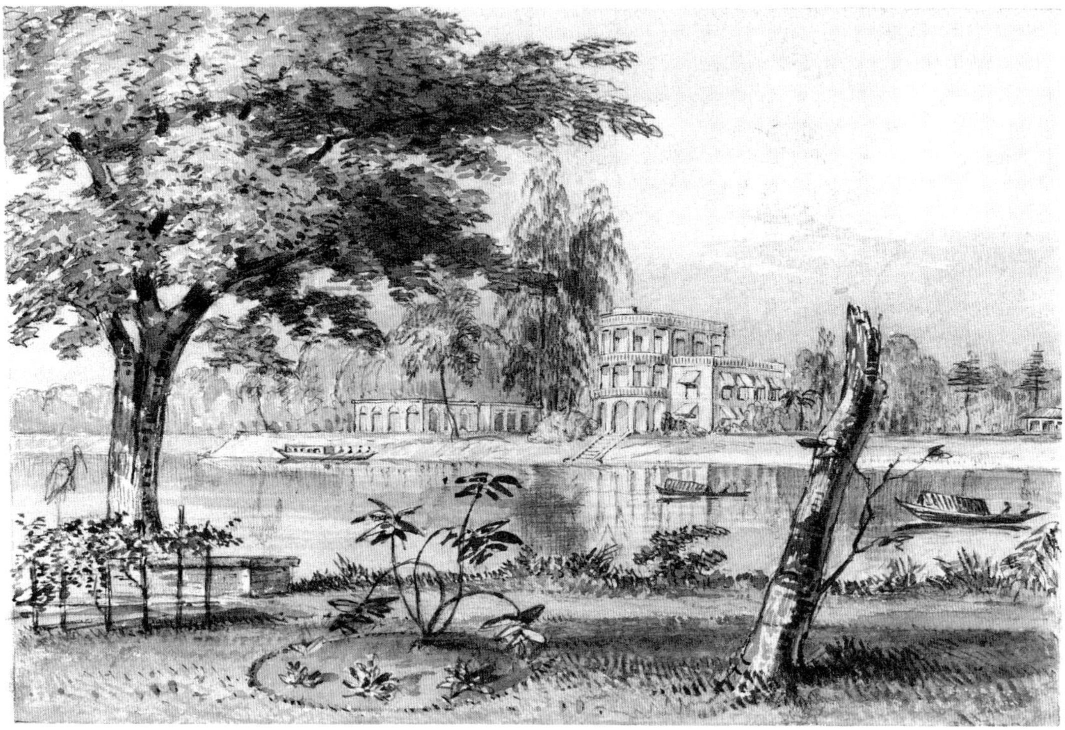

Figure 8.6 W. H. Fitch's sepia wash rendering of a sketch by (Sir) Joseph Hooker of the Superintendent's residence at the Calcutta Botanic Garden which had been built for William Roxburgh in 1795. (Kew Archives.)

the experimental tea nurseries established on sites he had selected. In 1837, as a member of a British mission to Afghanistan, he took the opportunity to collect plants in Kashmir. When he returned to England in 1842 on sick-leave he was accompanied by fifty cases of fossil bones and eighty of dried plants. The plants were consigned to the basement of India House, the fate of most collections despatched to London, and many had decayed when Joseph Hooker salvaged the contents of the notorious cellar in 1858. While he was in London writing up his researches on the Siwalik fossils, Falconer submitted botanical papers to the Linnean Society, including one on Costus (*Saussurea costus*), a native of the Punjab and Kashmir and an important spice and drug in the ancient world.

As Superintendent Falconer had a difficult role to perform, not only to rectify the damage perpetrated by Griffith and McClelland but also to convince the Government and many Calcutta residents that the Botanic Garden had a legitimate function in the economy of the country. The palms—among the first casualties of Griffith's reorganization—were reinstated in a palmetum of several acres; trees and shrubs, although grouped in related families, were massed so that trunks and stems, foliage and flowers provided contours and focal points; the tanks or water containers were filled with aquatic plants; and the great banyan tree and the *Amherstia* responded well to remedial care. In 1850 Falconer investigated the teak forests of Tenasserim, suffering from excessive felling and the absence of conservation. Two years later he recommended the introduction of cinchona from South America and the establishment of pilot nurseries in the hills of North Bengal and in the Nilgiris of South India to assess the feasibility of cultivating this quinine-yielding genus. During his term of office the Agricultural and Horticultural Society of India proposed and the Court of Directors approved that the free distribution of plants, a disproportionate drain on the resources of the Botanic Garden, would, with some exceptions, cease from April 1857. In the midst of this active and productive career, Falconer, now anxious to complete his reports on the Siwalik fossils, indicated his intention to retire no later than March 1855. When Thomas Thomson, on leave in London, learnt of Falconer's imminent departure, he persuaded the Company that he was an eminently suitable replacement.

Thomas Thomson (1817–78) was yet another Scot who, after studying medicine, sought a career in the service of the East India Company. He learnt his botany from Sir William Hooker at Glasgow University and in 1850 botanized in the Khasia Hills with his son, Joseph Hooker. A year after his installation at Calcutta in April 1855, Thomson submitted his first annual report in which he reviewed progress and future plans. The circular medicinal and natural gardens, made by Griffith and McClelland, were to be moved to a more fertile part of the garden. He wanted a glasshouse to propagate plants throughout the year. The library needed regular funding and a museum was desirable. The heterogeneous collections of dried plants accumulated since Wallich's time had been reorganized into a general herbarium[51]. Manning levels had to be improved if the garden were to fulfil its educational and recreational roles. Staff shortages had confined plant collectors to Assam, Khasia, Sikkim, and Malacca. The cessation in 1857 of the gratuitous distribution of plants to individuals, a service which had turned the place into 'a gigantic Nursery Garden', would release his hard-pressed staff for other duties. The last measure had the full support of Lady Canning, wife of the Governor-General. 'It is a fine establishment, but has been quite spoiled by the practice of giving plants to whoever wished for them. In ten months, nearly three hundred people were provided with, I think, above 50 000 plants.'[52] An entry in the diary of Lady Canning, a frequent visitor to the Botanic Garden, praised the blooms of a straggling *Beaumontia*. 'It is a sort of creeping Datura with branches of enormous white bells. I never saw anything so lovely. It was nearly wild, in a sort of boundary hedge of high bamboo.'[53] Following the proclamation in November 1858 announcing the transfer of the administration of British India from the discredited East India Company to the Crown, Thomson's institution was elevated to Her Majesty's Botanic Garden[54].

Thomas Anderson (1832–70), the third Scot in a row to occupy the Superintendent's desk, became Thomson's successor when he retired in 1861. His interest in the Indian flora had begun during his undergraduate days at Edinburgh while helping Hugh Cleghorn to organize the University's Indian herbarium. His report, covering the period April 1861 to April 1862, is the first official document, so far discovered, to add the epithet 'Royal' to the Botanic Garden's designation. He defended Griffith's plan to supersede the aesthetic planting of trees with a scientific display. He apportioned one-third of the oldest part of the garden to an arboretum and proposed a systematic arrangement in much of the remaining ground. An exception was made of climbing plants which had been trained up cast-iron columns for maximum effect. A few months after his appointment Anderson visited Java to study Dutch methods of cultivating cinchona and to collect plants and seeds. These were subsequently planted in Ceylon, the Nilgiri Hills, and northern Bengal. To his new responsibility for cinchona cultivation was added in 1864

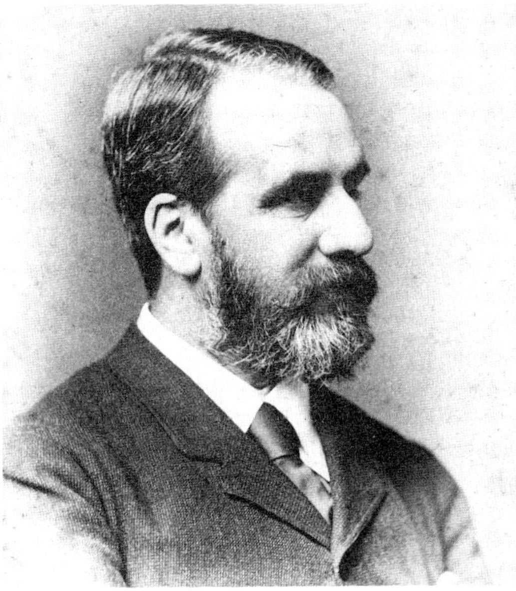

Figure 8.7 Sir George King (1840–1909).

the management of the Forest Department in Bengal. In 1864 a cyclone devastated the garden when tides over sixteen feet above normal flooded the grounds to a depth of six or seven feet, uprooting or severely damaging over a thousand trees, more than half of the total number. Three years later another cyclone almost completed the obliteration of the arboretum but the great banyan tree miraculously survived[55]. .Anderson thought advantage should be taken of this catastrophe to move the gardens nearer to Calcutta for the convenience of its residents—he had Kiddapore in mind—but he was invalided home in 1869 before he could pursue this objective. Charles Barron Clarke acted as locum until yet another native of Scotland, George King (1840–1909), was appointed Superintendent in 1871 (Fig. 8.7).

King, newly recruited into the Indian Medical Service, had delivered to the Botanic Garden in 1866 on behalf of Kew, India's first Ipecacuanha plant. When he accepted the appointment of Superintendent five years later, his only experience in managing a botanical garden had been a year in charge at Saharanpur. King inherited a garden starved of adequate financial support, wrecked by two cyclones and subsequently devoid of shade and exposed to sun and wind, much of it a wilderness of coarse grass and weeds. Large areas were marsh, remains of the rice swamps of the Gangetic delta. The general neglect was so depressing that Sir George Campbell, the Lieutenant-Governor of Bengal, was ashamed—so he told King—to bring visitors to see it. King set about rehabilitating and remodelling his domain with efficiency, imagination, and sensitivity. The swamps, converted into pools, formed ideal settings for the gigantic South American waterlily, *Victoria amazonica*; banks were raised to reduce the likelihood of future flooding, the monotonous flatness of the landscape was relieved by hillocks and gentle undulations. Abhorring the precision of formal planting, he massed trees and shrubs with a deliberate disregard for taxonomic relationships, except where kinship, as with bamboos and palms, naturally provided the diversity and irregularity he sought. Some formality, however, was present in an avenue of Royal Palms from the West Indies (*Roystonea regia*), a long row of tall Indian almonds (*Terminalia catappa*) and a parade of *Grevillea robusta* from Australia. His considerable talents in landscaping earned him commissions from Calcutta residents and an invitation to design the grounds of the new Zoological Gardens. He replaced the crude and insecure bamboo and mat shelter which had served as a conservatory with an iron structure covered by a thin thatch of grass—a shady refuge for plants during the intense heat of an Indian summer. He copied the internal lay-out of Victorian conservatories in England by introducing winding paths and rock-edged flower beds into his large octagonal Palm House with its high central dome. Orchids were protected during the cool season in nursery glasshouses. On the restoration in 1872 of the ground formerly occupied by the Agricultural and Horticultural Society, the Royal Botanic Garden expanded to just over 270 acres. In 1878 40 acres were acquired at Darjeeling, a hill station on the border with Sikkim, for the propagation of plants of the eastern Himalayas. A small botanical garden had been established at Rangarun a thousand feet below Darjeeling two years earlier but the damage caused by frequent hail storms led to its transfer to a site in Darjeeling presented by a local resident, William Lloyd. The Lloyd Botanical Garden, as it was known, was laid out by King as a distant annexe of the Calcutta garden. From about 1910 it became an important centre for the collection and distribution of Himalayan seeds, bulbs, and plants.

King never allowed his horticultural activities to conflict with his botanical researches. A fire-proof herbarium was built to house a rapidly growing collection of specimens (half a million by 1890), an indispensible resource for his monographic studies of Indian families and genera. In 1887 he launched the *Annals of the Royal Botanic Garden, Calcutta*, illustrated with drawings by Bengali artists. This was to become the vehicle for impressive and profusely illustrated botanical monographs. King also lectured on botany at the Medical College. Many demands were made of him by Government and yet he was still designated 'Superintendent', the same rank as that held by the gardener managing the less prestigious Lal Bagh garden in Bangalore. William Thiselton-Dyer, Director at Kew Gardens, drew the attention of the India Office to this anomaly and the equity of bringing Calcutta's post in line with the directorship of the smaller Peradeniya and Singapore Botanic Gardens[56]. Since this proposed change in nomenclature did not require any adjustment in salary or duties, the India Office was pleased to announce in 1891 that henceforth King and his successors would enjoy the status of 'Director'. In the same year he was made the first Director of the Botanical Survey of India.

The initiative for proposing to the India Office the co- ordination and intensification of plant collecting and botanical research in India came from Thiselton-Dyer, always acutely conscious of Kew's imperial role. As long ago as 1869, his father-in-law, Sir Joseph Hooker, had deplored the 'chaotic and disgraceful state' of Indian botany and regretted that nothing had been done 'to centralize and utilize the scientific efforts of the Indian Govt'[57]. Taking advantage of the presence of both George King and J. F. Duthie, Superintendent of the Saharanpur Botanic Garden, in London in 1884, Sir Joseph joined them in a visit to the India Office to discuss improvements in the organization of botany in India[58]. Botanical stations were being administered at a local level with variable and sometimes questionable records of achievement. In the Bombay Presidency the gardens were 'botanical' in name only, engaged in little more than the buying and selling of vegetable seeds while the botanical garden at Saharanpur devoted much of its effort to supplying plants for the gardens of military personnel. When Thiselton-Dyer prepared a memorandum in February 1885 at the request of Madras Government on the constitution of a Botanical Department for its Presidency, he took the opportunity to point out the desirability of a linkage between the various botanical departments in the Indian provinces. The Secretary of State for India's approval in 1887 led to the formation of the Botanical Survey of India with a remit to explore the flora of the subcontinent and to co-ordinate all botanical research. It was a loose federation of four centres with only good will rather than substantial financial backing coming from the Government of India. As the premier garden, Calcutta became the headquarters of the Survey with a regional responsibility for Bengal, Assam, North East

Frontier, Burma, and the Andaman and Nicobar Islands. The Government Botanist at Madras covered the South including Hyderabad and Mysore. The College of Science at Poona under its Principal, Dr T. Cooke, accepted responsibility for western India and Sind. Northern India became the domain of Saharanpur. The Revenue and Agricultural Department of the Government of India, though conceding it was a viable framework, had reservations.

> *The scheme for botanical survey, as it now stands, provides for botanical explorations over the whole area of our Indian possessions and is so far, satisfactory. But we are inclined to believe that it will not be necessary to maintain permanently so large an organization for botanical research, to which branch of scientific enquiry more expenditure and time have been devoted than to many other equally important subjects, and that it will be desirable to re-examinine the question when any of the appointments now held by botanical experts become vacant*[59].

The Survey with no staff of its own until about 1910 was largely dependent on the efforts of regional botanists in their own free time. Provincial governments, jealous of their local autonomy and forever conscious of the need for economy, were reluctant participants. When J. F. Duthie retired from Saharanpur in 1902, his post of Director of the Botanical Survey of North Western India was abolished; Bombay and Madras by the early years of this century had also severed their connections with the Survey, preferring to support the initiative of their own economic botanists. Although Survey work inevitably suffered, foundations had been laid during these formative years. Its *Records*, published from 1893, reporting botanical tours and taxonomic research, was both a forum and a link for professional and amateur botanists. A regeneration of the Botanical Survey of India had to wait the independence of India in 1947.

David Prain (1857–1944), Curator of the Calcutta herbarium in 1887, succeeded King as Director in 1898. After Prain became Director at Kew Gardens in 1905, Andrew Thomas Gage (1871–1945) held the post until 1923 when Charles Cumming Calder (1884–1962) was the last European to become Director. On his retirement in 1937, Kalipada Biswas was appointed the first Indian Director and had the privilege of

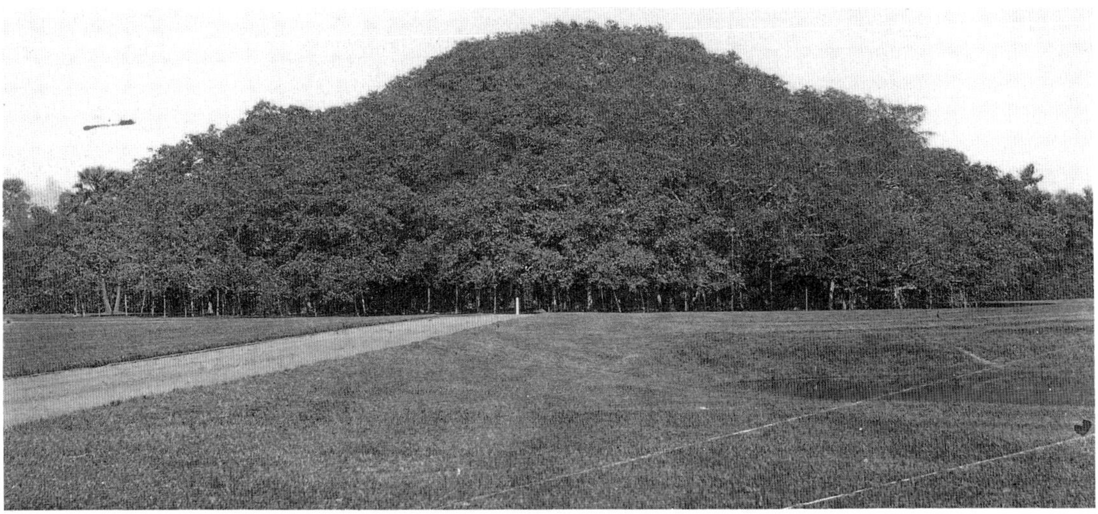

Figure 8.8 The great banyan tree, *Ficus benghalensis*, in the National Botanic Garden, Calcutta. Now more than two centuries old, it has a circumference of over 400 metres.

officiating at the 150th anniversary celebrations of the Royal Botanic Garden the following year. Renamed the Indian Botanic Garden in 1950, it was transferred from the control of Bengal to the Government of India in 1963. The most historic botanic garden in the whole of Asia has now passed its bicentenary, the great banyan tree (Fig. 8.8) continues to extend its welcoming shade and the monuments of Kyd, Roxburgh, Wallich, and Griffith inform the curious visitor of its indebtedness to the devoted service of a long line of superintendents and directors.

NOTES

[1] Wallich to Buchanan, 25 February 1815. S.R.O. GD 161/19/4.
[2] Memorial to Court of Directors, 28 December 1819. I.O.L.R. E/1/140, folios 300 *et seq.*
[3] Carey to Colebrooke, 14 June 1817. I.O.L.R. Eur. MSS B230.
[4] Hardwicke to Banks, 15 May 1817. B.L. Add. MSS 9869, folio 21.
[5] Carey to Colebrooke, 14 June 1817. I.O.L.R. Eur. MSS B230.
[6] Public letter from Government of Bengal, 21 July 1818. I.O.L.R. F/4/588, item 14270.
[7] Banks to Wallich, 26 August 1818. Mitchell Library, Sydney. MSS A300.
[8] Wallich to Banks, 20 January 1820. B.L. Add. MSS 33982, folio 201.
[9] Wallich to Government of Bengal, 23 December 1819. I.O.L.R. F/4/624, item 15931.
[10] V. Jacquemont (1835), Vol. 1, 114–15.
[11] Wallich to Government of Bengal, 14 June 1820. I.O.L.R. F/4/655, item 18040.
[12] There appears to be some confusion over George Potter's name. Some authorities refer to him as George Porter but the *East-India register and directory 1822* and E.I.C. archives confirm his name to be Potter.
[13] Wallich to Government of Bengal, 2 November 1822. I.O.L.R. F/4/760, item 20668; *Journal of Malaysian Branch of Royal Asiatic Society*, **54** (2), 1981, 35–7.
[14] Wallich to Government of Bengal, 5 November 1827. I.O.L.R. F/4/961, item 27345.
[15] See Radcliffe-Smith and Parsons (1970) for detailed analysis of Wallich's herbarium.
[16] Alphonse de Candolle to his father, Augustin Pyramus de Candolle, 24 April 1830. B.M. (N.H.) Roger de Candolle transcripts.
[17] In 1913 the Wallich herbarium was transferred to the Royal Botanic Gardens at Kew.
[18] Address from the Council of the Linnean Society to the E.I.C., 23 June 1832. L.S.
[19] Wallich to E.I.C., 3 April 1830. I.O.L.R. F/4/1139, item 30257.
[20] Bauer to W. J. Hooker, 3 May 1837. Kew Archives.
[21] *The Botanist*, no. 27, 1839, wrapper.
[22] *Gardener's Magazine*, **14**; 1838, 172.
[23] *Botanical Miscellany*, **2**, 1831, 94.
[24] Some sources say 254 copies were printed.
[25] *Rumphia* (1836–49), in which C. L. Blume examines the flora of the Indian Archipelago, uses the works of Reede tot Drakenstein, Rumphius, Roxburgh, and Wallich as a starting point.
[26] E.I.C. to Government of Bengal, 29 September 1830. I.O.L.R. E/4/729, folio 479.
[27] Wallich to Ward, 4 February 1834. L.S. Nathaniel B. Ward letters.
[28] W. Griffith, Private journal, 17 February 1836.
[29] Wallich to Government of Bengal, 1 October 1836. I.O.L.R. F/4/1761.
[30] E.I.C. to Government of Bengal, 21 August 1839. I.O.L.R. E/4/760, folios 237–48.
[31] Wallich to Government of Bengal, 21 December 1840.
[32] J. D. Hooker (1854), Vol. 1, 4.
[33] Wallich to W. J. Hooker, 27 December 1841. Kew Archives. Wallich letters, Vol. 1, 1841–9.
[34] Wallich to W. J. Hooker, 6 November 1842. Kew Archives. Wallich letters, Vol. 1, 1841–9.
[35] Voigt (1845), viii.
[36] Mabberley (1985), 325.
[37] E.I.C. to Government of Bengal, 3 January 1845. I.O.L.R. E/4/781, folios 771–2.
[38] J. D. Hooker (1854), Vol. 2, 244.
[39] Ibid., Vol. 1, 3.
[40] E.I.C. to Government of Bengal, 29 May 1844. I.O.L.R. E/4/778, folio 588.
[41] Wallich to W. J. Hooker, 8 February 1845. Kew Archives. Wallich letters, Vol. 1, 1841–9.
[42] *Calcutta Journal of Natural History*, **2**, 1842, 288–94.
[43] Ibid., **6**, 1846, 612–14.

[44] Wallich to W. J. Hooker, 2 October 1845. Kew Archives. Wallich letters, Vol. 1, 1841–9.
[45] Wallich to Swainson, 18 July 1819. L.S. Swainson letters.
[46] Wallich to W. J. Hooker, 8 February 1845. Kew Archives. Wallich letters, Vol. 1, 1841–9.
[47] Wallich to W. J. Hooker, 22 November 1845. Kew Archives. Wallich letters, Vol. 1, 1841–9.
[48] E.I.C. to Government of Bengal, 1 April 1846. I.O.L.R. E/4/787, folios 337–8.
[49] E.I.C. to Government of Bengal, 16 June 1847. I.O.L.R. E/4/792, folios 553–64.
[50] Memorial on Calcutta Botanic Garden (1846?). Kew Archives. Calcutta Botanic Garden, 1830–1928, folios 21–5.
[51] *Journal of Bengal Asiatic Society*, **25**, 1856, 405–18.
[52] A. Hare (1893), *Story of two noble lives*, Vol. 1, London, 116.
[53] Ibid., 144.
[54] It is not known when the epithet 'Royal' was conferred upon it and no Royal Charter has been found.
[55] Severe damage and the irreparable loss of historic and mature trees were inflicted by cyclones in 1910, 1926, and 1940.
[56] W. Thiselton Dyer to India Office, 29 October 1888. Kew Archives. Selection from Despatches to India, Vol. 2, 1888, folios 199–200.
[57] Huxley (1918), Vol. 2, 17.
[58] Ibid., 249.
[59] Government of India (Revenue and Agricultural Department) to India Office, 24 December 1889. Kew Archives. Indian Botanical Survey, 1884–1920, folio 48.

OTHER BOTANICAL GARDENS

By the end of the Victorian era many towns in British India boasted public parks and gardens administered by the municipal authority and often managed by gardeners trained at the Royal Botanic Gardens at Kew. Agri-horticultural societies also maintained private gardens for the use of their members. Some of these gardens purported to be 'botanical' but there were only four establishments in South Asia which strictly justified the epithet: Calcutta, Saharanpur, and Ootacamund in India, and Peradeniya in Ceylon.

The experimental plantation developed by William Roxburgh at Samalkot was referred to as a botanical garden in 1800 when the Madras Government decided to abolish it, having resolved to establish a new botanical garden in Mysore, the Company's most recent territorial acquisition. Benjamin Heyne, Roxburgh's successor at Samalkot, chose a former royal garden near Bangalore which had a relatively mild climate for much of the year, fertile soil, and abundant water. He was instructed to transfer valuable exotics from Samalkot to the new garden which would be integrated into the survey programme undertaken by Colin Mackenzie. Heyne recommended the cultivation of vegetables, spices, and medicinal plants on its forty acres. When its yield proved disappointing, its abandonment was seriously considered in 1803. Heyne defended its achievements in providing food for the regimental messes and in demonstrating the cultivation of European vegetables to the native population. When he failed to convince the Company, it passed to the Rajah of Mysore in 1810.

On his transfer from Bangalore in 1819 Major Waugh offered the Madras Government his garden which he had filled with European and Chinese plants over a period of 15 years. The Lal Bagh, as it was called, had been an orchard and formal garden started in 1760 by Haider Ali, the ruler of Mysore, and improved by his son, Tipu Sultan. Wallich at the Calcutta Botanic Garden urged the acceptance of this generous offer seeing the garden as a future nursery for acclimatizing English fruit-trees which invariably died in Bengal's uncongenial climate. In 1831 the Lal Bagh passed to the custody of the Chief Commissioner of Mysore who, eight years later, handed it to the newly-formed Agri-Horticultural Society of Mysore. On the dissolution of this short-lived society in 1842, the Lal Bagh once more became the Chief Commissioner's responsibility, receiving little attention apart from superficial tidying and routine sweeping of paths. It was Hugh Cleghorn, Professor of Botany and from 1856 Conservator of Forests, who campaigned for its role as a horticultural station under experienced gardeners. It already had a nucleus of fine trees which included Moreton Bay chestnuts, Spanish oaks, and olives from the Levant. Cleghorn saw it as a centre for improving indigenous crops and for the propagation of exotics and their distribution. He convinced the Company who appointed William New, a former Kew gardener, in 1858 to take charge. New created more walks, planted an avenue of fast-growing *Grevillea robusta*, and sowed seeds received from China, Mauritius, South Africa, and Kew. He was succeeded by John Cameron who had increased the number of named species and varieties to over 3000 by the early 1890s. The *Madras Mail* in 1880 described it as being essentially a pleasure-garden with little attention given to scientific research. It is now the headquarters of the Directorate of Horticulture of the State of Karnataka, attractively landscaped with dramatic outcrops of rock, a large lake, some splendid survivors of the original trees and with one of the finest glasshouses in the subcontinent. It has an impressive rose garden and is a venue for frequent flower shows.

One of Cleghorn's reasons for a horticultural station at Bangalore was the need for a link between the garden at Ootacamund, the hill station in the Nilgiri Hills, and Madras. A public garden in the hills was the inspiration of a local horticultural society and in the spring of 1848 William McIvor (d. 1876) arrived from Kew to landscape the steep slopes and to create a nursery. In 1857 the garden came under the Conservator of Forests for Madras until 1883 when the newly appointed Madras Government Botanist and Superintendent of Cinchona Plantations. M.A. Lawson, took charge of all government gardens and parks. Ootacamund became an important centre for cinchona cultivation in the south, and a reminder today of its Victorian prosperity are the Eucalyptus trees from Australia, grown in thousands for their timber, their glaucous leaves imposing distinctive smears of colour on neighbouring hills. McIvor's creation is now a public park with a display of mature trees from many parts of the world, an impressive arboretum deserving its present status of botanical garden. The only other garden of any horticultural consequence in the Madras Presidency was founded by the Agri-Horticultural Society of Madras in 1835 and owed its early success to its dynamic Secretary, Robert Wight. It pioneered the local cultivation of European vegetables and native and introduced fruit, and coffee cultivation in the south followed its researches in the 1840s.

After Calcutta, the most important botanical garden in the subcontinent was undoubtedly Saharanpur, over 800 miles northwest of the Bengal capital, south of the Siwalik Hills and the Himalayan range (Fig. 9.1). There, at little more than a thousand feet where summers and winters are not extreme, it is possible to grow plants from the more temperate regions of the world. The garden owed its existence to George Govan (1787–1865), appointed Civil Surgeon at Saharanpur in June 1815, after service in the Nepalese war of 1814. He perceived that the former pleasure grounds laid out by a Rohilla chief in the 1770s, the Farhatbaksh gardens, could be adapted as a botanical garden. The magistrate at Saharanpur had suggested its restoration in 1814 and even nominated the station surgeon, H. Hough, to take charge of it. Its 40 acres were choked with mango trees, apart from plots of vegetables in the few clearings, and the outbuildings were in a ruinous state. Govan, however, was not deterred from taking on its conversion in addition to his official medical duties. The submission he made in 1816 to Lord Hastings in Calcutta stressed the advantages of its situation between the great northern plains and the Himalayas making it conducive to 'the naturalization of many of the valuable vegetable productions of colder climates' and, at the same time, 'the more tender inhabitants of the south might be gradually inured to a colder atmosphere'[1]. Warming to his theme, he envisaged the garden stocked with the familiar fruits of Europe, the cinchona of South America, and all those vegetables and medicinal plants which persistently defied Calcutta's skills to grow them. Saharanpur's elevation and soil were similar to those parts of China where

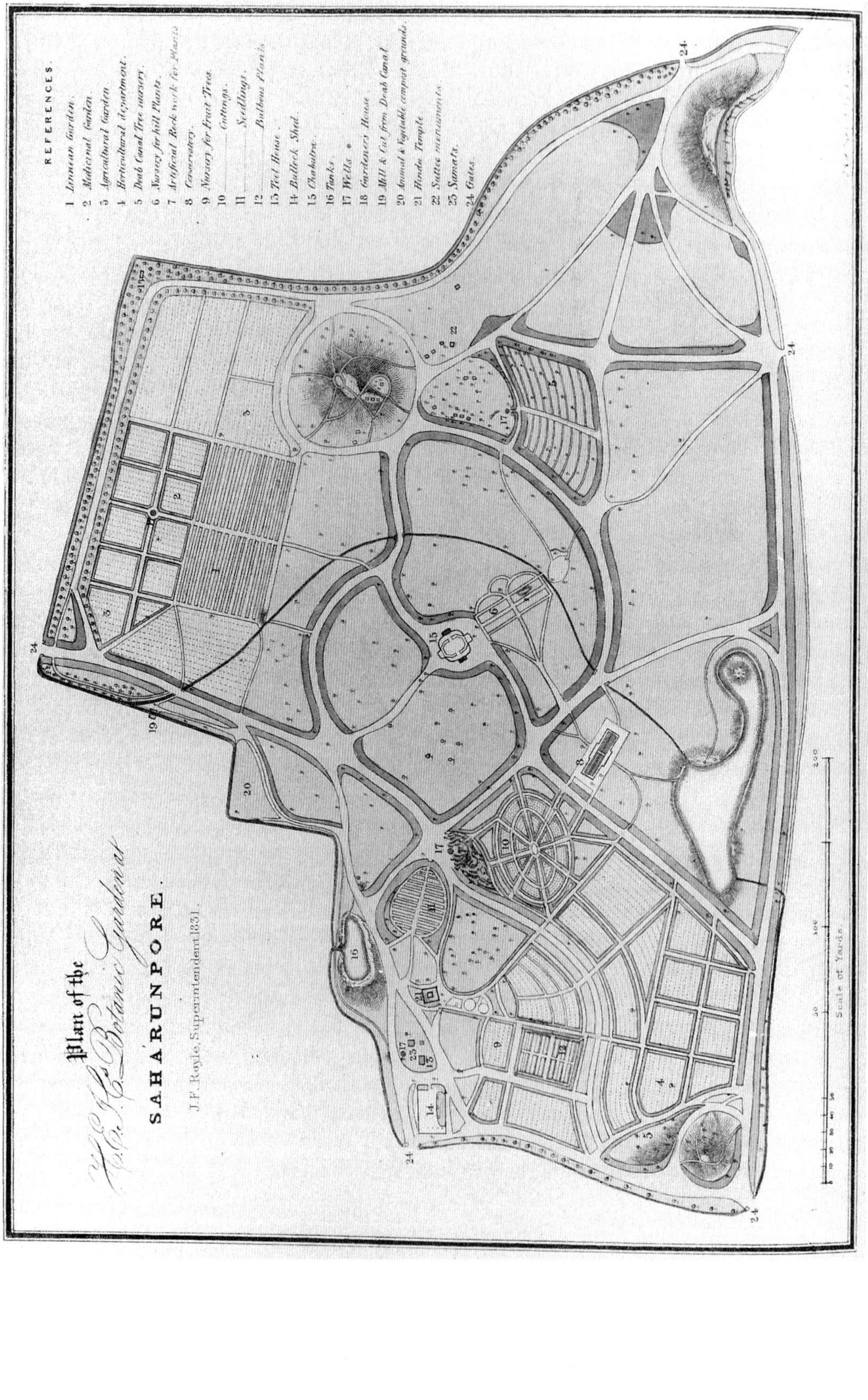

Figure 9.1 Plan of the Botanic Garden at Saharanpur in 1831.

tea plantations flourished; the successful cultivation of tea alone, it was argued, would justify the maintenance costs of this proposed garden. His persuasive presentation convinced the Governor-General who sanctioned his appointment as Superintendent in June 1819. Confident of official approval, Govan had already been busy clearing and levelling the neglected site, creating roads, initiating a planting programme, planning rock, medicinal, and Linnaean order beds and even building a small conservatory. His appointment being confirmed, Wallich at Calcutta immediately suggested an exchange of plants. Suffering from repeated bouts of fever, presumably malaria, Govan was granted European sick leave in 1821, never to return to Saharanpur. Shortly after the resumption of his Indian service, he joined an official geological survey in the Himalayas as its botanist and zoologist. He spent his periods of leave with his family at Simla where the French botanist, Jacquemont, met him and disparagingly dismissed him as 'Ecossais rigide, homme de bien, mais médiocre botaniste'[2]. Many of the crops Govan had confidently predicted would grow at Saharanpur failed to thrive but his optimism regarding tea was vindicated although not during his brief term of office.

A native gardener let the garden revert almost to a wilderness again before John Forbes Royle (1798–1858) took charge in 1823 as well as performing his duties as civil surgeon (Fig. 9.2). But he and his successors now came under the control of the Calcutta Botanic Garden. Once more the garden was cleared of invasive shrubs, the ground levelled or sloped to vary its contours, and a large pond dug for aquatics. In trying to create a fashionable English lay-out, assisted by an English gardener from 1827, Royle never lost sight of the garden's objectives in botany, agriculture, and horticulture. One desideratum being the production of drugs for the medical services, he established a

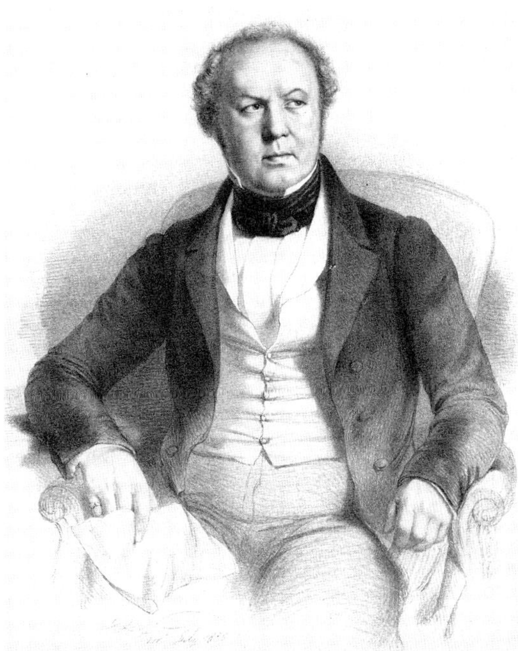

Figure 9.2 John Forbes Royle (1798–1858). Lithograph by G. H. Ford.

nursery for the cultivation of medicinal plants at the hill station at Mussoorie, nearly 7000 feet up in the Himalayas. At the same time he closed down Govan's inaccessible small garden at Nahan. He had great expectations of Mussoorie. 'The vegetable productions are such as to warrant the cultivation of European plants with every prospect of success', he reported to Wallich[3]. 'Rhododendrons and oaks covered with ivy are the most common trees, white roses and brambles, barberries and raspberries the most frequent shurbs, while violets and wild strawberries form the sward.' As well as medicinal plants—Belladonna, Senna, Digitalis, Valerian, Aconite—he grew alpines in his Himalayan nursery, acclimatizing them before moving them down to Saharanpur where he also successfully naturalized the fruit trees of Afghanistan, China, Europe, and America. Visitors such as Lieutenant Colonel Davidson, who had known the derelict garden of twenty years earlier, appreciated his efforts: 'had the air been a little damper, the sky more cloudy, and the walks better weeded, I might have fancied myself in the grounds of a first-rate connoisseur in England'[4]. Jacquemont, on the other hand, given to making acerbic comments, found it 'a wretched botanical garden'. Royle's regular attendance at two local hospitals limited his opportunities to botanize but he intensely enjoyed those rare occasions he was able to escape to the foothills.

Everything looked like a revival of spring, some of the trees and shrubs were putting forth new leaves, others like the Viburnums were in flower. The ground was covered with a profusion of beautiful flowers among which were Primula denticulata, Primula petiolaris, Potentilla atrosanguinea, and Potentilla splendens, with species of Trillium, Anemone, Ranunculus, Viola, Convallaria, Cordyalis, and Caltha. Higher up, patches of snow were seen, and the Himalayan bamboo, which had been levelled to the ground. Beyond this, everything had a wintry aspect, the snow lay in dense masses though detached, having melted away from around the trunks of many trees and the blocks of gneiss rocks. Few plants were even in leaf, but the Yew was in flower, and the Juniper seen straggling over the rocks, and Primula peeping up in the warmer situations. . . . Nothing could be more magnificent than the view from this culminating point, having the plains of India indefinitely extended to the southward, and on the north the snowy peaks of the Himalayas towering above this great height[5].

Such experiences were rare treats since he had to leave much of his plant collecting to his gardeners whom he sent to the Tibetan boundary of Kunawar (Kinnaur) and as far west as the mountains encircling the Vale of Kashmir. Some 30 000 plants representing some 4000 species, were under cultivation when he left the garden in 1831. He also bequeathed to his successor an herbarium for which he provided two systematic catalogues—one according to Linnaeus and the other to Jussieu. But he took home with him substantial collections and drawings for his projected account of the Himalayan flora. For a century Royle's herbarium—one of the major collections of the flora of North West India—was lost, to be discovered by chance in the possessions of the Liverpool Chemists Association in 1952. On examination it was found to include specimens from the botanists of Tranquebar—Rottler, for example—as well as gatherings by Roxburgh, Buchanan, and Wallich. It had suffered damage during the air-raids on Liverpool in the last war but eighty per cent had miraculously survived and some of the pressed flowers had even retained their delicate colouring[6].

The vacancy at Saharanpur was filled by Hugh Falconer who struggled to maintain the garden during a period of retrenchment imposed by the East India Company, alarmed by the heavy cost of their Burmese campaign. When the Governor-General, Lord Bentinck, made an official visit to Saharanpur, Falconer must have been very apprehensive. His lordship admired the display of flowers from the Himalayas and the weeping willows around the lake, confessing he had never seen a garden so beautiful and well-kept since his arrival in India[7]. Nevertheless, he reduced the establishment, closed the garden at Mussoorie, and cut the Superintendent's salary by half. The closure of the Mussoorie garden proved fortunately only temporary; in 1839 Falconer received permission from Lord Auckland to reopen it and began restocking it with fruit trees from Kashmir and Europe—Asafoetida and *Prangos pabularia* from Central Asia (the latter a potential fodder crop) and pines, hops, and rhubarb from England. The large contingent of British forces in Afghanistan prompted Falconer to recommend hop cultivation in the neighbourhood of Kabul to produce beer for the troops. At the same time he intensified seed collecting expeditions in the Himalayas, ranging from Kumaon to Kashmir, despatching a harvest of over 500 species to England, among them the Deodar (*Cedrus deodara*) and the Himalayan birch (*Betula utilis*). According to Professor Lindley, the successful planting of the Deodar alone justified the traffic in seeds[8]. Within a few years Scottish nurseries had raised as many as 5000 seedlings of this hardy conifer. British institutions and gardeners reciprocated with agricultural, medicinal, and ornamental plants suitable for trial cultivation at Saharanpur. Royle, now appointed Correspondent relating to the Vegetable Productions of India at India House in London, supervised this exchange of seeds and plants. British consuls in Egypt and Syria sent seeds of varieties of Egyptian wheat and the Carob tree which, it was hoped, would adapt to the arid conditions of northern India. From de Candolle in Switzerland came a present of rye and other cereals. Lord Auckland's relaxation of financial constraint was conditional upon Falconer giving priority to plants of 'practical utility' although he accepted that 'discoveries of a scientific nature will be made, and assistance afforded in furtherance of scientific enquiries in Europe and India'[9]. The future of the garden still remained in doubt. The Court of Directors agreed to send a gardener[10], to appoint a resident artist, and to grant an allowance for the purchase of English seeds. Falconer was to take charge of the public gardens at Agra and the tea plantations in Kumaon and Garhwal. 'We are willing to give the garden at Seharanpore a fair trial for a few years. .' At Falconer's suggestion, Hawal Bagh, a garden noted for its apples, pears, and plums, near Almora in Kumaon was acquired in 1841 as an additional nursery for European fruits.

Falconer retired in 1841, due to ill-health, and his successor, William Jameson (1815–82) remained in his post for over thirty years. Not being a botanist, Jameson devoted his energies to raising horticultural and agricultural plants. Private gardens, the gardens of military cantonments, and public parks as far away as Lahore, Lucknow, and Delhi were stocked with his flowers. Huge quantities of Deodar seeds went to Britain in fulfilment of a continuing demand. He developed tea plantations, introduced American maize to Kumaon, Garhwal, and the western hills, and persevered in the trial cultivation of flax.

With Jameson's departure in 1875 to become a tea planter, some officials wished to close the establishment and Duthie, the new Superintendent, was faced with indifference or, at best, lukewarm support of the Government of the North-Western Provinces. The former Professor of Natural History at Cirencester's Royal Agricultural College, John Firmer Duthie (1845–1922), strove for 27 years to reinstate and improve the institution's scientific role, neglected entirely by his predecessor. The garden he had inherited was thick with trees, solitary like the *Araucaria cunninghamii* (one of the largest specimens in the subcontinent) or gregarious like the palms and willows; there were several avenues of Casuarinas, some over 60 feet tall. The eye of the casual visitor was immediately arrested by the magenta bracts of a gigantic *Bougainvillea glabra* smothering a tree, the massive blue curtain of *Thunbergia laurifolia* hanging from another tree, and the smooth grey bark of a mature evergreen fig, *Ficus retusa*, planted by Royle over half a century earlier. The orchards were filled with fruits of the tropical and temperate zones and about 60 acres allocated to vegetables. Mussoorie still survived but the other small gardens had been sold and the responsibility for tea cultivation had passed to private management. Consequently, Duthie was able to give more time to botany. He lectured on taxonomy at the Forest School at Dehra Dun and travelled throughout the North-Western Provinces, the Punjab and Central Provinces, Kumaon and Garhwal, Simla and Kashmir, diligently seeking plants not only

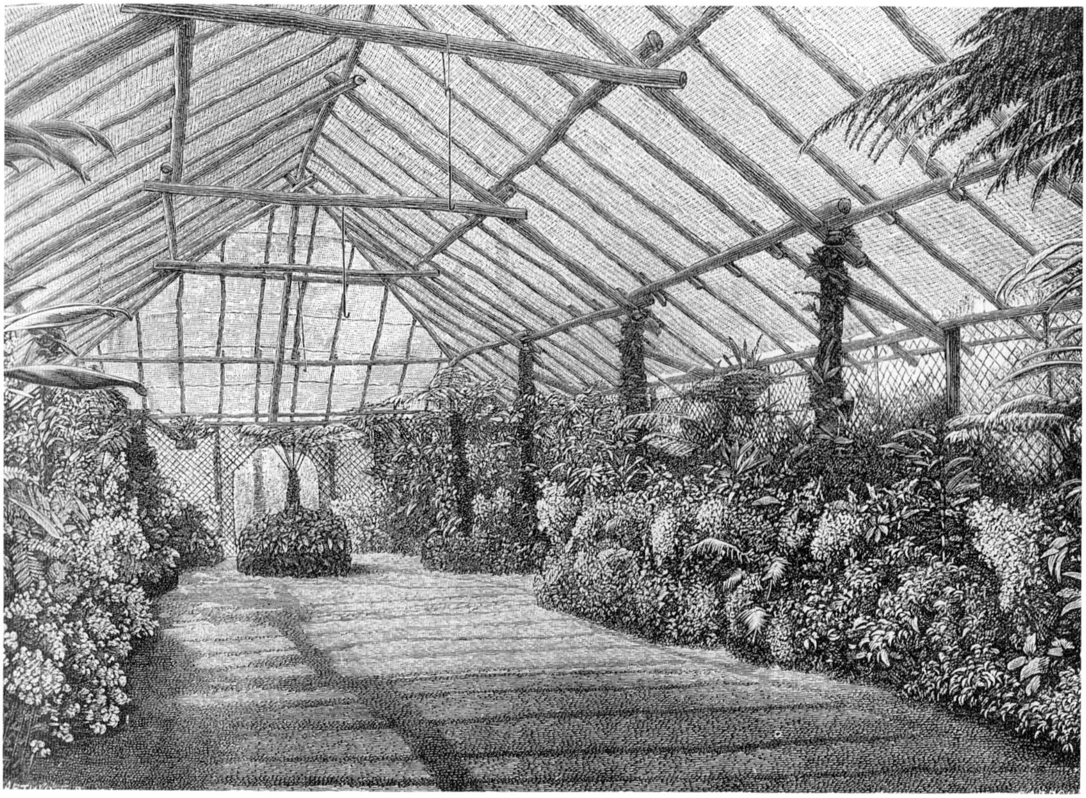

Figure 9.3 Conservatory at Government House, Ganeshkhind in 1888. Wood engraving. This sketch shows the crude construction of such conservatories whose roofs were lined with thin coconut-fibre matting. 'Such places are serviceable for sheltering the more delicate plants, which otherwise would be destroyed by the sun and the heavy rains; but from the want of sufficient light, the plants rarely thrive in them' (T. A. Firminger).

for Saharanpur's herbarium but also for Calcutta, the British Museum, Kew, and Edinburgh. The quest for fodder plants compelled him to pay particular attention to grasses and his collections were acknowledged by Sir Joseph Hooker as the finest ever made in northern India[11]. He became an authority on the vegetation of the northern plains and the first volume of his *Flora of the Upper Gangetic Plain and of the adjacent Siwalik and Sub-Himalayan Tracts* came out in 1903, the year in which he retired, accepting a post of Assistant for India in the Kew Herbarium. Before he left Kew in 1907 he had revised and supplemented Sir Richard Strachey's *Catalogue of the plants of Kumaon and of the adjacent portions of Garhwal and Tibet* (1882). Illness and a perfectionist's obsession with detail delayed the progress of his *Flora of the Upper Gangetic Plain*, never to be finished[12].

Duthie's herbarium was transferred from Saharanpur to the Forest School at Dehra Dun in 1908, the same year in which Mussoorie was closed and another garden opened at Naini Tal. In 1932 the post of Superintendent was abolished and responsibility for the garden passed to the Deputy Director of Gardens of the United Provinces. Old and historic trees were felled when Indian troops were stationed there during the Second World War. It is now a horticultural experimental and training centre for tropical and subtropical fruits, flowers, and vegetables.

In 1887 Theodore Cooke, Principal of the College of Science at Poona charged the

Government of Bombay with having done least for botany in the whole of British India; Bengal, the North-Western Provinces, and Madras all had botanical departments, only Bombay was without one. Horticultural and botanical gardens were also latecomers to western India. In 1828 the house and grounds of the late Colonel Forde at Dapoori near Poona were purchased for the Governor, Sir John Malcolm, who desired it as a residence for three or four months of the year. He authorized the creation of a botanical garden on 48 acres, an act which incurred the censure of the Court of Directors who had not been consulted, but who reluctantly gave their consent provided it was prudently managed. The first Superintendent, a Mr Williamson, died soon after his appointment, and Charles Lush (1797–1845) who had been a lecturer in botany at St Thomas's Hospital in London succeeded him. In 1837 the incumbency passed to Alexander Gibson (1800–67) who planted mulberries and encouraged the growing of arrow-root, tapioca, and potatoes. An ambitious scheme to create gardens for commercial crops in each collectorate in western India was rejected on the ground of cost but nurseries at Sheoneri Fort, Hirwe, and Nirgori were sanctioned. Gibson resigned in 1847 to become the first Conservator of Forests for the Bombay Government. He formed his own private arboretum on an 18 acre site close to his home near Junnar. Gibson's future collaborator on the *Bombay Flora*, Nicol Alexander Dalzell (1817–78), now took charge of the garden and supervised its transfer to the new Government House at Ganeshkhind in 1865 (Fig. 9.3). It was a botanical garden in name only; a few acres of dandelions were grown for the Medical Department but its *raison d'être* was its revenue from the sale of roses, crotons, and other ornamentals. Yet another move, this time to the Soldiers' Gardens in Poona marked the final stage in its decline; it was now no more than a public park where people promenaded while the band played.

NOTES

1 Hyde (1962), 49.

2 V. Jacquemont (1933), *Etat politique et social de l'Ind du Norde*, Paris, 240.

3 Royle to Wallich, 7 April 1826. I.O.L.R. F/4/995, item 27123 (2), folio 115.

4 C. J. C. Davidson (1843), *Diary of travels and adventures in Upper India*, Vol. 1, London, 133.

5 Stansfield (1953), 263–4.

6 Royle's herbarium is now in the Liverpool Museum.

7 A. C. Lowe (1894), *Diary of an officer of the 16th (Queen's) Lancers, 1822–1840*, Calcutta, 272.

8 *Calcutta Journal of Natural History*, **2**, 1842, 287.

9 Public despatch from London to Bengal, 15 July 1840. I.O.L.R. E/4/763, folios 391–404.

10 John Miller was appointed in September 1841.

11 Duthie's writings on grasses include *A list of the grasses of North-western India, indigenous and cultivated*, Roorkee, 1883; *Illustrations of the indigenous fodder grasses of the plains of North-western India*, Roorkee, 1886, *Fodder grasses of northern India*, Roorkee, 1888.

12 Volume 3, part 3 was published in Calcutta in 1929.

WILLIAM GRIFFITH AND ROBERT WIGHT

India's botanical gardens were the designated repositories for plants collected in the subcontinent and neighbouring countries or donated from more distant regions: living plants to be coaxed into flower in the garden and dried specimens to be systematically stored in the herbarium. The plant collectors were sometimes the superintendents themselves but more often their gardeners or locally recruited natives, or were officers in the army, surgeons in the medical service, administrators and planters on remote stations, or surveyors in the field, mostly amateurs with variable levels of botanical knowledge and competence but all motivated by curiosity and enthusiasm. I. H. Burkill's *Chapters on the history of botany in India* (1965) is dense with the names of men who assiduously collected seeds, bulbs, plants or herbarium specimens either in the vicinity of their homes or on their travels—military expeditions and diplomatic missions in particular provided unrivalled opportunities for collecting by the observant—but this account can do no more than identify the principal collectors and assess their importance. In this respect Griffith and Wight outstrip their contemporaries in achievement and discrimination.

Sir George King had no doubts about the intellectual stature of William Griffith (1810–45) (Fig. 10.1). 'His botanical predecessors and contemporaries were men of ability and devotion: Griffith was a man of genius.'[1] King's bold assertion sprang from his admiration for Griffith's familiarity with cryptogams as well as with flowering plants, his proficiency with the microscope, his taxonomic expertise, his artistic skills, and the abundance of his collections—mammals, fishes, and insects as well as plants. His talents were buttressed by abnormal energy, stamina, and fortitude. In just under 13 years Griffith's tours of duty took him from Afghanistan in the West to Malacca in the East and never at any time, despite considerable physical discomfort, even danger, did he neglect any opportunity to collect. It has been calculated that he amassed some 9000 specimens.

A precocious young man, he was the protégé of Professor John Lindley at University College, London, who introduced him to Wallich when he was dispersing the Indian

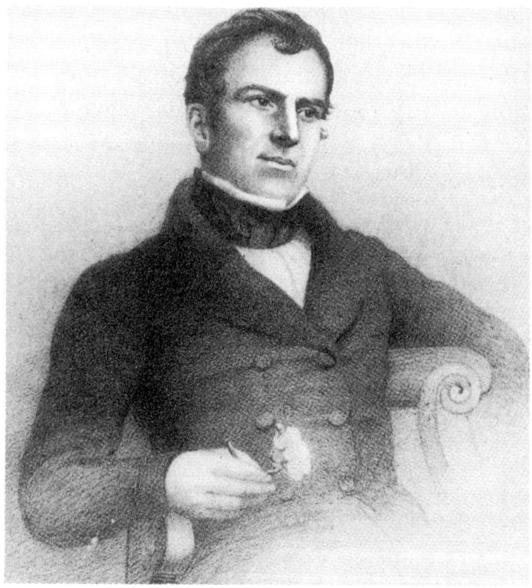

Figure 10.1 William Griffith (1810–45). Lithograph based on a daguerreotype.

herbaria in London. His floral analyses of *Humboldtia*, *Crypteronia* (*Henslowia*), and *Phytocrene* in Wallich's *Plantae Asiaticae rariores* were the first intimations of botanical promise. Appointed Assistant Surgeon on the Madras Establishment in 1832, his travels began in 1835 when he joined Wallich and McClelland on an expedition to Assam in search of indigenous tea. When later he got separated from a small British party reconnoitring the unsettled Burmese border his assassination was rumoured. Alone he descended the Irrawaddy river, his life in danger through the risk of his canoe capsizing; it was very low in the water because of its great burden of plants. In 1838 as a member of Major R. B. Pemberton's diplomatic mission to Bhutan, he became the first botanist to visit that reclusive country. The following year found him with the Army of the Indus on its punitive expedition to Afghanistan. He botanized as often as he could on the march to Kandahar and during bouts of sickness persuaded his fellow officers to collect for him. An excursion to Bamian some 50 miles north west of Kabul cropped nearly 200 alpine species. He chose the northern route back to Calcutta as he had predicted to his friend Robert Wight. 'From Cabul I shall return to India by as circuitous a way as I can select, and if Lord A[uckland] allows me to return via Simla, Mussoorie, etc. I shall then have seen the Himalayas at more points that any other person. And what a collection shall I not have. Imagine six months in the Himalayas!'[2]

His constant endeavour was the creation of a personal herbarium representative of the subcontinent's flora, a necessary prerequisite for his projected Indian flora. Wight sent him material from the South and from 1837 Griffith employed a native collector in the Khasia Hills. In 1841 Griffith was posted to Malacca where he subsequently died of hepatitis, tragically young, his promise unfulfilled. His collections of pressed specimens, his flowers and fruit preserved in spirits of wine, his botanical notes and drawings, were all bequeathed to the East India Company. Wight, apprehensive that 'the labours of the greatest botanist that ever set foot in India will be lost, perhaps for ever, swamped, midst the accumulated records of hundreds of men, that are daily being added'[3] to the India Museum in London, offered help and funds to get his manuscripts published. McClelland volunteered to edit Griffith's journals and the East India Company defrayed the cost of their publication, taking 250 copies for presentation to institutions and botanists[4]. Unfortunately, Griffith's papers needed expert editing and his friend, McClelland, was not even a botanist. Joseph Hooker had a low opinion of McClelland: 'He is a persevering Scotchman, without much ability, or power of perception; blinded by Griffith's extraordinary ability'[5]. But despite this incompetent editing Griffith's powers are manifest: his speculative mind, his independence of thought, his attention to morphological detail, his recognition of the relevance of phytogeography, and his interest in ecology. All these strands would eventually have been woven into his design for a scientific Indian flora, owing little to Roxburgh or Wallich. What is left of his herbarium now reposes in the Royal Botanic Gardens at Kew. In the Botanic Garden in Calcutta where he had been earmarked as Wallich's successor there is only a modest memorial. His premature death was an incalculable loss to the advancement of botany in India.

Griffith's friend, Robert Wight (1796–1872), was also an embodiment of phenomenal energy, endurance, and prodigious industry (Fig. 10.2). When he arrived in Madras in 1819 to take up an appointment as an Assistant Surgeon, his knowledge of botany amounted to no more than what he had learnt during his medical studies at Edinburgh University, and the scarcity of

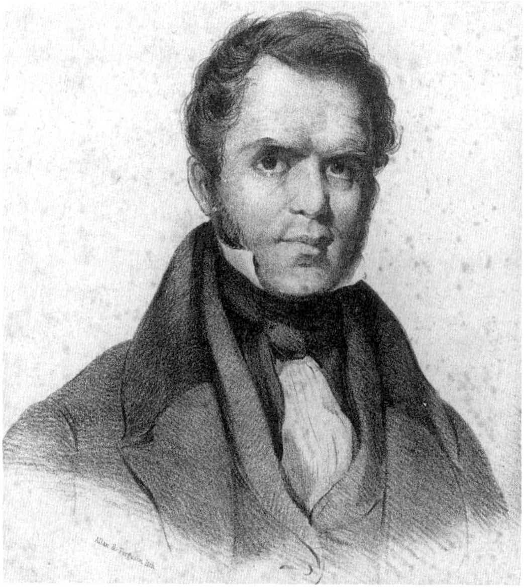

Figure 10.2 Robert Wight (1796–1872). Lithograph.

essential books in India at first hindered his progress. Nevertheless, he soon began botanizing in the vicinity of his military station and engaged natives to collect plants farther afield. William J. Hooker, Professor of Botany at Glasgow, received a donation of plants collected by Wight in Samalkot, Vellore, and Madras in 1826, the year in which Wight succeeded James Shuter as Naturalist in the Madras Presidency. This appointment provided both the stimulus and the opportunity to study botany more systematically. He returned from a nine-month tour in the South with a haul of about 2000 species of plants besides birds, insects, and minerals. He planned a more ambitious tour to last two years, beginning in January 1828. With the *Hortus Malabaricus* and Buchanan's account of Mysore, Canara, and Malabar as his guides, he intended to investigate the floristic diversity of the tropical rain forests of the west coast and the subtropical forests of the Nilgiris. He was confident of harvesting more than 5000 species. Unfortunately, the Madras Government, ever conscious of the need to curb expenditure, abolished the post of Naturalist at the beginning of 1828 and Wight found himself garrison surgeon at Negapatam on the coast. Disappointed but not deterred, he turned his attention to Tanjore and, at his own expense, employed native collectors and artists. Inspired by the plates in J. E. Smith's and J. Sowerby's *English botany* (1790–1814), he sent drawings to W. J. Hooker for publication in his *Botanical Miscellany*[6]. He left India in 1831 with a hoard of almost 100 000 specimens packed in six large cases, having been granted three years' sick leave. Wallich, frantically despatching the residue of duplicates still in his rooms in Frith Street, declined Wight's offer of yet more duplicates. So Wight enlisted the aid of George Arnott Walker-Arnott (1799–1868), a friend of his university days, to distribute the duplicates[7] and to collaborate in a work describing his herbarium of an estimated 3000–4000 species.

Koenig, Roxburgh, and Wallich had adopted Linnaeus's sexual classification which continued to have its devotees in India long after it had been superseded in Europe. But such a classification restricted to the characters of stamens and pistils often grouped together plants with no other affinity whereas a natural arrangement employing many more features, produced more rational relationships. By correlating a number of characters Antoine-Laurent de Jussieu formulated in his *Genera plantarum* (1789) a natural sequence of families and genera which many European botanists accepted, although traditionalists like Sir James Edward Smith were reluctant to abandon Linnaeus. Augustin Pyramus de Candolle elaborated Jussieu's system which was further modified by other taxonomists. When Wight returned to Britain in 1831, progressive botanists—Robert Brown, William Hooker, and John Lindley—were the leading advocates of natural classification. Wight and Walker-Arnott, willing converts, adopted de Candolle's system for their *Prodro-*

mus florae peninsulae Indiae Orientalis (1834), the first Indian flora to do so. Only the first volume which describes 1365 species was published since Wight was recalled to India. Joseph D. Hooker and Thomas Thomson, impressed by 'the accuracy of the diagnoses, the careful limitation of the species . . . the many improvements in the definition and limitation of genera and the higher groups of plants', pronounced the book 'as the most able and valuable contribution to Indian botany which has ever appeared'[8].

On his return to India in 1834, Wight was posted to Bellary, 300 miles north-west of Madras, to a regiment about to march 700 miles to the far South, a prospect which appealed to his collecting instincts. At Palamcottah near Cape Comorin, the regiment's destination, he recruited three collectors, two of whom were despatched to Courtallum and Malabar. Wight had with him tracings of Roxburgh's Coromandel plants and his unpublished drawings at Calcutta, systematically arranged in portfolios, to assist the identification of specimens which had reached 2000 species by 1835. The abundance of orchids especially pleased him but there were, inevitably, disappointments and frustrations, one such occasion being the receipt of a parcel from his Malabar collectors, its contents ruined by the monsoons. This agreeable routine of collecting and examining specimens was interrupted when he was temporarily relieved of military duties in 1836, and transferred to the Revenue Department, with instructions to investigate the state of agriculture in South India, in particular tobacco, coffee, senna, dye-stuffs, sugar, spices, and the failure of recent experiments with American cotton. He was to ascertain both scientific and vernacular names of native crops. 'It will be convenient to give this information in a separate report, with drawings of the most valuable plants, which may hereafter be lithographed for general use.'[9]

Before starting his fact-finding mission, Wight had made sure he left his two experienced plant collectors with sufficient work during his absence. From 1842 to 1850 he managed the experimental cotton farm at Coimbatore, his ever-expanding herbarium filling over 20 cases, each three feet high. Aware that collecting was very much a matter of personal initiative, seldom coordinated with other projects, he proposed in 1851 that the Madras Medical School should organize 'natural history classes in which youths taken from the East Indian and native communities could be trained as naturalists' to participate in a systematic exploration of the Presidency, submitting their collections to experts for identification[10]. When Wight retired in 1853 he took home with him an herbarium in excess of 4000 species, largely the efforts of his collectors in the hills. He presented it to Kew the year before his death.

The creation of a comprehensive herbarium was an essential tool for Wight's illustrated flora of South India. The project had been discussed with William Hooker in Glasgow during Wight's leave in the early 1830s. They both agreed that lithography was the ideal graphic process for any extensively illustrated publication. Lithography probably first came to India with the formation of the Government Lithographic Press in Calcutta in 1823 and Wallich was the first author in India to use it for floral illustration in his *Tentamen florae Napalensis illustratae* (1824–6) where the figures are rather primitive efforts compared with the sophisticated hand-coloured plates in his *Plantae Asiaticae rariores* (1830–2), printed in London. Wight visited lithographic firms in Glasgow, made himself familiar with the technicalities of the process, and purchased equipment to take back to India. There he wrestled with the warping of presses, the breaking of stones, the cracking of rollers, mould on paper, and the inexperience of his native operators before he produced any satisfactory prints.

By 1838 he had attained an acceptable competence in lithography to embark on two illustrated works almost simultaneously: *Illustrations of Indian botany* and *Icones plantarum Indiae Orientalis*. With an order for 50 copies from the Madras Government, he started printing *Illustrations of Indian botany* under considerable difficulties—the scarcity of good quality paints, the use of transfer paper and the climate resulted in substandard plates.

> *The drawing is copied on prepared paper, by being held against the light, with lithographic ink, which is communicated to the stone. No corrosion, and indeed very little assistance of any kind is given, so that the loss of tender parts which you observe arises not from corrosion, but from the fine lines coming out in printing, a not infrequent occurrence in this country where the great heat softens the ink too much. These explanations will show that we have done nearly all we can effect in lithographic improvement while opposed by climate*[11].

Many of the plates which were crudely hand-coloured by a team of Indians were drawn by Rungia, one of the artists he had trained. The work attracted only 235 subscriptions (including the 50 copies reserved for the Government) and Wight lost heavily. After 182 plates comprising two volumes, it ceased publication in 1850. In the companion work, *Icones plantarum Indiae Orientalis*, the descriptions are shorter and the plates uncoloured. Again without a guaranteed sale of 50 copies to the Government the book would soon have foundered. Wight's objective was a cheap and quickly produced publication by printing plain outline drawings, a decision which met with the approval of William Griffith who preferred them for floral analyses. About 400 of his plates were copied from Roxburgh's drawings—353 from the unpublished set at Calcutta so that they also illustrate species described in Roxburgh's *Flora Indica*. Most of the remaining plates were by Rungia and Govindoo. Wight's aim was to produce an Indian equivalent of Smith and Sowerby's *English botany* by illustrating the plants described in his *Prodromus*; 'to supply the Indian Botanical amateur with the means of acquiring a knowledge of the principles of the natural method of Botanical classification, by presenting him with a series of diagrams of the organs from which the characters of the orders are taken, to enable him to compare them with the written characters' (Prospectus). After the sixth and final volume in 1853 with 2101 plates printed, he relaxed for perhaps the first time in his chosen career, fully deserving the luxury of self-satisfaction as he calculated that his books had contributed figures of well over 2000 species to the existing corpus of illustrated Indian floras by Roxburgh, Wallich, and Royle. 'The Indian Flora can now, I believe, boast of being more fully illustrated than that of any other country under British sway, Great Britain alone excepted.'[12] His Indian colleagues loyally applauded his achievement, insisting that the plates in the work were 'equal to anything of the kind in botanical works in Europe, and certainly superior to any that have been produced in India'[13]. Wight selected 200 Nilgiri plants from his *Icones*, had them coloured and issued as *Spicilegium Neilgherrense* (1846–51) in a limited edition of a hundred copies (Fig. 10.3). Hooker and Thomson judged these three works 'the most important contributions, not only to botany, but to natural science, which have ever been published in India'[14].

The Nilgiris attracted many botanists who, after strenuous collecting in the hills, could relax at Ootacamund. Ill-health brought a German missionary, Bernard Schmid (1787–1857) to this hill station where he collected flowering plants, ferns, and mosses[15]. He was

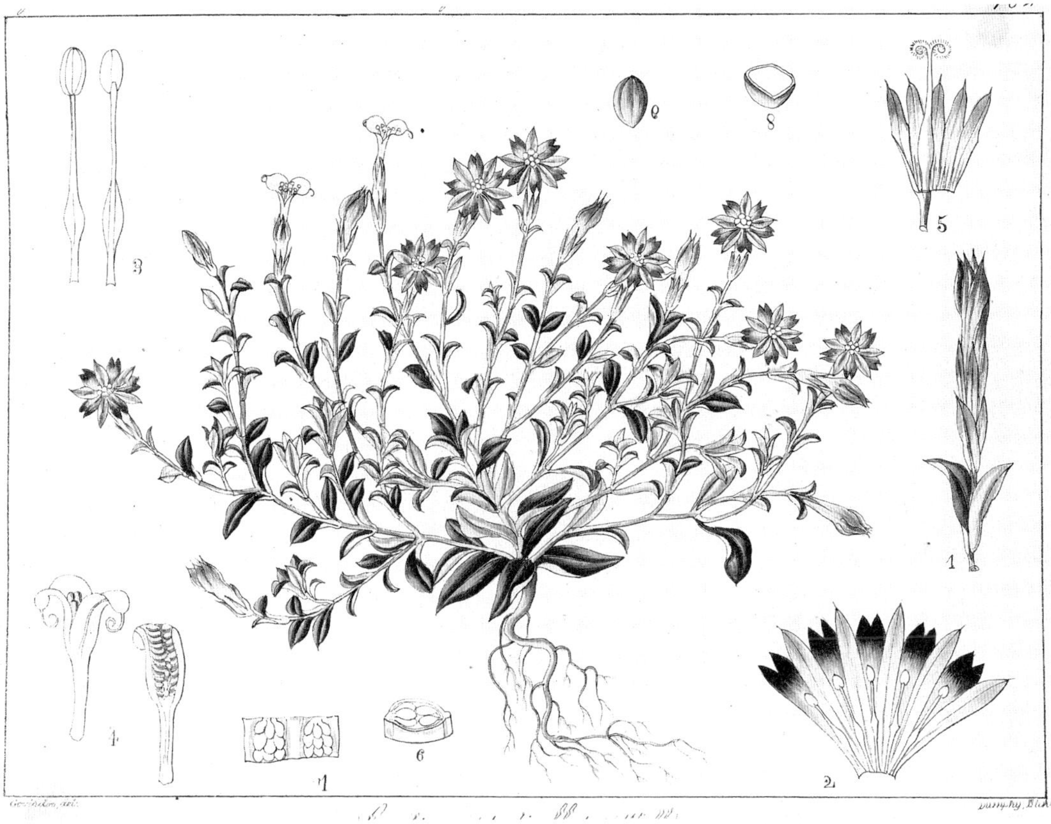

Figure 10.3 *Gentiana pedicellata.* (From R. Wight (1851), *Spicilegium Neilgherrense: or a selection of Neilgherry plants drawn and coloured from nature,* Vol. 2, Madras, plate 164.) Hand-coloured lithograph of a drawing by Govindoo, Wight's favourite Indian artist.

an hospitable host and guide to Karl von Huegel, the Austrian collector, who spent six weeks exploring the neighbourhood. Another missionary, F. Metz of the Basle Mission, who had collected at Mangalore for a German syndicate, made occasional forays into the Nilgiris. A distinguished foreign visitor was Jean Baptiste Louis Claude Theodore Leschenault de la Tour (1773–1826). He had been one of the two botanists on Nicolas Baudin's voyage in the southern hemisphere in 1801–4. In 1815 he was despatched to the French settlement at Pondicherry to establish a botanical garden there and to collect in India, Siam, Burma, and Malaya. The East India Company, under pressure from Sir Joseph Banks, instructed the Government in Bengal to offer him assistance including plants from the Botanic Garden at Calcutta. Banks independently wrote to Buchanan at Calcutta urging his full co-operation since Leschenault de la Tour was 'likely during his stay in India to visit those countries with which the English nation have the least intercourse & will no doubt discover articles of utility with which we are yet unacquainted'[16] His request to be allowed to explore the Western Ghats including the Nilgiris was reluctantly granted by the Court of Directors who usually suspected the motives of other nationals in British India. 'We are perfectly disposed to afford every facility to researches of this nature,' they informed the Madras Government; 'you will however bear in mind the propriety of using precautions to prevent as far as possible their being made the pretext for intrigue with the natives'[17]. Leschenault de la Tour's collections included animal and mineral specimens as well as plants for the National Museum in Paris.

He was one of a small band of French naturalists, active in India, the first being Pierre Sonnerat (1745–1814) who had seen service in Mauritius, Bourbon, and Madagascar with the naturalist Commerson, and had been a member of a successful expedition to the Pacific to collect spices and other useful plants for the French colony at Mauritius (Fig. 10.4). In 1774 he was naturalist on another voyage to the East during which his ship explored both east and west coasts of the Indian subcontinent. In 1786, now resident in Pondicherry, he directed his attention to both animal and plant kingdoms, engaging native artists to record his finds[18]. He was essentially a zoologist and William Roxburgh, who met him while he was stationed at Samalkot, had little regard for his botanical skills. 'Sonnerat was a miserable botanist. I know him well, collecting specimens was all the good he could do in that line. Indeed I am well informed that his publications were

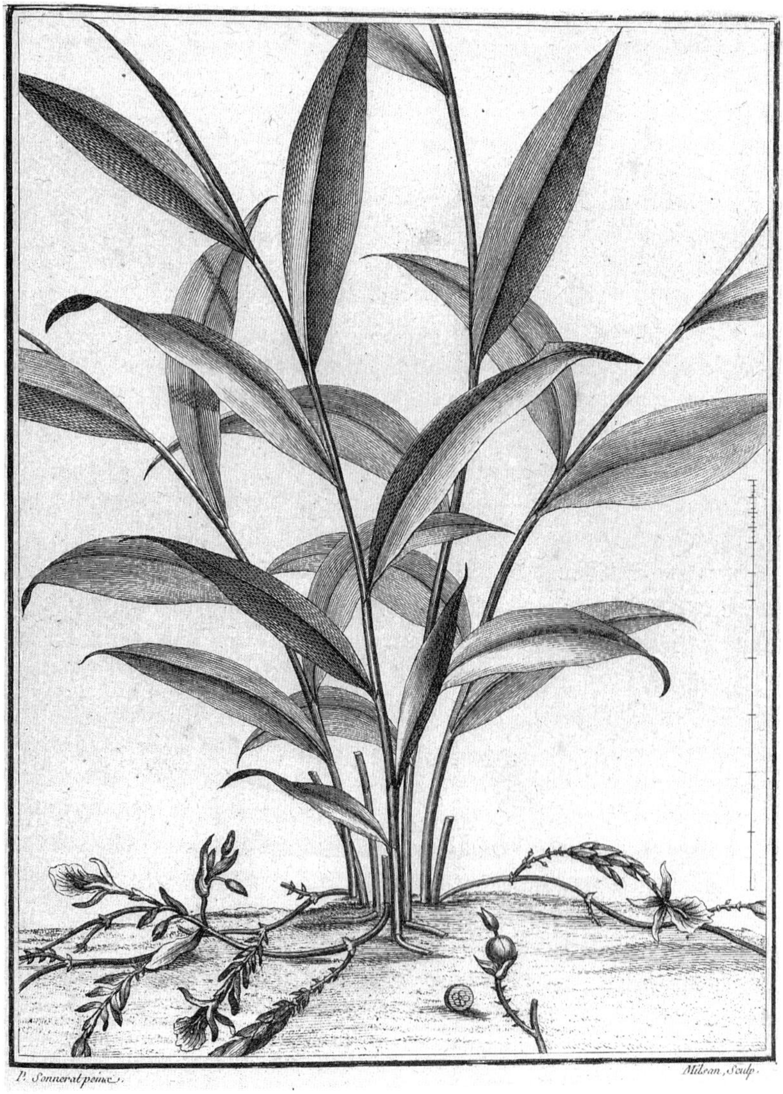

Figure 10.4 Cardamon, *Elettaria cardamomum*. Drawing by Pierre Sonnerat (from his *Voyages aux Indes orientales et à la Chine*... Paris, 1782, plate 136.) Engraving.

stolen from the missionaries at Pondicherry.'[19] George Samuel Perrottet (1793–1870) amassed a collection of South Indian plants representing some 1500 species. Charles Belanger (1805–81) collected at the French settlement at Mahé on the Malabar coast in 1825 before a leisurely journey across the peninsula to Pondicherry where he became Superintendent of the Botanic Garden and made the acquaintance of Robert Wight.

The productivity of Richard Henry Beddome (1830–1911) entitles him to be recognized as Wight's natural heir. His botanical career began when he transferred from the Madras Infantry to the newly-created Forest Department in the Madras Presidency in 1857. Forest conservation directed his attention to floristic studies. *The ferns of southern India* (1863–4) appeared at the same time as his *Trees of the Madras Presidency* (1864). The former included Ceylon and in his *Ferns of British India* (1866–70) he dealt with those species to be found elsewhere in the subcontinent. Before it was finished, Beddome had embarked on *Icones plantarum Indiae Orientalis*, a work clearly inspired by Wight's book with the same title. Like Wight he preferred outline drawings for clarity and cheapness, and for most of his publications he employed Wight's favourite artist, Govindoo. Only the first volume with 300 plates in 1874 was published. In his *Flora sylvatica for southern India* (1869–74) he described and illustrated the principal timber trees and large shrubs in the South and in Ceylon. In his last book before retirement, Beddome returned to his first love—ferns—with *Handbook to the ferns of British India, Ceylon and the Malay Peninsula* (1883).

For many years Western India was never subjected to the same intensity of botanical scrutiny as the South, the Gangetic Plains or the Himalayas. William Henry Sykes (1790–1872) as Statistical Reporter to the Bombay Government from 1824 to 1829 surveyed the Deccan when he collected about a thousand species of plants, mainly in the neighbourhood of Poona. His reports were illustrated with his own pen-and-ink sketches and with competent watercolours by his draughtsman, Bombardier Llewellyn L. Fidlor of the Bombay Artillery. Sykes paid tribute to Fidlor's 'considerable talent in coloring and in delineation. . . . The drawings are executed on a scale from Nature by actual measurements'. They were all scrupulously vetted and, where necessary, Fidlor's detail and colour were corrected by Sykes[20]. Joseph Nimmo (*fl.*, *c.* 1830–54) who sent plants to Sir William Hooker was 'the acknowledged head of the corps botanique of Bombay', according to Wight, 'a gentleman whose diligence in collecting is only equalled by his liberality in distributing the proceeds'. John Sutherland Law (1810–85) of the Bombay Civil Service divided his herbarium of some 1500 species between Oxford and Kew; Alexander Gibson and Nicol Dalzell collaborated in a *Bombay flora* (1861). Casual collecting was indulged in by a number of civilians and military personnel but their efforts lacked impetus and co-ordination. The Presidency desperately needed the initiative and resolve of a Wight or Beddome to involve the support of the Bombay Government. In 1888 W. Thiselton-Dyer, Director at Kew, deplored 'the absence of any competent botanical authority in this Presidency, or of any properly equipped herbarium, which would serve for the local identification of the indigenous vegetation'[21]. This complaint was prompted by a visit to Kew from a Bombay resident who wanted his collection of grasses named. 'It must be understood that, in a botanical sense, this establishment has the whole empire on its shoulders', wrote Thiselton-Dyer, rather pompously. 'With all the good will in the world, it is physically impossible that we can name collections for every private British subject who applies to us.' And he was a man who took pride in being blunt and to the point. 'There is no part of India where the interests of botanical sciences are more entirely neglected, no part from which it is more difficult to obtain specimens.' The India Office referred his plea for an official botanical survey of Western India to the Governor of Bombay, who was asked to investigate the feasibility of utilizing existing facilities in Bombay, i.e. locating an herbarium at the Victoria and Albert Museum and making the Professor of Botany at the Grant Medical College responsible for the project. But the India Office's refusal to countenance any additional expenditure effectively blocked any involvement by Bombay.

NOTES

[1] King (1899), 908.
[2] Griffith to Wight, 25 September 1839. W. Griffith (1848), *Posthumous papers*, Vol. 2, xiii.
[3] Wight to McClelland, 31 March 1845. *Calcutta Journal of Natural History*, **6**, 1846, 302.
[4] Griffith's posthumous publications: *Posthumous papers: journals of travels in Assam, Burma, Bootan and the neighbouring countries*, Calcutta, 1847. *Icones plantarum Asiaticarum*, Calcutta, 1847–54. 4 vols. *Notulae ad plantas Asiaticas*, Calcutta, 1847–54. *Itinerary notes of plants collected in the Khasyah and Bootan mountains, 1837–1838, in Afghanistan and neighbouring countries, 1839–1841*, Calcutta, 1848. *Palms of British East India*, Calcutta, 1850.
[5] Huxley (1918), Vol. 1, 234.
[6] *Botanical Miscellany*, **2**, 1831, 90–110, 344–60; **3**, 1832–3, 84–104, 291–302; *Hooker's Journal of Botany*, **1**, 1834, 62–7, 225–31; *Hooker's Companion to the Botanical Magazine*, 1835–7, 2 vols; *Annals of Natural History*, **1**, 1838, 395–6; **2**, 1839, 111–12.
[7] Walker-Arnott listed the duplicates in a manuscript that was subsequently lithographed, 1833–7.
[8] Hooker and Thomson (1855), 48–9.
[9] *Transactions Botanical Society of Edinburgh*, **11**, 1873, 371.
[10] Madras Central Committee for the Grand Exhibition of Industry and Art. Circular 8 January 1851. I.O.L.R. F/4/2492, item 141183.
[11] Wight to von Martius in Munich, 15 April 1845. *Transactions Botanical Society of Edinburgh*, **11**, 173, 377.
[12] R. Wight (1853), *Icones plantarum Indiae Orientalis*, Vol. 6, Madras, v.
[13] *Madras Journal of Literature and Science*, **8**, 1838, 210.
[14] Hooker and Thomson (1855), Vol. 1, 50.
[15] J. K. Zenker (1835-7), *Plantae Indicae, quas in montibus Coimbaturicis coeruleis Nilagiri seu Neilgherries dictis, collegit Rev. Bernhardus Schmidt*. Decas I–II. Jena.
[16] Banks to Buchanan, 10 January 1815. B.M. (N.H.) D.T.C., Vol. 19, folios 115–16.
[17] East India Company to Madras Government, 12 May 1819. I.O.L.R. E/4/923, folio 84.
[18] The State Library of New South Wales has a manuscript journal of Sonnerat, Nouveau voyage aux Indes Orientales in 3 volumes, and another volume of watercolours which includes 28 plant portraits by Indian artists.
[19] Roxburgh to A. B. Lambert, 21 February 1801. B.L. Add. MSS 28545, folio 173.
[20] W. H. Sykes (1826), First report on statistics of the Dukhun. I.O.L.R. MSS Eur. D141, nos 3 and 4.
[21] Thiselton-Dyer to Under-Secretary of State for India, 20 July 1888. Kew Archives. Selections from Despatches to India, Vol. 2, 1888, folio 227.

THE HIMALAYAS

While apathy or, at best, subdued activity prevailed in botanical circles in western India, the Himalayas beckoned the adventurous and the curious, among them surveyors and scientists who observed in a casual note or expansive report the indigenous plant and animal life of this complex mountain range which separates India from Central Asia, extending from Afghanistan for 1700 miles to Assam in the East, and continues into China. A diversity of soils and climates, especially the summer monsoons, support what the plant collector, Frank Kingdon Ward succinctly described as an 'immense reservoir of hardy plants'. A fertile jungle strip, ten to twenty miles wide—the terai—which skirts its eastern ranges is home to slender evergreens such as the sal (*Shorea robusta*) and tall bamboos. The abundant sal yields to conifers and oaks in a gradual ascent of four or five vertical miles through a progression of rain forest, temperate vegetation, conifers, and rhododendrons to the alpine zone above the tree-line at about 10 000 feet. Everywhere plants have adapted to their environment, a delicate ecological balance which man's intervention through cultivation, grazing and logging is swiftly destroying with consequent erosion and lowland flooding. The phytogeographical variations range between the arid zones of the Hindu Kush and the warmer, wetter regions of the east, with Kashmir, Nepal, and Sikkim offering bumper harvests for botanist and plant collector.

An extravagant palette of colour is provided by dense thickets of rhododendrons competing with the white of magnolias, the pink of daphnes or the blue, yellow, and crimson of Himalayan poppies. A bedazzled tourist to Simla complained that the rhododendron's 'rich tints throw out too much bright colour into the landscape'. Rhododendrons that rise to 60 feet and more on the lower slopes survive as miniature shrubs at higher altitudes—splashes of reds, pinks, purples, and yellows in crevices, and on rocks and screes. An alpine scrub of rhododendrons, willows, and junipers leads to a landscape of anemones, saxifrages, primulas, and gentians emerging from the melting snow. These colonies of alpine flowers continue to a height of 15 000 feet; on one expedition eidelweiss was found at over 18 000 feet and compact mats of *Stellaria decumbens* and *Arenaria bryophylla* have been seen even higher.

British official interest in the Himalayas probably began with George Bogle who led a mission to Lhasa in 1774. The Governor-General, Warren Hastings, instructed him to seek out minerals and to observe the fauna and flora during his passage through the mountains. Another mission, this time to the ruler of Garhwal at Srinagar, was undertaken in 1796 by Thomas Hardwicke who became the first European to collect plants in the north-western Himalayas. The East India Company never bestowed the dignity and authority of official status on one of its greatest explorers, William Moorcroft (1767–1825), impulsive and unpredictable, enthusiastic to the point of eccentricity, but unquestionably courageous. A qualified veterinary surgeon, he took charge of the Company's stud farm in Bihar in 1808. Four years later, disguised as an Hindu ascetic, and with an Anglo-Indian companion, he made the first of his great journeys, climbing the Niti Pass in Garwhal in the Great Himalayan Range to the sacred lakes of the Sutlej river. Since he had absented himself without leave from his station in Bihar to penetrate territory beyond the Company's jurisdiction, one wonders whether the Bengal Government were at all placated when he returned with a flock of shawl-wool goats and a bundle of plants (which was despatched to Robert Brown in London). He was soon organizing the delivery of seeds of trees, flowers, and fruit through the contacts he had made on this expedition.

The objective of his next major exploration in 1819 was Central Asia to purchase stallions for his stud. Although this time he was given an armed guard, the Company denied him any official authorization, fearful of antagonizing local tribes. Two plant collectors provided by Wallich were part of his escort of 300 men. He pioneered a route over the Lesser Himalayas to Leh in Ladakh and proceeded to Kashmir where he studied shawl manufacture in incredible detail and painstakingly noted every aspect of agricultural practice in the valley. Victor Jacquemont, the next European to visit Kashmir, maliciously observed that Moorcroft's 'chief occupation was making love and, if his friends are surprised that his travels were so unproductive, they may ascribe it to this cause'. This caustic comment from a reputed philanderer is hardly borne out by the immense amount of data Moorcroft crammed into his letters and journals. He reported enthusiastically to Wallich on fruit-laden orchards, piquantly distinguishing the subtle flavours of Kashmiri apples, pears, peaches, apricots, quinces, plums, and cherries[1]. With a philistine insensitivity he recommended the conversion of the terraced garden at Pinjore in the Punjab, the last of the great Mughal landscapes, into a fruit nursery for British settlements. Moorcroft made a special journey to Dras in Kashmir to investigate reports of a nutritious bush called Prangos (*Prangos pabularia*), grown by local farmers as a winter fodder for their sheep, goats, and cattle, credited with remarkable fattening properties and, as a bonus, able to destroy liver fluke in sheep. As it was easily cultivated even on poor soil, Moorcroft saw it as a promising new crop for British livestock. In 1822 and 1823 he despatched consignments of the hay, fruit, and seeds of Prangos to the Calcutta Botanic Garden and Wallich sent seeds to London. Both Griffith, who encountered it in Afghanistan, and Falconer, who grew it experimentally in his Mussoorie garden, thought Moorcroft had exaggerated its agricultural potential. Several attempts to introduce it into British agriculture failed. Moorcroft also saw Ladakh rhubarb competing commercially with Chinese exports to the West. Throughout his travels, his voluminous journals seldom depart from a conscientious observation of local husbandry and industries. There is never a hint of his difficulties and dangers but occasionally the grandeur of the Himalayas leavens his pedestrian prose. An uncharacteristic purple passage intrudes uneasily into his description of the fertile Kulu Valley.

Vast slopes of grass studded with clumps of cypress, cedar and fir go off from the summit in various lines . . . The bottom of the valleys glitter in the sun with ribbands of water. The mountains near at hand thrusting their green peaks into the sky sometimes are wholly visible, clear and defined and at others are hidden . . . by dense clouds or mists . . . The surface of the ground is literally enamelled with small asters, anemones and great var-

ieties of the sterile strawberry as scarlet, crimson, white, and yellow.

Three years after reaching Ladakh and Kashmir, Moorcroft's party continued its trek, prudently avoiding contact with the suspicious Sikhs, passing safely through the unsettled regions of Afghanistan and the Hindu Kush, only to be detained for six months by the rapacious chieftan of Kunduz before eventually reaching the domes and minarets of secretive Bukhara in 1825. Moorcroft died of fever on his return journey to India.

In 1822 while Moorcroft was still in Ladakh, three brothers—Alexander, James, and Patrick Gerard—led one of the first parties of 'tourists' to the Spiti Valley north east of Simla. They walked entranced through belts of birch and pine and carpets of alpines at 13 000 feet; near the Sutlej river they picked blackcurrants and raspberries and indulged a longing for England with gooseberries at Nisangwere[2].

Godfrey Thomas Vigne (1801–63), an independent traveller—engaged, it has been suggested, in espionage—met the fraternal trio at Simla in the summer of 1834. He was a talented artist, a good shot, and had an amateur's interest in geology and botany. During the three years he resided in Kashmir he got to know the valley more intimately than Moorcroft, who had disliked the Kashmiris. He acclaimed Gulmarg, a hill resort in the valley, 'so highly picturesque that little is wanting but a mansion and a herd of deer to complete its resemblance to the perfect English Park'[3]. While exploring the Lesser and Greater Himalayas, he found time to press a few plants which he presented to Royle—too late, unfortunately, to be included in *Illustrations of the botany . . . of the Himalayan Mountains* (1833–40).

While he was in Srinagar in 1835 Vigne was introduced to an Austrian army officer, Baron Karl Alexander Anselm Freiherr von Huegel (1795–1870) and together they left the Vale of Kashmir for the Punjab to be entertained by the Sikh leader, Ranjit Singh. An indifferent explorer but an industrious collector, von Huegel's labours enriched the Vienna Museum. He worked diligently, wrote copiously[4], and reaped a satisfactory burden of plants. 'My booty in new plants and seeds exceeded my expectation, and the autumn had even spared a few rare flowers.'

The way for the Baron's and Vigne's entry to Kashmir had been prepared by the charismatic Frenchman, Jacquemont (Fig. 11.1), the first European allowed into the state after the tactless and seemingly aggressive behaviour of Moorcroft. Jacquemont had enthusiastically recommended Kashmir when he met von Huegel in Poona in 1832. Victor Vinceslas Jacquemont (1801–32), with the support of the Royal Asiatic Society in London and the permission of the East India Company, had departed for India in 1829 in order to study the ethnology, geology, and botany of the country and to collect plants for the Jardin des Plantes in Paris. His tour was scheduled to last seven years and, in the event of his not being able to reach the

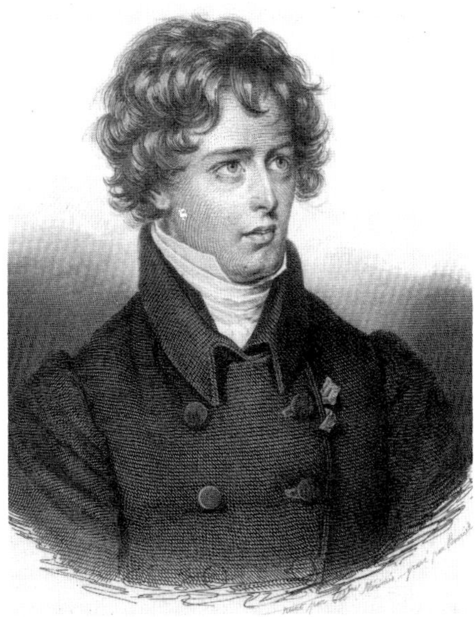

Figure 11.1 Victor Jacquemont (1801–32). Engraving after a portrait by Mérimée.

Himalayas, he proposed concentrating his efforts on the natural history of Malabar and the Western Ghats. A cultured young man, the friend of Prosper Mérimée and Stendhal, he brought the sophistication of the Paris salon to Indian society. His considerable charm gained him the friendship of the Governor-General and other influential officials who provided him with introductions during his peregrinations in northern India. The unstinted hospitality he enjoyed made him an Anglophile, immensely impressed by what he perceived to be the enlightened paternalism of British administrators. His posthumously published letters, acutely observant, seasoned with humour and a dash of malice, were usually written in the solitude of his tent after a day's exhausting trek. They flicker with banter and not infrequently with self-ridicule. 'I am very brown, have large mustachios of a distressing colour, no whiskers, long hair and a very small, light and flexible hat of palm leaves.' He was very proud of his Pondicherry hat which he relined with black silk every three months. 'In the morning I pull it like a cap over my ears, and find it very warm; it assumes any shape I wish.' He had an appreciative eye for pretty women whether they were to be found in an English household, an Indian nautch, or bathing on the banks of the Ganges.

He left Calcutta in November 1829, after an agreeable stay of nearly six months, visited Royle at Saharanpur, paused at Simla, and crossed the Tibetan frontier at Tashigong. 'These wilds have been travelled over by a great many English, and I have reason to believe that their Flora is sufficiently well-known, although, no doubt, they have left something to be done on a close inspection.'[5] He owed his introduction to Ranjit Singh to General Jean-François Allard, a former Napoleonic officer serving in the Sikh army. Ranjit Singh's empire embraced the whole of the old Punjab, from the Sutlej to the Indus, and since it included Kangra and Kashmir no access to the western Himalayas was possible without this ruler's permission. Jacquemont formed an instant rapport with the old roué—having a mutual interest in women doubtless fostered the relationship. The Frenchman's shrewd appraisal of people as well as his literary fluency are abundantly evident in his portrait of this illiterate leader who was the scourge of the British.

He is a small thin man of a fine shape although one-eyed through the effect of small-pox; there are however very few marks of it left on his face. The right eye which has been spared to him is very large, the nose is fine and slightly raised, the mouth is well-shaped, with the superb teeth and little moustaches which are constantly passed through his fingers; a long white beard falls on his breast. His appearance gives vent to a constant nobility of thought, a great finesse and a profound penetration[6].

Jacquemont spent four months during 1831 in Kashmir collecting plants, animals, and minerals (Fig. 11.2.), all the time pondering a book which would not be a conventional flora but rather an excursion into phytogeographical relationships. Like botanical travellers before him and Joseph Hooker after him he was struck by the similarities between Asian and European plants. 'I shall compare the vegetation of the Himalayas with that of the Alps, the rocky mountains to the west of the Missouri, and the lofty Cordilleras of equinoctial America.'[7]

Jacquemont's interest in plant geography may well have been stimulated through meeting Royle, the Superintendent at Saharanpur. Royle was the first botanist to publish observations on floristic affinities between Europe and Asia, a thesis he pursued through all his researches. Royle's plant collectors accompanied merchants trading with Kashmir. His friend Lieutenant E. Maxwell brought him plants from Kunawar. Royle

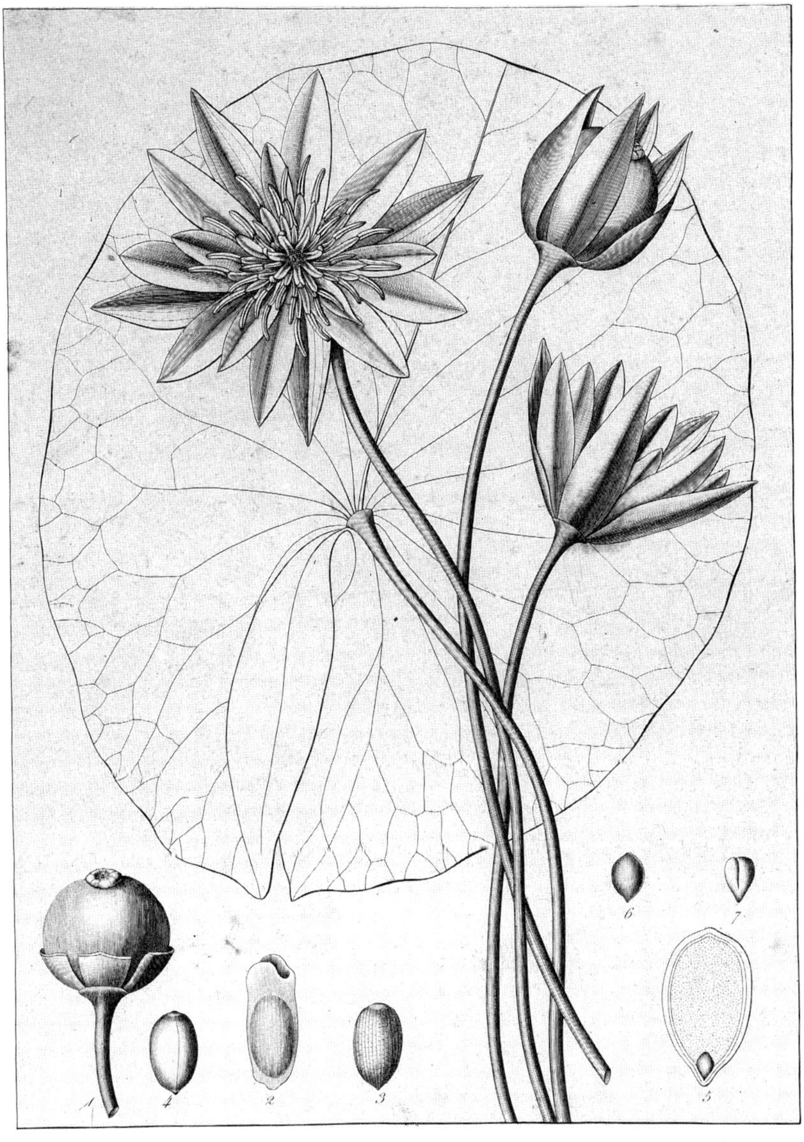

Figure 11.2 *Nymphaea alba*. (From V. Jacquemont, *Voyage dans l'Inde pendant les années 1828–32*, Vol. 4: *Botanique* by J. Cambessèdes, Paris, 1844, plate 10.) Engraving of drawing by E. Delile.

concentrated on Kashmir and Kunawar because they were densely populated regions, providing ample evidence of the native use of plants in agriculture, medicine, and commerce. Royle's triple approach to plants—their description and classification, their geographical distribution and their usefulness to man—underlies the presentation of his *Illustrations of the botany and other branches of the natural history of the Himalaya mountains and of the flora of Cashmere* (1833–40). In writing this pioneer mountain flora, Royle consulted not only his own herbarium but also the plants of Wallich's collectors in Kumaon and the collections of Govan and Gerard made in Garhwal and Sirmur. Hardwicke lent him drawings done some 30 years earlier on his visit to Kashmir. While Wallich was on extended leave in England, Royle was allowed to borrow the Calcutta team of artists and 44 of the hundred published plates were lithographed from

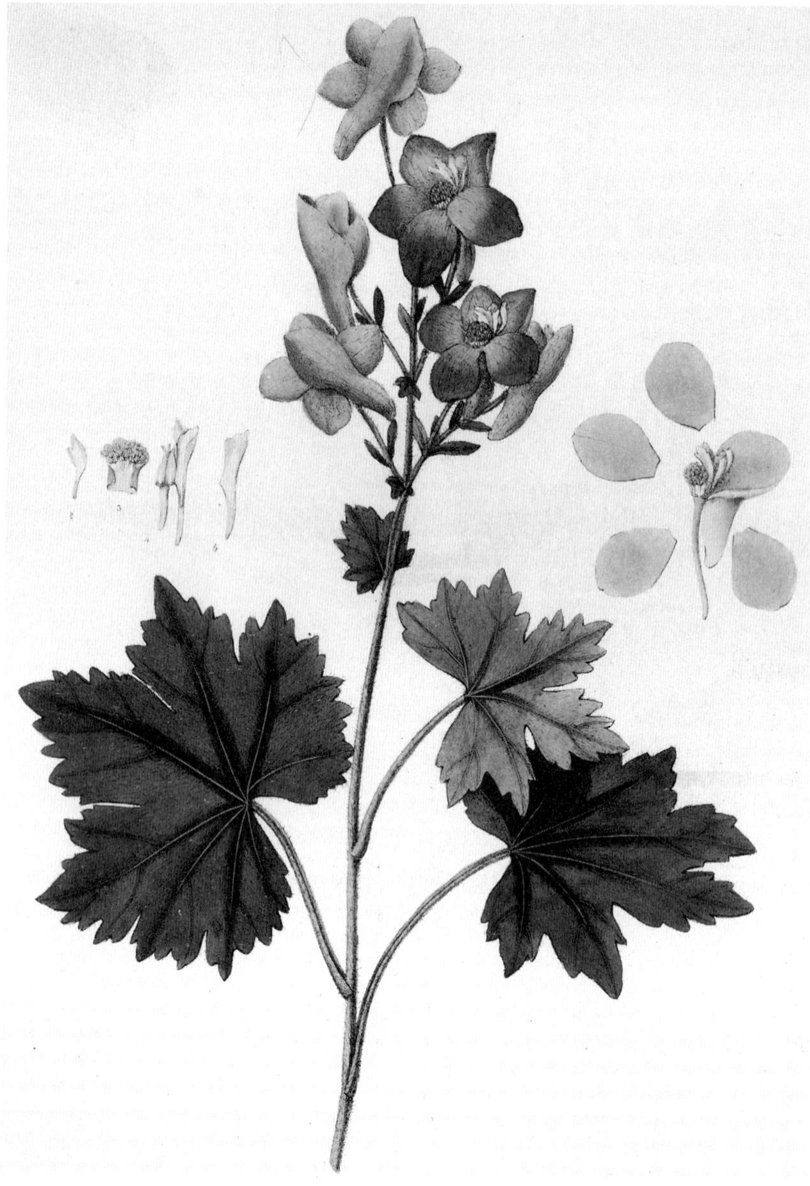

Figure 11.3 *Delphinium cashmerianum.* (From J. F. Royle (1833), *Illustrations of the botany and other branches of the natural history of the Himalayan mountains and of the flora of Cashmere,* London, plate 12.) Hand-coloured lithograph of a drawing by Vishnupersaud whom Royle had borrowed from the Calcutta Botanic Garden while Wallich was in England.

drawings by Vishnupersaud (Fig. 11.3) who had illustrated an earlier Himalayan flora, Wallich's *Tentamen florae Napalensis illustrate* (1824–6). Plants with medicinal properties feature prominently in his selection and there is, for instance, an extensive account of the history, distribution and cultivation of native and introduced cotton. He recalled the astonishment of the seventeenth century traveller, François Bernier, when he viewed a mountain range in Kashmir. 'The side facing the south, that is, looking towards

Hindoustan, is full of *Indian* and *European* plants mingled together; but the side exposed to the north is crowded exclusively with the vegetable productions of *Europe*.'[8] Royle considered this observation. The aspen, willow, poplar, and walnut were some of the trees familiar to the European traveller and his own plant collectors had brought back species belonging to European genera.

Royle continued his speculations on the geographical dispersal of plants in the comfort of his London home while, in India, Ranjit's death in 1839 led to warring factions in the Punjab and the first Sikh conflict with British forces in 1845. At the critical battle of Ferozshah, Dr Werner Hoffmeister, physician to Prince Waldemar of Prussia was killed by grape-shot from the Sikh lines. Hoffmeister, a botanist, had added to his personal herbarium as they had progressed on a sight-seeing tour through Nepal and Kumaon, reaching Shipke on the Tibetan frontier[9]. Both Edward Madden (1805–56) and (Sir) Richard Strachey (1817–1908) served in the First Sikh War. Madden sent seeds of plants collected on his Himalayan travels to the botanical garden at Glasnevin in his native Ireland. At the end of the Sikh campaign in 1846 he met Strachey about to commence a scientific survey of Kumaon. Strachey was joined in 1848 by James Edward Winterbottom (1803–54) who had botanized in Kashmir and they collected together until 1849, amassing some 2000 species in Kumaon and Garwhal and adjacent Tibet[10].

While Winterbottom was in Kashmir he met Thomas Thomson (1817–78) who had joined the medical service of the East India Company in 1839 and was taken prisoner during the First Afghan War of 1839–42, narrowly escaping being sold in the Bukhara slave market. The political instability in the western Himalayas prompted the Government in Calcutta to establish the precise boundary between Ladakh and Tibet by appointing in 1847 a border commission consisting of Alexander Cunningham, Henry Strachey (brother of Richard), and Thomas Thomson. They worked independently in the territory they were to explore, Thomson's region being north of Leh and in the valleys of Nubra and Shyok. He was the first European to winter in Baltistan; he pioneered a new route into Ladakh over the Pir Panjal when the rhododendrons were at their best and potentillas and primulas were in flower. In the arid landscape of Ladakh, with little opportunity for botanizing, he collected minerals. He stood weak from lack of oxygen on the high desolate ridge of the Karakoram. His specimens eventually exceeded 2500[11]. In 1866 the Royal Geographical Society recognized his achievements by presenting him with its Founder's Medal. In making the award, the President observed that Thomson had 'brought back collections that link the labours of the Russian botanists in the north with those of the English in the south'.

Some six years after Thomson had climbed the peaks of the Karakoram, precisely the same route was followed by Hermann Schlagintweit (1826–82) and Robert Schlagintweit (1833–85) who succeeded in crossing the Karakoram pass, heading east to Khotan. With their brother Adolphe (1829–57), they had arrived in India in 1854 to carry out a magnetic survey of the subcontinent with the backing of the King of Prussia, the distinguished geographer Baron Alexander von Humboldt, and the Royal Asiatic Society. Their survey work, executed with Teutonic thoroughness, encompassed natural history, geology, and ethnology. Adolphe, travelling alone in disguise in Central Asia, was beheaded during disturbances at Kashgar in Turkestan (now Sinkiang). Hermann and Robert continued exploring the Tibetan plateau and the Karakoram until 1858 when a selection of their heterogeneous collections was presented to the India Museum in London: specimens of rocks, soils, fossils, skeletons, and skins of animals, over 300 plaster casts of faces, feet, and hands and a plethora of native artifacts. Unfortunately, the bulk of their collections was destroyed during an air-raid on Berlin in 1943; smaller collections survive in many repositories including 2000 Himalayan and Tibetan plants in the British Museum (Natural History).

Due largely to inaccessibility and difficult terrain, fewer plants have been collected in the north-western Himalaya than farther east in Nepal, Bhutan, and Sikkim. The Kingdom of Nepal, over 500 miles long and 150 miles at its widest, incorporates about one-third of the Himalayan chain and some of its highest peaks. It is a transitional region between the West and the East for within its boundaries are to be found elements of the flora of western Asia and of China and Japan. Along its southern border lies the narrow strip of forested terai, much of which has been progressively cleared for cultivation. The landscape of the foothills and deep valleys is subtropical but zones of temperate broad-leaved trees, evergreen oaks, tree rhododendrons, and conifers climb to the snow-line and the alpine meadows. It has been calculated that Nepal with its monsoon summers and dry winters is home to some 6800 species of flowering plants.

The country has had a turbulent history; in 1768 Gurkhas overthrew the ancient Hindu dynasty and embarked on a campaign of conquest. They over-ran Kumaon in 1789 and annexed Garwhal in 1804, threatened Sikkim and boldly crossed the border into British India. Attempts at appeasement by the British were followed by persuasive force and in 1816 the Ghurkas were compelled to cede Kumaon and Garwhal and to accept the ultimate humiliation, the permanent presence of a British Resident in Khatmandu. The first Englishmen in Nepal were Colonel William Kirkpatrick and two army officers sent by the East India Company in 1793 to mediate between the Nepalese and invading Chinese forces. Kirkpatrick's *Account of the Kingdom of Nepaul* (1811) refers only casually to the vegetation. He assessed forests of sal and pine with commercial astuteness; he dismissed the reputed cinnamon as being the inferior 'Cassia lignea' or *Cinnamomum aromaticum*; he noted the dependence of native medicine and dyeing on indigenous plants; approved the walnuts, oranges, and honey harvested from a countryside 'so over-run as this is with aromatic and sweet flowering shrubs'. No doubt his reports of Nepal's flora determined the East India Company to engage in 1795 'a native of Napaul well skilled in the knowledge of herbs'[12] to bring back seeds and plants, an operation which Dr John Fleming in Calcutta regretfully reported a year later had failed miserably.

The first opportunity to indulge in botanical espionage offered itself after the signing of a treaty between the British and the Gurkhas in October 1801 which allowed Captain Knox to proceed to Khatmandu as Resident. His surgeon was Francis Buchanan who brought with him a native artist and all the equipment necessary to make a collection. Buchanan's first impression of Khatmandu which he entered in April 1802 was unfavourable: 'a bare, ugly valley resembling many of those in Scotland before the introduction of fences and other improvements'.[13] Knox and his small mission had been instructed to gather discreetly information useful to British interests—the production of drugs and dyes, for instance. At an opportune moment, Knox was to negotiate access rights to the sal and pine forests of the terai. Buchanan used low-caste gardeners to collect plants but on one occasion all his local labour was withdrawn by Nepalese officials who suspected they were spying for the British. He found flowers and trees similar to European types and questioned Sir William Jones's identification of spikenard. 'All I can say is, that, if this root was the spikenard of the Roman ladies, their lovers must have had a very different taste from the youth of modern Europe.'[14] He noted that potatoes were grown reasonably successfully in the hills and that the rich enjoyed European vegetables. By the time the mission was recalled in April 1803 due to the increasing hostility of the Nepalese, Buchanan had formed an herbarium of between 1500 and 2000 specimens, a portfolio of drawings and had sent 117 consignments of seeds, bulbs, and rhizomes to the Calcutta Botanic Garden. Almost two decades were to pass before another European botanist continued the exploration of Nepal's flora.

When Buchanan returned to Britain in 1805 he presented his specimens, 'a strange mixture of European, Siberian, Japanese and Indian genera',[15] 114 drawings, principally of

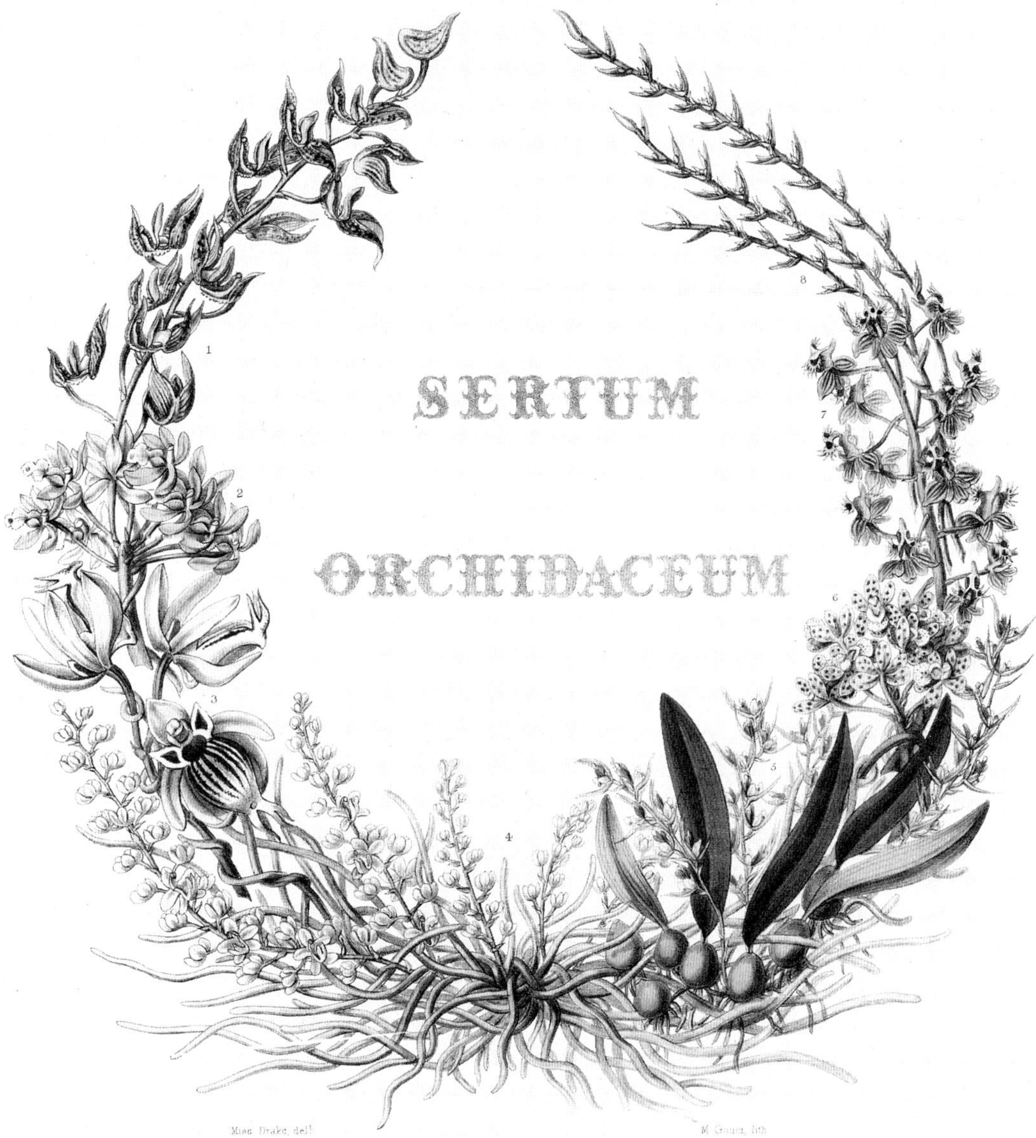

Figure 11.4 This wreath of orchids (From the title-page of J. Lindley (1838), *Sertum orchidaceum*, London) was composed from drawings in the East India Company's library in London. With a single exception, the orchids had all been collected in Nepal, mainly by Nathaniel Wallich.

orchids which were difficult to preserve, to Sir James Edward Smith. He even included his descriptive notes on the new species he had collected. Not having the time to work on the collection himself, he was confident that Smith would be eager to use them in his *Exotic Botany*. In the event he was to be gravely disappointed; only 12 were selected by Smith for his short-lived publication[16], and a few descriptions for Rees's *Cyclopaedia*. Buchanan had already lost interest when he politely declined Wallich's offer in 1821 of collaboration in a work on Nepalese plants. 'I have not a single note respecting any of the plants I brought with me from Nepal. Smith has the whole.'[17] He was further dissuaded by learning that David Don was at that moment immersed in a similar work with the aid of duplicate specimens that Buchanan had given to A. B. Lambert and material sent by Wallich.

Soon after Wallich became Superintendent in 1817, some officers of the Rajah of Nepal returning to their country allowed two of the Botanic Garden staff to accompany them. Bharat Singh who had previously worked for Buchanan was to collect plants and Francis de Silva, son of Roxburgh's Portuguese servant, was to press them adding whatever data he could discover about their native names, habitat, and uses[18]. In under a year their labours had rewarded the Botanic Garden at Calcutta with nearly a thousand species. Quite independently, Edward Gardner, the British Resident at Khatmandu, employed at least six collectors, sending them to sites far from the capital. Wallich fulsomely acknowledged his indebtedness to Gardner who additionally helped to procure his admission to Nepal in 1820.

Arrangements had been made for Wallich to go to Nepal in December 1814 and Buchanan, in anticipation, had formulated instructions for his guidance. Medicinal plants were to be given his first consideration. 'And as the natives employed to collect are apt to bring anything on which they can conveniently lay their hand, instead of the real production wanted, Dr Wallich will be allowed to engage a native doctor, who may be well acquainted with the medicines used, and who will be able to prevent imposition.'[19] Since Buchanan's collecting had been confined to the Khatmandu Valley and neighbouring hills, Wallich was to concentrate on higher altitudes and the plants beyond the snow-line. For whatever reason, presumably political, the expedition was cancelled, and it was not until December 1820 that Wallich at last achieved his ambition (Fig. 11.4). Because his own movements were restricted to the valley, he sent his collectors, sometimes prevailing as well on pilgrims, to the mountains around the sacred lake of Gossain Than, some three to four days' march north of Khatmandu. Soon his haul of specimens exceeded Buchanan's. He was eager to introduce *Edgeworthia gardneri* as a new shrub to English gardens and believed that *Rhododendron arboreum*—now the national flower of Nepal—would be hardy in Britain. Several hundred basket loads of plants, roots, cuttings, and seeds found their way to his Botanic Garden, 'If you were to see his last letter home it would make you laugh', wrote Hardwicke to A. B. Lambert. 'He is raised to the highest degree of earthly happiness. His room he describes as so filled with bundles of plants, boxes, bottles, growing plants, dead birds, quadrupeds, insects, etc. that he can scarcely find his way to his writing table when he leaves it, and the division of his attention to all these subjects hardly allows him time to sleep.'[20] After his return to Calcutta in November 1821, a fulfilled and contented man, Wallich presented an inventory of his collections to the Governor-General: 12 portfolios of descriptions of the Nepalese flora, 364 drawings in 8 folios, catalogues of wood specimens and a spirit collection of flowers and fruit—all raw material for his *Tentamen florae Napalensis illustratae*

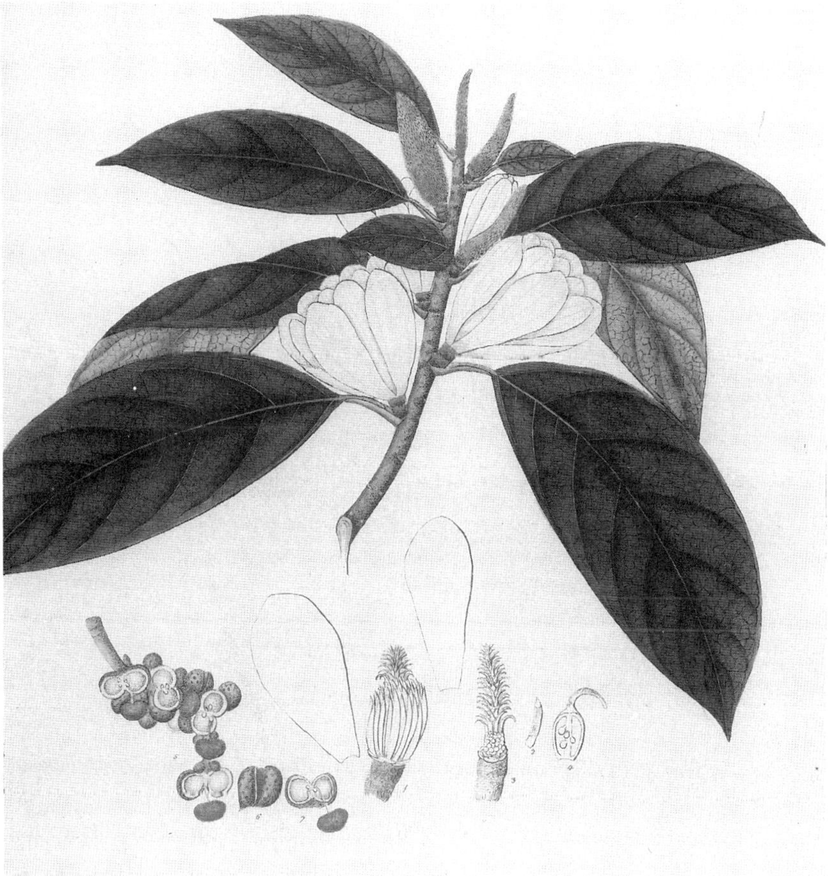

Figure 11.5 *Michelia doltsopa.* (From N. Wallich (1824), *Tentamen florae Napalensis illustratae,* Calcutta, plate 2.) Hand-coloured lithograph of drawing by Gorachand.

(1824–6) (Fig. 11.5). In the interval between the two parts of this work, David Don's *Prodromus florae Nepalensis* (1825) appeared, a compilation based on Buchanan's and Wallich's collections. Buchanan found it full of errors and John Lindley damned the author's Latin descriptions.

Joseph Dalton Hooker, temporarily denied access to Sikkim, spent three months of the autumn of 1848 in the eastern region of Nepal—the first European to be permitted to botanize outside the Khatmandu Valley. With a vigilant escort of Nepalese soldiers he followed the Tambur river and explored the Wallanchoon and Kanglachem passes into Tibet. Impressed by the number of different species of rhododendron, although they were not in flower, Hooker hopefully harvested their seed. He left Nepal at the Islumbo pass, able to enter Sikkim at last. Sikkim and Nepal were visited by Hermann and Robert Schlagintweit in 1857, J. Scully was there in 1876, J. F. Duthie four years later, and in 1907 I. H. Burkill collected plants where Wallich had been some 86 years earlier. Foreigners were never particularly welcome, one notable exception being the sportsman and naturalist, Henry John Elwes who, invited by the Maharajah in 1914, arrived with an original if unusual present of a pedigree Aberdeen-Angus bull. Kew received the plants collected by Alexander F. R. Wollaston, medical officer and naturalist to the Mount Everest expedition in 1921, who survived the perils of exploration only to be murdered in 1930 by a student at Cambridge. Thereafter the country was effectively closed to the outside world and the

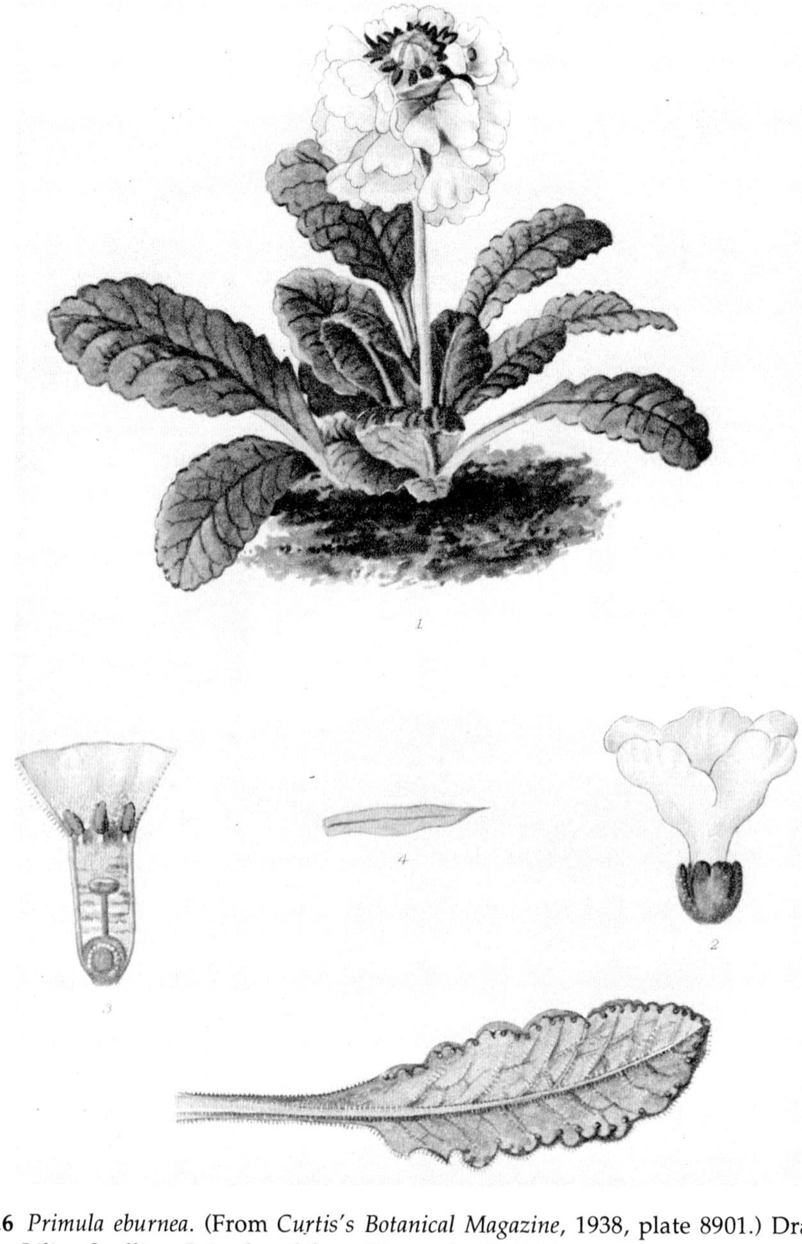

Figure 11.6 *Primula eburnea.* (From *Curtis's Botanical Magazine,* 1938, plate 8901.) Drawn and lithographed by Lilian Snelling. Introduced from Bhutan by R. E. Cooper.

only plants to reach the West from 1927 to 1937 were collected by a Nepalese army officer, Lall Dhwoj (*d.* 1931), and Professor N. K. Sharma; among them was *Meconopsis regia*. This prohibition on foreigners was relaxed only in 1949 when Major H. W. Tilman led a mountaineering party to Langtang Himal; his party included the botanist Oleg Polunin[21]. The following year another group of mountaineers included D. G. Lowndes, formerly a colonel in the Indian Army, who was able to make a collection. Polunin was back again in 1952, this time in West Nepal, with two other botanists, L. H. J. Williams and W. Sykes; in 1954 J. D. A. Stainton made the first of many visits which provided him with the experience and material for his *Forests of Nepal* (1972). The first Japanese party arrived in 1952, the aim of their subsequent botanical expeditions being to investigate the botanical affinities between the eastern Himalayas and Japan[22]. Unrestricted tourism, now essential to Nepal's economy, today poses a threat to its flora, but far more serious is the irreversible damage to the ecology perpetrated by the Nepalese themselves. In just ten years, one-third of its forests has disappeared, cut down for fuel or cleared for terraces of grain crops. Nepal, botanically in 1950 the least explored Himalayan region, has now become one of the best known. More than 100 000 specimens have been collected there and the *Enumeration of the flowering plants of Nepal* (1978–82) records some 6800 native species of flowering plants.

The Kingdom of Bhutan, still as reclusive as Nepal used to be, has through its isolation and an enlightened conservation policy preserved much of its vegetation. Over 60 per cent of the country, about the size of Switzerland, is still covered by temperate and subtropical forest, a rich reservoir of the Himalayan flora. The East India Company for a long time had wanted to investigate the commercial potential of Bhutan, so when George Bogle passed through the country in 1774 on his way to Tibet, Warren Hastings asked him to bring back whatever he judged of interest. In 1783 yet another diplomatic mission, this time conducted by Samuel Turner, headed for Lhasa. The surgeon accompanying the party, Robert Saunders, had a casual interest in natural history and wrote the first account of the geology and botany of Tibet and Bhutan[23]. His observations, however, are superficial: soft fruits such as raspberries, strawberries, and bilberries were prolific; 'scarce a plant to be seen that is not of European growth'; rhubarb grew on snow-covered hills; the principal food crop was the turnip—potatoes, cabbages, and lettuces left by Bogle had been neglected. Saunders was clearly not competent to be more discerning. The first botanist to report on the country's flora was William Griffith, a member of R. B. Pemberton's mission in 1837–8. During his four and a half months' stay in Bhutan, he collected some 1200 specimens including many new species which, unfortunately, reposed in obscurity until others had collected and published their findings. The famous Bhutan turnip which Warren Hastings had been anxious to introduce into cultivation in Britain, Griffith dismissed as no more than an outsize coarse radish[24].

Although Griffith found plants of horticultural interest, it was Roland Edgar Cooper (1890–1962) who supplied seed to English growers. Cooper's mission was to find plants suitable for British gardens for his sponsor, A.K. Bulley, the proprietor of Bees Nursery at Ness in Cheshire. During two tours in 1914 and 1915, he collected 19 species of primulas including his favourite, 'the gem of the whole of the Eastern Himalaya', *Primula eburnea* (Fig. 11.6). This triumph he owed to an assiduous Lepcha collector who had returned to the remote monastery where it had been seen by Cooper, and had garnered the precious seed just before the snows came. Unfortunately, many of his primulas have proved not to be hardy in the British Isles. Although an herbarium was incidental to his seed collecting, he nevertheless accumulated nearly 4000 dried specimens during these two trips. The ubiquitous Frank Kingdon Ward crossed into eastern Bhutan in 1925; F.M. Bailey, Political Officer in Sikkim, was in the country in 1927; another Sikkim Political Officer, Basil Gould, collected over 2000 specimens during official visits in 1938 and 1939; but the extent of the floral riches of Bhutan was only fully appreciated as a result of the seven visits by Ludlow and Sherriff from 1933 to 1950 (Fig. 11.7).

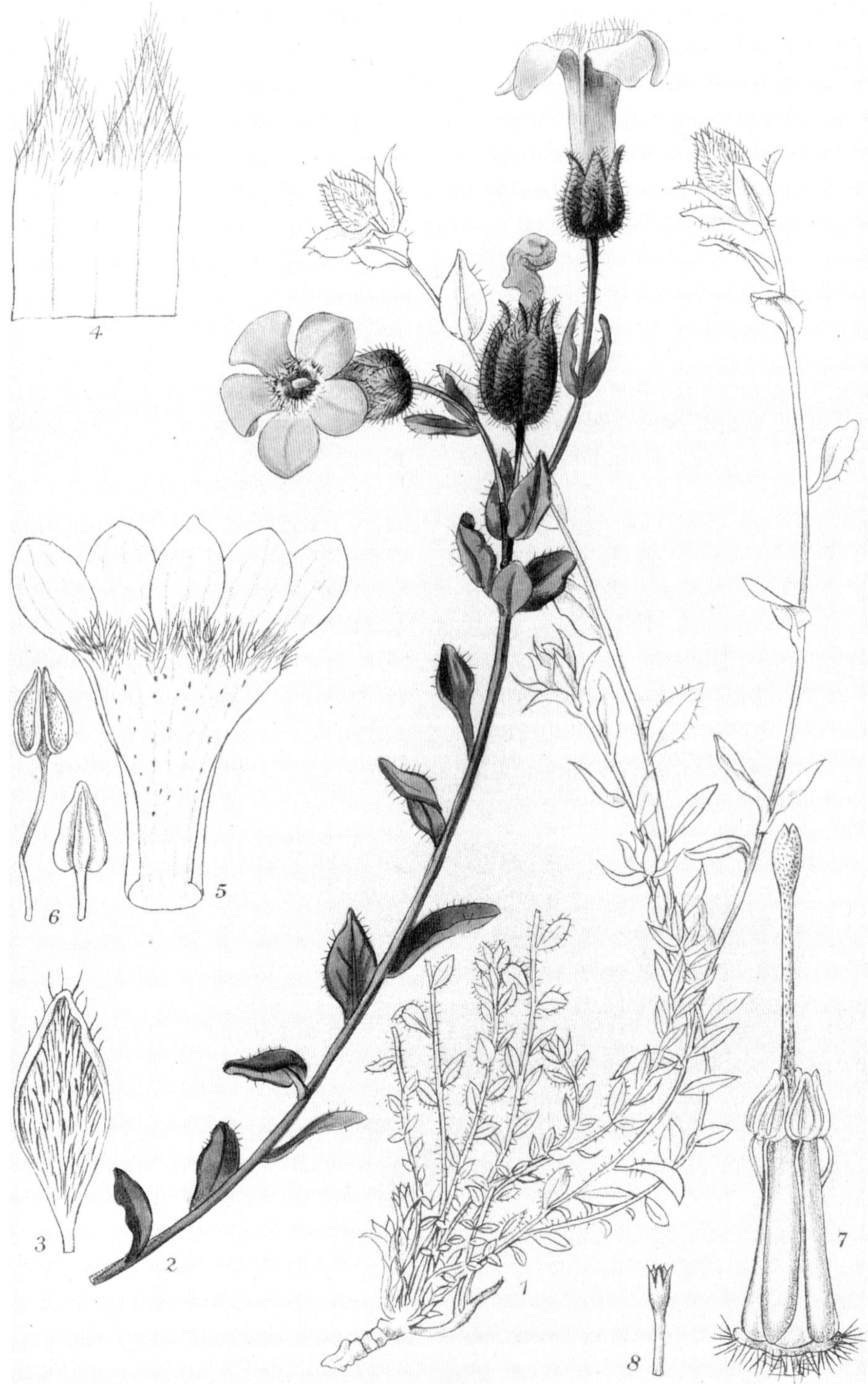

Figure 11.7 *Cyananthus sherriffii.* (From *Curtis's Botanical Magazine,* 1943, plate 9655.) Drawn and lithographed by Lilian Snelling. Collected in Bhutan by Ludlow and Sherriff in 1936.

Frank Ludlow (1885–1972) spent much of his working life in South Asia as Vice-Principal of Sind College in Karachi, then as an inspector in the Indian Education Service, and finally as headmaster at Gyangtse in Tibet. In 1929 he met the British Vice-Consul at Kashgar in Chinese Turkestan, George Sherriff (1898–1967). Both men were experienced travellers and with a common love of flowers and birds, the ensuing friendship developed into a collecting partnership. Each expedition was made memorable by some spectacular discovery. On their first tour in 1933 they found the diminutive *Primula kingii*, its bell- like flowers glowing ruby-red in the sun and *Meconopsis simplicifolia* nodding on long stems. They rediscovered *Meconopsis superba*, previously known only from a herbarium specimen collected in Tibet in 1884. The profusion of flowers at Me La, or the Pass of Flowers, on Bhutan's eastern border, enticed them back on two subsequent tours. There they made some of their best hauls: the blue *Meconopsis grandis*, potentillas in a spectrum of yellows, a cluster of saxifragas but, above all, primulas. Fields densely colonized by the yellow Himalayan cowslip, *Primula sikkimensis*, were an unforgettable sight. Ludlow judged their 1936 expedition as horticulturally the most rewarding—65 species of primulas of which 14 were new and a number of rhododendron novelties. On Black Mountain in Central Bhutan in 1937 they climbed through carpets of primulas and other alpines and were awed by a specimen of *Rhododendron grande*, at least 70 feet high. During their last expedition in 1949 they found a new lily, which W. T. Stearn named in 1950 *Lilium sherriffiae* after Sherriff's wife, Betty, who described her flower as 'maroon with inside of corolla chequered with gold'. There were, inevitably, periods of frustration and disappointment. One such occasion for Sherriff was Lubsing La at over 14 000 feet. 'This place seems completely dead. I don't know where the flowers are. We've gone high and we've gone low, and we can't find anything.'[25] Ludlow and Sherriff may have lacked the discipline of a botanical education but they had a discriminating eye for plants of horticultural merit, a talent for which gardeners everywhere have been infinitely grateful. The Royal Botanic Garden at Edinburgh, the recipient together with the British Museum (Natural History) of the bulk of their collections, is now engaged on a flora of Bhutan, primarily for the use of native botanists and foresters but also a timely reminder to the Bhutanese of their floral heritage[26].

Jonas Thomas Booth (*b.* 1829), or Thomas J. Booth as he preferred to be known, was thought by his contemporaries to have collected in Bhutan. Both Hooker's *Journal of Botany* and *Curtis's Botanical Magazine* confirmed this belief which presumably arose from the ambiguous appellation, 'Bhutan Himalaya'. Booth was the nephew of Thomas Nuttall, the celebrated botanical explorer in North America, who, excited by the flamboyant plates in the first number of Joseph Hooker's *Rhododendrons of Sikkim-Himalaya* (1849), resolved to acquire living plants for his Lancashire garden. After a brief initiation at Kew, his nephew, a complete innocent in botany and the tropics, sailed to India in 1849 to collect rhododendrons and other horticulturally interesting plants for his uncle. The news of the rough treatment meted out to Joseph Hooker by the Sikkimese authorities put paid to his plans for collecting in that country. On the advice of the Calcutta Botanic Garden he headed instead for the Himalayan range in northern Assam and the Balipara Frontier Tract on the eastern frontier of Bhutan. Despite his inexperience, appalling hardships in the mountains during the monsoons, and being 'bled to death by myriad of leaches', Booth served his uncle well. When he left India in 1851 he had despatched three substantial consignments of plants, seeds, and dried specimens. They included orchids and begonias but the choicest were his rhododendrons. Thomas Nuttall described 22 species discovered by his nephew[27] and the star was undoubtedly the snow-white *Rhododendron nuttallii* which Sir William Hooker ecstatically greeted as 'the finest of all the Rhododendrons . . . As *Victoria regia* [i.e. *V. amazonica*] is justly considered the Queen of Water-lilies, so the plant here represented may with equal justice be called the Prince of Rhododendrons'.[28] Unfortunately, many of his species proved too tender for outdoors in Britain. After a spell as a nurseryman at Rainhill near Prescott, Booth left England never to be heard of again.

The atrocious climate, the difficult terrain, and hostile tribes hindered the exploration of the mountain ranges passing through Assam and Burma. William Griffith, a member of the Assam Deputation in 1835–6 searching for the elusive tea plant, made a brief collecting foray in the Mishmi Hills in north- eastern Assam. Here where the rainfall is over 300 inches a year and slippery rock faces make climbing hazardous, Kingdon Ward concentrated most of his efforts between 1926 and 1956. He viewed the north-east frontier of India as 'the real treasure house of the world, the last stronghold of the collector in Asia'. Frank Kingdon Ward (1885–1958), explorer and geographer as well as plant hunter, was also at one time another of A. K. Bulley's collectors. He was the doyen of the plant collectors of his day—knowledgeable, self-reliant, and tenacious. Familiar flowers excited him as much as new discoveries. Alpines, especially primulas, and rhododendrons, perhaps his favourites, unfailingly stimulated this fluent writer to passages that rival the lushness of the prose of another plant collector, Reginald Farrer. *Rhododendron tanastylum* has flowers of 'daring purple, rasped with clusters of darker stars, giving it a sort of blackberry-and-apple colour, or in some lights, crushed strawberry'.[29] He enjoyed the contrived perfection of a garden but preferred to see flowers in their natural habitat.

> *Rhododendrons in the wild, lashed by the mountain gales, caressed by the sun or stung by the driven rain, in winter buried under 10 or 20 feet of snow, are unforgettable. Man can assemble in close embrace a far greater range of rhododendrons than nature ever knew; but never can he reproduce the drama of living plants on the high and lonely passes of Sino-Himalaya*[30].

Wedged between Nepal and Bhutan is the tiny state of Sikkim, at one time constantly in conflict with its aggressive neighbours. Since William Griffith's collectors were there in 1843 it has been a place of pilgrimage for scores of botanists and collectors, and preeminent among them is Sir Joseph Hooker, the most distinguished scientist to explore its mountains. His visit was of momentous significance, producing not only a superbly illustrated flower book and a classic work of travel but also serving as an apprenticeship for the incomplete *Flora Indica* and the still unsuperseded *Flora of British India*.

Joseph Dalton Hooker (1817–1911), the son of Sir William Jackson Hooker, Director of the Royal Botanic Gardens at Kew, began his career as Assistant Surgeon and Naturalist on HMS *Erebus* on an Antarctic voyage that lasted three years. The *Botany of the Antarctic voyage . . . 1839–43* (1844–60) revealed his talent as a taxonomist, but it was his first venture into print in 1837—descriptions of three new Himalayan mosses—that presaged his long attachment to India. In 1847 Hooker was ready for another overseas tour, preferably to the Andes or the Himalayas, in order to acquire first-hand knowledge of tropical vegetation. He preferred India where Hugh Falconer at the Calcutta Botanic Garden had promised assistance, and where his friend, Thomas Thomson, was about to join a boundary commission in

Kashmir. His father obtained permission to employ him as a plant collector in India for two years on the grounds that the Indian flora was poorly represented at Kew. Originally, the Admiralty had intended him to proceed afterwards to Borneo to investigate agricultural prospects on the island of Labuan, but when the Borneo expedition was cancelled, Hooker was allowed a third year in India. His ship docked in Calcutta in January 1848, too early to depart for little-known Sikkim, recommended by Falconer. He acclimatized himself during the next three months with local tours in Bengal before proceeding to Darjeeling which became the base of his operations for the next two years. This hill station had been ceded in 1835 by the Rajah of Sikkim to the British who had perceived its potential development as a health resort. There Hooker was the guest of the reclusive orientalist and naturalist and the former Resident at Khatmandu, Brian Hodgson. Hodgson's connections with the Nepalese government enabled Hooker to spend three months in East Nepal. In the neighbourhood of Darjeeling Hooker, for the very first time, pushed and scrambled and climbed through the lush, dark vegetation of a Himalayan forest.

It is difficult to conceive a grander mass of vegetation—the straight shafts of the timber trees shooting aloft, some naked and clean, with grey, pale, or brown bark; others literally clothed for yards with a continuous garment of epiphytes, one mass of blossom, especially the white Orchids Coelogynes, which bloom in a profuse manner, whitening their trunks like snow. More bulky trunks bore masses of interlacing climbers, Araliaceae, Leguminosae, Vines, and Menispermeae, Hydrangea, and Peppers, enclosing a hollow, once filled by the now strangled supporting tree, which had long ago decayed away. From the sides and summit of these, supple branches hung forth, either leafy or naked; the latter resembling cables flung from one tree to another, swinging in the breeze, their rocking motion increased by the weight of great bunches of ferns or Orchids, which were perched aloft in the loops. Perpetual moisture nourishes this dripping forest, and pendulous mosses and lichens are met with in profusion[31].

Archibald Campbell, Superintendent of Darjeeling and the Governor-General's representative in matters concerning Sikkim, recruited a team of local Lepchas to collect plants for him (Fig. 11.8). Sometimes Hooker had as many as 20 of them out in the terai and the valleys and hills while he sheltered from the rains in Hodgson's bungalow, drawing, mounting, and labelling their finds.

On his first foray into Sikkim Hooker discovered three species of rhododendrons, a modest beginning to a collection that was to establish him as one of the greatest of all rhododendron hunters. Apart from a large escort of porters to carry his baggage, he usually collected and surveyed on his own but late in 1849 Campbell decided to join him. The two men indiscreetly crossed the Sikkim frontier into Tibet only to be promptly turned back by Chinese troops. The ineffectual Rajah who feared the incident might exacerbate already tense relations with China did not prevent his Anglophobic Chief Minister seizing Campbell who was beaten and imprisoned. Hooker, having no political value, was free to continue his tour but refused to desert his companion. 'I kept as near as I was allowed, quietly gathering rhododendron-seeds by the way.' Punitive action by British troops secured their release by the end of the year and part of Sikkim was annexed as retribution for the offence.

Arriving back in Darjeeling early in 1850, Hooker met up with Thomas Thomson whom he had known from his university days at Glasgow. As neither Bhutan nor

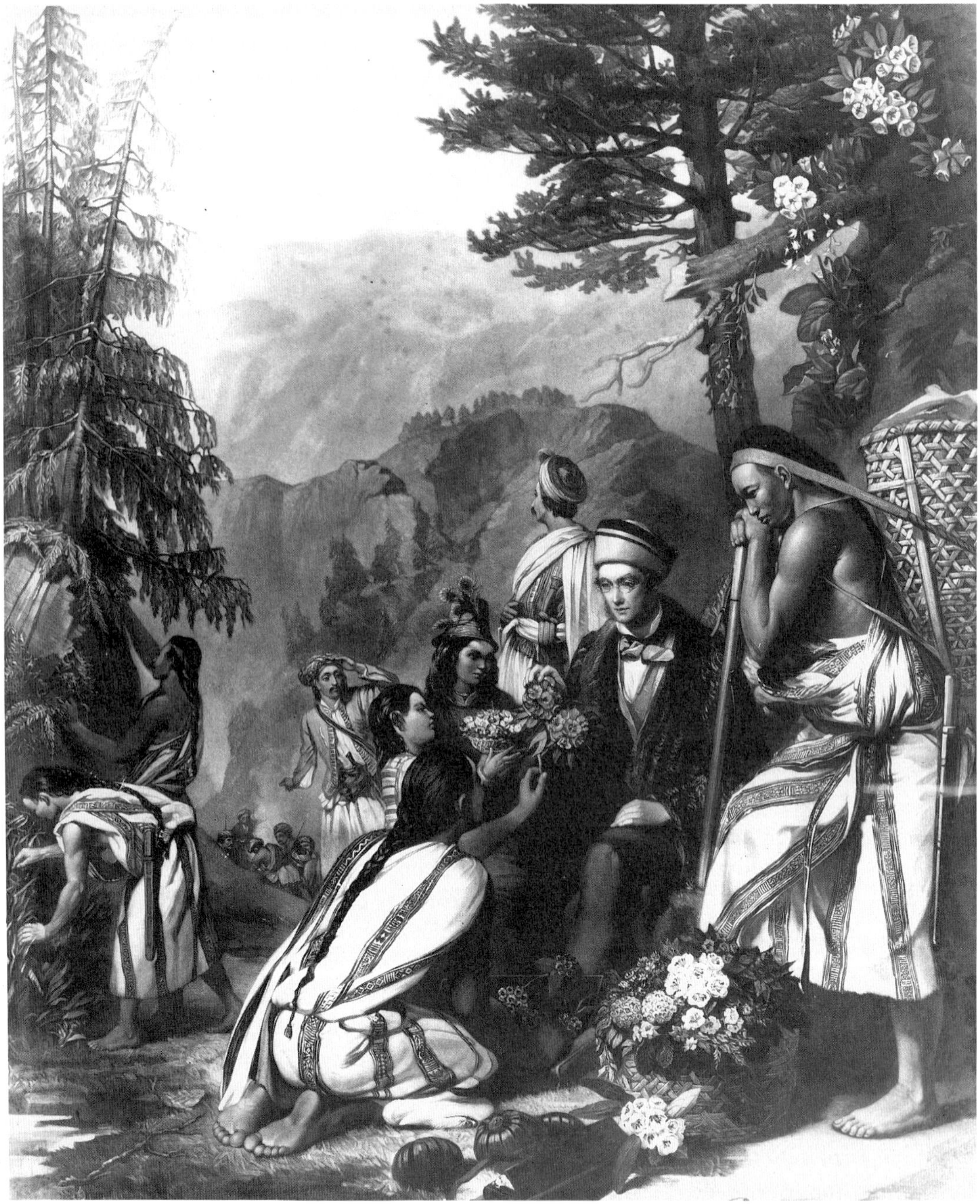

Figure 11.8 Joseph Hooker with his Lepchas and Gurkha guard inspecting the plants collected at the end of a day's march. This whimsical scene by Frank Stone was inspired by a sketch by William Tayler of the Bengal Civil Service. Mezzotint. In a letter to his future wife, Hooker describes posing for Tayler: 'My dress was . . . Thibetan in the main with just so much of English peeping out as should proclaim me no Bhotea . . . So I have a large loose worsted Bhotea cloak . . . Enough of it is thrown back to show English pantaloons & my lower extremities cased in Bhotea boots—my shirt collar is romantically loose and open, with a blue neckerchief, which, & my projecting shirt-wrists, shew the English man'.

Nepal was accessible, they spent the remainder of the year in the Khasia Hills in eastern Bengal. Hooker found he preferred the humid forests of Bengal to the subtropical vegetation of Sikkim on discovering they had a more diverse flora—the richest in India, he asserted, if not in all Asia. He and Thomson collected over 3000 flowering plants and several hundred ferns, mainly in Khasia, and sampled the vegetation of Sylhet, Cachar, and Chittagong. Sir William at Kew, well content with his son's efforts, featured the more spectacular horticultural discoveries in *Curtis's Botanical Magazine* which he edited: *Primula sikkimensis* and *Meconopsis napaulensis*, for instance, and a generous selection of rhododendrons, many of which were successfully raised at Kew. Joseph Hooker always had a pencil tied to his jacket and a notebook ready in his pocket and a thousand or more sketches of scenery, people, and botanical subjects testified to his industry. The first European to survey a number of the passes into Tibet, his accurate mapping was utilized by the Indian Trigonometrical Survey (Fig. 11.9).

Hooker's collections in the east and Thomson's in the north-west together amounted to some 150 000 specimens. Back in England they dedicated themselves to the formidable task of identifying and labelling them, working in a shed at the rear of the Orangery in Kew Gardens. The disposal of their duplicates was of a magnitude comparable to Wallich's operation in 1828 to 1832, but on this occasion the botanic gardens at Kew and Calcutta were not omitted from the list of recipients. When Robert Wight returned from India in 1853, his duplicates, too, were distributed from Kew.

For years botanists had been aware that herbaria and small collections in universities,

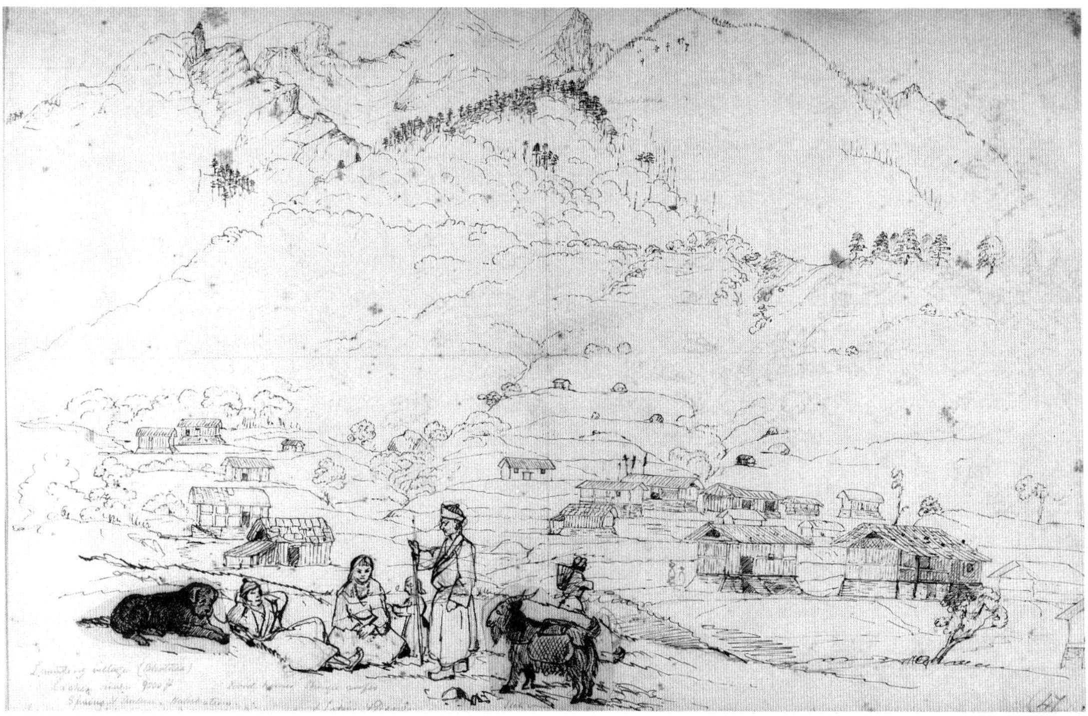

Figure 11.9 Sketch by Joseph Hooker of Lamteng village, Sikkim, 1849. (Kew Archives.) 'My drawings will be considered tame compared with most mountain landscapes, though the subjects comprise some of the grandest scenes in nature' (Hooker (1854), Himalayan journals, Vol. 1, p. xvii).

private ownership and, more especially, in the India Museum in London provided a reasonable selection of species for the preparation of an Indian flora. It was a proposition that Hooker and Thomson had seriously discussed as they botanized in the Khasia Hills. Thomson was convinced that 'no flora in the world is more interesting than that of India'[32], and in Hooker he had a collaborator who could match his enthusiasm, knowledge, and industry. When their intentions became known, offers of help came from all quarters; from Wallich in retirement, from botanists in India who sent collections, and from the Jardin des Plantes in Paris who gave them a set of Jacquemont's specimens. Everything went to Kew where the collaborators could consult Sir William Hooker's excellent herbarium and library. Citing the precedent of Hugh Falconer whose pay and allowances were not suspended while he was in England cataloguing the Siwalik Hills fossils, Thomson sought, in 1851, from the Court of Directors, a similar dispensation for himself. He tactfully reminded the Company that the numerous individual collections stored in the India Museum were of little use to science until they had been sorted and identified, a task he volunteered to undertake but for which he required some financial assistance[33]. He did not ask for any subvention for eventual publication. In this application he was supported by a resolution passed at the annual meeting of the British Association for the Advancement of Science in 1851. The Company, however, weakened by successive financial crises and emasculated by Government control, declined, but, not wishing to appear completely unsympathetic, offered to consider some financial recompense on the completion of the work. Hooker was more fortunate; the Admiralty agreed to pay him an annual salary of £400 for three years while he was engaged on the project.

The first volume of *Flora Indica*, the cost of which was covered by a small legacy to Thomson, appeared in 1855. Almost half the book was taken up by an Introductory Essay of 280 pages, also available separately. The Essay was a perceptive resumé of the current state of Indian botany: it gave a cautious rather than a radical appraisal of the species concept; it considered the relationships between the Indian vegetation and that of China and Japan, Malaya, Siberia, Europe, and Africa; it proposed a division of India into phytogeographical regions; it stressed the importance of climate in the geographical distribution of plants; it provided an excellent historical summary. When the *Flora Indica* was published, Thomson was back in India, this time as Superintendent of the Botanic Garden in Calcutta and Hooker was Assistant Director at Kew. Their new duties hindered the continuation of the work but a series of *Praecursores* reviewing families of Indian plants was published between 1858 and 1861. The *Flora Indica* never proceeded beyond the first volume but what was urgently needed at that time was a general flora, selective in details, rather than the exhaustive description of species embarked upon by Hooker and Thomson. Had they continued with the same density of information, it would have resulted in a compilation of some 12 000 pages and an indefinite number of years in gestation. Since the *Flora Indica* was never completed, the Court of Directors felt under no obligation to recompense Thomson but they did purchase a hundred copies, 60 of which went to India.

While Joseph Hooker was in India he entertained his father with long informative letters, a selection of which duly appeared in *Hooker's Journal of Botany and Kew Garden Miscellany*, providing much of the material for his *Himalayan journals*, published in 1854. He dedicated it to his friend Charles Darwin who constantly consulted Hooker on botanical matters in the course of his researches. Hooker's interest in native culture and his expertise in geology and meteorology

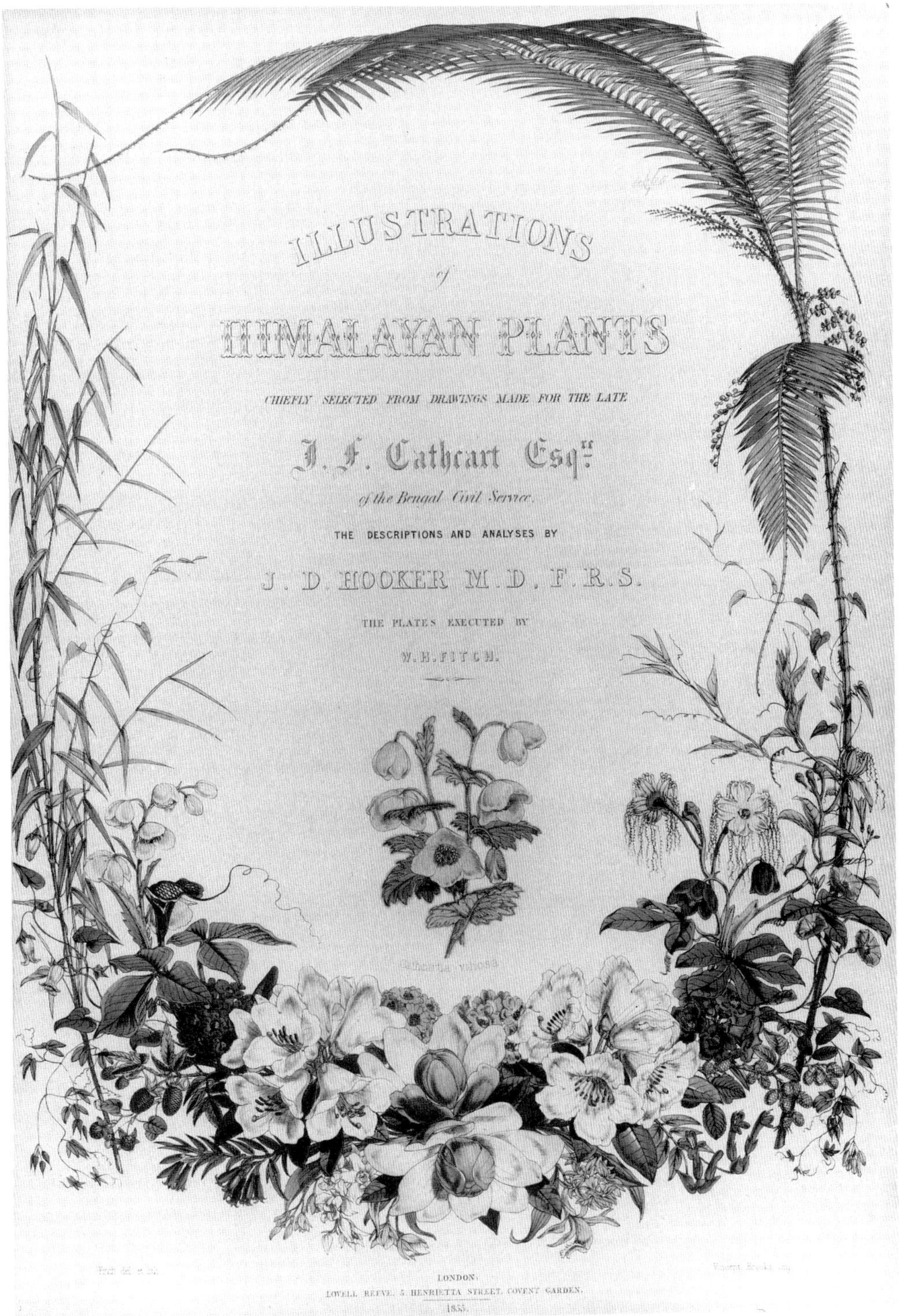

Figure 11.10 (From Joseph Hooker's *Illustrations of Himalayan plants*, London, 1855.) Title-page decorated by Kew's artist, W. H. Fitch, who redrew for this book some of the flower drawings by J. F. Cathcart's Indian artists.

as well as botany both instruct and beguile the reader. His heightened awareness of exotic surroundings evokes striking phrases and passages: Lepcha porters running in the rain 'look like snails with their shells on their backs'; butterflies poised with erect wings on the sand remind him of 'a crowded fleet of yachts on a calm day'; the bark of *Rhododendron hodgsonii* is 'as delicate as tissue-paper'. Spectacular sunsets staining sky and clouds with bands of glowing colour recall water-colours of Turner. The book was an instant success and a worthy addition to the classics of travel literature.

The third book to come out of his Indian tour was *Rhododendrons of Sikkim-Himalaya*. The first ten lithographed plates which were issued in 1849 earned the admiration of *The Athenaeum*. 'That he [Hooker] should have ascended the Himalaya, discovered a number of plants, and that they should be published in an almost unequalled style of magnificent illustration, in less than eighteen months—is one of the marvels of our time.' The drawings, based on Hooker's dried specimens and field sketches, were by Kew's official botanical artist, Walter Hood Fitch. Fitch's vigorous, confident lines and bold colouring miraculously recreated exotic blooms—29 had been delineated by the time the last plate appeared in 1851—which excited the botanist and were to provide the gardener with some of our best garden species.

Fitch's services were again enlisted to reinterpret a selection of drawings by Indian artists for *Illustrations of Himalayan plants* (1855) (Fig. 11.10). They had been commissioned by a Bengal civil servant, John Ferguson Cathcart (1802–51), convalescing at Darjeeling when Hooker made his acquaintance. An all-round amateur naturalist, he employed Lepchas to collect specimens and native artists to draw them. At one time he had a team of five artists who between them could produce three portraits a week. He had intended publishing an illustrated flower book similar in presentation to *Rhododendrons of Sikkim-Himalaya*. When he died on his return to England, his sister donated nearly a thousand drawings to Kew from which Hooker made a small selection for publication. Some were fully finished drawings, many were pencilled outlines with just an inflorescence and leaf in colour. Hooker found fault with their perspective and diagnoses and had Fitch redraw them to correct 'the stiffness and want of botanical knowledge displayed by the native artists who executed most of the originals' (Introduction). Hooker always had reservations about the botanical accuracy of Indian flower drawings. Similar objections are repeated in his preface to *A century of Indian orchids selected from the drawings of plants, in the herbarium of the Royal Botanic Garden Calcutta* (1895).

> *I should call attention to the fact that, excellent as the drawings are in many respects, as representatives of the plants portrayed, they err in manifesting that tendency to enlarge, which is a besetting sin of Indian botanical artists; and the analyses leave much to be desired in the matter of proportion and accuracy . . . In no case do these defects appear to me to detract materially from the value of the illustrations as a means of identifying the plants represented.*

This is perhaps an appropriate moment to consider the status of native artists who worked for Europeans, to examine cursorily their traditional background and the cultural heritage which they had to abandon or to modify in order to meet the exacting scientific requirements of their employers. When Humayun (1508–56), the son of the first Mughal emperor, Babur, returned from exile in India, he brought with him two painters from the Safavi court, the founders of a new school of Mughal art. They introduced the Persian miniature with its themes of romance and epic stories set in a crowded composition of figures and animals in a landscape of contorted rocks, pendulous trees, and indeterminate flowers. Opaque watercolours were applied to paper with brushes fashioned from the fine hairs of cats and squirrels or the fibres of particular trees. Colours were painstakingly built up in layers, bonded together by burnishing the reverse side of a painting with a smooth crystal or polished agate. During the reign of Akbar (1542–1605), Humayun's son, Mughal painting, now imbued with a robust realism, moved gradually away from its Persian origins. Akbar ordered the memoirs of his grandfather, Babur, to be written and illustrated, and of the two versions and fragments which survive, the one in the British Library is probably the finest. Eleven of its 96 miniatures by some 40 artists depict individual plants drawn with a factual honesty that makes identification relatively easy. A celebrated early seventeenth-century painting of squirrels in a Chenar tree (*Platanus orientalis*), its leaves autumnal red and gold, invites comparison with the best of contemporary European botanical art. It was executed during the reign of Jahangir (1569–1627) who continued his father's royal patronage of the arts. An amateur naturalist who appreciated the importance of dispassionate and precise observation, Jahangir encouraged his artists to paint with a critical eye, employing, whenever appropriate, the techniques of western art.

Not infrequently, Indians had the opportunity to examine and copy European paintings—the huge devotional canvases in the Catholic churches of Portuguese Goa, portrait miniatures or illustrated books and individual engravings. Mughal artists were intrigued by western rendering of anatomy and drapery and the treatment of perspective and nowhere is their indebtedness more apparent than in their animal and flower studies. They copied engravings in European herbals and florilegia with the same verve and dexterity that also facilitated their imitation of western portraiture and the introduction of western landscapes into their miniatures. Plants whose linear qualities are so easily adapted and formalized are an ubiquitous motif in Islamic art. In early Mughal miniatures trees and flowers gave cohesion and colour to a composition. They eventually migrated to enclosing borders, spaced out as solitary blossoms of favourite garden flowers—roses, tulips, narcissi, poppies, and irises—never the central theme or focal point of a painting.

Experts on oriental art now suspect that the individual flowers studies that suddenly emerged during the reign of Jahangir may have been inspired by European flower books; their pose was similar, even to the inclusion of a hovering insect and, more conclusively, some of the flowers portrayed were neither native to India nor to Western Asia. This supposition has now been positively confirmed by the discovery that a seventeenth century Mughal miniature of a Martagon lily (Figs 11.11, 11.12) in the Victoria and Albert Museum was copied from Pierre Vallet's *Le jardin du Roy tres Chrestien Henry IV* (1608)[34]. Ustad Mansur unquestionably deserved the title 'Wonder of the Age', bestowed on him by Jahangir. He painted animals and birds with clinical clarity against a minimal background, never allowing himself to be seduced by the decorative appeal of line or colour. His animal studies are zoological statements that had to satisfy the demanding standards of Jahangir the naturalist. He is reported by the emperor as having painted a hundred flowers during a royal visit to Kashmir but, unfortunately, only three have survived—an iris, narcissus, and tulip, all delineated with botanical precision. This scientific discipline was relaxed during the reign of Shah Jahan (1592–1666) when floral borders to miniatures were fused in undulating arabesques, and flowers formed stylized patterns on carpets, textiles, and the marble cladding of buildings. Conformity to objective truth was no longer a requirement. In the album[35] prepared for Darah Shikoh, eldest son of Shah Jahan,

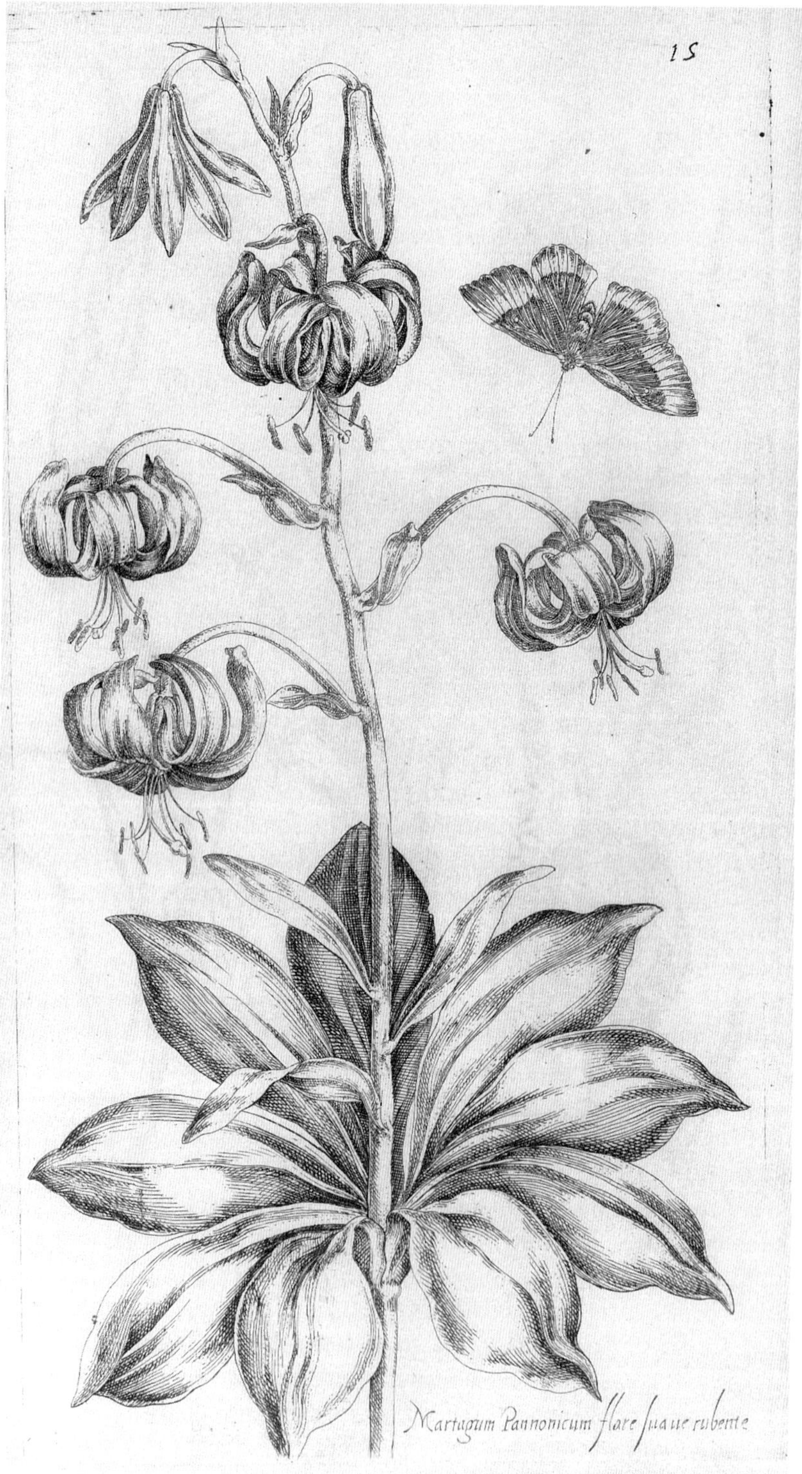

Figure 11.11 Martagon Lily, *Lilium martagon*. (From P. Vallet (1623), *Le jardin du Roy tres Chrestien Loys XIII Roy de France*, Paris, plate 15.) Engraving. (A later version of 1608 edition).

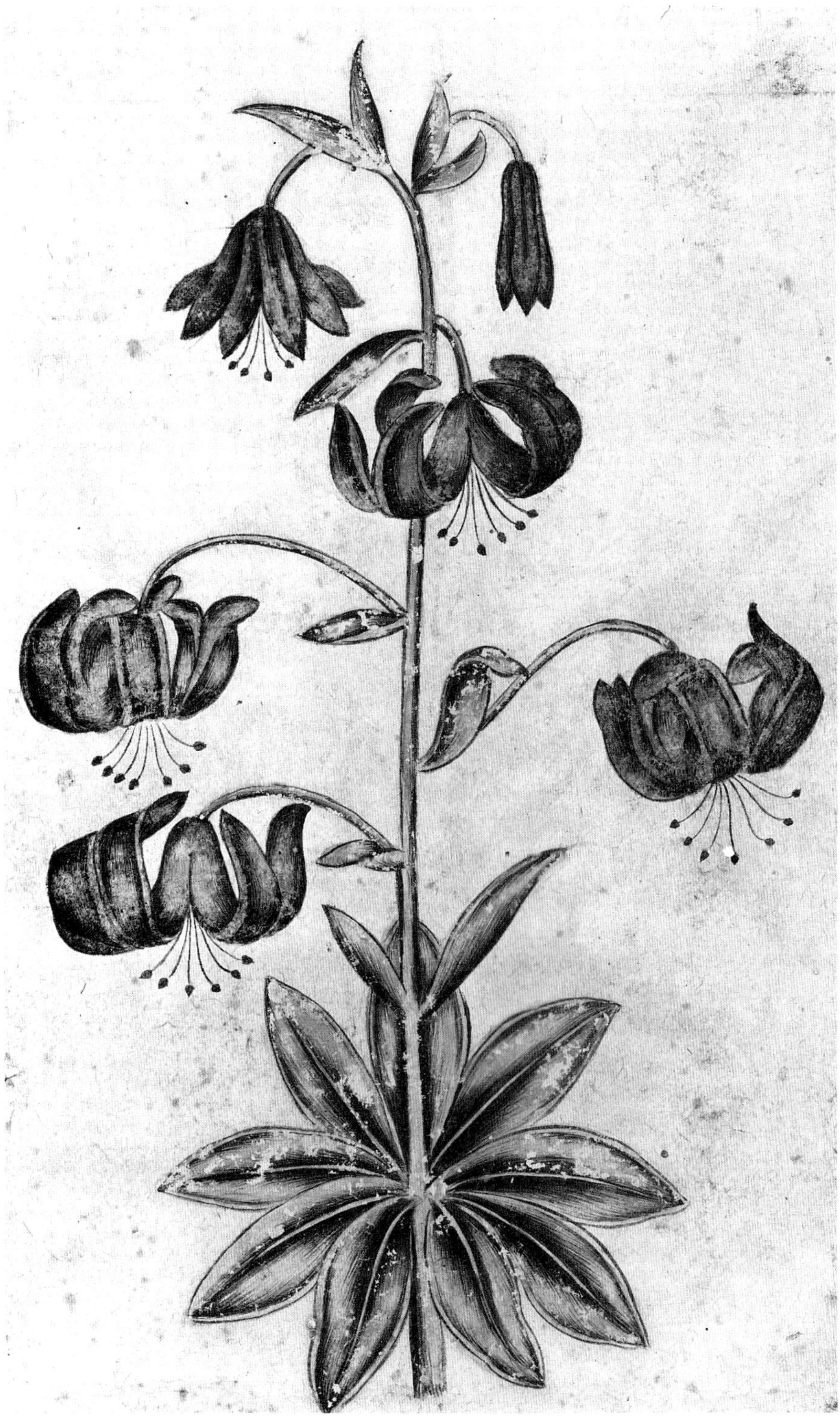

Figure 11.12 Mughal copy of the Martagon Lily in P. Vallet's *Le jardin du Roy* . . . , 1623. (Victoria and Albert Museum.)

some miniatures depict plants that combine features from monocotyledons and dicotyledons and colours chosen solely for decorative effect, demonstrating a flippant disregard for fidelity to nature.

With the disintegration of the Mughal empire in the eighteenth century, those Indian artists who sought European patronage readily adapted their traditional techniques and style to suit the needs of their new clients. Their efforts, however, did not always meet with unqualified approval; some were considered weak in perspective and inept in shading. Thomas Twining thought that 'the merit of their drawing is almost confined to a very accurate imitation of flowers and birds. I never saw a tolerable landscape or portrait of their execution.'[36] Michael Symes during a visit to Burma in 1795 lamented that his Bengali artist 'though skilful in copying figures and making botanical drawings, was unacquainted with landscape painting and perspective'.[37] Despite such reservations, Indian artists found employment as draughtsmen to engineers and surveyors in the East India Company and received commissions from individuals who wanted a pictorial souvenir of their stay in India. But Robert Kyd, able to find only one suitably talented native artist for the newly-founded Botanic Garden in Calcutta, sought permission to recruit two from China[38]. There was a substantial export of Chinese watercolours of flowers and birds to the West, but Cantonese artists could not be tempted to India by a generous salary and expenses and, in the event of the climate proving 'prejudicial', the guarantee of an immediate return to China.

There were European patrons who were more fortunate in their artists or less exacting in their requirements. Lady Mary Impey, wife of the Chief Justice of Bengal, was well content with the drawings, mainly of birds, by the three artists she engaged while residing in Calcutta from 1777 to 1783. Inevitably, artists manifested variable standards of proficiency since they were expected to paint in a naturalistic idiom completely alien to their tradition of fastidious miniatures, to renounce opaque gouache for transparent watercolour, to modify their palette by using more subdued colours, and to draw in sepia ink and pencil. Francis Buchanan was exasperated by the sluggish pace of the artist who drew fish for him. Benjamin Heyne achieved a reasonable output only by getting his artist to draw plants in outline, adding colour to just a single flower, fruit or leaf. Cathcart used the same procedure for many of his plant portraits. Botanists like Joseph Hooker insisted not only on the inclusion of minute morphological characters but also scientifically correct floral dissections. Indian artists found difficulty in rendering the fragility of petals and the texture of leaves. A drawing of '*Musa nepalensis*', probably *Ensete glaucum*, incurred the displeasure of one botanist who indignantly scribbled on it: 'Most abominable leaves for which master painter shall be duly cut with reference to his month's wages.' Colonel Hardwicke found fault with a drawing of *Rhododendron arboreum* by one of Roxburgh's artists, declaring it to be nothing like the drawing he had of the same species. Botanists were not always so uncharitable. Buchanan praised the engravings in Roxburgh's *Plants of the coast of Coromandel*: 'The workmanship does great credit to the natives and the whole is highly scientific.'[39] Amateur naturalists on the whole were more tolerant. William Tennant, a Company chaplain, admired 'the laborious exactness with which they imitate every feather of a bird, or the smallest fibre on the leaf of a plant'. Lord Valentia perceived 'great merit' in Heyne's botanical drawings and when Maria Graham examined Roxburgh's portfolios, she was certain that 'they are the most beautiful and correct delineations of flowers I ever saw. Indeed, the Hindoos excel in all minute works of this kind.'[40] It was precisely this facility for applying microscopical detail, an accomplishment that came easily to artists familiar with miniatures, that provoked Marianne North's envy. The paintings of butterflies and moths in the botanical garden at Peradeniya were, she thought, executed in 'watercolours so exquisitely that one could see almost every hair in the insect's wing'.[41] It is this marriage of tradition and western scientific discipline, an instinctive urge to create patterns in the context of static repose that endow these drawings of plants with uniquely Indian sensibility.

Indian artists, seeking employment, were drawn to the principal British settlements like Madras and Calcutta. In 1800 Henry Colebrooke was unable to persuade any competent artist to leave the northern towns to work for him in the Deccan. Lady Impey's three artists were all inhabitants of Patna—Bhawani Das and Ram Das, both Hindus, and Shaikh Zain al-Din, a Muslim. Shaikh Zain al Din was also employed by Sir William Jones. It was not unusual for an artist to have more than one client; Manu Lal worked for both Wellesley and Richard Parry and the experienced painters at the Calcutta Botanic Garden never lacked private commissions. The best known of the Calcutta team was Vishnupersaud (or Vishnu Prasad), attached to the Botanic Garden in 1815 after the departure of Buchanan who had employed him on his statistical surveys. His published output was considerable: his drawings appeared in Wallich's *Tentamen florae Napalensis illustratae* (1824–6) and *Plantae Asiaticae rariores* (1829–32) and Royle's *Illustrations of the botany . . . of the Himalaya mountains* (1833–40). Wallich's two books also include examples of the work of another of his official artists, Gorachand.

The only other contemporary artist who could match Vishnupersaud in productivity and competence was Govindoo. The botanist, Robert Wight, felt so indebted to him that he named a genus of orchid after him (*Govindovia nervosa* now *Tropidia angulosa*).

> *I have dedicated it to the artist whose facile pencil produced the drawings for the greater part of the plates of the last three volumes of this work* [Icones plantarum Indiae Orientalis] *and whose skill in analytical delineation is, I believe, as yet quite unrivalled among his countrymen, and, but for his imperfect knowledge of perspective, rarely excelled by European artists*[42].

Wight's disciple in the South, Richard H. Beddome, had no hesitation in retaining Govindoo to prepare clean outline drawings with dissections for his *Ferns of southern India* (1863–4), *Icones plantarum Indiae Orientalis* (1869–74), and *Flora sylvatica for southern India* (1869–74).

How often, one wonders, was the high regard in which some of these artists were held reflected in their salaries? When Roxburgh referred a bearer with promising artistic talent to Wallich at Serampore in 1808, he suggested that he should 'pay him according to his ability'.[43] In the 1790s Roxburgh was paying about three rupees a drawing[44] which was approximately what Lady Anne Monson paid each of her two painters in 1775: 'A painter will earn about 35 rupees a month and he will draw about 12 plates taking one with another'.[45] In 1787 Colonel Kyd calculated that an artist could earn between 50 or 60 rupees a month but gave no indication of optimum output[46]. In 1804 Roxburgh allocated a monthly disbursement of 100 rupees to pay the wages and all other expenses incurred by the native painters at the Botanic Garden. When the formation of a botanical garden at Saharanpur was under consideration in 1817 the *ad hoc* local committee rejected the recruitment of two painters 'as the wages of such men are high'[47]. It was such reluctance, typical of officialdom, that goaded Wallich in 1827 to appeal against the 100 rupees monthly allowance for painters that had prevailed at the Calcutta Botanic Garden since its inception[48]. He

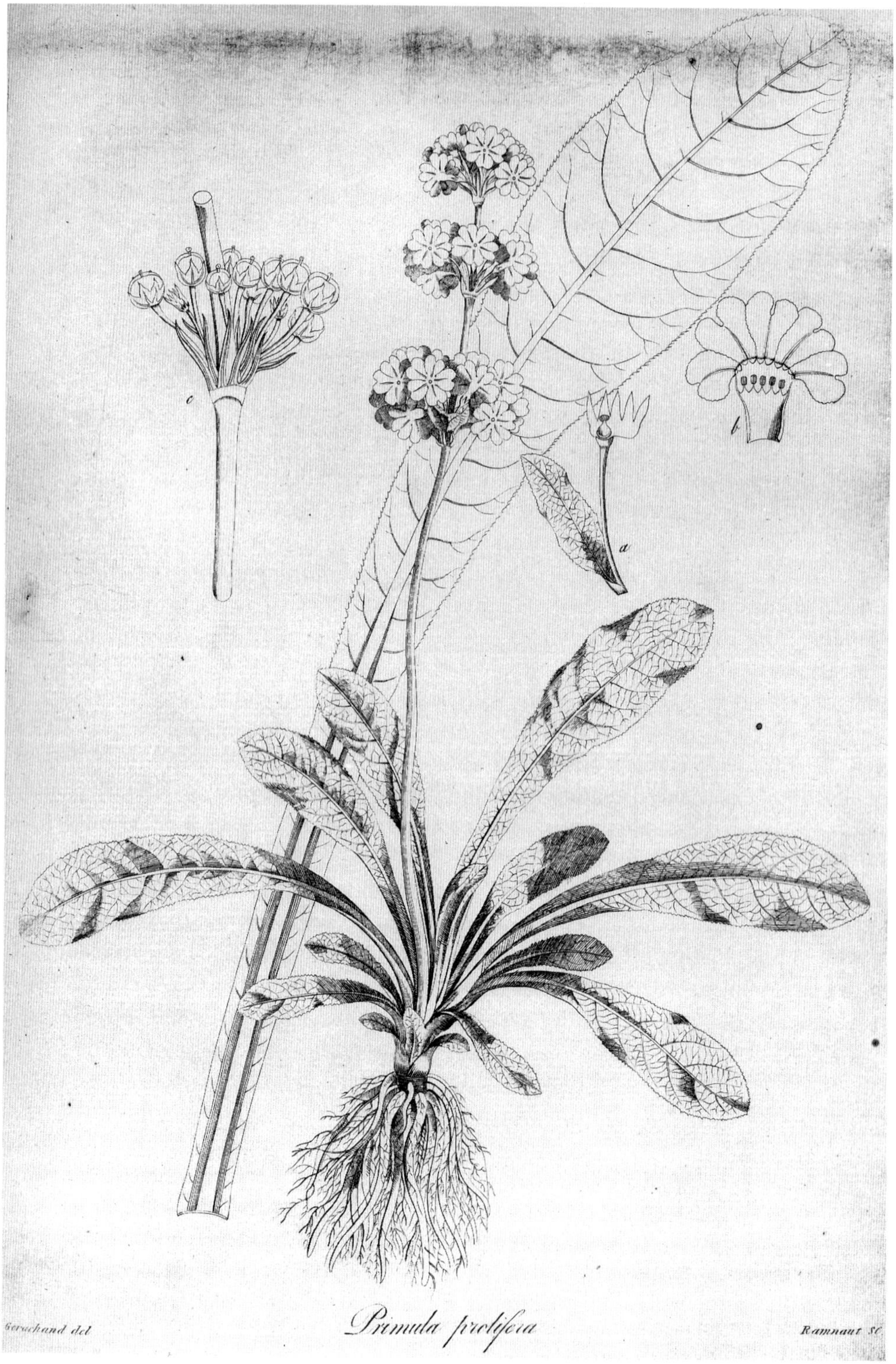

Figure 11.13 *Primula prolifera*. (From *Asiatick Researches*, Vol. 13, 1820, p. 372.) Drawn by Gorachand and engraved by Ramnaut under Nathaniel Wallich's supervision.

tried to circumvent this limitation by diverting 45 rupees a month from a contingency fund to recruit extra artists. In another submission he pointed out to the Government of Bengal that his head painter, Vishnupersaud, who had earned 50 rupees when he worked for Buchanan on his statistical surveys now received only 35 rupees 'which . . . is entirely inadequate as a reward for his skill which I would venture to say is not equalled among the natives of this country and rarely exceeded by any botanical draftsmen in Europe'. As a result of his petition Vishnupersaud's salary was increased to 60 rupees a month, as recommended, and his three assistants also received modest increases[49]. In 1840 the Court of Directors agreed to appoint a painter at Calcutta at 75 rupees a month[50] and in 1845 to increase an apprentice painter's monthly wage from 10 to 20 rupees[51]. The equity of these rates of pay can only be judged by some scale of comparison. In 1845 a plant collector at Calcutta earned 25 rupees a month, a little more than the apprentice painter got. In 1807 the Governor-General recommended the following complement of staff for Buchanan's impending statistical survey in Bengal.

One native officer	100 rupees a month
One learned Brahmin	100 rupees a month
Two Writers [clerks]	50 rupees a month each
Two Painters	50 rupees a month each
Persons employed in collecting natural curiosities	40 rupees a month[52]

So a painter was considered to merit a clerk's pay, an improvement on the scale Heyne had proposed for the staff to support him as Naturalist in the Madras Presidency in 1802.

One writer	12 rupees
One European draftsman	12 rupees
One native draftsman	6 rupees
Two plant collectors	6 rupees[53]

Heyne evidently believed that a native artist was worth half of his European counterpart. It must, however, be remembered that an Indian in his own country could live more cheaply than a European.

Not only were Indians taught to draw in the European style but they were also initiated in the technicalities of western graphic processes. When Wallich succeeded Roxburgh at the Botanic Garden, he set about instructing Indians in the craft of copper engraving. 'The workmanship does great credit to the natives', commented Buchanan approvingly[54]. Raffles thought that the specimens of engravings Wallich sent him 'more than equal [his] expectation'[55]. The first published results of their apprenticeship appear to be the plates in *Asiatick Researches* for 1820 where drawings by Vishnupersaud and Gorachand (Fig. 11.13) are competently engraved by Kasinaut and Ramnaut[56]. Three years later the first lithographic press was imported into India[57]; the Government Lithographic Press, shortly to be followed by the Asiatic Lithographic Press, was formed and native printers were taught the new process. Thereupon Wallich abandoned engraving and his *Tentamen florae Napalensis illustratae* became the first botanical work in India to have lithographically reproduced plates. The first part in 1824 was printed by the Government Lithographic Press at Calcutta, the second part in 1826 by the Asiatic Lithographic Press. Simple washes of colour were applied to some of the 50 plates to produce a work which was commendable considering the lack of local expertise. Nevertheless, Wallich was very conscious of its inferior presentation compared with sophisticated European lithographs.

The state of infancy in which the graphic art still continues in India has invariably made me feel sorrow and shame on comparing the plates attached to my humble efforts with the smallest and most pretensionless production of a similar description, which has issued from the press at home[58].

One of his reasons for seeking official leave in England in 1828 was to have the drawings for his projected *Plantae Asiaticae rariores* executed by skilled lithographers. By the end of the nineteenth century Indian printers had acquired a comparable skill and the plates in the *Annals of the Royal Botanic Garden, Calcutta* were lithographed by the Government School of Art in Calcutta. The plates in *The Orchids of the Sikkim-Himalaya* which appeared in volume 8 (1898) of the *Annals* were coloured by 'the sons of Nepalese coolies, who had never, until [the author] Mr Pantling took them in hand, been accustomed to use any implement more delicate than a hoe. Mr Pantling's perseverance and skill in drilling these boys into accurate colourists was considered marvellous by all who had seen them at work.'[59] This is yet another instance of that remarkable adaptability that has always distinguished Indian artists and craftsmen.

NOTES

[1] Moorcroft to Wallich, 8 July 1823. *Transactions of Agricultural and Horticultural Society of India,* **1**, 1838, 62–73.
[2] A. and P. Gerard, 'Account of a journey through the Himalayan Mountains,' in *Edinburgh Philosophical Journal,* **10**, 1824, 295–305.
[3] G. T. Vigne (1842), *Travels in Kashmir, Ladak, Iskardo, the countries adjoining the mountain-course of the Indus, and the Himalaya, north of the Panjab,* London, 460–2: J. Forbes Royle 'Observations on the vegetation and products of Afghanistan, Kashmir and Tibet'.
[4] Huegel (1845).
[5] Jacquemont (1835), Vol. 1, 147.
[6] *Calcutta Review,* **56**, 1935, 56.
[7] Jacquemont (1835), Vol. 1, 328.
[8] F. Bernier (1934), 406.
[9] Hoffmeister's collections were published by J. F. Klotzsch but many of his names were subsequently invalidated. J. F. Klotzsch and A. Garcke (1862), *Die botanische Ergebnisse der Reise. . . des Prinzen Waldemar von Preussen, 1845–46, durch W. Hoffmeister,* Berlin.
[10] Their collections are now in the B.M. (N.H.) and Kew Gardens. J. F. Duthie (1906), *Catalogue of the plants of Kumaon and of the adjacent portions of Garwhal and Tibet, based on the collections made by Strachey and Winterbottom during the years 1846 to 1849, and on the catalogue originally prepared in 1852 by Lt. Gen. Sir Richard Strachey*, London.
[11] T. Thomson (1852), *Western Himalaya and Tibet; a narrative of a journey through the mountains of Northern India during the years 1847–8*, London.
[12] E.I.C. to Government of Bengal, 3 July 1795. I.O.L.R. E/4/643, folios 310–11.
[13] *Annals of Royal Botanic Garden, Calcutta,* **10** (2), 1905, xvi.
[14] Buchanan (1819), *An account of the kingdom of Nepal*, Edinburgh, 98.
[15] Buchanan to J. E. Smith, 17 March 1806. L.S. Smith letters, Vol. 2, folio 211.
[16] Figures and descriptions of Nepalese plants in *Exotic Botany,* **2**, 1805, plates 97–101, 107, 108, 113–17.
[17] Buchanan to Wallich, 16 October 1821. B.M. (N.H.) Buchanan letters, folio 157.
[18] Bengal public letter to E.I.C., 5 August 1819. I.O.L.R. Board's collection 1820–30, item 15534.
[19] 'Memorandum respecting the instructions to be given to Dr Wallich.' B.M. (N.H.) Buchanan letters.
[20] Hardwicke to Lambert, 5 February 1821. B.L. Add MSS 9869, folio 53v.
[21] O. Polunin (1952), Natural history of the Langtang Valley, in *Nepal Himalaya* (ed. H.W. Tilman), 242–63.
[22] H. Hara and H. Ohashi (1966, 1971, 1975), *Flora of Eastern Himalaya*. Tokyo.
[23] R. Saunders (1979). Some account of the vegetable and mineral productions of Boutan and Thibet. *Philosophical Transactions of the Royal Society*, **79**, 79–111.
[24] W. Griffith, Journal of the mission which visited Bootan in 1837–38 *Annals of Natural History*, **4**, 1840, 424–9; **5**, 1840, 119–25, 205–11, 405–9.
[25] Fletcher (1975), 336.
[26] A. J. C. Grierson and D. G. Long (1983–), *Flora of Bhutan* (in progress).

[27] T. Nuttall (1853), Description of and observations on some species of Rhododendron, collected in Assam and Bootan by Thomas J. Booth. *Hooker's Journal of Botany and Kew Garden Miscellany*, **5**, 353–67.
[28] *Curtis's Botanical Magazine*, 1859, plate 5146.
[29] F. Kingdon Ward (1930), *Plant hunting on the edge of the world*, London, 199.
[30] B. L. Urquhart (1958), *The Rhododendron*, 23.
[31] J. D. Hooker (1854), Vol. 1, 110.
[32] *Journal of Asiatic Society of Bengal*, **25**, 1856, 414.
[33] Thomson to E.I.C., 19 April and 26 July 1851. I.O.L.R. L/F/2/143.
[34] Skelton (1972).
[35] Now in I.O.L.R.
[36] T. Twining (1893), *Travels in India a hundred years ago*, London, 459.
[37] M. Symes (1800), *An account of an embassy to the kingdom of Ava sent by the Governor-General of India in the year 1795*, London, 252.
[38] Kyd to Government of Bengal, 7 May 1787. I.O.L.R. H/Misc/799, folio 89.
[39] Buchanan to Roxburgh, 4 February 1817. B.M. (N.H.) Buchanan letters to Roxburgh, 1795–1812.
[40] M. Graham (1812), *Journal of a residence in India*, Edinburgh, 145.
[41] M. North (1892), Vol. 1, 309.
[42] R. Wight (1853), *Icones plantarum Indiae Orientalis*, Vol. 6, Madras 34.
[43] *Bulletin of Botanical Survey of India*, **25**, 1983, 350.
[44] Carter (1988), 268.
[45] James Robertson to Banks, 17 May 1775. B.L. Add. MSS 33977, folio 48.
[46] Kyd to Government of Bengal, 7 May 1787. I.O.L.R. H/Misc/799, folio 89.
[47] Submission from three local agents to Secretary of Board of Commissioners, 26 April 1817, I.O.L.R. F/4/587, item 14218.
[48] Wallich to Government of Bengal, 5 September 1827. I.O.L.R. F/4/1068, item 21980, folios 70–7.
[49] Bengal Public Consultations, 10 April 1828. I.O.L.R. F/4/1191, item 30877.
[50] E.I.C. to Government of India, 15 July 1840. I.O.L.R. E/4/763, folios 391–404.
[51] Government of India to E.I.C., 3 January 1845. I.O.L.R. E/4/781, folio 746.
[52] Bengal Public Consultations, 11 September 1807. I.O.L.R.
[53] Fort St George Public Consultations, 24 September 1802. I.O.L.R. F/4/152, item 2601.
[54] Buchanan to Wallich, 4 February 1817. *Annals of Royal Botanic Garden, Calcutta,* **10** (2), 1905, xxviii.
[55] Raffles to Wallich, November 1819. *Journal of Malay Branch of Royal Asiatic Society*, **54**, 1981, 11.
[56] Wallich (1820), Descriptions of some rare Indian plants. *Asiatick Researches*, **13**, 369–415.
[57] *Gleanings in Science*, **1**, 1829, 294.
[58] Wallich to Government of Bengal, 5 November 1827. I.O.L.R. F/4/961, item 27345, folio 45.
[59] *Orchid Review*, **31**, 1923, 171.

Ceylon

Hugging the southern tip of the Indian subcontinent is Sri Lanka or Ceylon. The Ancient Greek geographers knew the island as Taprobane, hence the botanical specific epithet *Taprobanicus*. Strategically situated on the principal sea route between East and West, it was a meeting place, an international market, for merchants from Africa and Arabia, the Malay Archipelago, and China. One of its best-known landmarks, Adam's Peak, was revered by Muslims as the spot where Adam took his last look at the Garden of Eden; Chinese monks made pilgrimages to the island to worship Buddhist relics; Victorian settlers, enchanted by its lush vegetation and equable climate, christened it the Pearl of the Indian Ocean. Europe first learned of the existence of the island through officers in Alexander's army. Megasthenes at the Maurya court in the Fourth century BC was told that Taprobane 'has palm-groves where the trees are planted with wonderful regularity all in a row, in the way we see the keepers of pleasure-parks plant out shady trees in the choicest spots'[1]. Classical authorities gave contradictory dimensions for the island which covers about 25 000 square miles; it is not as large, in fact, as Ireland. They believed it to be the largest island in the known world; both Ptolemy and Pliny had it stretching into the southern hemisphere. Even as late as the sixth century the Alexandrian monk, Cosmas, calculated its length to be 900 miles (it is, in fact, 290 miles at its longest). Ibn Batuta (1304–*c*.1378), perhaps the greatest of all medieval travellers, covered some 75 000 miles during 30 years wandering through Asia and Africa, visiting every Muslim state. His pilgrimage, punctuated by frequent marriages and desertions, brought him to Ceylon in 1344. As a devout Muslim, he climbed Adam's Peak to pray at the blessed footprint of Adam. On his ascent he admired 'flowers of various colours, with the red rose, about the size of the palm of the hand, upon the leaves of which they think they can read the name of God and of his prophet'[2]. This has been identified as *Rhododendron arboreum* subsp. *nilagiricum*[3] and must be the earliest mention of any rhododendron in South Asia although it was not until 1817 that a John Davy recognized it as being a member of that genus.

The profitable cinnamon trade (Fig. 12.1) and Ceylon's pivotal position in the Indian Ocean lured the Portuguese to its shores during the first decade of the sixteenth century. They enforced a treaty with its King, built forts, and introduced plants from their other overseas possessions, but even after 150 years their occupation extended to little more than the west coast. The *Itinerario* of Jan Huygen van Linschoten, a Dutchman in the service of the Portuguese, was published in Holland in 1595 and recounts five years' sojourn in Goa. This book which disclosed Portuguese maritime routes was consulted by the leaders of the first Dutch expedition to India in 1595. In 1602 the United East India

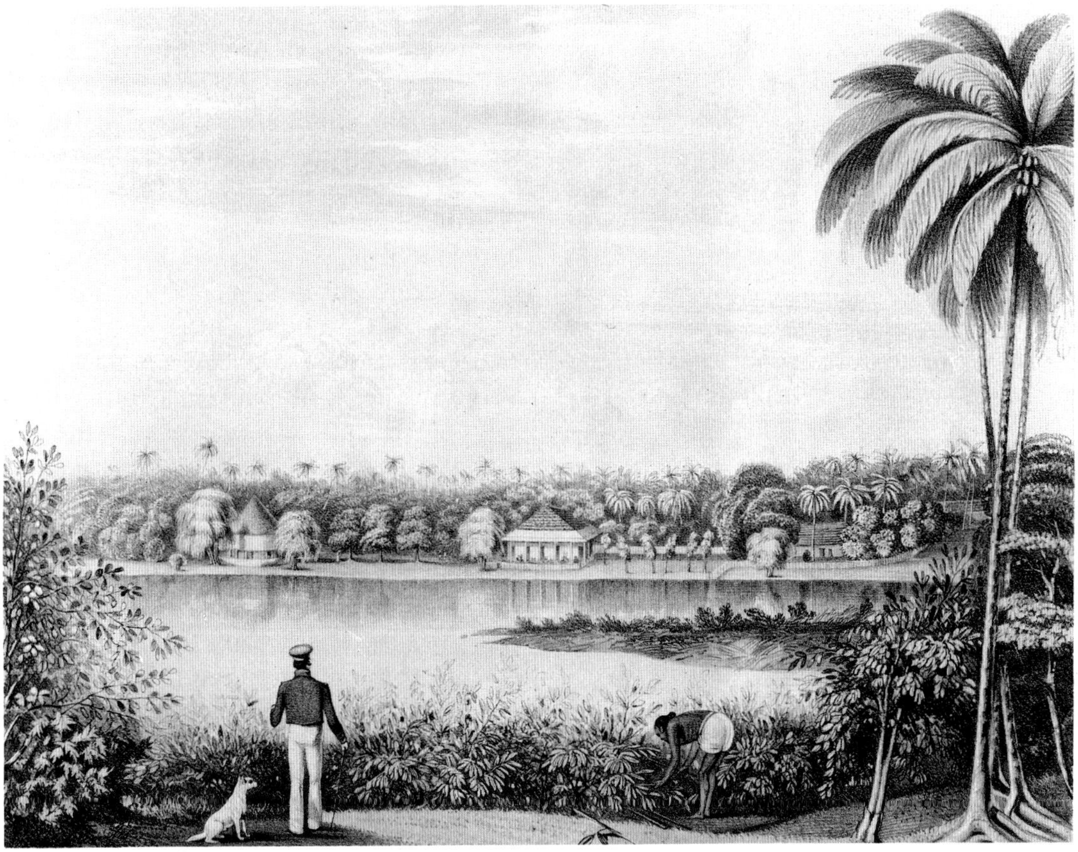

Figure 12.1 The cinnamon gardens at Kollupitiya, now a fashionable residential district of Colombo (From J. Deschamps (1845), *Scenery and reminiscences of Ceylon*, London, plate 6.) Lithograph.

Company represented the interests of former Dutch companies but, unlike the British East India Company in its early days, it was granted powers by the Government to levy troops, establish fortified bases and to elect overseas governors. Its first objective, the monopoly of Ceylon's cinnamon trade, was achieved through superior sea power forcing the Portuguese on the defensive. With the surrender of their fort at Jaffna in 1658, Portuguese dominance in the island was at last at an end. The Dutch ruthlessly enforced regulations to protect their interests in cinnamon, making the unauthorized sale of its oil a serious offence.

Phillipus Baldaeus, a minister of the Dutch Reformed Church, came to Ceylon after the capture of Colombo in 1656. His account of the island, published in Amsterdam in 1672[4], enthused about the vegetation: citrus fruit superior to Portuguese and Spanish varieties, an abundance of grapes, the efficacy of tamarinds as an anti-scorbutic, snakewood (*Strychnos colubrina*) as an ointment for skin complaints or as a recipe for colic and fever. Robert Knox, a British sailor, captive in the Kingdom of Kandy from 1660 to 1679, had ample opportunity to observe and respect the horticultural expertise of the natives. It was, he learnt, an offence to cut down a healthy coconut palm, the source of food, drink, clothes, and materials for utensils and ropes. The large flexible leaves of the talipot palm were fashioned into sunshades and tents. European vegetables introduced by the Portuguese and the Dutch grew well. 'The woods are their apothecaries shops, where with herbs, leaves, and the rinds of trees they make all their physic and plaisters, with which sometimes they will do notable cures.'[5] Both sexes dressed their hair with wild flowers but surprisingly never cultivated them in gardens. A Dutch translation of Knox's book was published in Utrecht in 1692 and subsequently plagiarized by François Valentijn in the fifth volume (1726) of his *Oud en nieuw Oost-Indien*. Valentijn who had served in the Dutch East Indies, never visited Ceylon and depended on contemporary writers such as Baldaeus and Linschoten and interviews with travellers for his lengthy compilation on the island. He relied almost exclusively on Knox for his descriptions of cinnamon, the talipot and the kittul palms, *Pandanus thwaitesii* and *Croton laccifer*. Even Linnaeus cited Knox in his *Flora Zeylanica*.

Van Reede tot Drakenstein, who had distinguished himself both as a soldier and diplomat during the Dutch conquest of Malabar, was honoured in 1667 by being appointed First Captain and Sergeant-Major of Ceylon. In 1669 he received a request for local medicinal plants from the Dutch settlement in Batavia which saw Ceylon as a cheap alternative source of supply. Van Reede was required to encourage 'master surgeons and others who have knowledge of medicinal herbs to speculate and satisfy their curiosity about these things in order to discover in Ceylon all that might be found to be a great comfort there'[6]. Dr Robert Padbrugge, then investigating the natural resources of the island, had the previous year inspired optimism by claiming to have identified some thirty species of medicinal plants recorded in classical texts, confidently predicting that many more new drugs were awaiting discovery. The High Government was convinced that Ceylon 'is so blessed a land of all sorts of valuable and splendid medicines as any land on the earth can be, all this so plentifully that before long only very few medicaments need be sent from the home country'[7]. The momentum in floristic studies in Ceylon diminished with the departure of Paul Hermann to the Netherlands in 1680 to the level of routine collecting of drug plants for Batavia. Van Reede, now Commissioner-General of the Western Quarters, sought to revitalize botanical research throughout the Dutch possessions in India and Ceylon by ordering his Governors, Directors, and Commanders in 1691 to ship regular consignments of seeds, bulbs, and roots to the Netherlands 'not only to satisfy the curiosity of many wise men and amateurs, but also in order that the world might enjoy and be served by the medicinal virtues which the above-mentioned wise men might discover and find out'[8]. He expected enterprise and initiative in their selection. 'You should also know that by the names of seeds we do not mean seeds of garden plants which are tame and known in Europe, but exotic, wild, and unesteemed seeds which occur and can be found without exception in the forests'. Ceylon, already the depôt for the despatch of Asian goods to Europe, was a natural centre for this co-operative collecting. Of all the Dutch overseas territories, Ceylon alone continued donating seeds and herbarium specimens to the botanical gardens in Amsterdam and Leiden until well into the second half of the eighteenth century.

It was not until Paul Hermann (1646–95) arrived as a physician in 1672 that the island's vegetation was subjected to serious and intensive study. He had been recommended to the Dutch East India Company by Arnold Syen, Professor of Medicine and Botany at Leiden University and editor of the first volume of Van Reede's *Hortus Malabaricus*. His collecting, it is believed, was limited mainly to the coastal region; plants, living and dried, were sent to Syen, to Jan Commelin at Amsterdam and Jacob Breyne, a merchant in Danzig who subsequently passed on specimens to James Petiver in London. Hermann's surviving herbaria in

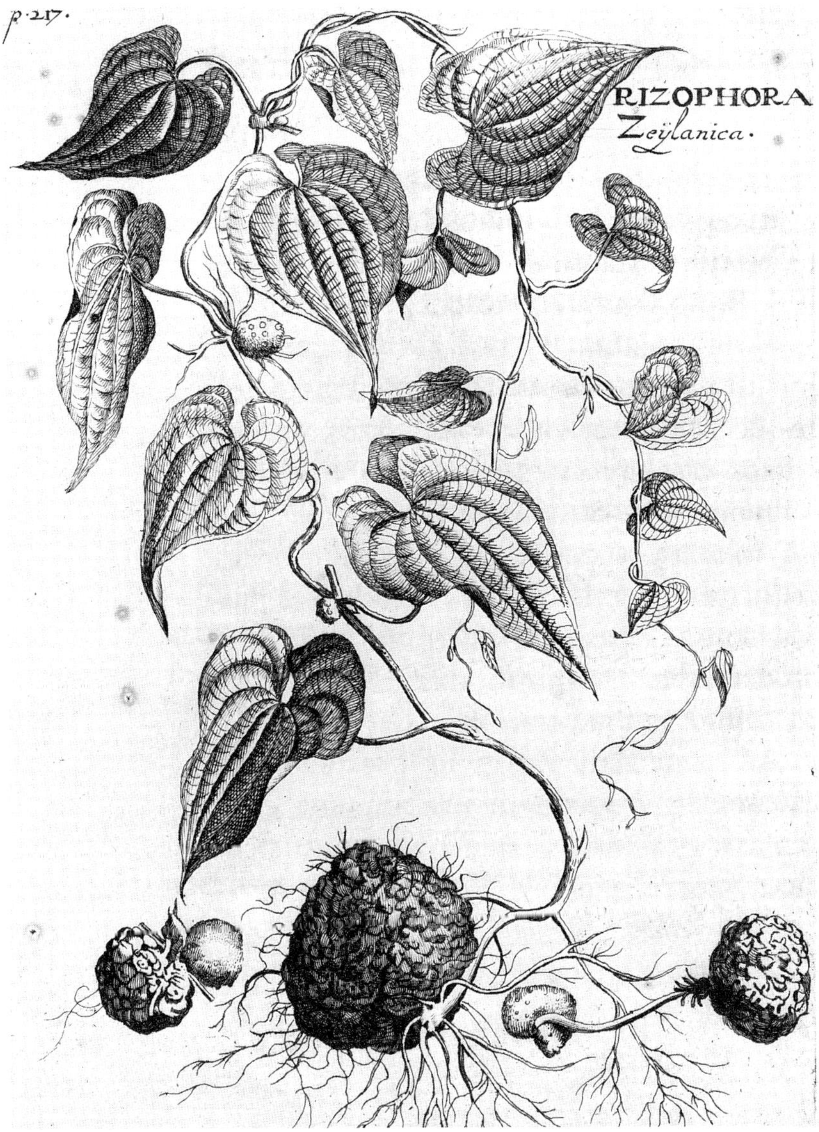

Figure 12.2 *Dioscorea bulbifera*. (From P. Hermann (1698), *Paradisus Batavus*, Leiden, p. 217.) Engraving. One of a number of plants from Ceylon described and illustrated in this book.

Leiden, London, and Paris reveal that he was acquainted with about a thousand species. His recall to the Netherlands in 1679 dismayed the Governor, reluctant to 'be deprived here of an expert in the knowledge of the native herbs and [to] be obliged to order many drugs from elsewhere, which for the rest are available here in plenty'[9].

His departure was occasioned by his appointment as Syen's successor at Leiden in 1680. When he was in England two years later he met William Sherard who was to edit Hermann's papers after his premature death in 1695, published as *Paradisus Batavus* (1698) (Fig. 12.2). The book includes some Ceylonese plants, illustrated by small figures; Sherard may also have edited a much more significant work—*Musaeum Zeylanicum* (1717). A slim volume of only 71 pages, it lists Hermann's herbarium which was lost for nearly 30 years until Linnaeus identified it.

Johannes Burman (1707–78) (Fig. 12.3), Professor of Botany and Director of the Municipal Garden at Amsterdam, utilized

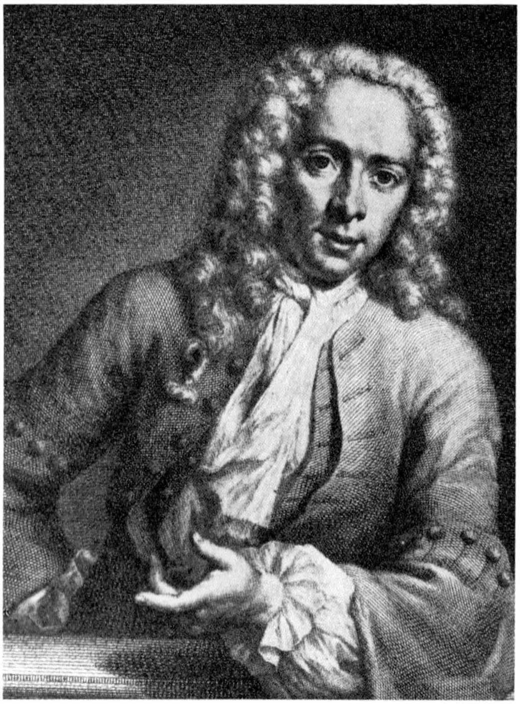

Figure 12.3 Johannes Burman (1707–78). Engraving by J. Houbraken of a portrait by J. M. Quinkhard.

Musaeum Zeylanicum and Hermann's specimens to compile his *Thesaurus Zeylanicus* (1737), an alphabetical list of Ceylonese plants, complemented by 110 plates of competent drawings. He became the first author to link the floras of Ceylon and Malabar through identifications in the *Hortus Malabaricus*.

During the preparation of his *Thesaurus* Burman briefly enjoyed the help of the young Swedish botanist, Linnaeus, who laboured under the handicap of not being able to consult Hermann's herbarium which had last been seen at an auction in 1711. It reappeared, almost miraculously, in 1744 when August Günther, the Danish Apothecary-Royal, sent Linnaeus a parcel containing four volumes of herbarium specimens and an album of about 400 plant drawings. Linnaeus recognized three of the volumes as the source of *Musaeum Zeylanicum*. Now he was presented with an irresistible opportunity to work on a tropical herbarium and within a few years his *Flora Zeylanica* (1747) emerged. Linnaeus named 429 plants arranged under genera; he also listed another 228 he could not allocate to genera. Here the essential characteristics of the species are expressed in carefully drafted diagnostic polynomials; it was not until the publication of his *Species plantarum* (1753) that he provided alternative concise binomials for everyday use. Thus his *Eupatorium foliis ovato-hastatis petiolatis dentatis* in the *Flora Zeylanica* received the concise alternative but less informative name *Eupatorium zeylanicum*. These polynomial names served the same purpose as a key in a modern flora for identifying plants. In this first tropical flora arranged according to his own sexual classification, Linnaeus, by citing Burman's *Thesaurus Zeylanicus*, connected all extant Ceylon herbaria with the *Hortus Malabaricus*. He prefaced his *Flora Zeylanica* with a survey of the island's natural history, extolling 'the abundance of its aromatic productions', the groves of cinnamon, nature's larder of coconuts, date palms, bananas, mangos, and jack fruit, 'and flowers of the most exquisite colour, structure and fragrance . . . as well as those plants which saturate the night air with their delicious scent'[10]. Living in Sweden with its short summers and depressing winters, distant Ceylon must have seemed an idyllic island.

William Curtis, the London apothecary, nurseryman and author, dedicated his book, *Linnaeus's system of botany* (1777) to his friend Joan Gideon Loten (1710–89), formerly Dutch Governor of Ceylon from 1752 to 1755. Like Van Reede, Loten's relaxation from official duties was natural history—more specifically, ornithology. He employed Pieter C. de Bevere to draw the mammals, birds, insects, fishes, and plants of the Malay Archipelago and Ceylon. Some of these drawings, now in the Natural History Museum in London were copied in Thomas

Pennant's *Indian zoology* (1769) and *Synopsis of quadrupeds* (1771) and George Edwards's *Gleanings of natural history*. Loten temporarily retired to England where he acquired the fellowship of the Royal Society and a second wife, and made the acquaintance of Sir Joseph Banks who commissioned the young artist, Sydney Parkinson, to copy 40 of Loten's Indian animal studies.

When Linnaeus died in 1778, his former pupil, Carl Peter Thunberg (1743–1828), became his worthy apostle. As surgeon in the Dutch East India Company, he visited Java and the Dutch station at Nagasaki in Japan, and from mid-1777 to February 1778 lingered in Ceylon. Lord Valentia who paused briefly in the island in 1804 during his aristocratic tour of the subcontinent had little regard for Thunberg's published account of his travels. He dismissed Thunberg's botanical activities in Ceylon as minimal and inaccurate. There is no evidence that Thunberg tried to contact the botanical fraternity in South India but it is conceivable, however, that he met J. G. Koenig, who was on the island during 1777.

There is a possibility that Koenig was the victim of a botanical hoax when he sent Banks a specimen of 'genuine' gamboge from Ceylon. Since it differed from another specimen of gamboge (*Garcinia morella*), collected on the island some years later by General Walker's wife, Robert Brown subjected it to microscopical examination. He discovered that the inflorescence of one species had been expertly stuck with sealing wax to the branch of another![11] It is inconceivable that Koenig, whom Sir William Jones eulogized as 'having as much simplicity as nature herself', could have perpetrated such a deception. The Chinese were known to have practised this type of fraud.

Koenig had been one of the first of the botanical missionaries at Tranquebar and Johann Peter Rottler was one of the last. Rottler visited Ceylon in 1788 and 1795, on the second occasion as interpreter and naturalist to Hugh Cleghorn, father of Hugh F. C. Cleghorn, the distinguished forester. He tactfully reconciled official duties and pastoral responsibilities with botanical opportunism without neglecting or offending Cleghorn who recorded in his diary that

> *My most worthy friend Rottler has found many curious and undescribed plants; and the classifying and preserving them, together with his pious exhortations to such Malabar Christians as he meets with, innocently and profitably fills up the time of this truly primitive and good man*[12].

Hugh Cleghorn was an academic with a penchant for the drama of 'cloak and dagger' diplomacy. He secretly persuaded Swiss mercenaries, constituting the major part of the Dutch garrison in Ceylon, to defect to the British. When the Netherlands were invaded by the French in 1794, the British Government temporarily annexed the Dutch Asian colonies with the promise of their restoration on the conclusion of peace. The natural harbour at Trincomalee, capable of sheltering the largest ships, and occupied in turn by the Portuguese, Dutch, and French, was the primary objective of the expeditionary force sent by the Government of Madras in 1795. The signing of a treaty between the British East India Company and the King of Kandy in 1796 precipitated the collapse of Dutch power in the island and the emergence of British supremacy in the Indian Ocean. Due to mismanagement by Company officials, Ceylon became a Crown Colony in 1798 and Cleghorn was appointed Secretary to its first Governor, the eccentric Honourable Frederick North, the younger son of Lord North, George III's disastrous Prime Minister. The Governor requested permission to employ two or three young people for special duties, one of them a Frenchman, Eudelin de Jonville, as 'Clerk for Natural History and Agriculture'. At first Jonville had little time for natural history, his attention having been directed to surveying the cinnamon plantations. Nevertheless, in February 1800 Jonville was able to report to the Governor that he had made a collection of specimens and preliminary sketches of the fauna and flora. In March he accompanied Major-General MacDowall's mission to the Court of Kandy as interpreter. He was subsequently made Surveyor-General of Ceylon and then suddenly no more is heard of him. A volume of his flower drawings, another of animals and a third of notes entitled 'Quelques notions sur l'Isle de Ceylan', presumably intended eventually for publication, are now in the India Office Library and Records, the official repository of the records of the East India Company and the India Office[13].

One of Jonville's drawings—the temple of Buddha at Arandera near Idalmalpany—was reproduced in Cordiner's *A description of Ceylon* (1807). The Reverend James Cordiner (1775–1836) embarked on his ministry as army chaplain at Colombo in 1799; before his departure in 1804 he had chronicled the life of the European garrisons, observed villagers at work and worship, witnessed an elephant hunt, sketched the landscape and studied the wildlife and flowers.

> *Nothing about Colombo is more apt to excite admiration than the flourishing state of the vegetable world. So much beauty and variety are in few countries equalled, and no where excelled. The thick shade of majestic trees, the open prospects, the lively verdure, the flowering shrubs, and parasitic creepers, unite their charms to render the morning rides delightful*[14].

With only a superficial acquaintance with botany, he gave a cursory description of the flora, cited scientific names when he could, approvingly quoted from Robert Knox and referred to Burman's *Thesaurus Zeylanicus*. He was particularly impressed that the coconut and palmyra palms and the jack tree between them provided the native population with food, timber for houses, furniture, boats, and even writing material. That so many of the flamboyant flowers lacked perfume disappointed him; he was distinctly uneasy about the insectivorous plant, *Nepenthes*, which he rejected as 'a curious and indecent plant'. He admired the Colombo garden of Major-General MacDowall who received boxes of exotics from Roxburgh at Calcutta by almost every ship that called at Ceylon. The cinnamon, its description and cultivation, got a whole chapter to itself. He was the first amateur naturalist to publish his observations during the British era in Ceylon.

Under the terms of the Peace of Amiens in 1802 Ceylon was ceded to Britain; North remained Governor until 1805, when as a passionate philhellene he retired to Corfu where he renounced European dress for classical costume. It has been suggested that when the Chief Justice of Ceylon, Sir Alexander Johnston, returned to England in 1809 he proposed to Lord Liverpool the creation of a botanical garden on the island. Sir Joseph Banks who was consulted nominated William Kerr for the post of 'Superintendant and Chief Gardener', an appointment he did not take up until 1812. Kerr had been at Canton since 1803 collecting plants for the royal gardens at Kew and this transfer was his reward for loyal and diligent service. The garden was sited on Slave Island, Colombo, and nostalgically called 'Kew'. Kerr's mission was to pay particular attention to the introduction of Chinese plants[15], 'to make discoveries in botany' and 'to form a Herbarium of dried plants for the future use of the establishment in Ceylon'.

William Jackson Hooker (1785–1865), on the threshold of his distinguished career as a botanist, yearned to travel. Hearing about the proposed botanical garden in Ceylon, he obtained Banks's support and was ready to join Sir Robert Brownrigg, the new

Governor of Ceylon, in 1811 when domestic crises delayed his departure. In the meanwhile Sir Joseph advised him to study Indian plants in cultivation in gardens in the neighbourhood of London.

> *To be well acquainted with Indian plants before you go will enable you to do in one year that without it could not be effected in two or three, to find on your return that a considerable number of plants that have taken up much precious time in delineating & describing are well known & already properly published is the necessary consequence of not being well prepared before hand*[16].

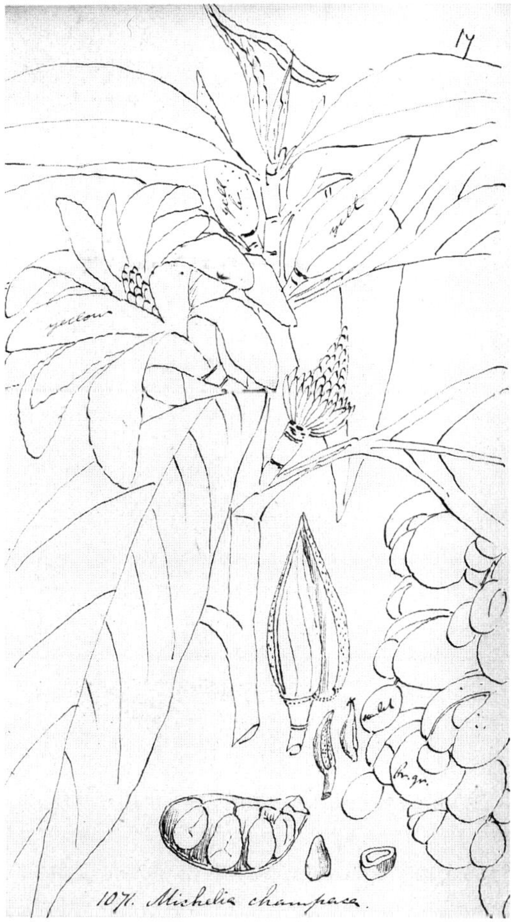

Figure 12.4 *Michelia champaca.* Pen and ink drawing by William Jackson Hooker. Ten pocket volumes of copies of Roxburgh's collection of drawings were made by Hooker preparatory to his projected journey to Ceylon. (Kew Archives, Vol. 1, folio 17.)

Hooker responded to this sound counsel by patiently making about 2000 pen and ink sketches of the Roxburgh drawings stored in the India House Library (Fig. 12.4). The Ceylon venture had to be abandoned when it became known that the conflict with the King of Kandy would make travelling in the island hazardous. In 1813 Banks proposed Java as an alternative destination but the concern of his parents and father-in-law, who all feared for his health, prompted Hooker to decline. Sir Joseph, who had sailed around the world in Captain Cook's *Endeavour*, was disappointed by this display of caution and timidity influenced by

> *the importunities of those who think they can guide you to a more serene, quiet, calm & sober mode of slumbering away life than that you proposed for yourself. . . . I was about 23 when I began my peregrinations. You are somewhat older but you may be assured that if I had listened to a multitude of voices that were raised up to dissuade me from my enterprise I should have been now a quiet country gentleman ignorant of a multitude of matters I am now acquainted with & probably never attained to no [sic] higher rank in life than that of country Justice of the Peace*[17].

William Kerr's sudden and unexpected death in November 1814[18] would appear timely confirmation of Hooker's circumspection. During his brief tenure of office Kerr had done little more than prepare plans and establish another garden at Caltura (now Kalutara) some 28 miles south of Colombo. The site of the main garden was an ill-conceived choice as Hardwicke reported to Banks when he saw it in 1815.

> *It is a small spot on the borders of an extensive lake but in a situation very flat and not sufficiently elevated above the surface of the water for the purpose of a garden. For a short time plants placed here thrive, but when the*

roots extend below the insufficient thickness of dry soils their growth is checked and the greater part gradually decay and perish[19].

Banks recommended another Kew gardener as Kerr's replacement—Alexander Moon, 'a smart young man', competent in both horticulture and botany[20]. Four years after his arrival in 1817, Moon was directed by the Governor to find a new location for the garden near Kandy where European vegetables were cultivated for the Colombo market. The availability of water and an abundance of trees to provide shade for tender exotics made the former royal palace at Peradeniya a desirable site to develop. Moon transferred to Peradeniya in 1822, at first concentrating on cinnamon and coffee; the removal of exotics from Caltura progressed slowly right up to the 1840s. He botanized, started a herbarium, sent seeds to Kew, and compiled a floristic guide in both English and Singhalese. *A catalogue of the indigenous and exotic plants growing in Ceylon* (1824) lists 366 species cultivated in the botanical garden and another 761 found in the fields and forests. Its limitations were largely due to the absence of an adequate library to assist Moon. Lacking descriptions of the plants it listed, the work was of little use to botanists. Moon's assistant was a native clerk with artistic talent, Harmanis de Alwis Seneviratne (Fig. 12.5). Taught how to draw plants to western standards, de Alwis was made the official draughtsman at Peradeniya in 1823, a post he held for 38 years.

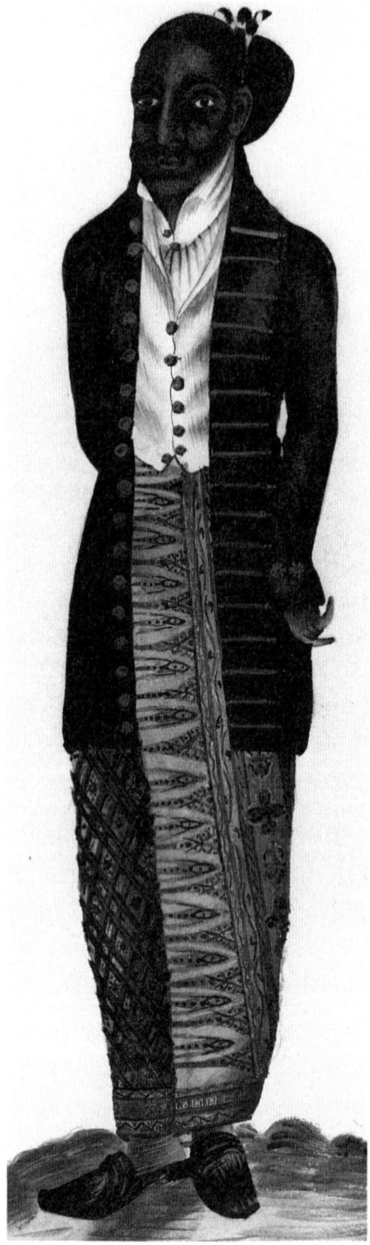

Figure 12.5 Harmanis de Alwis Seneviratne (*d.* 1894). Watercolour. Artist employed at the Botanic Garden, Peradeniya, 1823–61. (Royal Horticultural Society.)

For nearly twenty years after Moon's death in 1825 the botanical garden endured a succession of superintendents, gradually degenerating into little more than a market garden. Andrew Walker temporarily filled the post until the appointment of James Macrae in 1827, an eminently well-qualified candidate with first-hand experience of tropical plants; he had worked at the St Vincent Botanic Garden in the West Indies and had collected for the Horticultural Society of London in the Hawaiian Islands and South America. Before he died in 1830 he added to the herbarium started by Moon and sent a few plants to England. A former apprentice gardener at the Botanic Garden at Calcutta who had accompanied Wallich to England in 1828 to widen his horticultural skills, took charge of Peradeniya from G. Bird in 1832. This was James George Watson, an Anglo-Indian whom the Adjutant-General, General George Warren Walker, castigated as 'an ignoramus who could not read the language of botany'. Walker and his wife appear to have been the only people on the island with any botanical expertise. He sent specimens to Robert Wight who visited Ceylon in 1836, to Robert Graham at Edinburgh and to William Hooker at Glasgow, sometimes accompanied by Mrs Walker's neat flower sketches.

When Watson died in 1838, J. G. Lear who happened to be in Ceylon collecting plants for Joseph Knight of the Exotic Nursery in Chelsea filled the vacancy for two years. Orchids were his speciality. While he was there, Robert Wight had de Alwis seconded to Madras for three months to be taught how to dissect flowers; and Wight used some of his outline drawings in his *Icones plantarum Indiae orientalis*. Lear was followed in turn by two surgeons—H. T. Normansell (*d.* 1843) and W. C. Ondaatje, later to be elected a Fellow of the Linnean Society of London.

General Walker's plea for a professional botanist to run Peradeniya was at last realized by the selection of George Gardner (1812–49) in 1843. He had been a pupil of William Hooker at Glasgow where he studied medicine before devoting five years to collecting plants in Brazil on a prodigious scale. His first annual report, presented in August 1844, examined in a thorough and business-like manner the state of the garden and its services, its defects and his remedies. He resisted a suggestion that the jungle which encroached over a great deal of the garden should be planted with coffee, and although he favoured a diversification of economic crops he also wanted Ceylon's indigenous flora to be represented. His warning that the rapid clearance of virgin forest would result in the extinction of some species has a familiar ring for us today. 'Botanists of future times will look in vain for many of the species which their predecessors had recorded in the annals of science as natives of the Island'[21]. Believing that a botanical garden should also be a place for relaxation and enjoyment, he disapproved of a scientific presentation of plants. He launched an onslaught on weeds, smoothed irregularities in the ground caused by white ants' nests and gravelled the paths. The market garden activities, occupying one-third of his labour force, was stopped forthwith. He firmly insisted on the return of Moon's herbarium which Wight had borrowed. He intended intensifying the collecting of herbarium specimens with duplicates to be sent to Kew and the British Museum. The library, too, came under scrutiny as did de Alwis's botanical drawings which he considered as being 'rather pictures than scientific drawings', and unsuitable for publication. As it was Gardner's ambition to publish an illustrated flora of Ceylon, de Alwis thenceforth always drew under his rigorous supervision, producing drawings of uniform size with obligatory dissections. In 1845 he petitioned for an increase in the draughtsman's salary which was only $3 per month; 'his equal as a native botanical artist I am certain does not exist in India'[22]. He was determined to raise botanical science in the island from its habitual obscurity and as an earnest of his intentions published six papers on Ceylon's flora in the *Calcutta Journal of Natural History*. In these aims he was warmly supported by other naturalists and sympathizers: the Colonial Secretary, Sir James Emerson Tennent who looked forward to the promised 'Singhalese Flora'; Captain John Champion (1815–54), an entomologist who became a botanist under Gardner's tuition; and William Ferguson (1820–87), a surveyor whose researches on grasses, ferns, and economic plants were regular features in the *Ceylon Observer* and *Tropical Agriculturist*. Another congenial resident was John Whitworth Bennett, the author of the *Coco-nut palm* (1831), *Ceylon and its capabilities* (1843), and *Selection of rare and curious fruits indigenous to Ceylon* (*c*.1850), illustrated with his own drawings. Gardner was convalescing in Madras when Prince Waldemar of Prussia and his botanist physician, Werner Hoffmeister, toured the island in 1844. They were escorted around the botanical garden by de Alwis 'whom we found well qualified to act as a most efficient cicerone, and even acquainted with the botanical names of the plants, and their classification'[23].

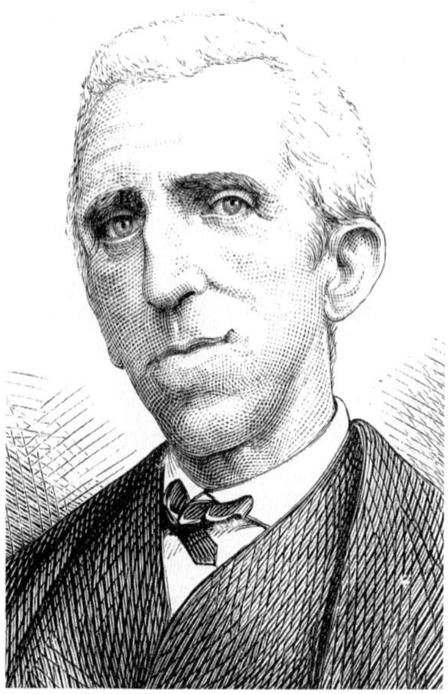

Figure 12.6 George Henry Kendrick Thwaites (1812–82).

When Gardner died suddenly of apoplexy at the early age of 37, it seemed that Ceylon botany might relapse into its former somnolence. Joseph Hooker, collecting in the Himalayas, refused the post, preferring to stake his chances on a suitable appointment at Kew. George Gardner's installation had established the precedent for the post to be filled by a botanist. George Henry Kendrick Thwaites (1812–82) (Fig. 12.6) abandoned accountancy for botany when he became a botanical lecturer at the School of Pharmacy and also at the Medical School in Bristol in the mid-1840s. His disappointment in failing to get an academic post in Ireland was assuaged by a successful application for the vacancy at Peradeniya. When he saw the botanical garden for the first time in 1849, jungle and scrub still covered much of the ground. Progressive clearance allowed the planting of more economic crops: varieties of cotton, vanilla, Shiraz tobacco, West Indies ginger, and Manila hemp. Nor was ornamental gardening neglected (Fig. 12.7); the native flora competed with flamboyant newcomers—*Amherstia nobilis* from Burma, *Saraca declinata* from Sumatra, *Brownea grandiceps* from Brazil, and *Delonix regia* from Madagascar. Decorative clumps of palms were added. In 1857 William Cameron arrived from Kew Gardens to be the first Curator and, at the same time, the designation of Thwaites's post was elevated from Superintendent to Director.

For 31 years, never once taking a holiday or leaving the island, Thwaites dedicated himself to the improvement of Peradeniya and the revelation of the island's flora. Among his initial priorities was the formation of a herbarium since Gardner's family had removed his predecessor's specimens. He collected, exchanged specimens with European herbaria and published about 25 new genera in Hooker's *London Journal of Botany*. Denying a rumour that Gardner had left notes for a Ceylon flora, he told Joseph Hooker that he intended compiling an illustrated flora on the lines of Wight's *Icones*. The last attempt had been Moon's inadequate *Catalogue* (1824). The *Calcutta Journal of Natural History* had published the little that Gardner had written, Hooker had included Ceylon plants in his *Icones Plantarum* and *Journal of Botany*, and many were illus-

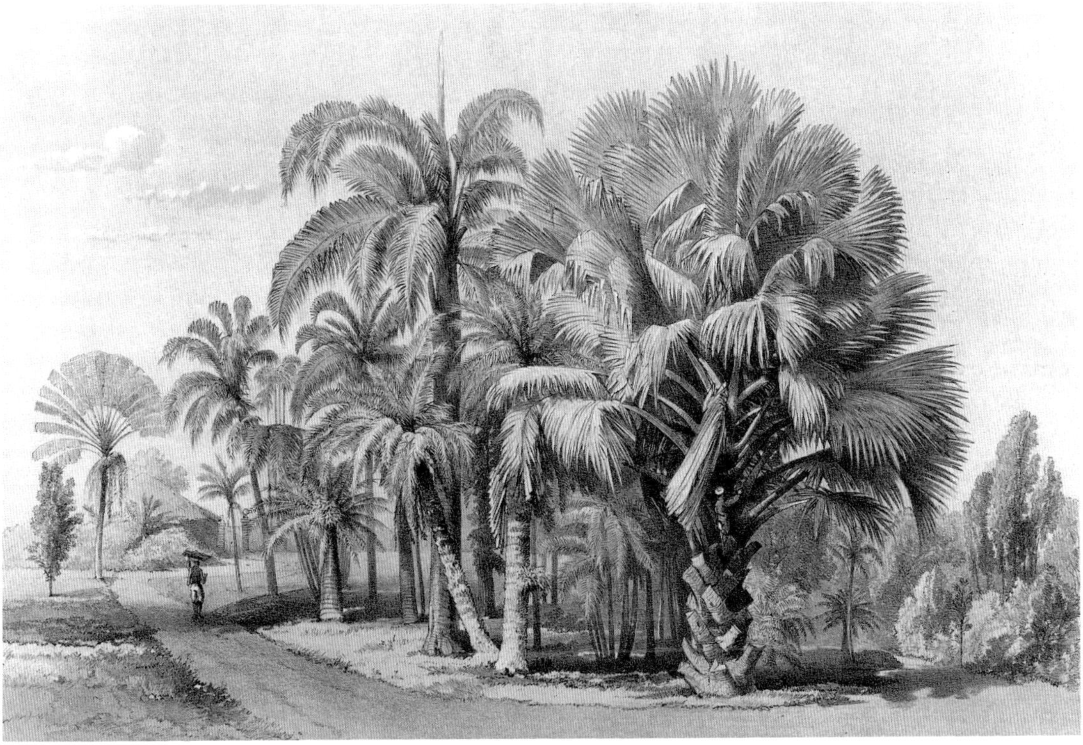

Figure 12.7 Group of palms at the entrance to the Botanic Garden at Peradeniya. (From C. O'Brien (1864), *A series of fifteen views in Ceylon*, London, plate 9.) Lithograph.

trated in Wight's *Icones*. Thwaites's plan was to consolidate these fragments, adding his own contributions to form a single work. Fully aware that this would be no easy task, he gladly accepted Joseph Hooker's offer in 1857 of help in identification and synonomy. *Enumeratio plantarum Zeylaniae* came out in five parts from 1858 to 1864, a competent precursor to Trimen's more thorough investigation. In its preface the author acknowledges his indebtedness to his principal plant collector and his artist, de Alwis who, when he retired in 1861, had the satisfaction of seeing his son, William, succeeding him. Thwaites had hoped to follow the *Enumeratio* with a popular flora adopting the formula laid down by Sir William Hooker for a projected series of colonial floras. It never appeared, perhaps for the reason that Thwaites became absorbed by his first love, cryptogamic plants.

Ceylon is a pteridologist's paradise and Thwaites shared his enthusiasm for ferns with the Chief Justice, Sir William Norris (1793–1859) and the businessman, George Wall (*c*.1821–94), author of *A catalogue of ferns indigenous to Ceylon* (1873) and a *Checklist* (1879). The many boxes of fungi sent to Kew were identified by the mycologists, M. J. Berkeley and C. E. Broome. The coffee-leaf fungus, *Hemileia vastatrix*, which devastated the island's coffee plantations, was identified by Berkeley and led to the appointment of a mycologist by the Ceylon Government. The algologist, William Henry Harvey (1811–66), *en route* to Australia, was Thwaites's guest in 1853. He spent a profitable fortnight dredging algae and exploring the mountains with his host. His first brief introduction to the tropics alerted him 'to the troubles of collecting in rude countries'. After the initial *frisson* of pleasure at seeing rampant *Gloriosa superba* from the shore-line up to 6000 feet and colonies of palms and breadfruit trees, Harvey complained of the lack of variety. 'The gem of the island is *Medinilla walkeri* which covers, like ivy, the trunks of large trees, connecting them in the flowering season into a most gorgeous *may-pole* or flowering columns.'

Another distinguished visitor was the peripatetic artist, Marianne North, on her way home from South East Asia (Fig. 12.8). Thwaites, whom she fondly referred to as a 'dear old gentleman', insisted on showing her Peradeniya despite the rain.

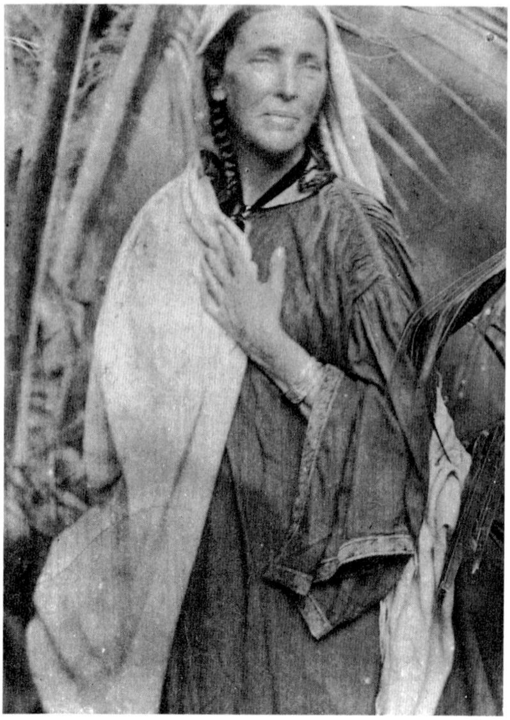

Figure 12.8 Marianne North (1830–90). According to Miss North, Julia Cameron, who took the photograph, 'dressed me up in flowing draperies of cashmere wool, let down my hair and made me stand with spiky cocoa-nut branches running into my head . . . Then she tried me with a background of breadfruit leaves and fruit'.

He had planted half the trees himself, and had seldom been out of it for forty years, steadily refusing to cut vistas, or make ribband-borders and other inventions of the modern gardener. The trees were massed together most picturesquely, with creepers growing over them in a natural and enchanting tangle[25].

Thwaites's stewardship can be considered under three categories: his floristic surveys, his study of cryptogams, and his propagation of economic plants. A mountain garden was established at Hakgala in 1860 for cultivating cinchona, the source of quinine, and a lowland garden at Heneratgoda was opened in 1876 as a rubber plantation. Thwaites always accepted the commercial exploitation of plants as a legitimate function of a botanical garden yet some of his most persistent critics were the coffee planters, convinced that Peradeniya had failed to provide an adequate service. An attempt in 1854 by the Legislative Council to disband the garden was, fortunately, vetoed by the Governer.

Thwaites's retirement in 1880 brought Henry Trimen (1843–96) to Ceylon as the next Director. Joint author of the *Flora of Middlesex* (1869) and *Medicinal plants* (1875–80), Assistant in the Department of Botany at the British Museum, and editor of the *Journal of Botany*, he was an eminently well-qualified candidate for the post (Fig. 12.9). Having this academic background, it comes as no surprise that he abhorred his predecessor's informal planting with its penchant for aesthetic perception. Trimen preferred a systematic and informative display with all the plants neatly labelled with their Latin names. He introduced his reforms in the South Garden, the last remaining uncultivated area. The Great Lawn, a miracle of smooth turf covering some seven acres, took shape in 1885. In 1894 a metal octagon house replaced flimsy structures. During the 1880s the cinchona nursery at Hakgala became a refuge for plants needing the cooler conditions of higher altitudes. About the same time more satellite gardens were opened at Anuradhapura and Badulla.

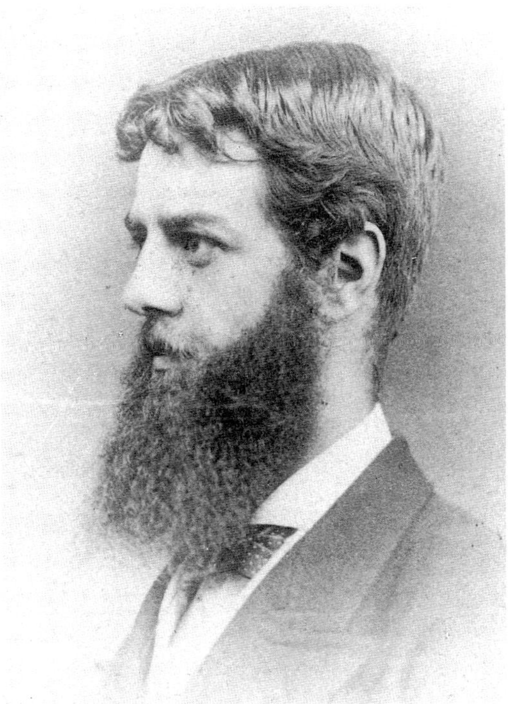

Figure 12.9 Henry Trimen (1843–96).

As the era of new economic crops was nearly over, Trimen concentrated on those already in cultivation such as cinchona, cacao, and cardamoms. The former Director's bungalow in the middle of the garden was converted into a museum of economic botany.

But Trimen's great work and his memorial is not his changes to Peradeniya but his *Handbook to the flora of Ceylon*, in its day a model of a tropical flora. He supplemented the resources of his garden's herbarium with those at Kew and the British Museum, the latter housing Paul Hermann's historic collections. With Sir Joseph Hooker's *Flora of British India* as his examplar, three volumes were published between 1893 and 1895 before paralysis and deafness compelled him to abandon the project. Hooker, a sprightly eighty years old, who was asked to complete it, confidently predicted that 'It should be easy as I have already described all the Ceylon plants in Flora of British India'. Two further volumes appeared in 1898 and 1900, an appropriate finale to Hooker's monumental treatment of South Asia. Most of the plates were drawn by William de Alwis, the garden's artist (Fig. 12.10). A volume incorporating additions and amendments, edited by A. H. G. Alston, appeared in 1931, and a revision of the entire work was undertaken in 1968 by the Smithsonian Institution, the Ceylon Department of Agriculture, and the University of Ceylon. Four editions of the *Hand-guide to the Royal Botanic Gardens* were edited by Trimen between 1883 and 1894 and his *Hortus Zeylanicus* (1888) listed all the plants, native and foreign, in cultivation at Peradeniya.

John Christopher Willis (1868–1958) finished the manuscript of his classic *Dictionary of flowering plants and ferns* on the voyage out to Ceylon in 1896 to become the new Director. Tea was now the predominant crop, the staple of the island's prosperity. The Government, having witnessed the economic catastrophe wreaked by the coffee-leaf disease readily agreed to the appointment of John Bennett Carruthers as mycologist in 1900. In 1898 they had sanctioned Willis's request for a post of scientific assistant to be filled by overseas botanists on a two-year secondment and in 1900 a small laboratory was equipped for European botanists. In 1901 Willis launched the *Annals of The Royal Botanic Gardens, Peradeniya*. He himself made a special study of Podostemaceae in Ceylon and India, continuing his researches when he went to Brazil on this family of plants which inhabit rocks in streams.

He gave free reign to Hugh Fraser Macmillan (1869–1958), the Curator since 1895, to re-landscape the grounds. A flower garden was planned, the lake enlarged, the arboretum thinned out, and formal avenues of Cabbage and Royal Palms, *Brownea grandiceps* and Coco-de-mer extended the spectacular geometry of walks and drives. Notwithstanding all these achievements and the frenetic

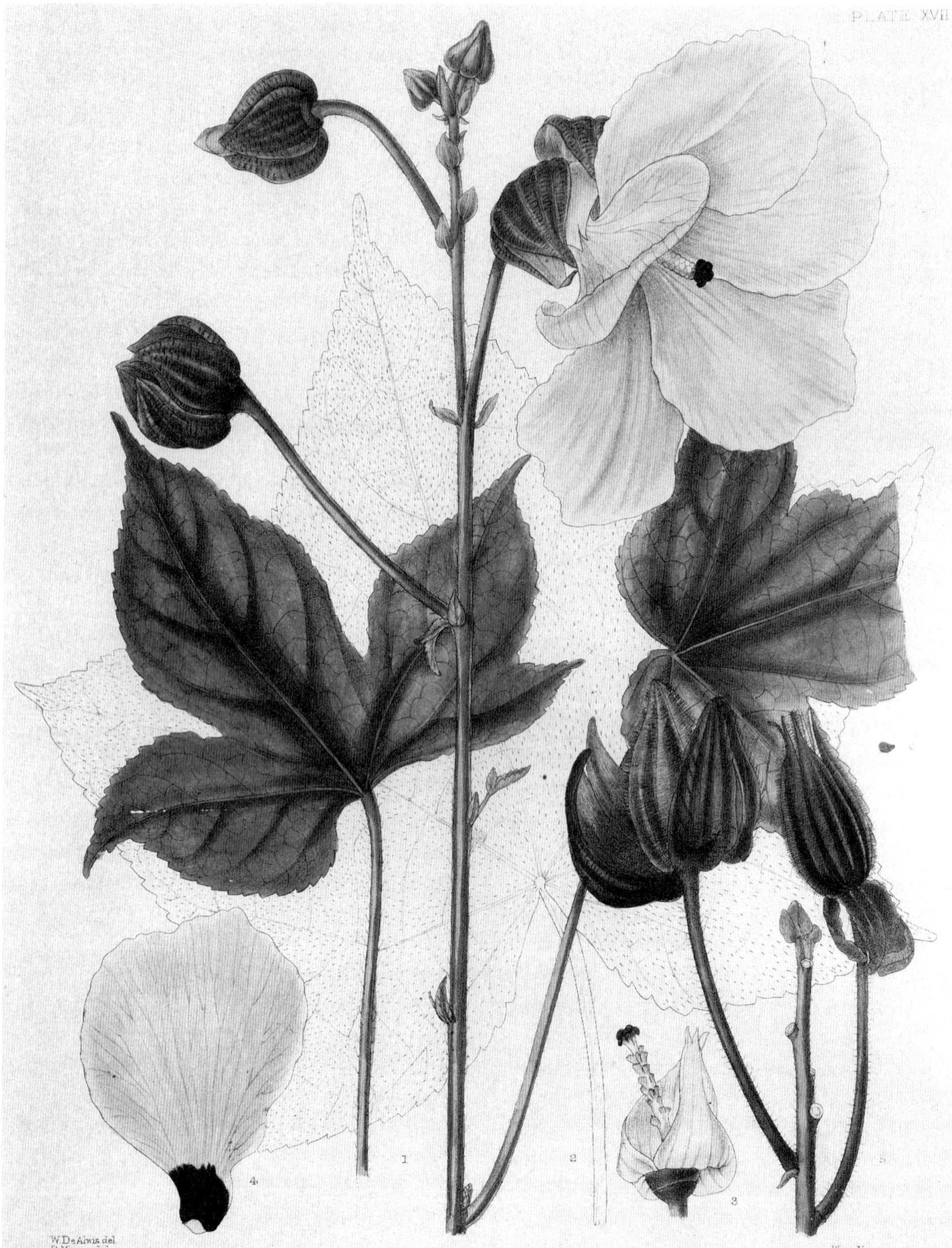

Figure 12.10 *Abelmoschus angulosus* var. *grandiflorus*. (From H. Trimen (1893), *Illustrations to Handbook to flora of Ceylon*, London, plate 17.) Lithograph of a drawing by William de Alwis, who succeeded his father as botanical artist at Peradeniya.

activity he enjoyed, Willis resigned in 1911 when it became abundantly clear that the Royal Botanic Gardens were to be subordinated to a new Department of Agriculture, a development almost certain to restrict his independence. He applied his dynamism to re-organizing the botanical gardens at Rio de Janeiro for the next three years before returning to Cambridge to teach. Calcutta and Peradeniya are the two great botanical gardens of South Asia. Calcutta is older and historically central to the development of botanical research but Peradeniya offers more to delight the tourist: regiments of tall trees, clumps of gigantic bamboos and the chromatic brilliance of exotic blossoms on branches and creepers and in flower-beds.

Ceylon may have been the goal of many botanists and plant collectors but it was often their final resting place. William Kerr died after only two years in the island, and Alexander Moon and his successors, James Macrae and James George Watson, also died prematurely. Few people fully appreciate the trials and tribulations, the dangers and the frustrations attendant on that ostensibly gentle pursuit of plant collecting. Victor Jacquemont succumbed to a liver complaint, Adolphe Schlagintweit was murdered, Joseph Hooker briefly suffered imprisonment, and Anton Hove was robbed several times. Personal isolation in remote regions was sometimes disliked more than physical hardships. Captain J. D. Herbert, leading a mineralogical survey in the Himalayas, pleaded companionship as well as professional expertise in urging the appointment of George Govan as his zoologist and botanist.

The presence of a companion in these excursions is very desirable both on public and private grounds nor is it to be expected that a solitary individual will feel the same ardour in penetrating to those distant and inclement spots where alone the most interesting observations are to be made[26].

Yet Jacquemont, the most sociable of men, who enjoyed comfort, conversation, and the company of women, never minded the long stretches alone with his porters and bodyguard. Until his second marriage when his wife joined him on expeditions, Frank Kingdon Ward mostly travelled alone. On the other hand, the bond of friendship and shared enthusiasms between Ludlow and Sheriff contributed to the success of their collecting missions.

Isolation and deprivation could induce doubts about the purpose and validity of collecting. William Griffith, having just narrowly avoided an ambush by Afghan tribesmen, wondered 'what business have I here collecting plants, with so many in Calcutta demanding attention? How I am living! alone, without a table, chair, wine or spirits, with a miserable beard, and in native clothes'[27]. This spasm of self-pity was uncharacteristic; Griffith was inured to danger and hardship. He had faced enraged buffalo and hostile natives in Assam, had travelled alone in Burma, and knew how to survive in the jungle.

Griffith was prepared to walk or to paddle a canoe; Buchanan chose a lumbering elephant or the rhythmic discomfort of a palanquin. Von Huegel always rode or endured the palanquin consoled by the ministrations of three cooks and a butler. Benjamin Heyne, thankful for the protection against sun and rain afforded by the palanquin, usually covered 25 to 30 miles a day with fresh relays of bearers. His entourage on his trip to Hyderabad included sepoys, six boys, four bearers, a personal servant, a draughtsman, and two plant collectors. William Moorcroft's retinue of muleteers and porters was so large that nervous tribesmen feared a British invasion. Jacquemont was proud of his self-restraint when he reduced his complement of porters to 36 and made do with only five servants. He told a correspondent to imagine him

Mounted on a white horse (I am predestined to have white horses, pistols in good order, etc., etc.) I shall open the march, followed by two poor devils, who will cost me twenty-four or thirty francs a month; one of whom, the syce, is properly the groom; and the other, the gassyara, or grass-cutter, is laden with my horse's food. Each will carry one of my guns loaded with ball or shot according to circumstances. When I gallop they will run; this is the custom[28].

A bamboo cart, hauled by a couple of oxen, carried his baggage. Before setting out he breakfasted on rice boiled in milk and dined at the end of the day's journey on fowl garnished with rice and rancid butter, drank water with a blush of brandy or maybe a cup of hot tea if the temperature justified it. Joseph Hooker in Nepal maintained, as far as he could, an agreeable domestic comfort surrounded by his instruments and collections.

My tent was made of a blanket thrown over the limb of a tree; to this others were attached, and the whole was supported on a frame like a house. One half was occupied by my bedstead, beneath was stored my box of clothes, while my books and writing materials were placed under the table. The barometer hung in the most out-of-the-way corner, and my other instruments all around. A small candle was burning in a glass shade, to keep the draught and insects from the light, and I had the comfort of seeing the knife, fork and spoon laid on a white napkin, as I entered my snug little house, and flung myself on the elastic couch to ruminate on the proceedings of the day, and speculate on those of the morrow, while waiting for my meal, which usually consisted of stewed meat and rice, with biscuits and tea. . . . After dinner my occupations were to ticket and put away the plants collected during the day, write up journals, plot maps, and take observations till 10 p.m. As soon as I was in bed, one of the Nepal soldiers was accustomed to enter, spread his blanket on the ground, and sleep there as my guard. In the morning the collectors were sent to change the plant-papers, while I explored the neighbourhood, and having taken observations, all were ready to start at 10 a.m.[29].

Hooker's elaborate organization and generous provisions ran counter to Edward Madden's advice to would-be Himalayan travellers to travel lightly with the minimum of native helpers. Madden recommended lightweight tents, bedding wrapped in double-waxed cloth, an abundance of footwear, metal eating utensils, and a green gauze veil or goggles for protection against snow glare[30].

Madden's experience had taught him that it was preferable to recruit native labour on a daily basis. Hooker employed natives specifically to carry his store of paper for drying plants and to climb trees. He had nothing but praise for his Lepchas; his Bhutanese porters, on the other hand, were rebellious and unco-operative. Buchanan found that his native collectors were reluctant to climb trees where they might be bitten by ants. Tigers were understandably a greater deterrent.

Our woods are so much infested with tygers that it is not easy to get men to go far into them in search of plants or seeds[31].

Nathaniel Wallich's Bengali collectors, disliking the cold of the higher reaches of Nepal's mountains, gave scant attention to alpine plants. When Govan collected during Herbert's expedition in the Himalayas he was allowed just one plant collector and two other natives 'for the preparation of subjects in natural history'. Hooker and Thomson collecting in the Khasia Hills seldom had fewer than twelve. R. E. Cooper established such a good working relationship with his Lepcha collector that the man volunteered to join a second expedition despite having been seriously ill during the first tour. Benjamin Heyne was impressed by the botanical knowledge of his collectors.

I have some collectors who have made such progress in the Linnaean system as to be able to distinguish male flowers from female in the Dioecious class in plants which they have never seen before[32].

Frank Kingdon Ward, the prince of plant hunters, distinguished two types of collectors—the botanist and the horticulturist. The latter is concerned not only with novelty but also beauty, hardiness, and ease of propagation; he must be able to assess a plant's acceptability by the gardening fraternity. He and the botanist may admire a plant in bloom but both have to wait patiently for the seed to set before collecting it. Robert Ross, head gardener at Calcutta during the mid-nineteenth century, proposed one way round this problem by suggesting that cuttings heeled in a box of friable mould would survive a journey of several days. The rediscovery of the location of a specific plant in order to harvest its seed demands an unerring visual memory. Kingdon Ward said of a pygmy rhododendron, *R. imperator*, that 'to get seed of it I had to find, not only the same gully again (which was comparatively easy) but the identical tiny tuffet'. When he returned to one of his earlier sightings of *Primula alta* in Assam, he was dismayed to find it had set no seed. Another destination had been obliterated by an early blanket of snow. 'I scraped away the snow, and revealed the familiar foliage; next minute I was digging like a badger, until the whole plant was exposed. And not one single capsule did it bear'. His experience corrects the romantic notion of an idyllic life, a carefree holiday. Plant collecting can be arduous, sometimes dangerous, often lonely, the weather appalling and the business of collecting seed, drying and pressing plants, a monotonous routine. The rewards are independence, the anticipation of new discoveries and the privilege of intimacy with the natural world. As an old man Sir Joseph Hooker nostalgically recalled an early encounter with the vegetation of the tropics.

I was skirting the edge of the magnificent forest that then clothed the base of the Sikkim Himalaya (now, I believe, replaced by tea plantations); twilight had just commenced, and I scarcely realized the scene, when the cicadas burst into full cry (it is impossible to call it song) with startling effect, sound and scene combining to herald my advance into, to me, a new world of interest and botanical excitement.

NOTES

1 McCrindle (1877), 170.
2 Ibn Batuta (1829), *Travels of Ibn Batuta* (trans. Rev. Samuel Lee), London, 189.
3 *Curtis's Botanical Magazine*, plate 9323.
4 A true and exact description of the great island of Ceylon. *Ceylon Historical Journal*, **8**, 1958–9.
5 Knox (1911), 31.
6 Heniger (1986), 29.
7 Ibid., 61.
8 Ibid., 269.
9 Ibid., 85.
10 *Hooker's Journal of Botany*, **1**, 1834, 180–1.
11 *Kew Magazine*, **4** (4), 1987, 201.
12 *Madras Journal of Literature and Science*, N.S. no. 11, May 1861.
13 I.O.L.R. MSS Eur E.80–2. These volumes include 184 flower drawings and nearly 150 pages of notes on the botany of Ceylon.
14 Cordiner (1983), 35.
15 Banks to Kerr, 30 June 1810. B.M. (N.H.) D.T.C. Vol. 18, folios 45–6.
16 Banks to Hooker, 1 April 1811. Kew Archives. Banks letters, Vol. 1, folio 33.
17 Banks to Hooker, 19 June 1813. Kew Archives. Banks letters, Vol. 1, folio 39.
18 Sir Alexander Johnston to Banks, 12 January 815. B.M. (N.H.) D.T.C., Vol. 19, folios 121–8.
19 Hardwicke to Banks, 28 October 1815. B.L. Add. MSS 9869, folio 18.

[20] Banks to Johnston, 25 June 1816. B.M. (N.H.) D.T.C., Vol. 19, folios 294–6.
[21] *Report on the Royal Botanic Garden of Peradenia. . . .* , August 1844, viii.
[22] Ibid., August 1845, 7.
[23] W. Hoffmeister (1848), 131.
[24] Harvey to W. J. Hooker, 20 November 1853. Kew Archives. Hooker letters, Vol. 33, folio 196.
[25] M. North (1892), Vol. 1, 305.
[26] 30 November 1826. I.O.L.R. F/4/957, item 27123(24).
[27] 30 January 1840. *Posthumous papers*, 1847, xviii.
[28] Jacquemont (1935), Vol. 1, 147.
[29] J. D. Hooker (1854), Vol. 1, 132–3.
[30] *Journal of Asiatic Society of Bengal*, **15**, 1846, 79–81.
[31] Sir John Murray to Banks, 13 April 1790. B. L. Add. MSS 33979, folios 24–5.
[32] B. Heyne (1814), *Tracts, historical and statistical on India*, 248.
[33] *Curtis's Botanical Magazine*, 1893, plate 7282.

Pursuit of the Picturesque

Until Hodges's paintings were published in London towards the end of the eighteenth century, the British visual perception of India relied upon engravings inspired by travellers' descriptions or amateurs' drawings. William Hodges (1744–97) was the first professional artist of any merit to record his impressions of the subcontinent. Trained at William Shipley's drawing school in the premises of the Society of Arts, he was apprenticed to the landscape painter Richard Wilson, a devotee of the classical landscapes of Claude and Poussin. Hodges learnt from Wilson how to compose poetic landscapes in the manner of these French painters, adroitly manipulating atmospheric light and shade on canvas and paper. The Admiralty engaged him as a topographical artist on Captain Cook's second voyage to the Pacific in 1772. During his three years on HMS *Resolution*, he had to subordinate his romantic vision to the prosaic requirements of topographical accuracy. He returned to the tropics, selecting India as a subject for his brush after the death of his wife in 1777. Landing in 1780 on the broad strand skirting Madras, his trained eye appraised approvingly 'the clear, blue, cloudless sky, the polished white buildings, the bright sandy beach, and the dark green sea'. Such light and strong colours tested his technical skills, the country and its inhabitants proffered picturesque compositions—temples and forts, parades of soldiers, caravans of merchants, and the 'fine antique figures' of women emerging from the river 'with wet drapery which perfectly displays the whole person'. He was attracted by the extraordinary configuration of the banyan tree. He rejoiced in the rebirth of the countryside during the monsoons, everywhere 'such a freshness of verdure . . . such vigour and fulness of foliage'. The powerful aquatints in his *Select views of India* (1785–8) enticed other artists to the country.

> *Many other tours in that interesting country might be undertaken by the enterprising artist. We know that the whole coast of Malabar possesses picturesque beauty equal to any country on earth; and how valuable would be the representation of that scenery, whether as a natural object, or as connected with the history of the country, and the manners of the people*[1].

For most visitors to India, the term 'picturesque' conjured up ruins in an ancient landscape, pastoral life in remote villages, exotic animals and birds in the jungle, and flamboyant flowers everywhere (Fig. 13.1). The gospel of the picturesque had been preached by the Reverend William Gilpin who tirelessly toured England, Scotland, and Wales lauding irregularity, ruggedness, and variety in the landscape. His accolade was always bestowed on 'that kind of beauty which would look well in a picture'. Eighteenth-century tourists were encouraged to view a landscape through paintings they remembered, the canvases of Claude Lorrain, Gaspar Poussin, and Salvator Rosa.

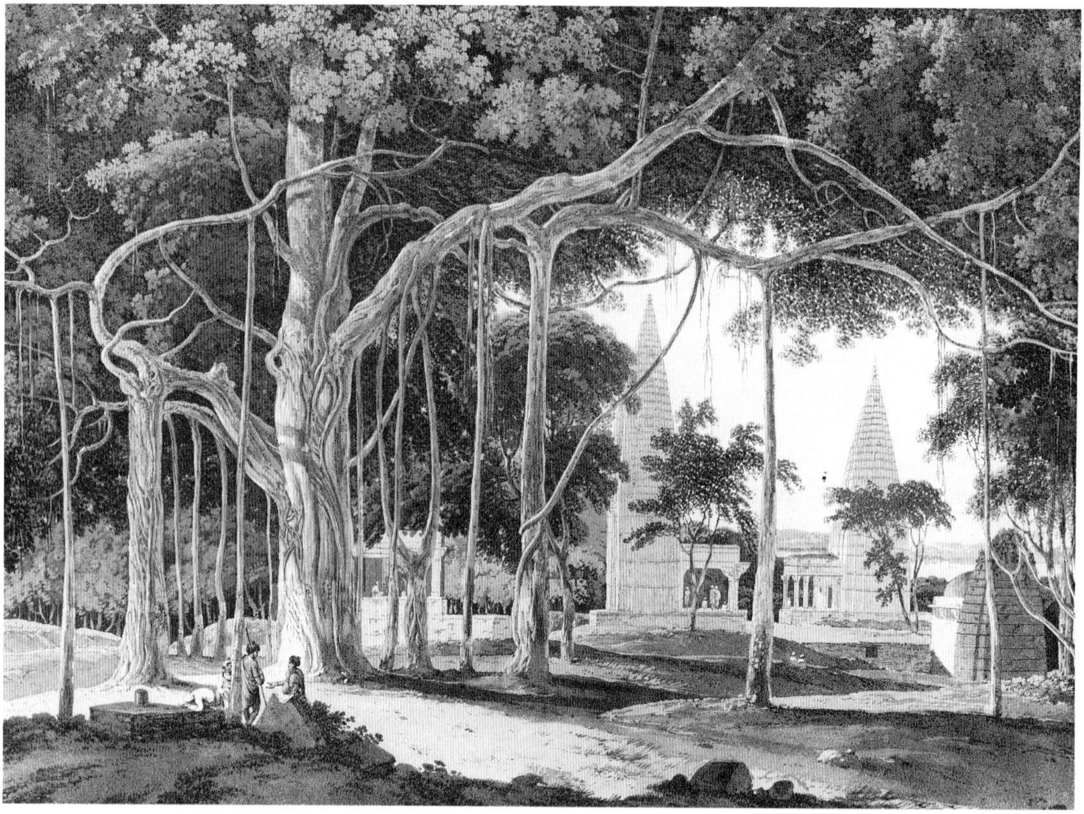

Figure 13.1 Banyan tree and Hindu temples at Agouree on the River Soane. (From T. and W. Daniell, (1795–7) *Oriental scenery*, London, plate 19.) Coloured aquatint. The Daniells' engravings confirmed the European romantic vision of India as an exotic landscape with a surfeit of ancient buildings.

Bishop Heber inspecting his Indian diocese exclaimed that 'Salvator Rosa never painted glens more wild and remote than today in our path to Dikkalee'[2]. Joseph Hooker, climbing and stumbling in the Himalayas, paused to invoke the name of this Italian artist:

> *The scenery is as grand as any pictured by Salvator Rosa; a river roaring in sheets of foam, sombre woods, crags of gneiss and tier upon tier of lofty mountains flanked and crested with groves of black firs, terminating in snow-sprinkled rocky peaks*[3].

It was not long before a stream of artists followed Hodges to India in search of people and places to paint—Tilly Kettle, Johann Zoffany, Thomas and William Daniell, Arthur Devis, and Francesco Renaldi. In 1804 William Westall urgently solicited Sir Joseph Banks's intervention to get a passage to Ceylon,

> *a country where I could scarcely fail of success: for the rich picturesque appearance of that island; every part affording infinite variety, must produce many subjects to a painter extremely valuable*[4]

François Balthazar Solvyns (1760–1824) practised both as a painter and engraver in Antwerp and Brussels before fleeing the country during the upheavals of 1789. Reaching India the following year, he stayed in Calcutta, studied Hindi and settled down to drawing representatives of the confusing hierarchy of castes in the local community.

A tangible expression of his industry is a substantial volume, *Collection of two hundred and fifty coloured etchings descriptive of the manners, customs and dresses of the Hindoos*, published in Calcutta in 1799, naive in execution and crudely coloured. It was subsequently expanded into the more ambitious *Les Hindous*, published in Paris, 1808–12. Its title is misleading since it also embraces animal studies and 12 engravings of useful plants, such as the mango, banana, sugar cane, rice, cotton, and indigo (Fig. 13.2). He reminded his readers that his drawings

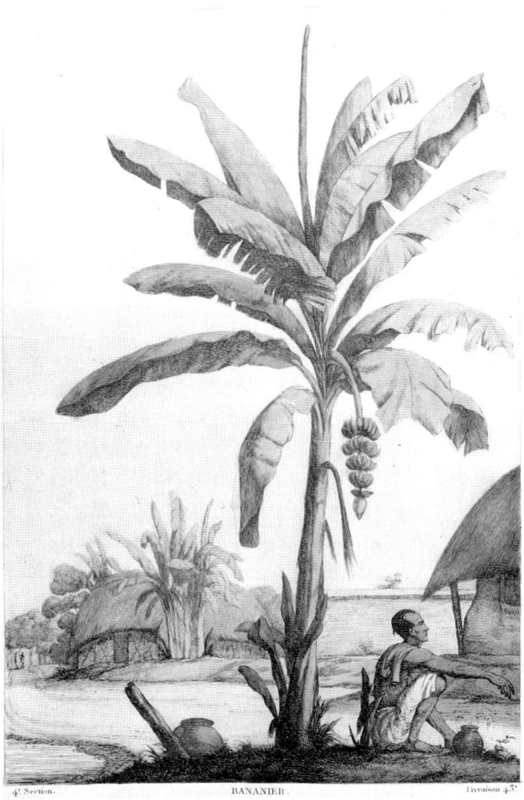

Figure 13.2 Banana trees. (From F. B. Solvyns (1808–12), *Les Hindous*, Paris.) Hand-coloured etching.

> *were taken by myself upon the spot. Instead of trusting to the works of others, or remaining satisfied with the knowledge contained in preceding authors, I have spared neither time, nor pains, nor expense, to see and examine with my own eyes, and to delineate every object with the most minute accuracy*[5].

His plant portraits were placed in the communal context of people and their homes. His insistence on self-verification extended also to the text which contained

> *strictly no more than what I myself was witness to. This is an advantage which I flatter myself to possess over the writers of natural history who have not travelled. All that I have read in Europe since my return from India, is a proof to me that the natural history of this country is still but little known, and that those who have written upon it are often ill-informed*[6].

But natural history plays a minor role in *Les Hindous* which exalts costume as the quintessence of the picturesque. Lieutenant Colonel Charles Forrest saw nature as an incomparable theatrical setting for his *Picturesque tour along the River Ganges and Jumna in India*, published with 24 aquatint plates in 1824. He made many sketches of villages 'with their straw-thatched bamboo cabins, backed by a rich group of wood, in which the palmyra reared its towering height and fan-like leaves, the bamboo waved its graceful and feathery branches, and the plantain threw its immense leaves of the most vivid green, green in every fantastic form' (Preface). His book, published by Rudolph Ackermann in the Strand in London, one of a group of publishers who specialized in extravagantly produced colour-plate books, was bought by a public eager to indulge in armchair travelling. For most of them no region of the subcontinent was more picturesque than the remote Himalayas.

James Baillie Fraser (1783–1856) who joined the family firm in Calcutta in 1813 was fortunate to have his *Views in the Himala Mountains* (1820) engraved by Robert Havell, one of England's leading aquatinters. His *Journal of a tour through part of the snowy range of the Himala Mountains, and to*

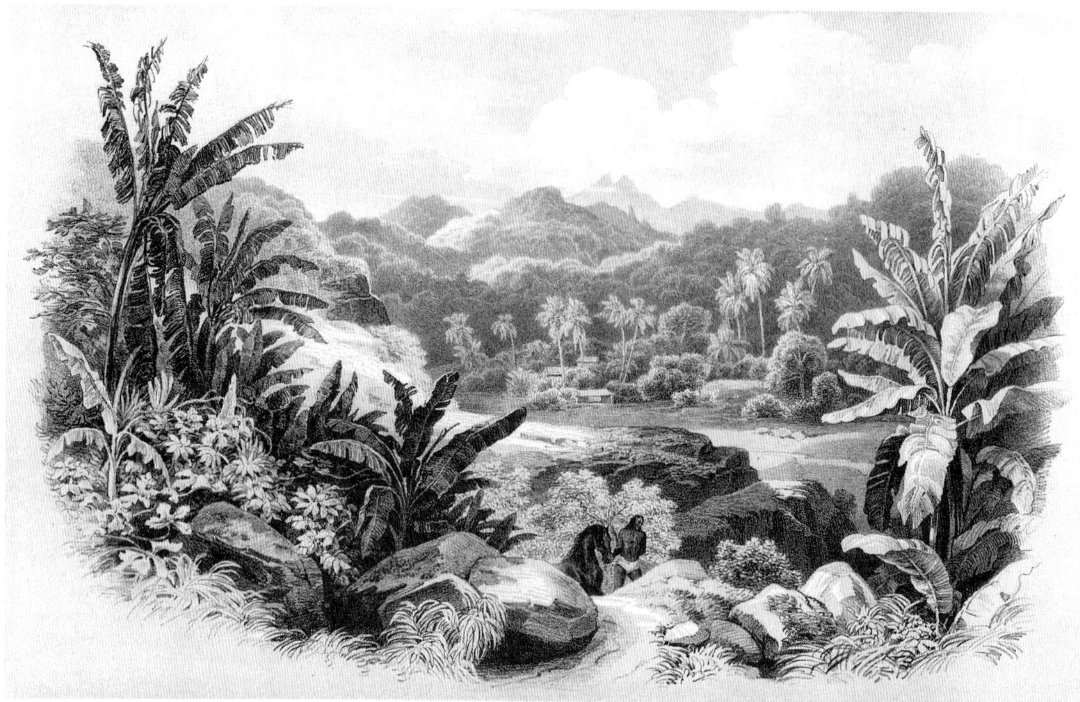

Figure 13.3 Rampant vegetation demonstrating 'disorder' and 'irregularity'—desirable attributes in any picturesque composition. (From F. W. Waldemar (1853), *Zur Erinnerung an die Reise den Prinzen Waldemar von Preussen nach Indien in den Jahren 1844–46*, Berlin, plate 25.) Lithograph after drawing by Prince Waldemar.

the sources of the rivers Jamna and Ganges, published in the same year, is punctuated with ecstatic cries of pleasure when his inexpert eye thinks it recognizes birch, oak, holly, poplar, strawberries, columbines, buttercups, and cowslips.

> *The variety of the trees and shrubs and smaller herbs was astonishing. At each step some new plants attracted the eye; and many of those meadow flowers which are common in Europe were hailed with great satisfaction*[7].

Fraser's magnificent plates were for the privileged wealthy; George Francis White's *Views in India chiefly among the Himalaya Mountains* (1836–8) was aimed at a larger, less affluent and perhaps less discriminating audience. The author's language was calculated to give his readers a vicarious thrill:

> *above our heads crag piled itself upon crag, the interstices being richly clothed with foliage, forest trees springing from the rifts, while creepers threw down their wild garlands to our feet*[8].

More restrained are the laconic notes in the diary of Edward Lear who explored the neighbourhood of Simla during 1874:

> *Quiet and immensity of the scenery, silence of forest. Lovely creepers and cordage, and jungle; bamboos, many dead: hay coloured. Astonishing variety of colours, mostly deep greens, but mixed with yellow green, yellows, orange, red and brown.*

Familiar flowers were always a comforting reminder to the home-sick expatriate, an invitation to indulgent nostalgia. It was a

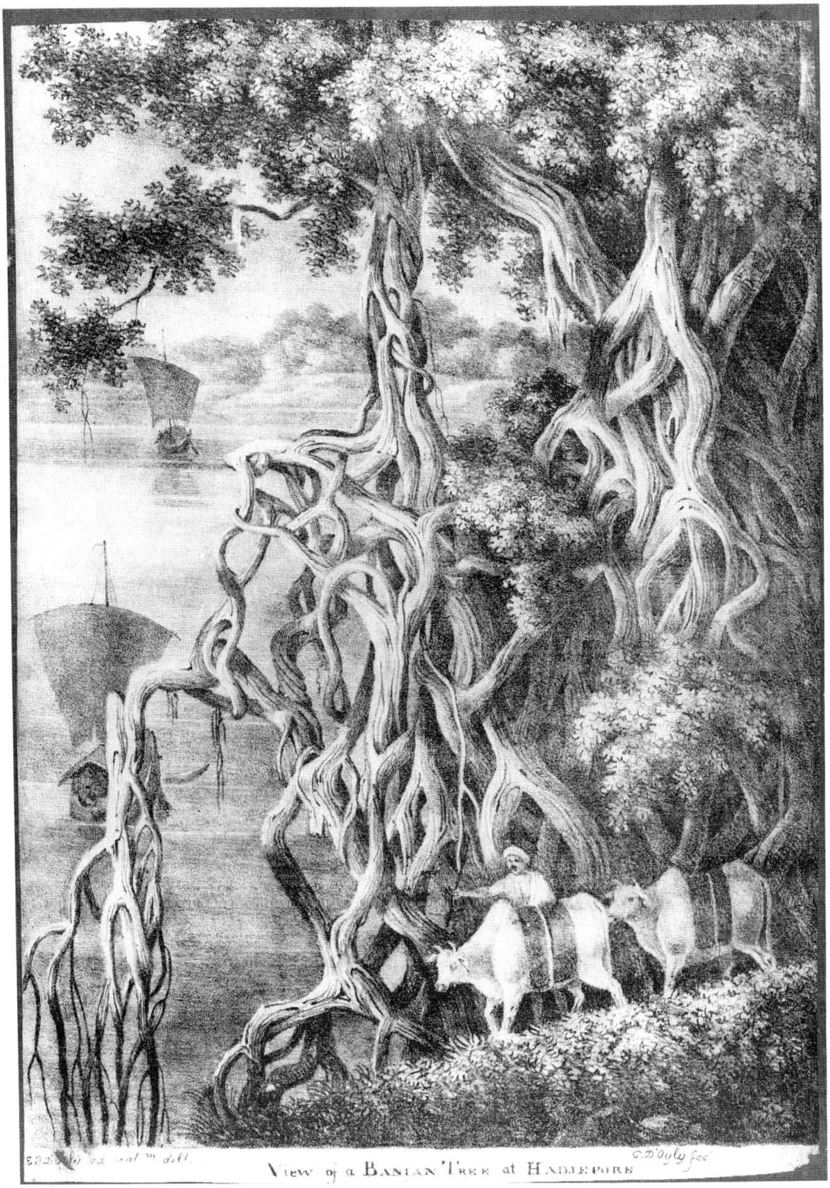

Figure 13.4 The convoluted aerial roots of a banyan tree was an irresistible subject for Sir Charles D'Oyly's lithographic skills.

memorable day for Fraser when he discovered a plant in the foothills that reminded him of the heather of Scotland.

> *And no one who acknowledges the lively feeling of attachment to his country . . . by the sight of a flower, of a scene, however rude, that he has seen in the days of his youth, but will sympathise in the joy with which I hailed this much loved and long looked for remembrancer of the 'land of brown heath and shaggy wood.' I plucked it and placed it in my hat, with a delight that I believed amused my attendants*[9].

Joseph Hooker was surprised to identify on a rocky hillside path

> *the two commonest of all British weeds, a grass (Poa annua) and the shepherd's purse! They had evidently been imported by man and*

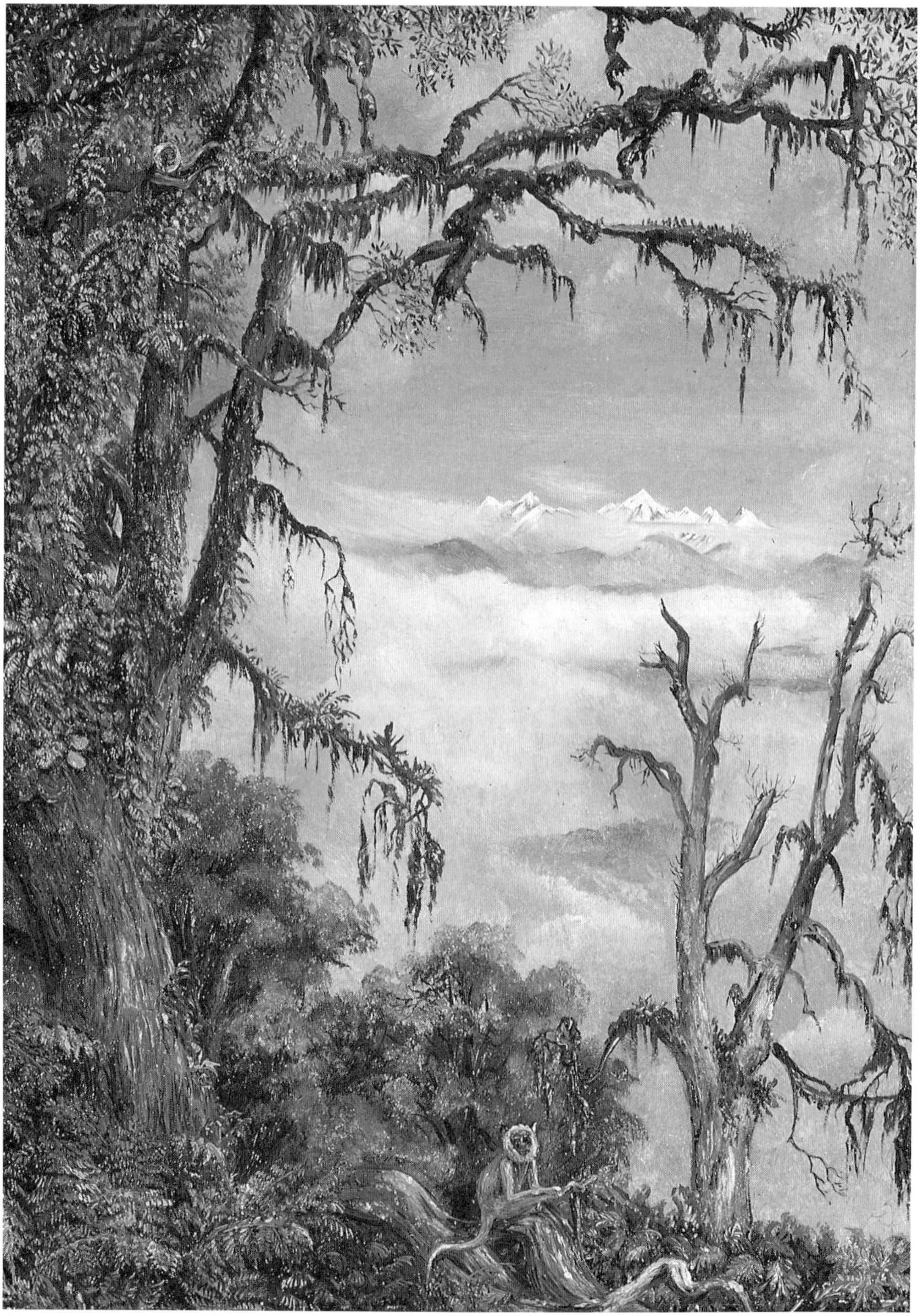

Figure 13.5 Oak trees against a backdrop of the Himalayas. Oil painting by Marianne North. 'The oak-trees had their branches hung with lichens and mosses, while the rhododendrons were almost as big, with pink bark. There were endless studies everywhere' (North (1892), *Recollections*, Vol. 2, p. 16).

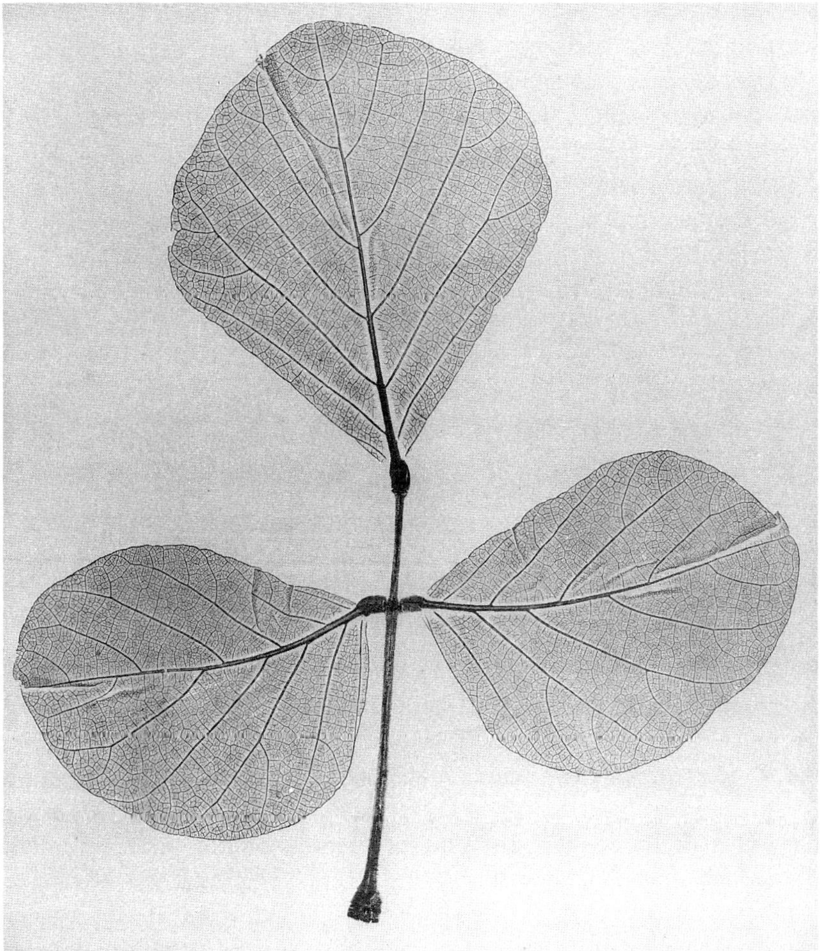

Figure 13.6 From H. Smith (1857), *Specimens of nature printing from unprepared plants*, Madras.

yaks, and as they do not occur in India, I could not but regard these little wanderers from the north with the deepest interest[10].

By Victorian times the word 'picturesque' had been trivialized to imply the pretty and the sentimental, used indiscriminately to describe any scene that pleased the often uncritical eye. There were those—a tiny minority—who perversely refused to admit any picturesque attributes in the Indian landscape. Jacquemont, having just reached Simla, was scornful. 'With regard to the picturesque, India is very poorly gifted.' He denigrated the landscape in the foothills as 'poor and monotonous. In the highest mountains in the world there is necessarily grandeur; but it is grandeur without beauty'. He preferred the Alps. Emily Eden, a shrewd observer of social and domestic life, told one of her correspondents that India had 'the most picturesque population with the ugliest scenery, that ever was put together'.

One of her relaxations was sketching flowers in the Calcutta Botanic Garden. Drawing was the pastime of many Europeans who were inspired to record in brush and pencil the Indian scene around them. Their level of achievement was remarkably high, often equal in competence to the work of professional artists. Many were military personnel who had received instruction during their training. The Royal Military academy at Woolwich appointed its first drawing-master in 1743; one of their most eminent instructors was the topographical artist, Paul Sandby. Drawing was taught to the cadets of the East India Company military college at Addiscombe and to its future administrators at Haileybury College. Since the reason for these drawing skills was to assist surveying, it is not surprising that many officers were more at ease sketching landscape than figures. Thomas Bacon, a lieutenant in the Bengal Horse Artillery, drew 'the graceful palm and lofty cocoa-nut, the thickly clustered pepal, the feathery jhyut and spreading mango'. Bengal captivated him. 'The scenery in many parts of the river is highly picturesque, and furnishes to the artist the most beautiful examples of form, of colouring and of light and shade.'[11] He would have agreed with Emma Roberts that 'there never can be wanting subjects for the pencil in a country, and amid a people so truly picturesque'[12].

Artistic skill is a talent displayed by many naturalists. Thomas Hardwicke and Brian Hodgson could draw well enough to meet their own strict standards; William Griffith's rapid pen and ink floral dissections satisfied most of his colleagues; and Joseph Hooker's field sketches provided all the essential details for W. H. Fitch's polished versions. For those who lacked any facility whatsoever there was always nature printing, a process with an honourable lineage—even Leonardo da Vinci tested its capabilities. A silhouette image of a plant or an insect can be obtained by an application of ink or paint to the object which is then gently pressed on a sheet of paper. Readers of the *Madras Athenaeum* in 1848 were instructed in 'easy and expeditious methods of taking facsimiles of plants or of forming ornamental devices from creepers'. The South appears to have had a monopoly of interest in nature printing. The *Madras Journal of Literature and Science* discussed the process in its October/December 1857 issue, and Henry Smith, Superintendent of the Madras Government Press, produced *Specimens of nature printing from unprepared plants* (1857) (Fig. 13.6). A few impressions were printed in green ink but the majority are a lifeless black. Far superior in execution is *Nature's self-printing: a series of useful and ornamental plants of the South Indian flora* (1862), published by the Basel Mission Press at Mangalore. It was the work of the Reverend J. Hunzikar who counterfeited in natural colours the images of plants collected around Mangalore, the coastal region and as far away as Ootacamund. He reminded the purchasers of his book that 'the process of printing from fresh plants was found to be a difficult undertaking requiring much time and attention, therefore only a small number of copies have been compiled'. The limited number of impressions that could be taken before a plant became sodden and shapeless prevented the process from being commercially viable. In any case a new toy was competing for the amateur's attention—the camera. The long exposures required for early photographs posed problems for the photographer of vegetation whose image could be blurred by the slightest breeze. Thomas Biggs, a professional photographer working in the Bombay Presidency, found that

> *foliage is difficult to get in this country, on account of its never being still between 9 am and 5 pm, there being always a strong breeze from east or west, consequently pictures must be taken before 9 am, and then the time of exposure is considerably increased. On the coast this difficulty is not so great, and five minutes is sufficient for foliage; but though the island of Bombay presents some most beautiful subjects in foliage, I have never seen a good picture of any taken there*[13].

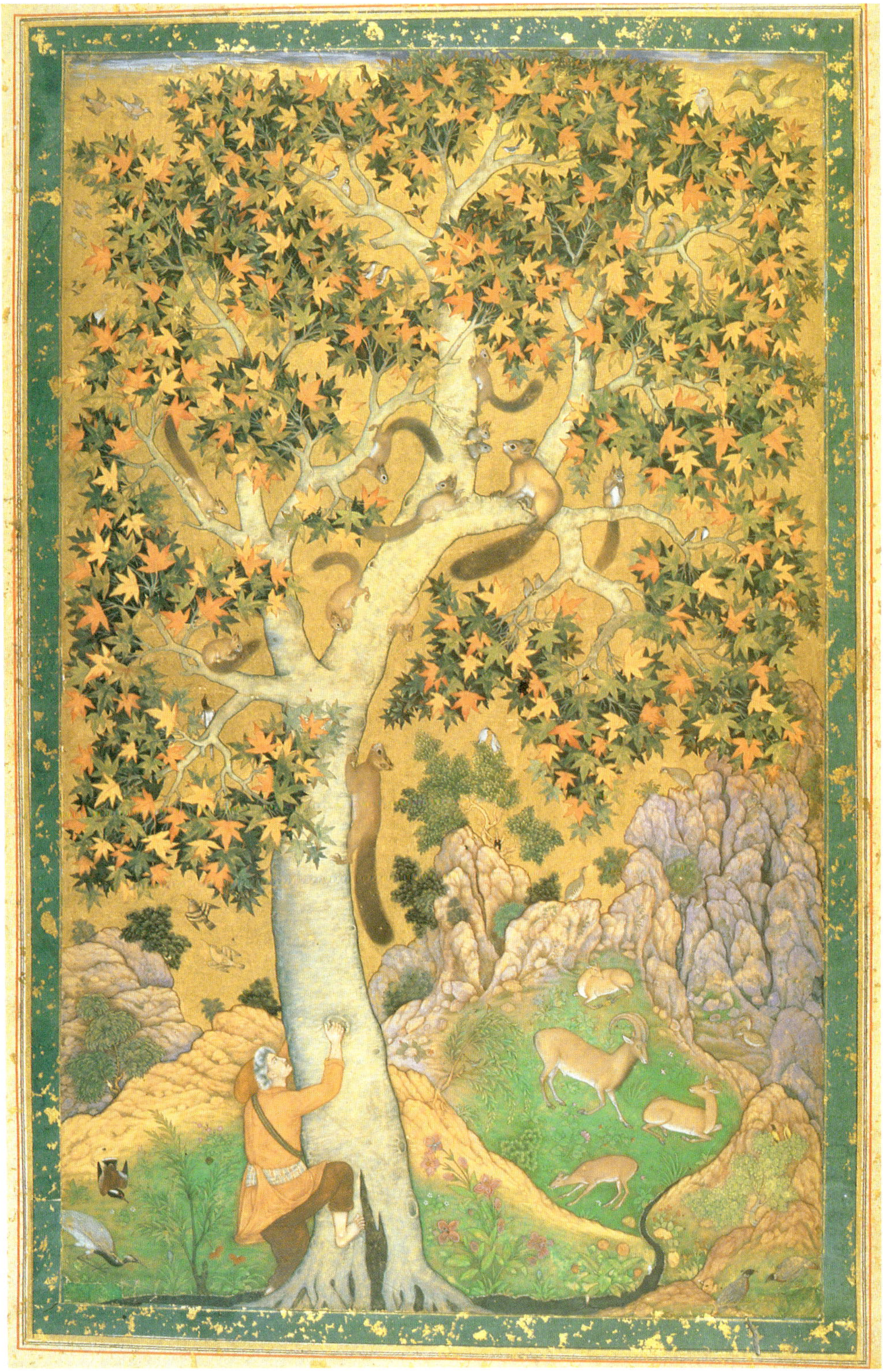

Plate 1 Chenar tree, *Platanus orientalis*, in its autumn tints. Mughal miniature by Abu'l-Hasan and Mansur, *c*.1610. During the reign of Jahangir, Mughal painters of plants and animals excelled in their meticulous observation of nature. (India Office Library and Records.)

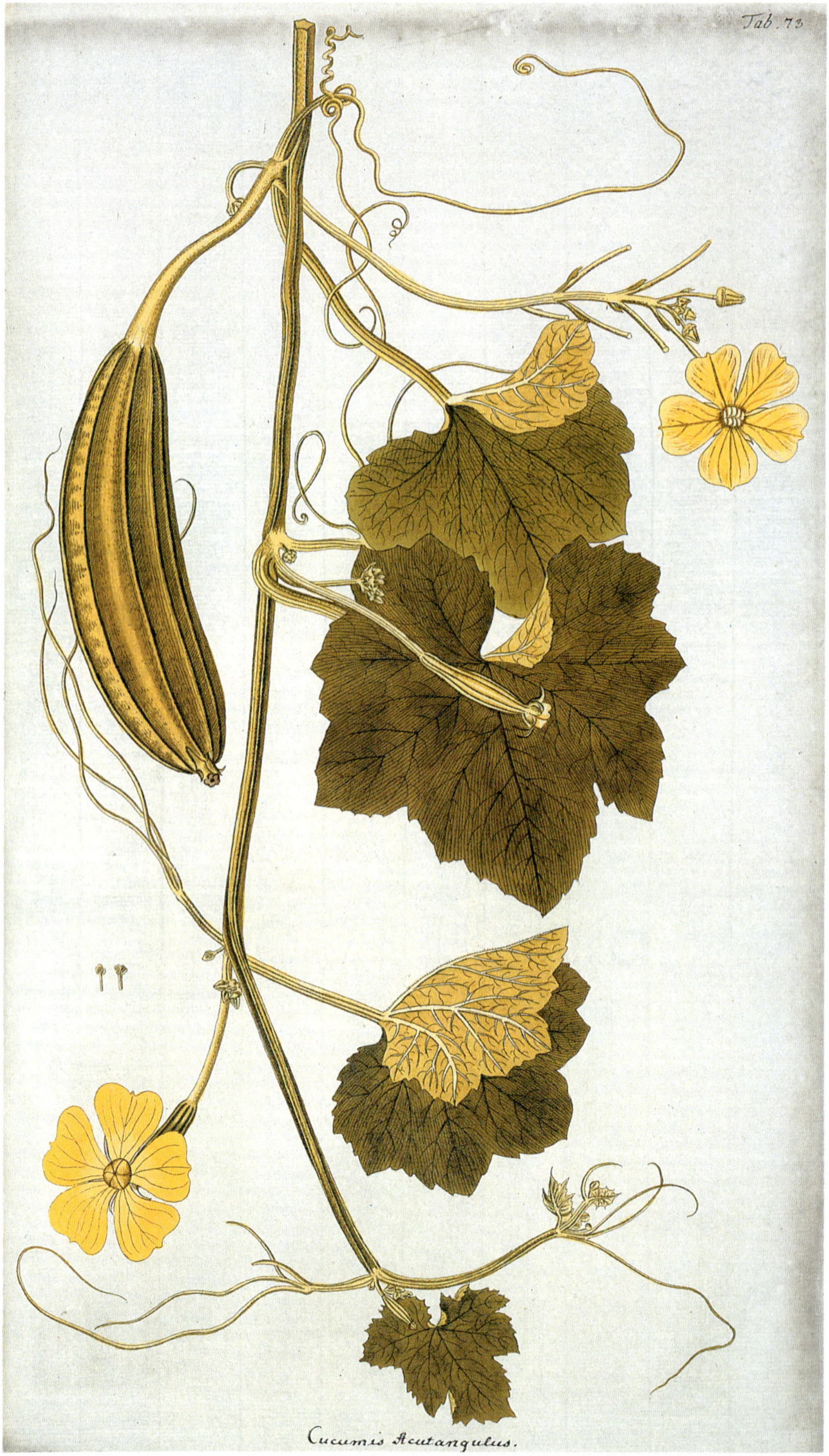

PLATE 2 *Luffa acutangula*. (From N. J. von Jacquin (1776), *Hortus Botanicus Vindobonensis*, Vol. 3, Vienna, plate 73.) Hand-coloured engraving. Jacquin was Professor of Botany and Chemistry at Vienna and Director of the University botanical garden. This work, issued under the patronage of the Empress Maria Theresa, records both ornamental and economic plants growing in the garden—some, like this Luffa, of Indian origin.

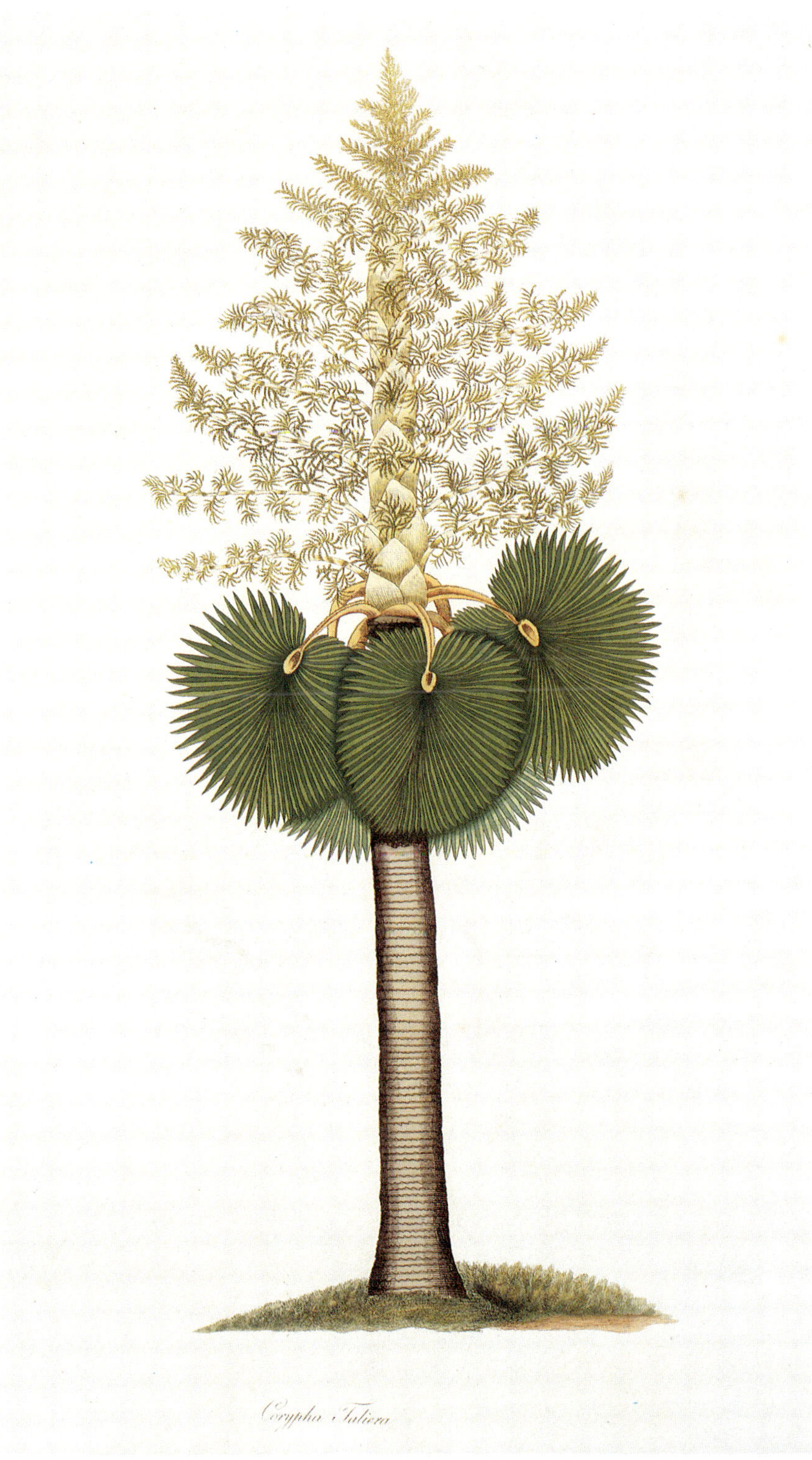

PLATE 3 *Coryphea taliera*. (From W. Roxburgh (1820), *Plants of the coast of Coromandel*, Vol. 3, London, plate 255.) Hand-coloured engraving. This palm, cultivated in Bengal, was once very common in Calcutta but in 1943 the only living specimen was in the Botanic Garden there.

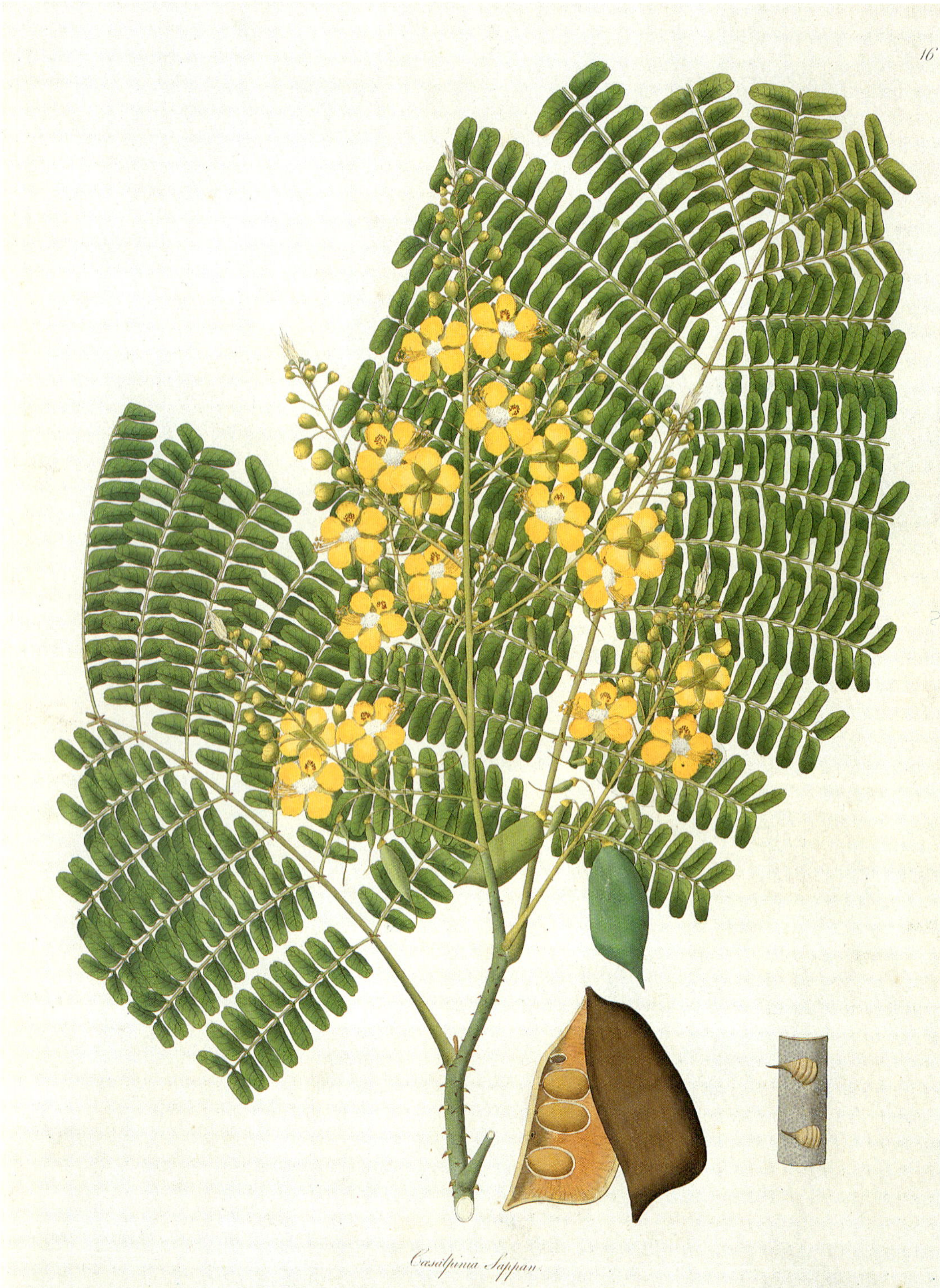

PLATE 4 *Caesalpinia sappan*. (From W. Roxburgh (1795), *Plants of the coast of Coromandel*, Vol. 1, London, plate 16.) Hand-coloured engraving. It was found in the Circars by William Roxburgh but it is also native elsewhere in India. The red heart-wood yields a dye used on cotton, silk, and woollen fabrics. Its abundance of thorns makes it a useful hedge plant.

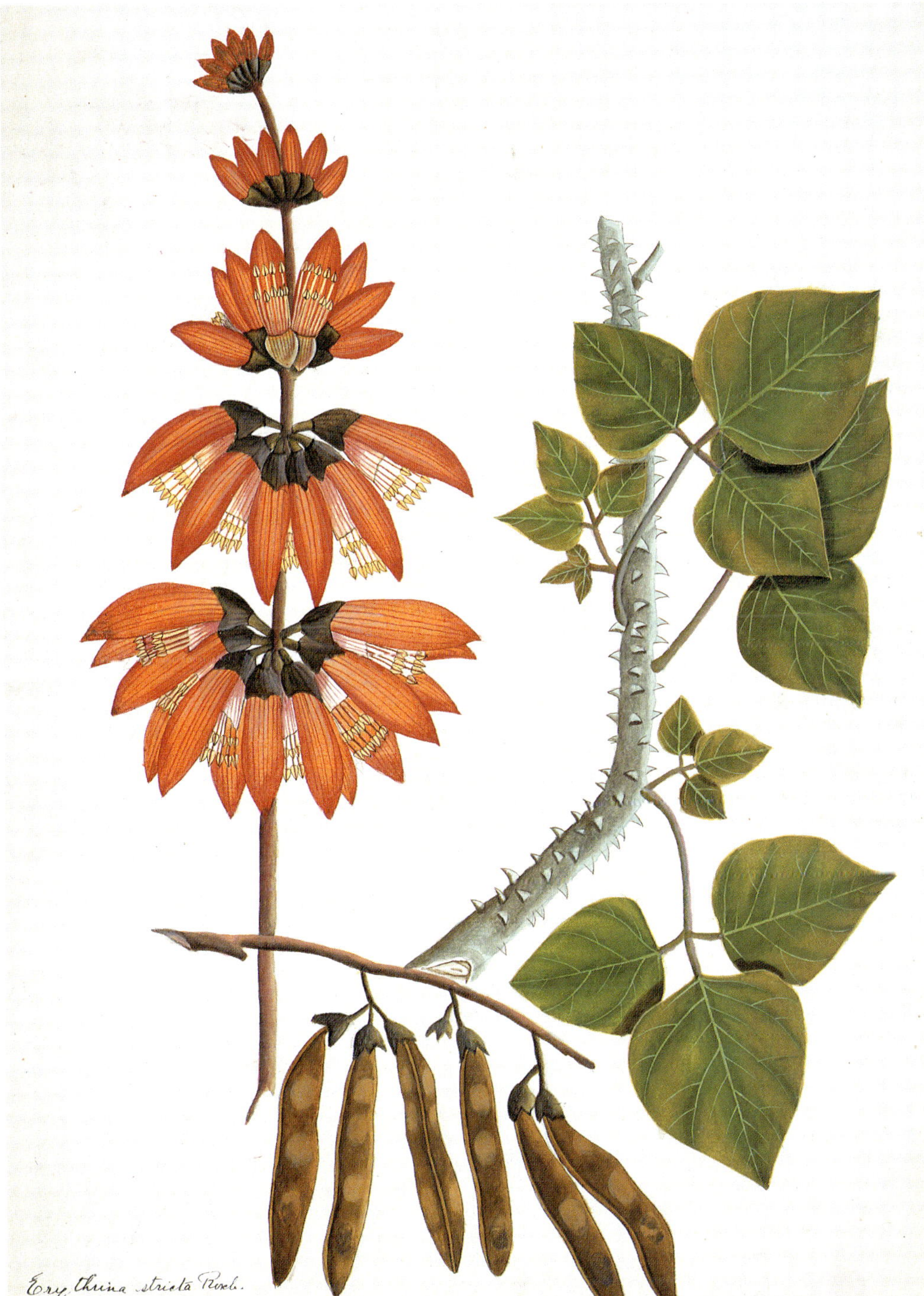

PLATE 5 *Erythrina stricta*. Watercolour from a collection of flower paintings by Indian artists believed to have belonged to Claude Martin (1732/5–1800). Originally a soldier in the French army, he transferred his allegiance to the British, eventually becoming a Major-General in the East India Company's forces. He was a discriminating collector of books, manuscripts, and paintings which he housed in his extraordinary palace at Lucknow, built to his own design.

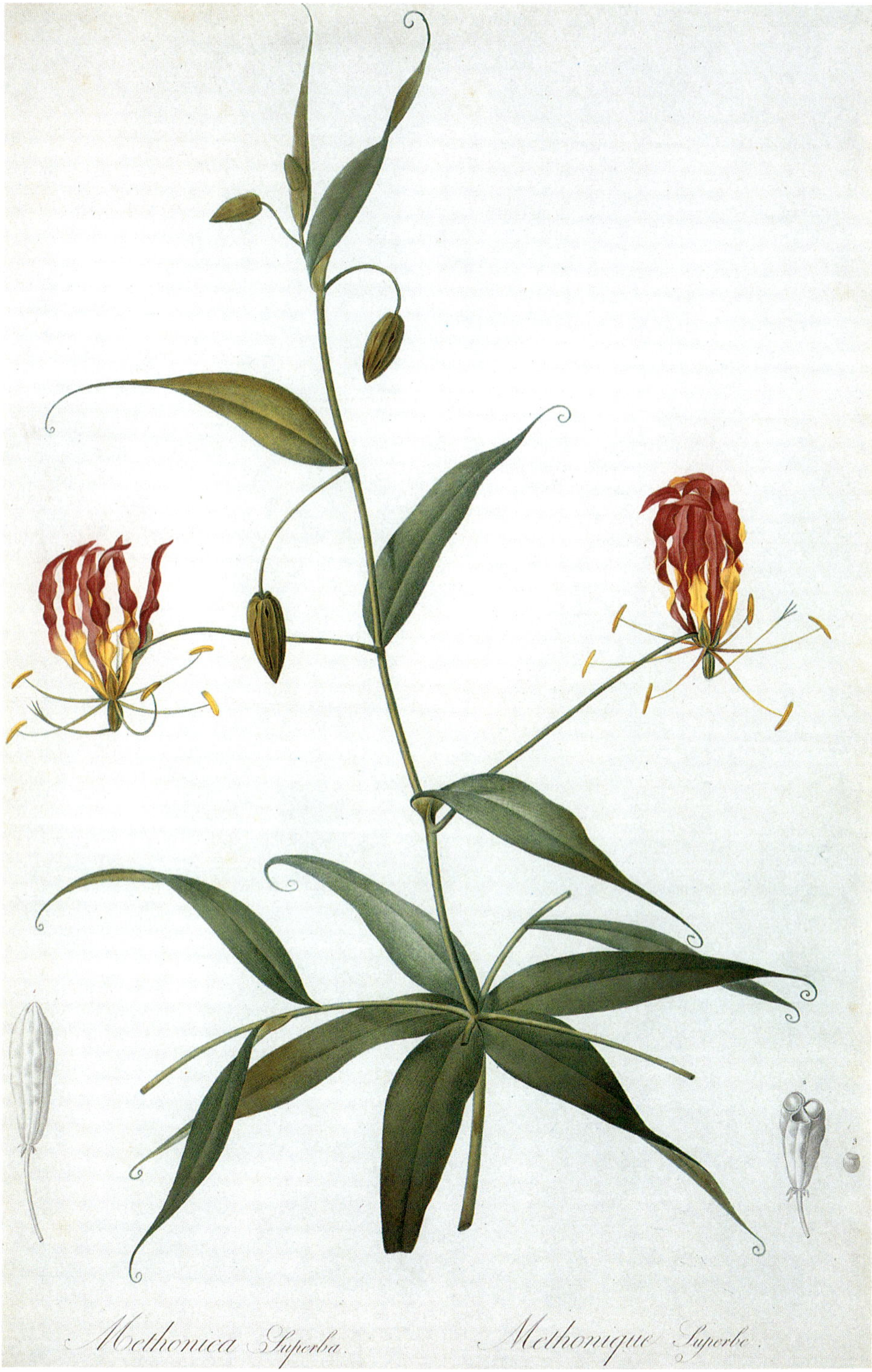

Plate 6 *Gloriosa superba.* (From P.-J. Redouté (1802), *Les Liliacées*, Vol. 1, Paris, plate 26.) Stipple engraving, colour-printed à la poupée. Native to Malabar, this plant is a vigorous climber which flowers during the early monsoons.

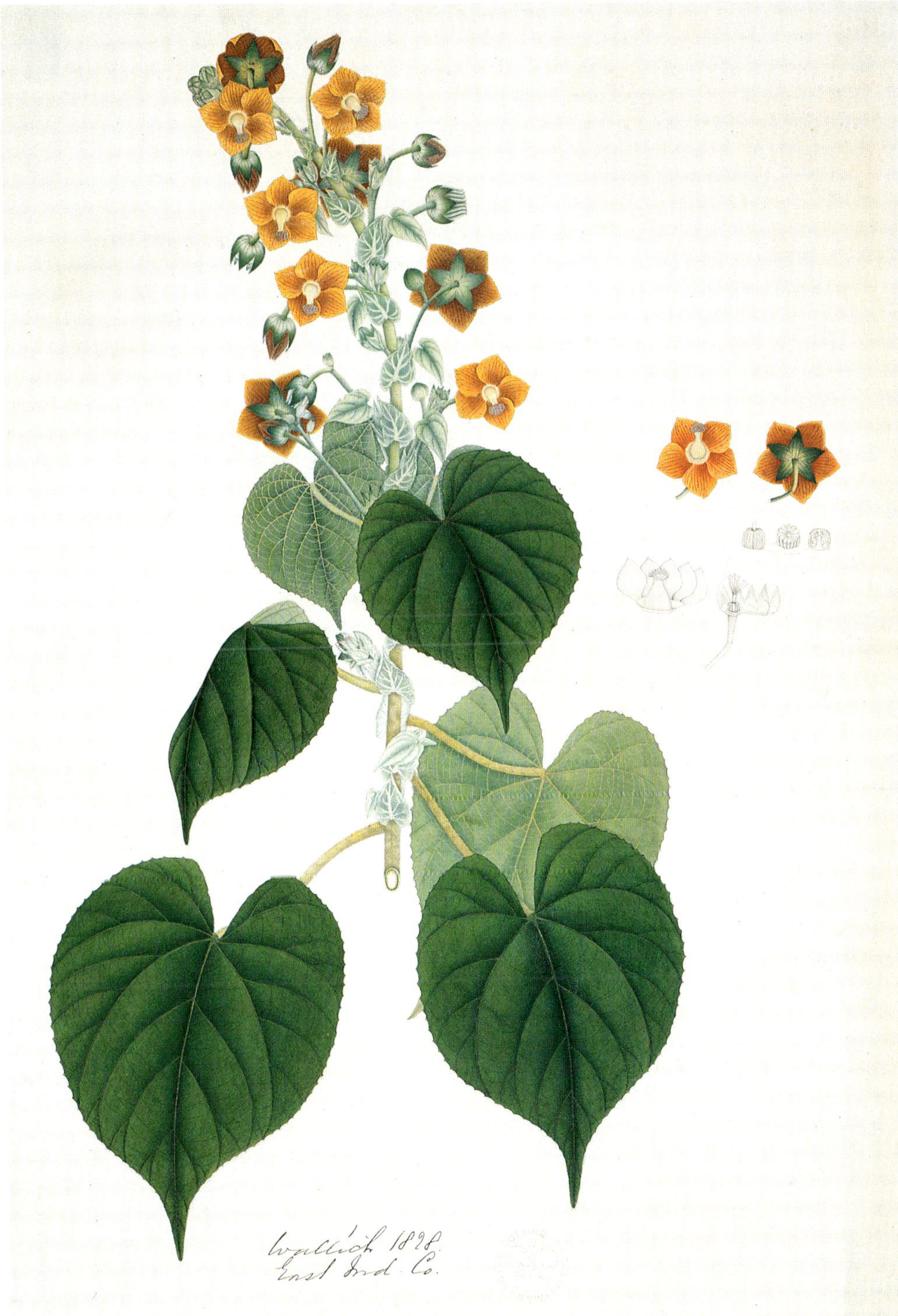

PLATE 7 *Abutilon auritum*. One of the many watercolour portraits of plants done for Nathaniel Wallich while he was Superintendent of the Calcutta Botanic Garden. The formal pose of the plant, decorative rather than natural, suggests the artist was an Indian, probably a member of the team of native artists employed at the garden.

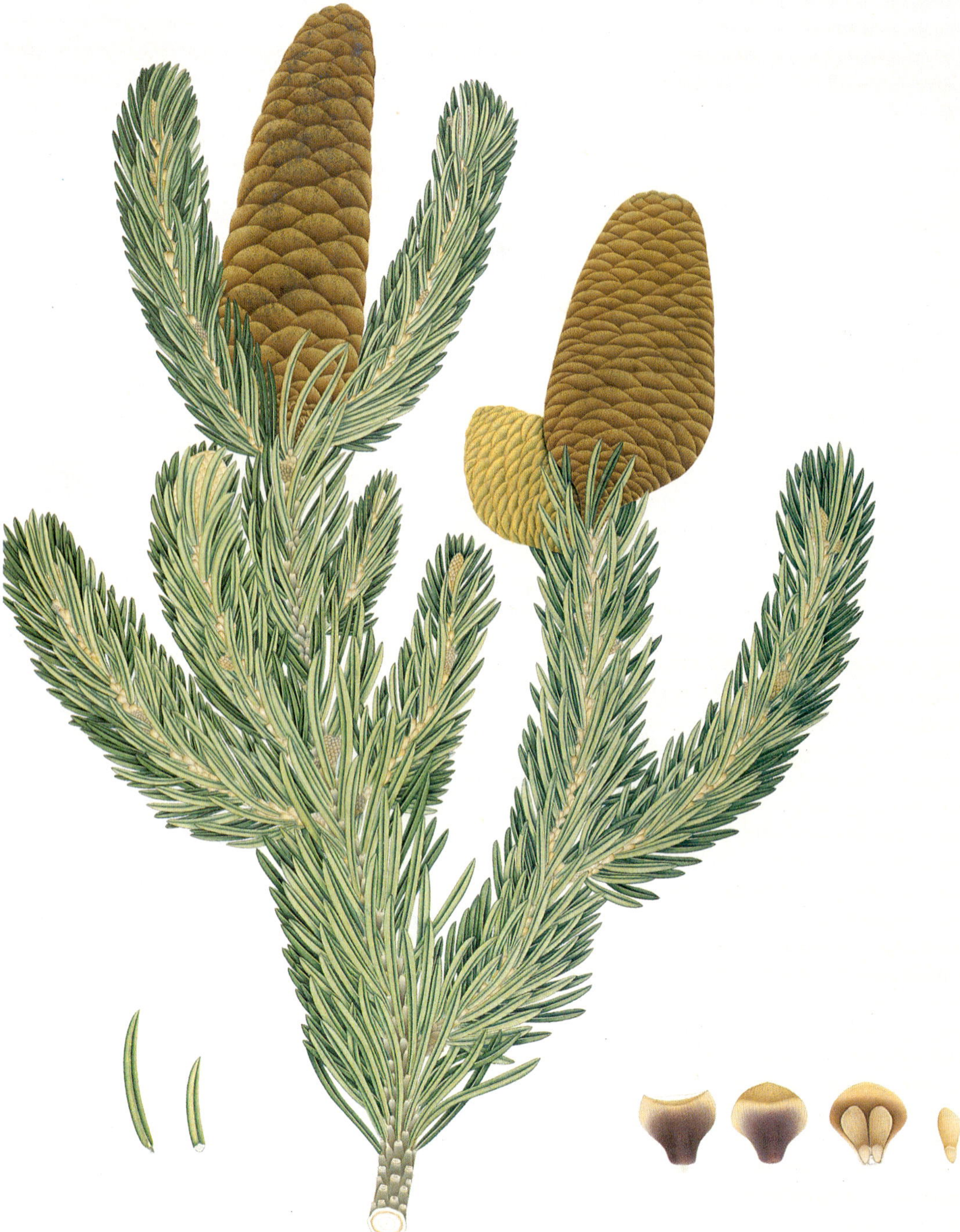

PLATE 8 *Picea smithiana*. This watercolour by Gorachand was selected by Wallich for reproduction in his *Plantae Asiaticae rariores* Vol. 3, 1832, plate 246. Wallich named this conifer, native to the Western Himalayas, after Sir James Edward Smith. George Govan, Superintendent of the newly founded Botanic Garden at Saharanpur, sent seed in 1818 to Scotland.

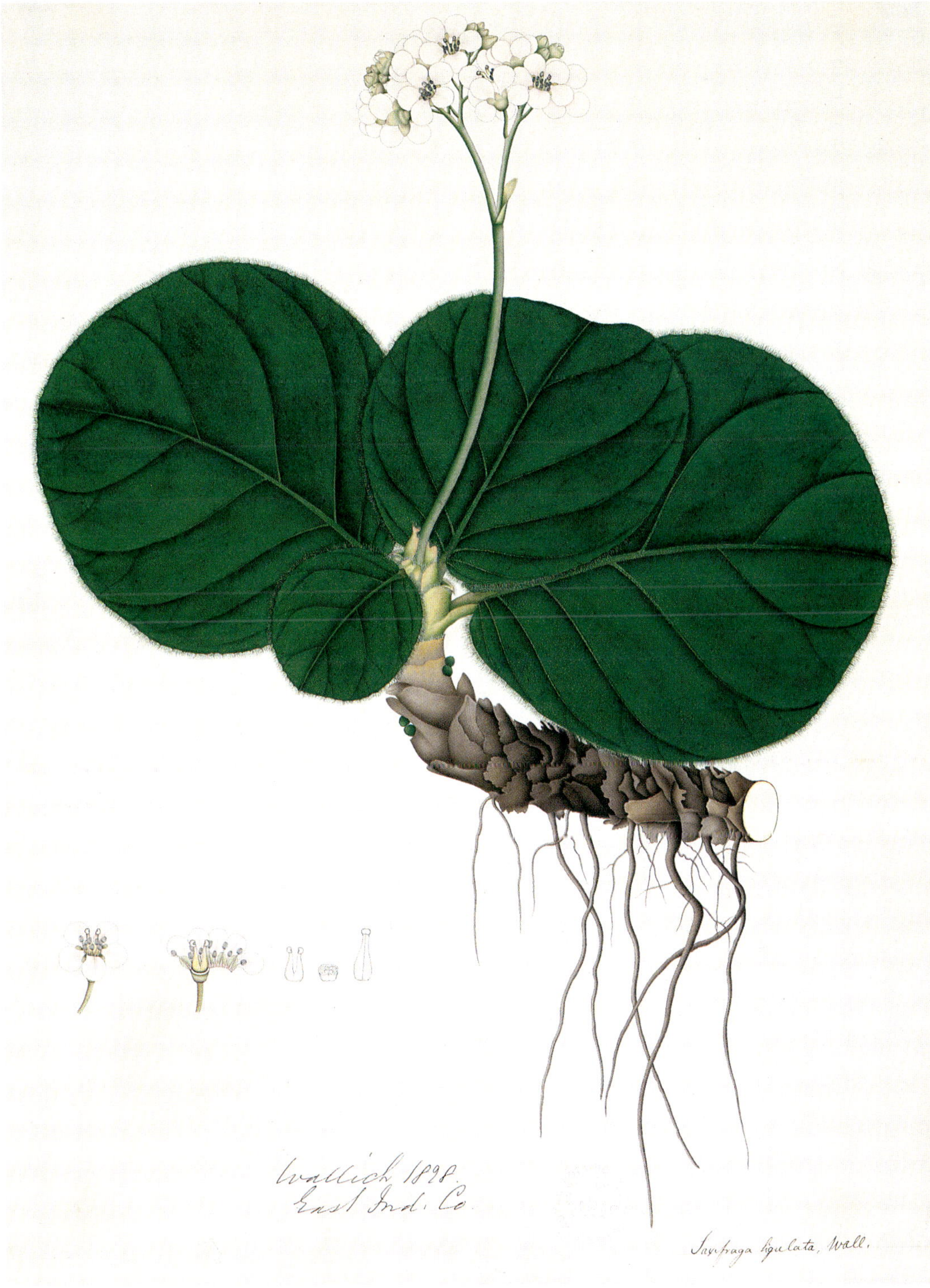

Plate 9 *Bergenia ciliata*. One of the large collection of drawings Wallich brought with him to England in 1828. Although now cultivated out of doors, in 1835 W. J. Hooker was recommending it as a greenhouse plant.

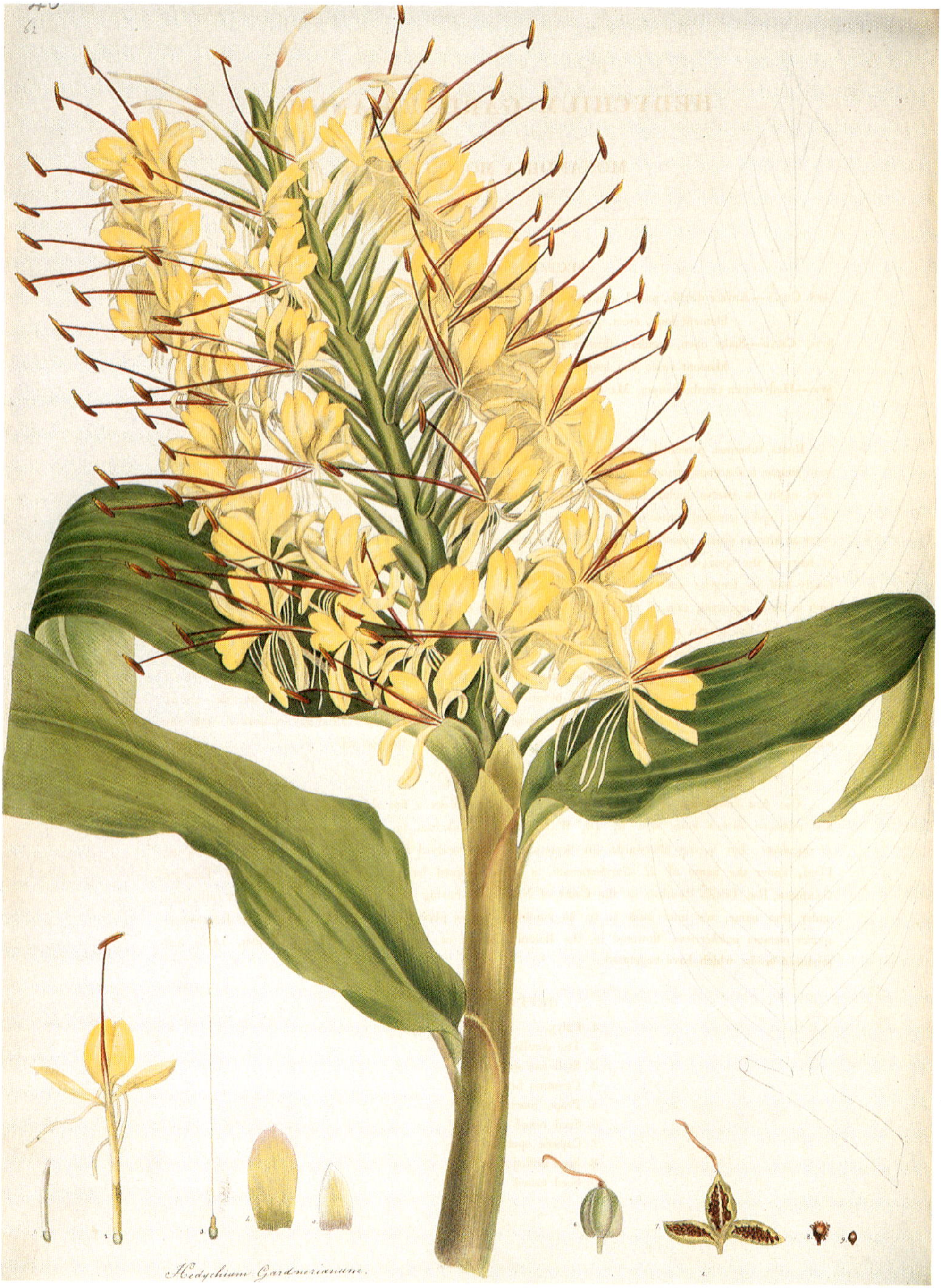

PLATE 10 *Hedychium gardnerianum*. (From W. Roscoe (1828), *Monandrian plants of the order Scitamineae chiefly drawn from living specimens in the Botanic Garden at Liverpool*, plate 62.) Lithograph of a drawing, probably by Thomas Allport. Nathaniel Wallich, who named this plant in honour of the British Resident in Nepal, Edward Gardner, declared it was 'the queen of the genus if not of the whole order'. In 1819 he despatched a specimen to Liverpool where it flowered the following year.

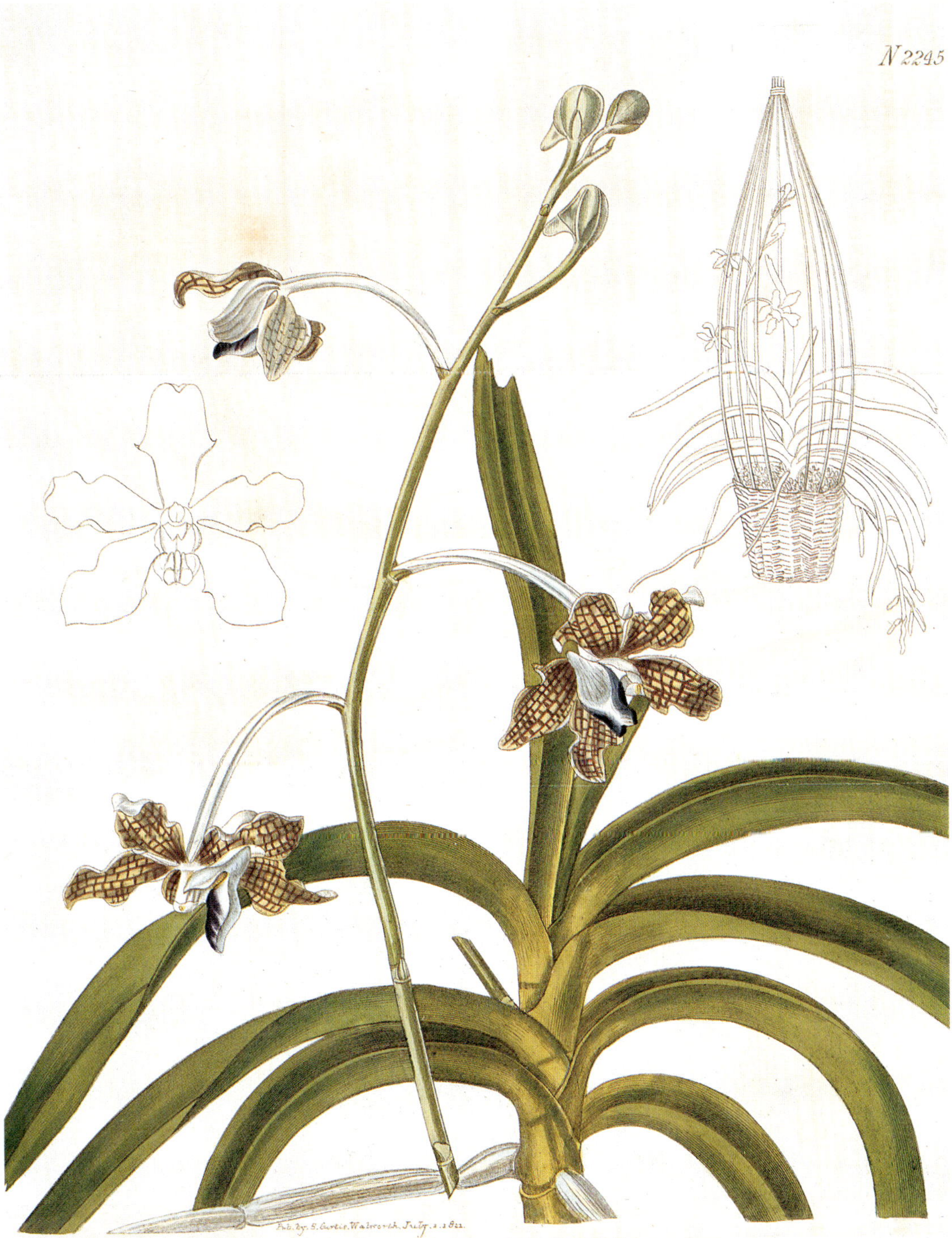

Plate 11 *Vanda tessellata*. (From *Curtis's Botanical Magazine*, 1820–1, plate 2245.) Hand-coloured engraving. This Indian orchid flowered in November 1820 at Spring Grove, the home of Lady Banks, suspended in the stove in a basket similar to the one illustrated here.

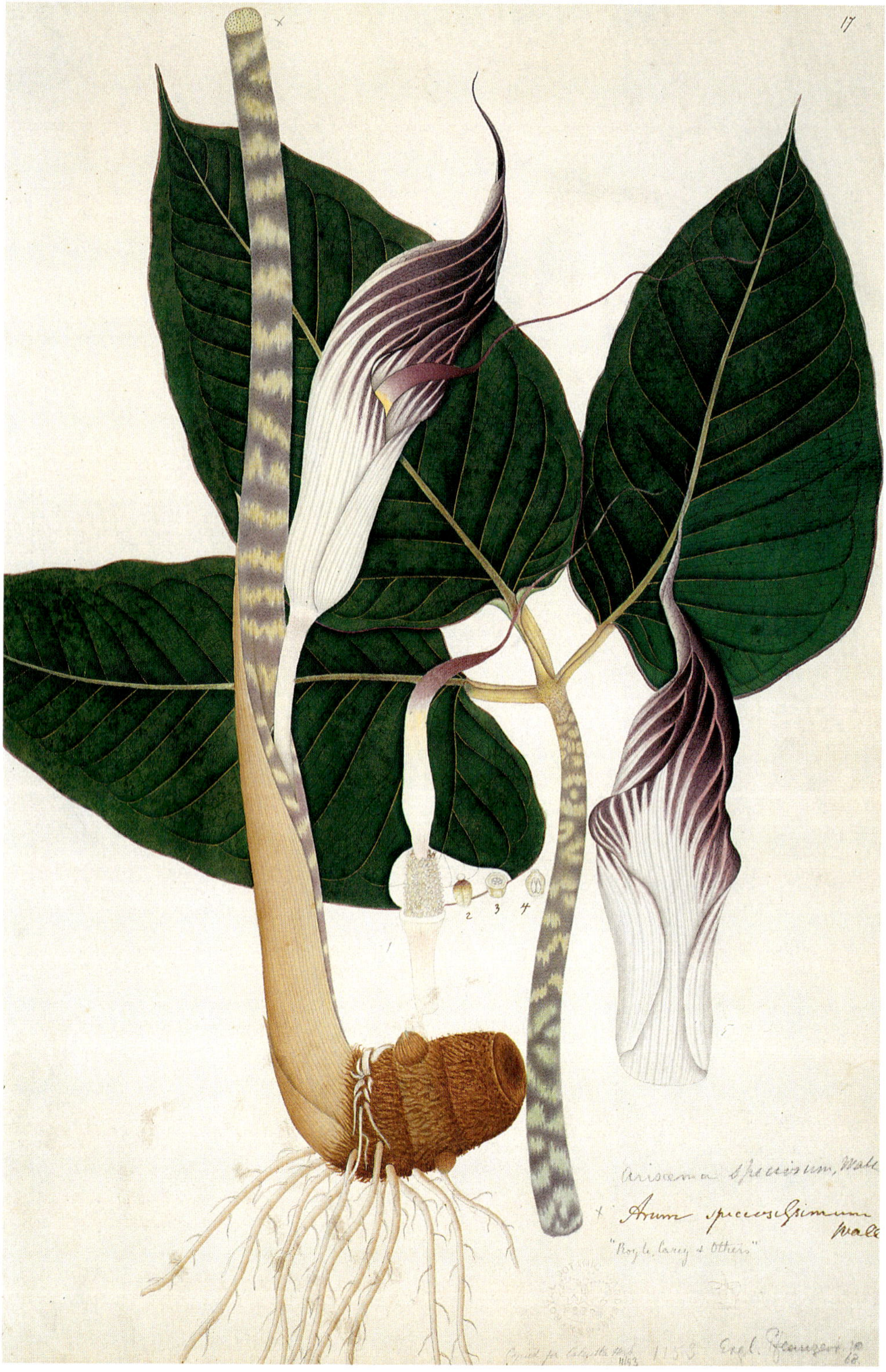

Plate 12 *Arisaema speciosum*. Watercolour drawing by an Indian artist. One of several thousand botanical drawings transferred to Kew Gardens on the dispersal of the contents of the India Museum in 1879, originally founded by the East India Company in 1801.

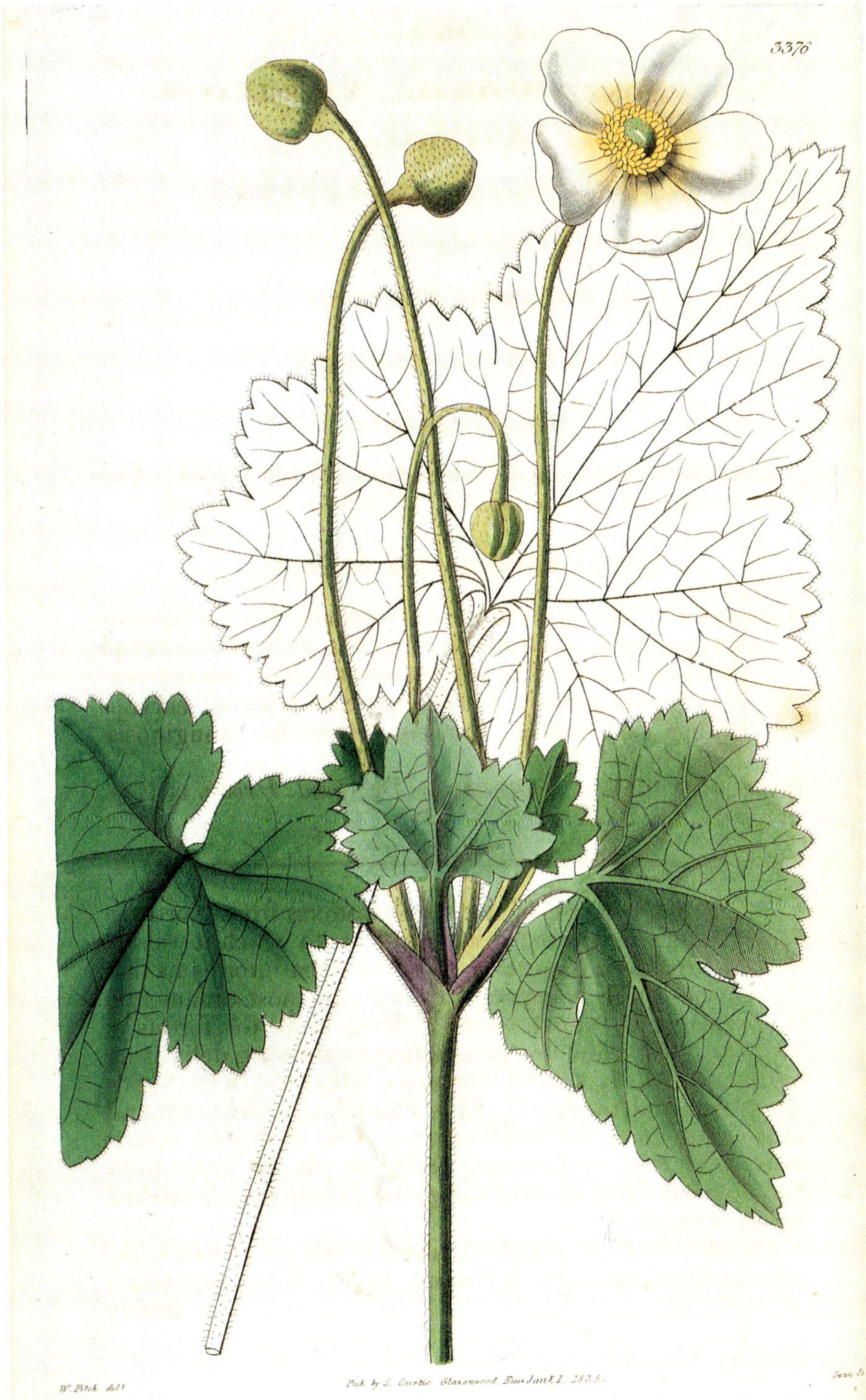

Plate 13 *Anemone vitifolia.* (From *Curtis's Botanical Magazine*, 1835, plate 3376.) Hand-coloured engraving of a drawing by W. H. Fitch. Wallich praised this plant 'as one of the commonest as well as the most ornamental flower-plants in Nepal'. In 1834 it was flowering in open borders in James Bateman's garden at Knypersley.

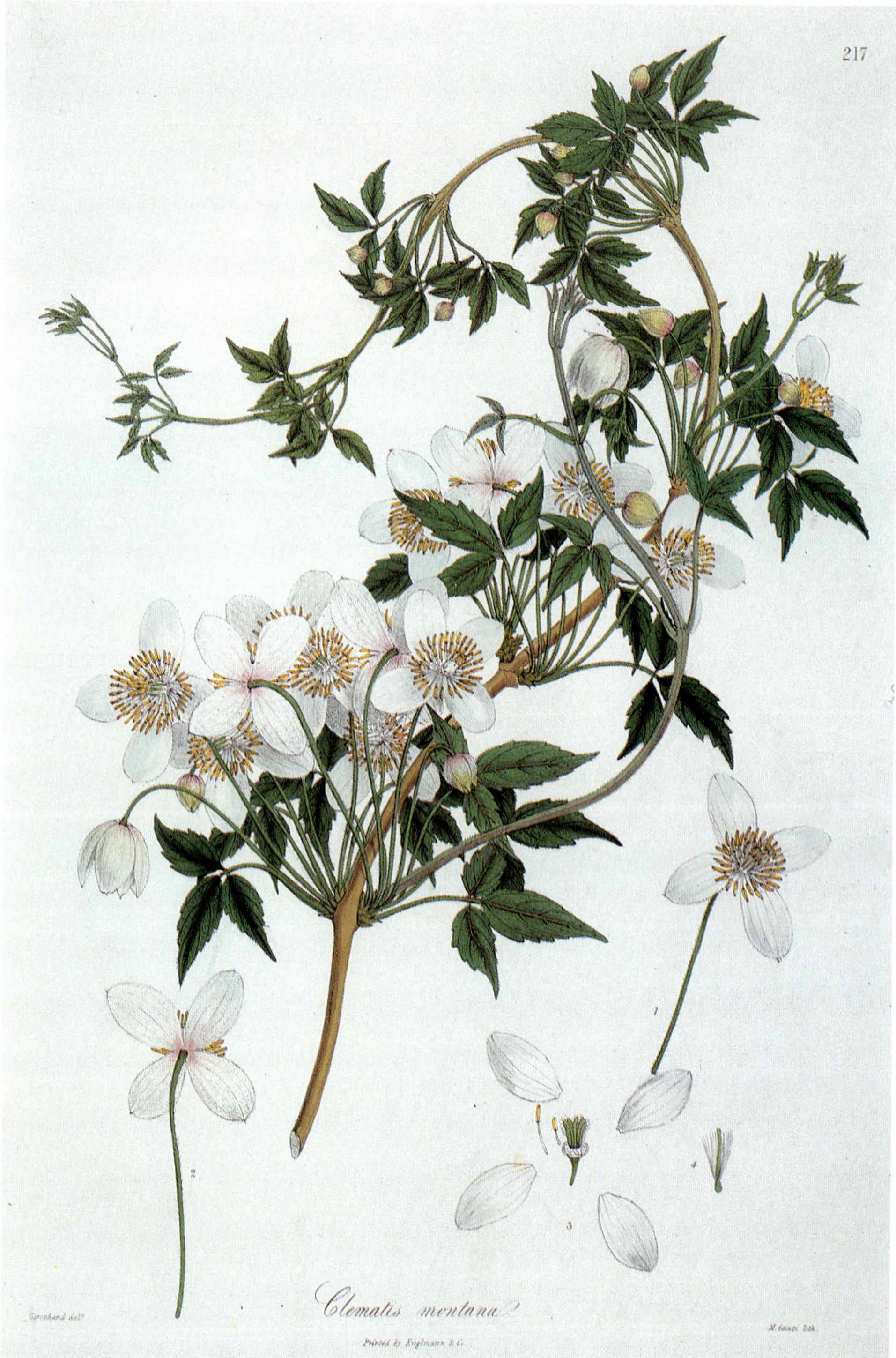

PLATE 14 *Clematis montana.* (From N. Wallich (1831), *Plantae Asiaticae rariories*, Vol. 3., London, plate 217.) Hand-coloured lithograph of a drawing by Gorachand, one of the team of artists employed at the Calcutta Botanic Garden. Both this hardy climber and *Anemone vitifolia* were introduced by Lady Amherst, whose husband was Governor-General from 1823 to 1828. J. C. Loudon recommended it as 'a delightful plant for placing at the root of unsightly tall trees'.

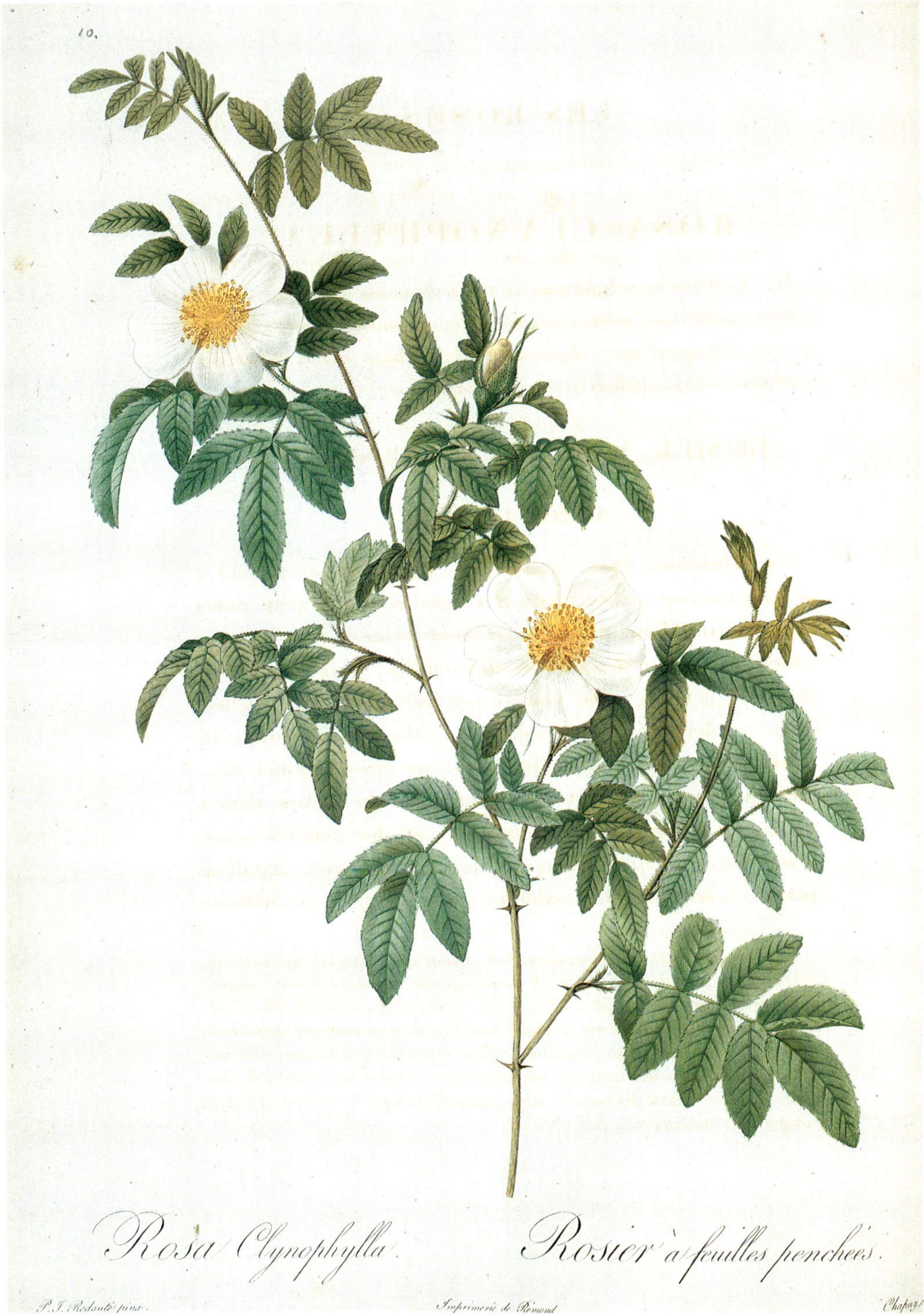

PLATE 15 *Rosa clinophylla*. (From P.–J. Redouté (1817), *Les roses*, Vol. 1, Paris, plate 43.) Stipple engraving, colour-printed à la poupée. This rose, which grows in the north of the Indian subcontinent, was first collected by Francis Buchanan in Nepal about 1803. It was described by William Roxburgh and introduced to England before 1818. J. C. Loudon reported in 1834 seeing a specimen 11 feet high against a wall in Loddiges's nursery in Hackney.

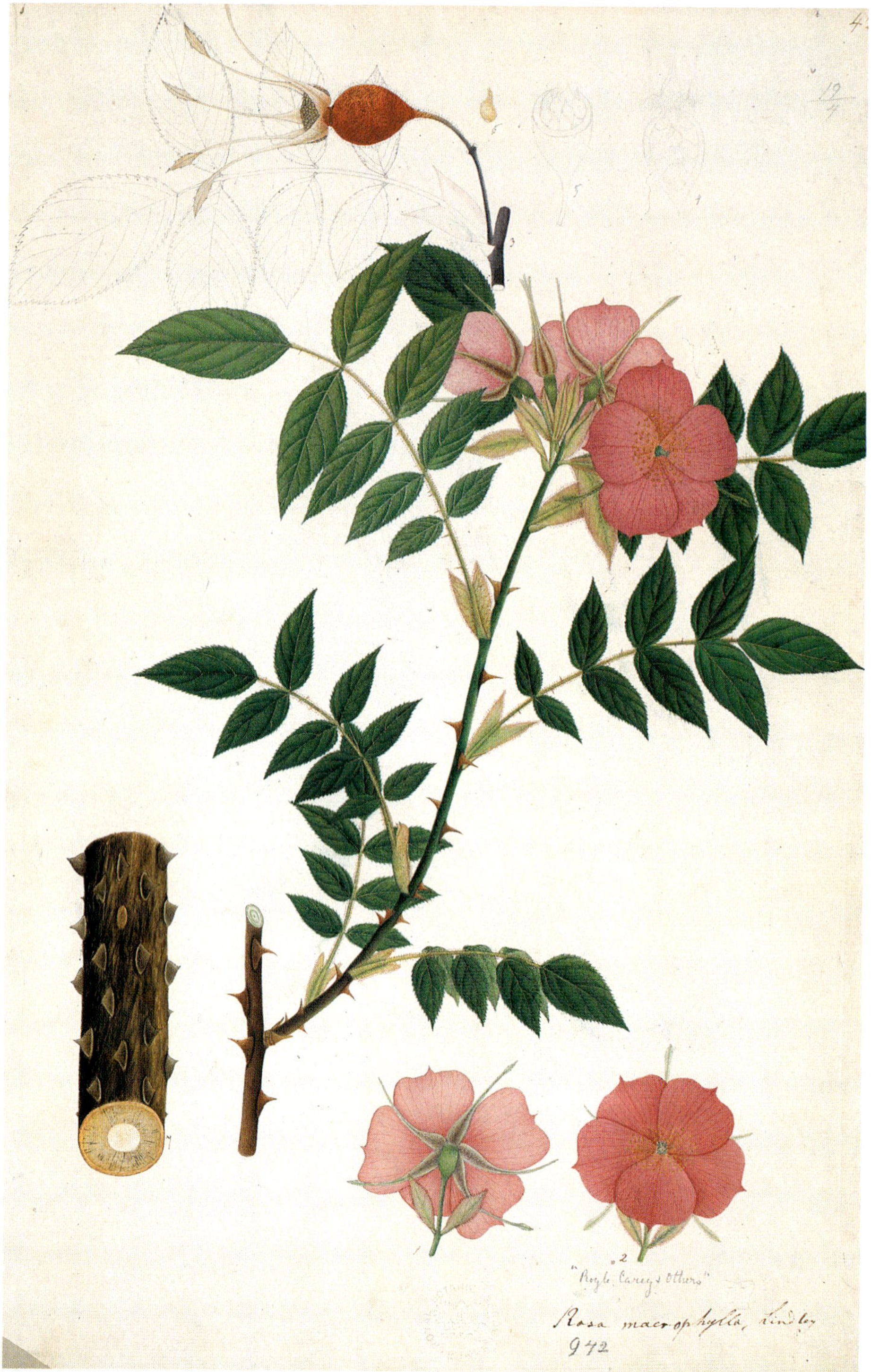

Plate 16 *Rosa macrophylla*. Watercolour drawing by an Indian artist. The rose is a native of the Himalayas at altitudes between 4200 and 10 800 feet. It was introduced into cultivation in England early in the nineteenth century.

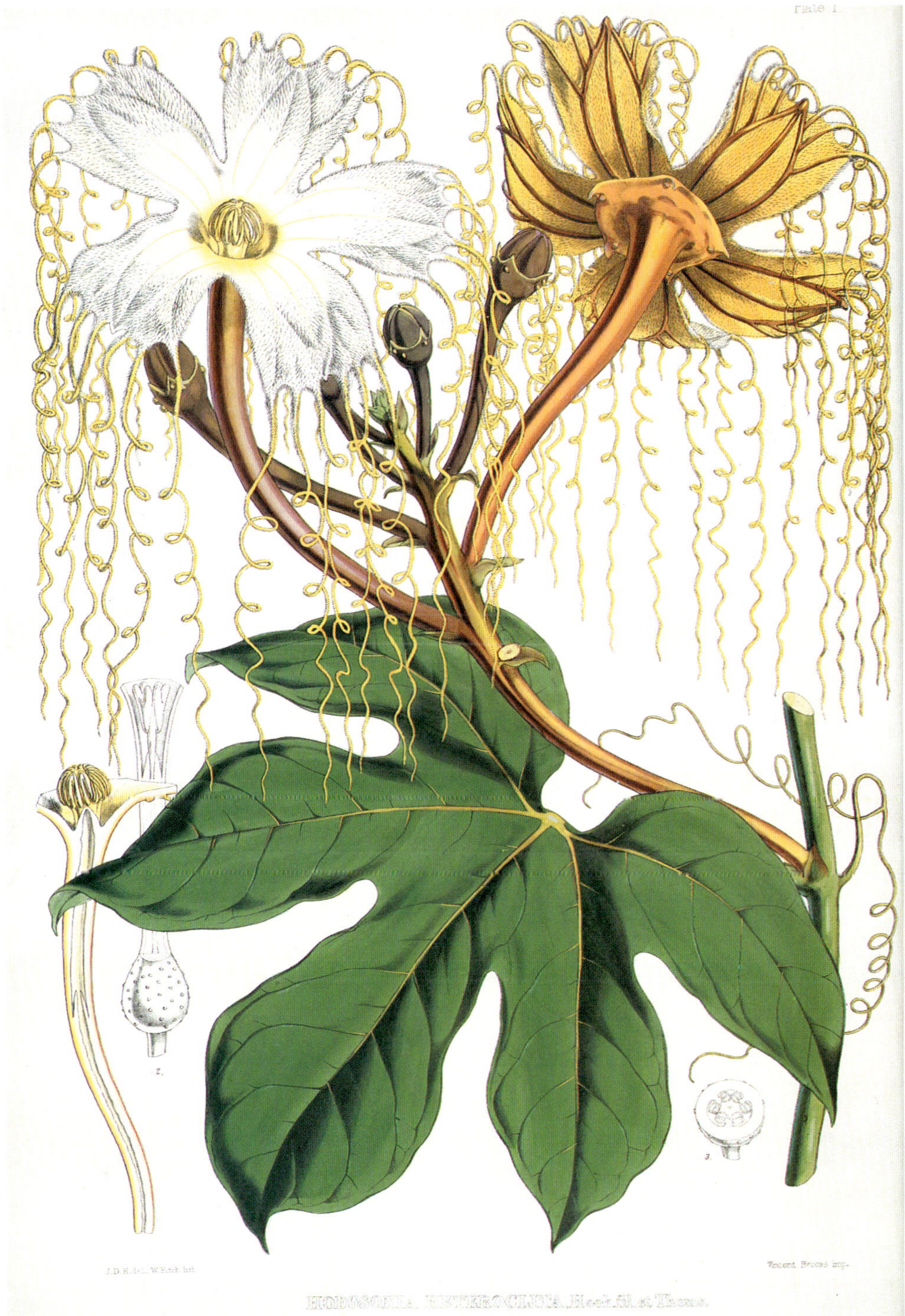

PLATE 17 *Hodgsonia heteroclita*. (From J. D. Hooker (1855), *Illustrations of Himalayan plants*, London, plate 1.) Lithograph by W. H. Fitch, after a drawing by Hooker who named the plant after Brian Hodgson, diplomat and naturalist in Nepal before retiring to Darjeeling. Hooker, in his *Himalayan journals*, acclaimed it the 'most magnificent plant of these jungles'.

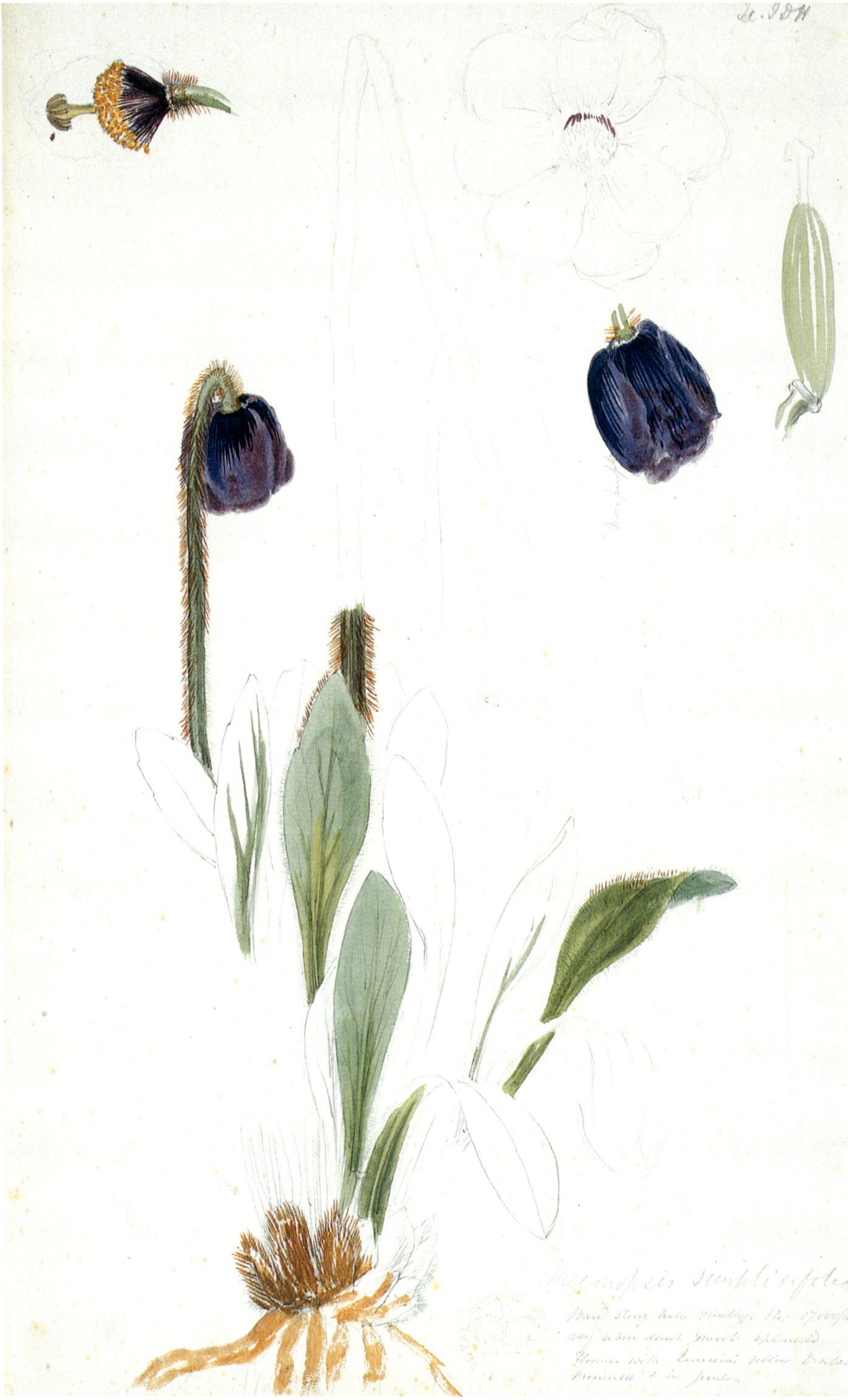

PLATE 18 *Meconopsis simplicifolia*. Watercolour sketch by Joseph Dalton Hooker. Wallich's collectors were the first to discover the plant in Nepal. Hooker, who collected it on his Indian visit, rated it 'the most beautiful and conspicuous of all the alpine flowers of Sikkim'.

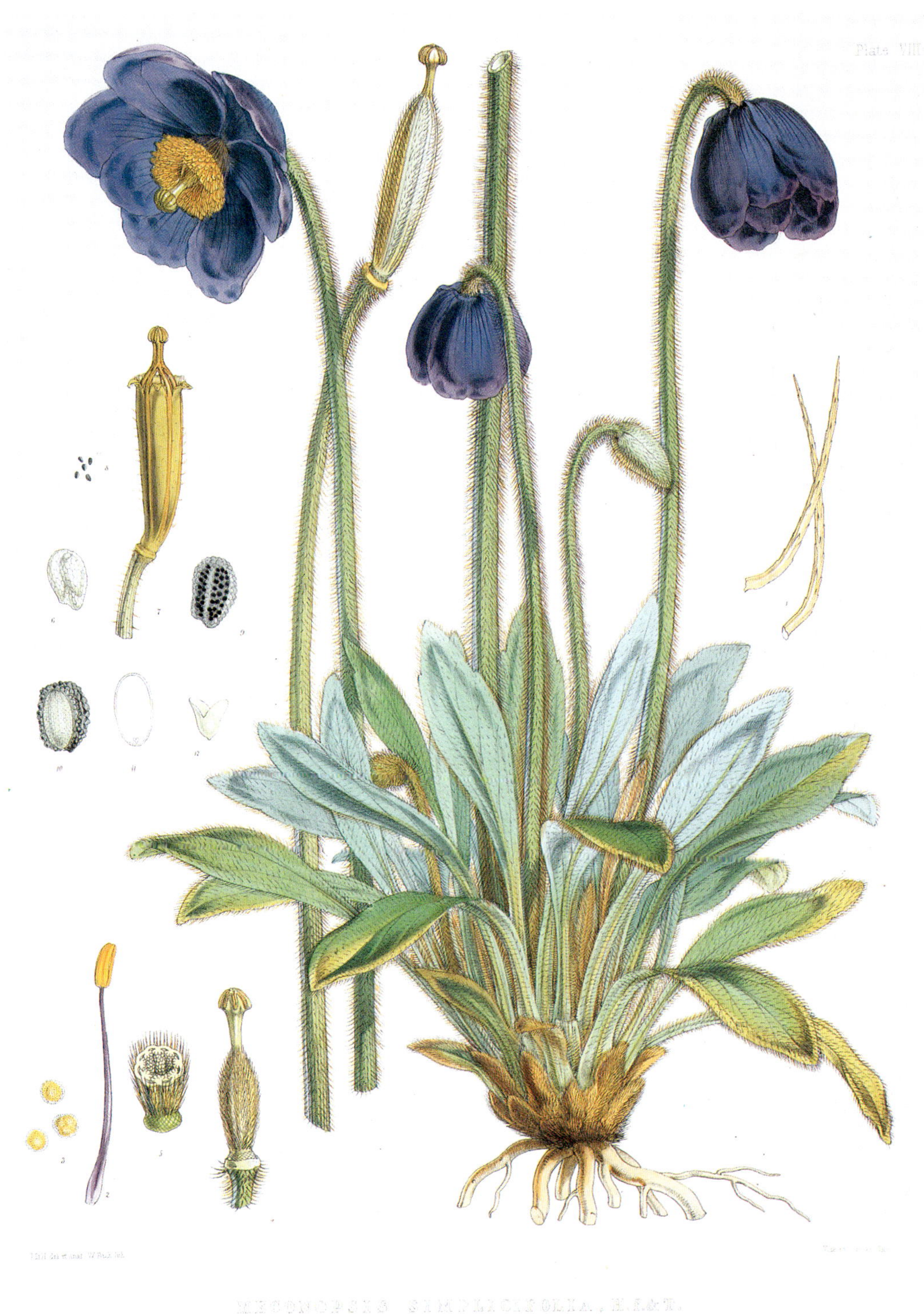

Plate 19 *Meconopsis simplicifolia*. (From J. D. Hooker (1855), *Illustrations of Himalayan plants*, London, plate 8.) Lithograph by W. H. Fitch, Kew's botanical artist, based upon Hooker's sketch (see Plate 18). An intimate knowledge of plant morphology together with confident artistic ability enabled Fitch to create accurate plant portraits from rudimentary sketches and dried specimens.

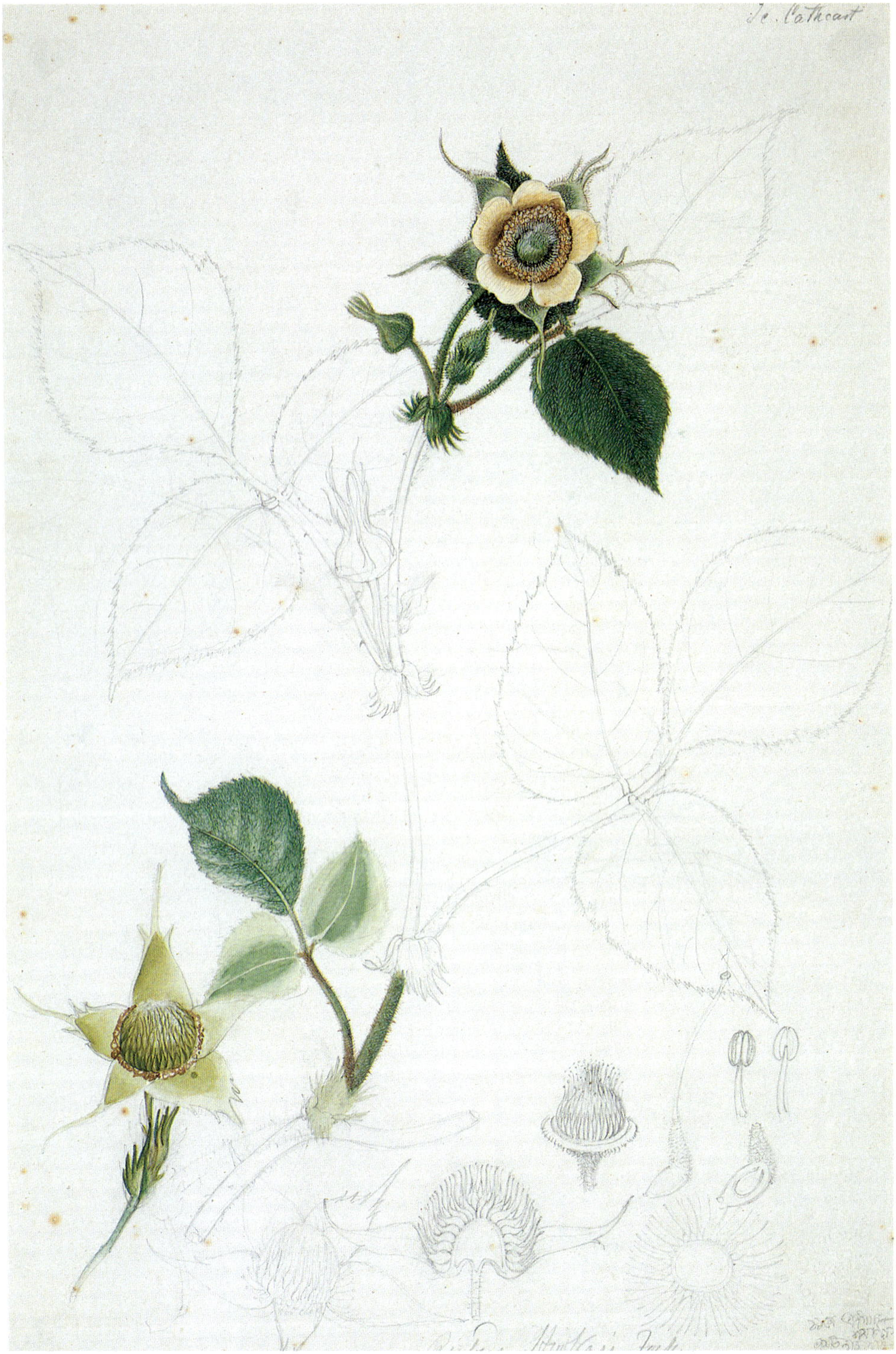

PLATE 20 *Rubus wardii*. Drawing by one of John Ferguson Cathcart's team of artists. At one time he employed as many as five who among them executed three drawings a week. Many of the drawings were pencil sketches with just a leaf and flower rendered in watercolour.

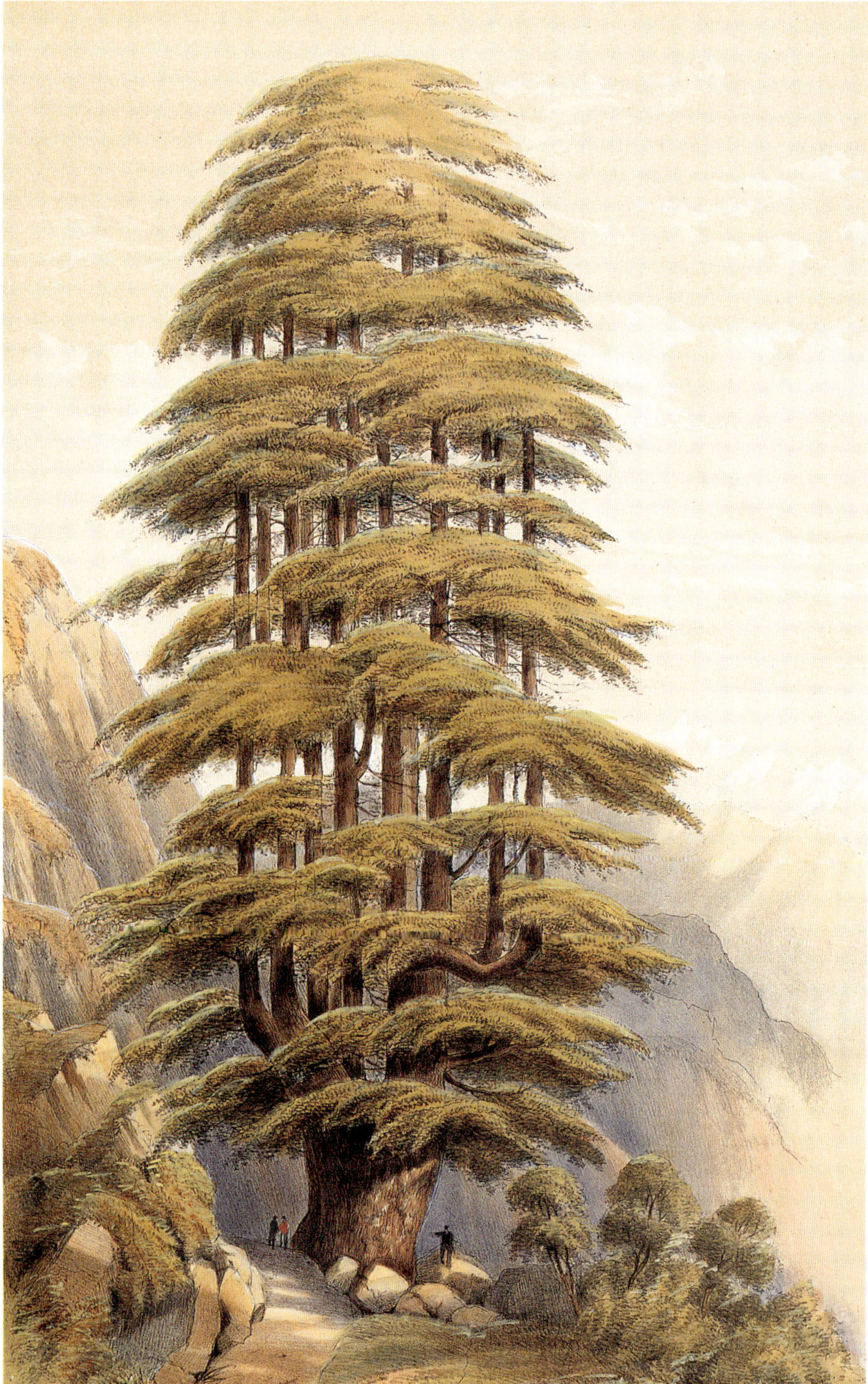

PLATE 21 *Cedrus deodara*. (From E. Ravenscroft (1884), *Pinetum Britannicum*, Vol. 3, Edinburgh, plate 43.) Lithograph of a drawing by the wife of the Governor-General, Lady Canning, who was a talented artist. This cedar, a native of the Himalayas from Afghanistan to Garwhal, was introduced to Britain in 1831, and through the agency of the East India Company a plentiful supply of seed was available by the mid-nineteenth century

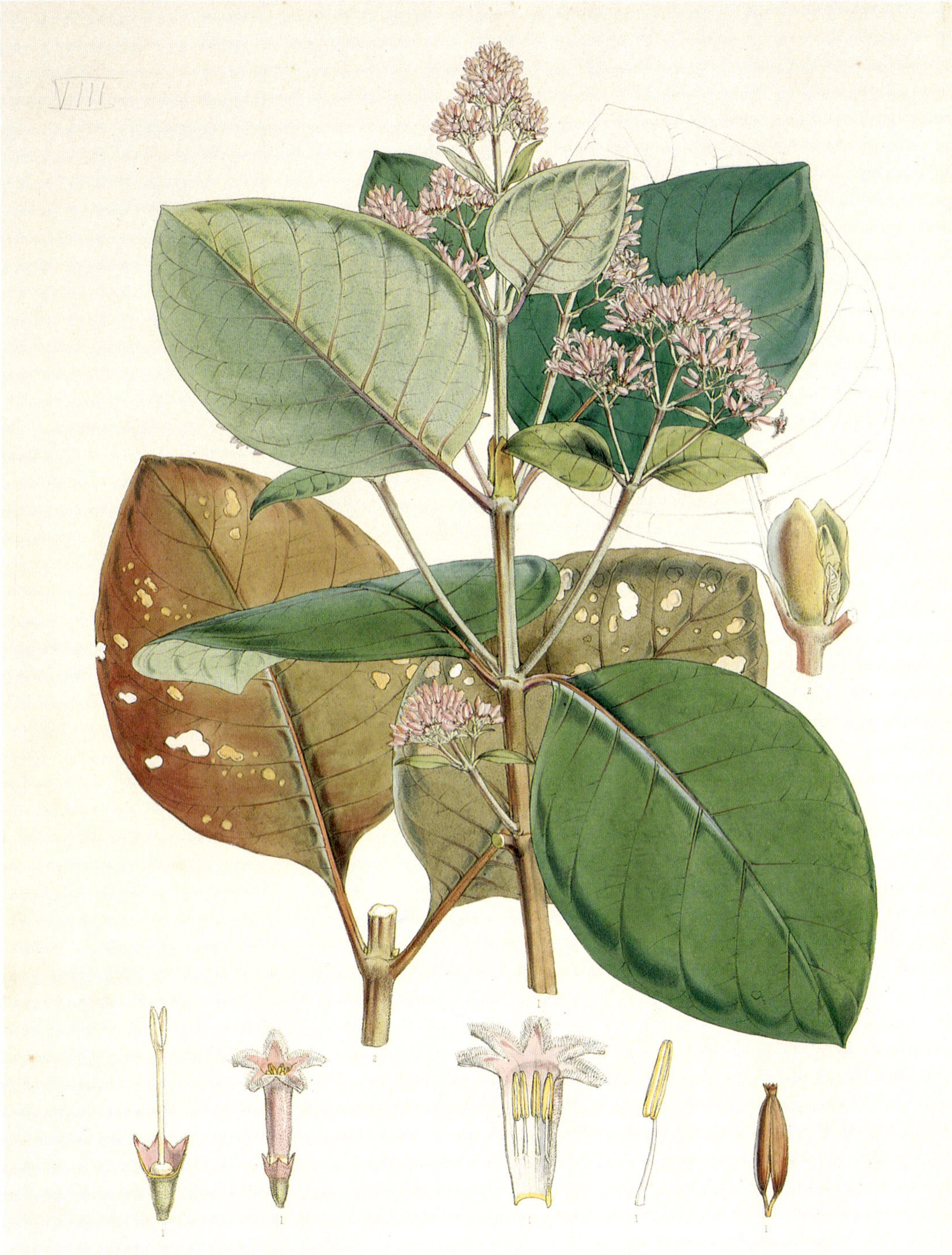

Plate 22 *Cinchona succirubra*. (From J. E. Howard (1862), *Illustrations of the nueva quinologie of Pavon*, London, plate 8.) Drawn and lithographed by W. H. Fitch. Howard, a leading quinine manufacturer, having purchased in 1858 in Madrid some cinchona specimens collected by J. Pavon in South America in 1777, sent Fitch to the Royal Museum in Madrid to draw Pavon's original specimens for this book. Howard also wrote another authoritative work on cinchonas—*Quinology of East Indian plantations* London, 1869–76—and advised C. R. Markham on his cinchona project.

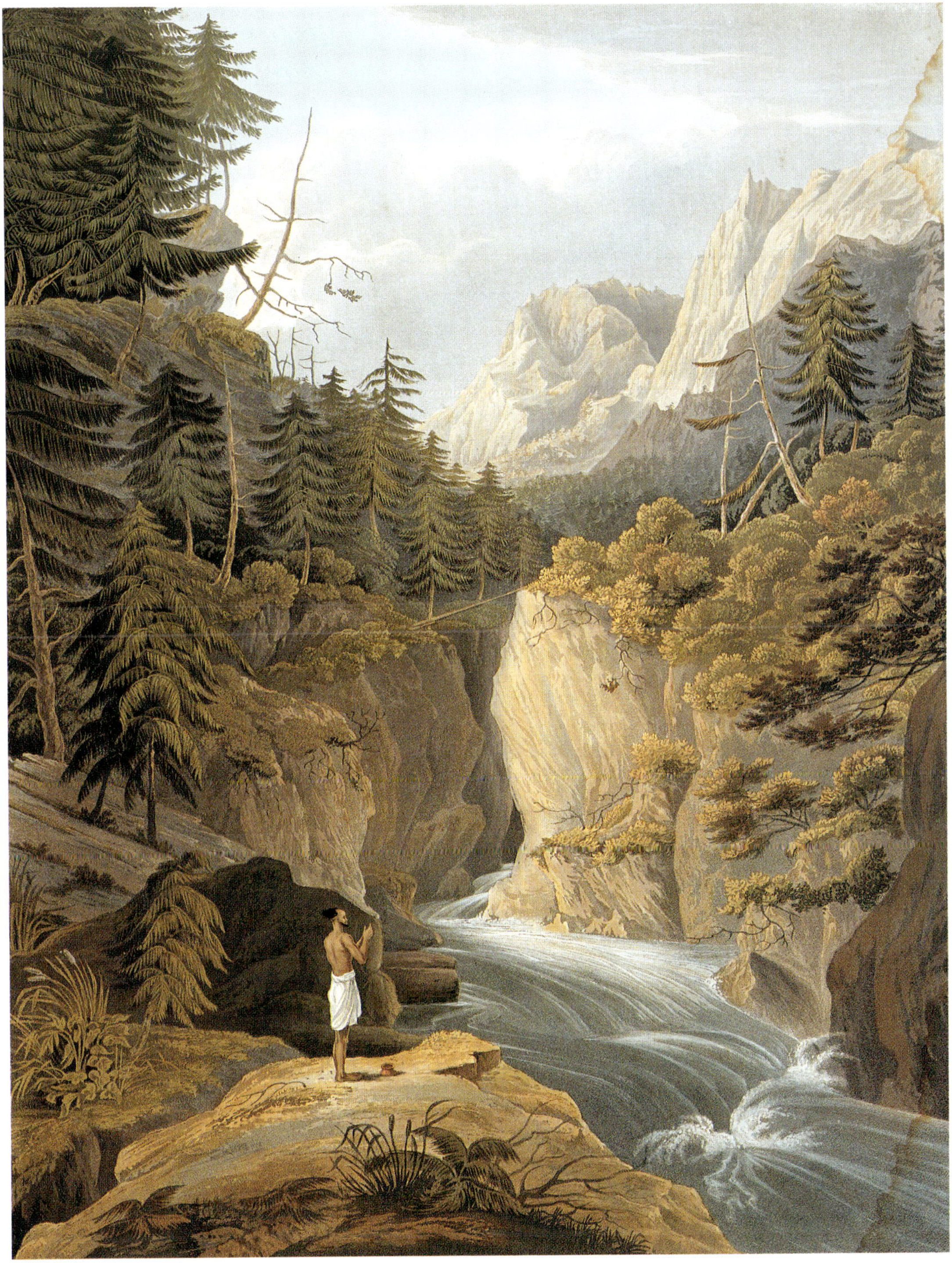

Plate 23 Bhairamghattee in the Lower Himalayas. (From J. B. Fraser (1820), *Views in the Himala mountains*, London, plate 34.) Coloured aquatint. Fraser had an eye for the picturesque in the landscape. 'Around us the fantastic forms of the old trees, their rich masses of foliage contrasting with the grey bare crags, and the blasted pines and withered oaks, formed a foreground for a picture worthy of the pencil of a Salvator' (Fraser, *Journal*, 1820, p. 158).

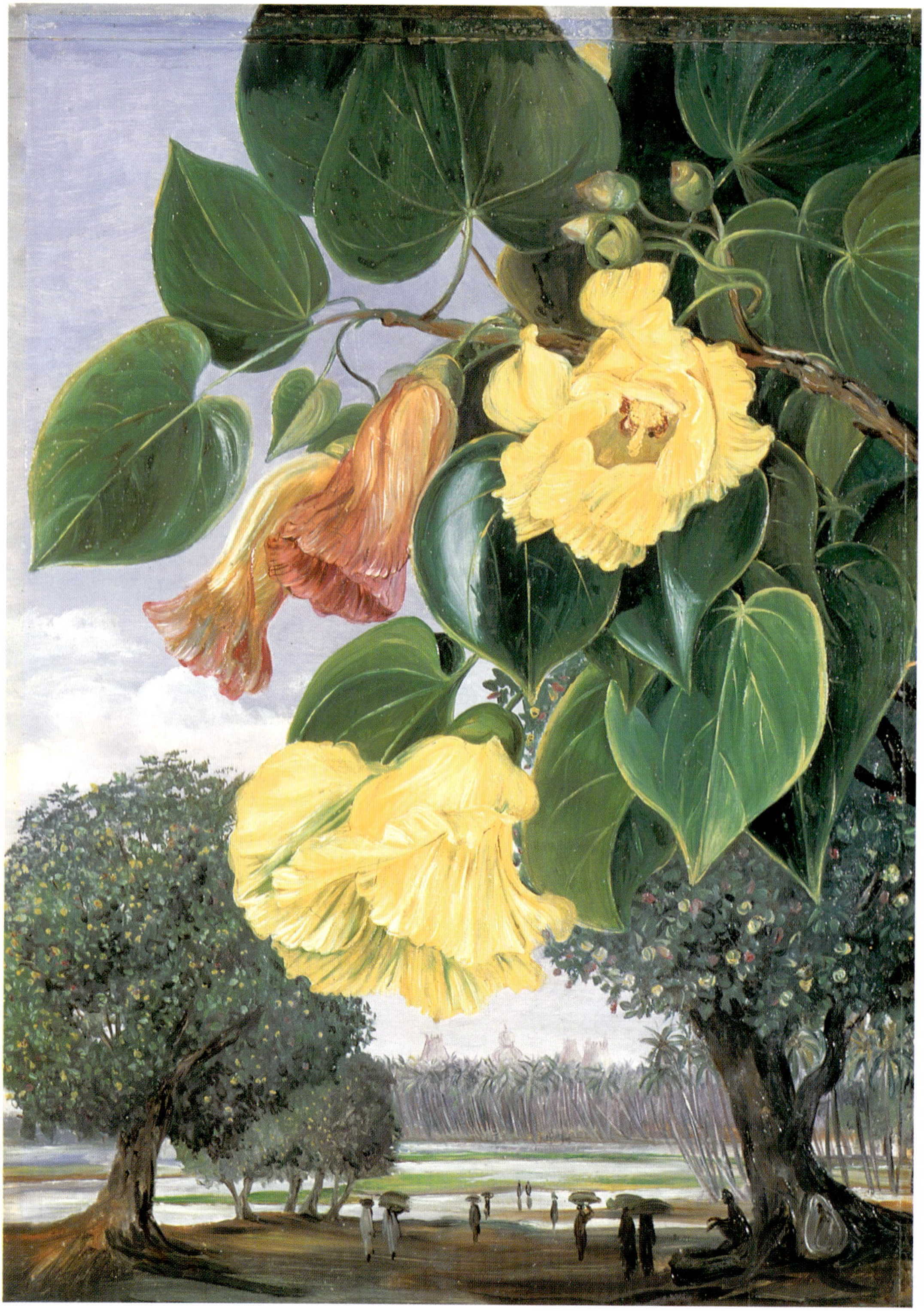

PLATE 24 *Thespesia populnea* with the temples of Madura in southern India in the distance. Oil painting by Marianne North, *c.*1877, who travelled the world painting with speed and manifest pleasure plants in their natural habitats. Her collection of over 800 studies are permanently displayed in Kew Gardens in a special gallery she generously provided.

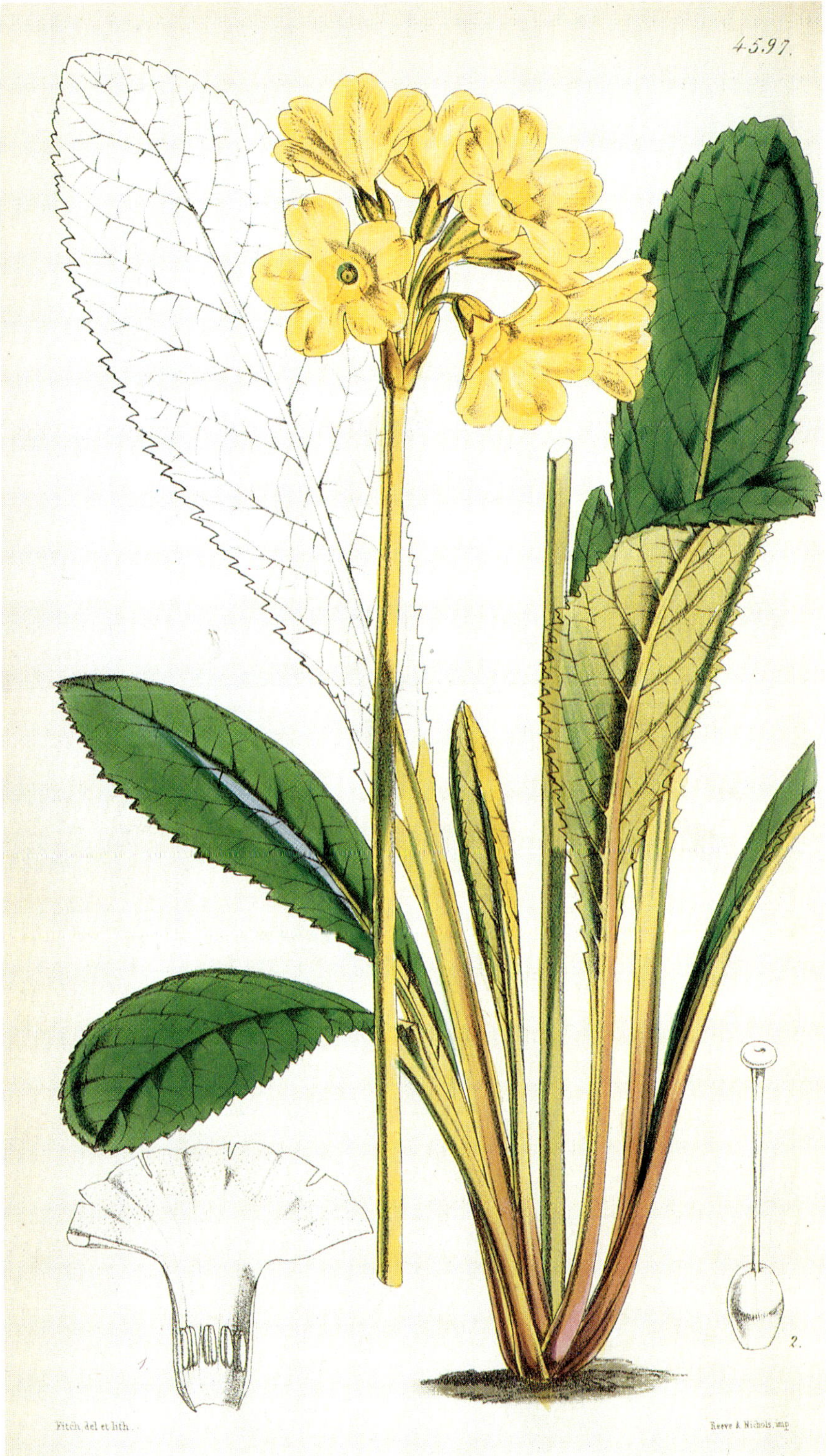

PLATE 25 *Primula sikkimensis*. (From *Curtis's Botanical Magazine*, 1851, plate 4597.) Drawn and lithographed by W. H. Fitch. One of the most popular of Joseph Hooker's introductions from Sikkim which flowered at Kew from his seed in 1851.

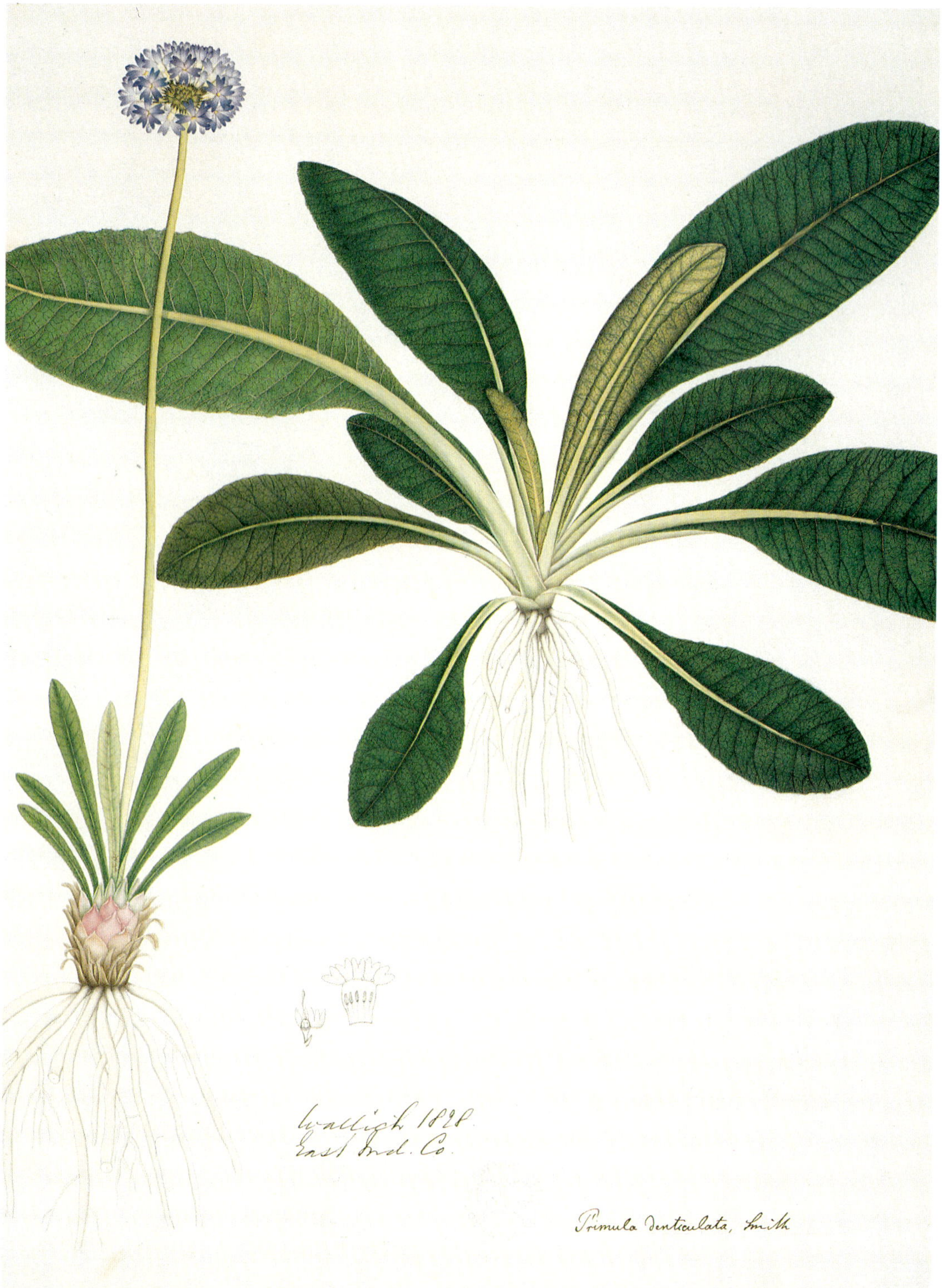

PLATE 26 *Primula denticulata.* Watercolour drawing by one of Wallich's Indian artists. The Veitch nursery at Exeter exhibited it at a Horticultural Society of London show in 1842.

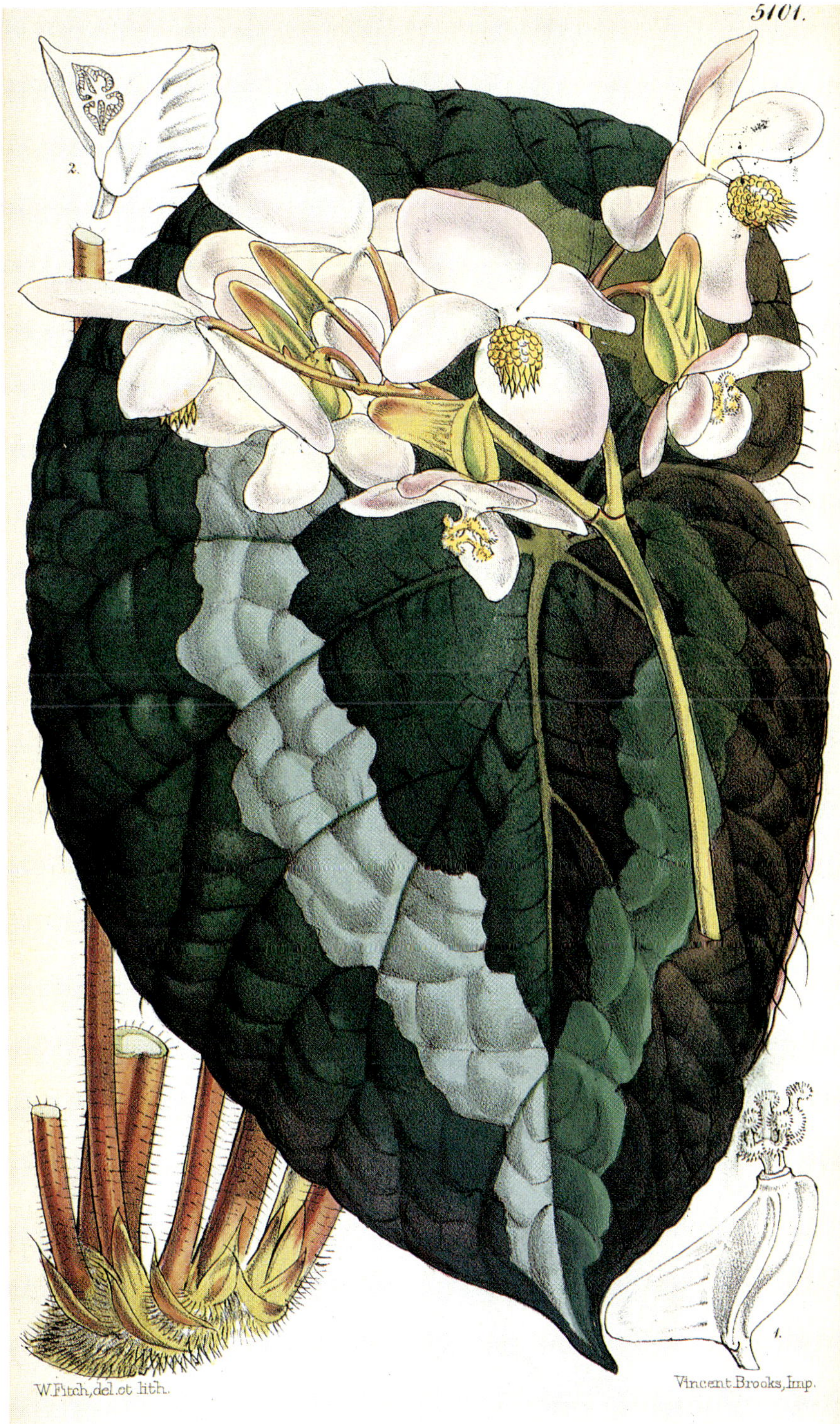

Plate 27 *Begonia rex*. (From *Curtis's Botanical Magazine*, 1859, plate 5101.) Drawn and lithographed by W. H. Fitch. Discovered in Assam, it was hailed by Sir William Hooker as 'the most lovely of the many lovely species of Begonia'. It was displayed in Victorian homes and conservatories and was the parent of many successful hybrids.

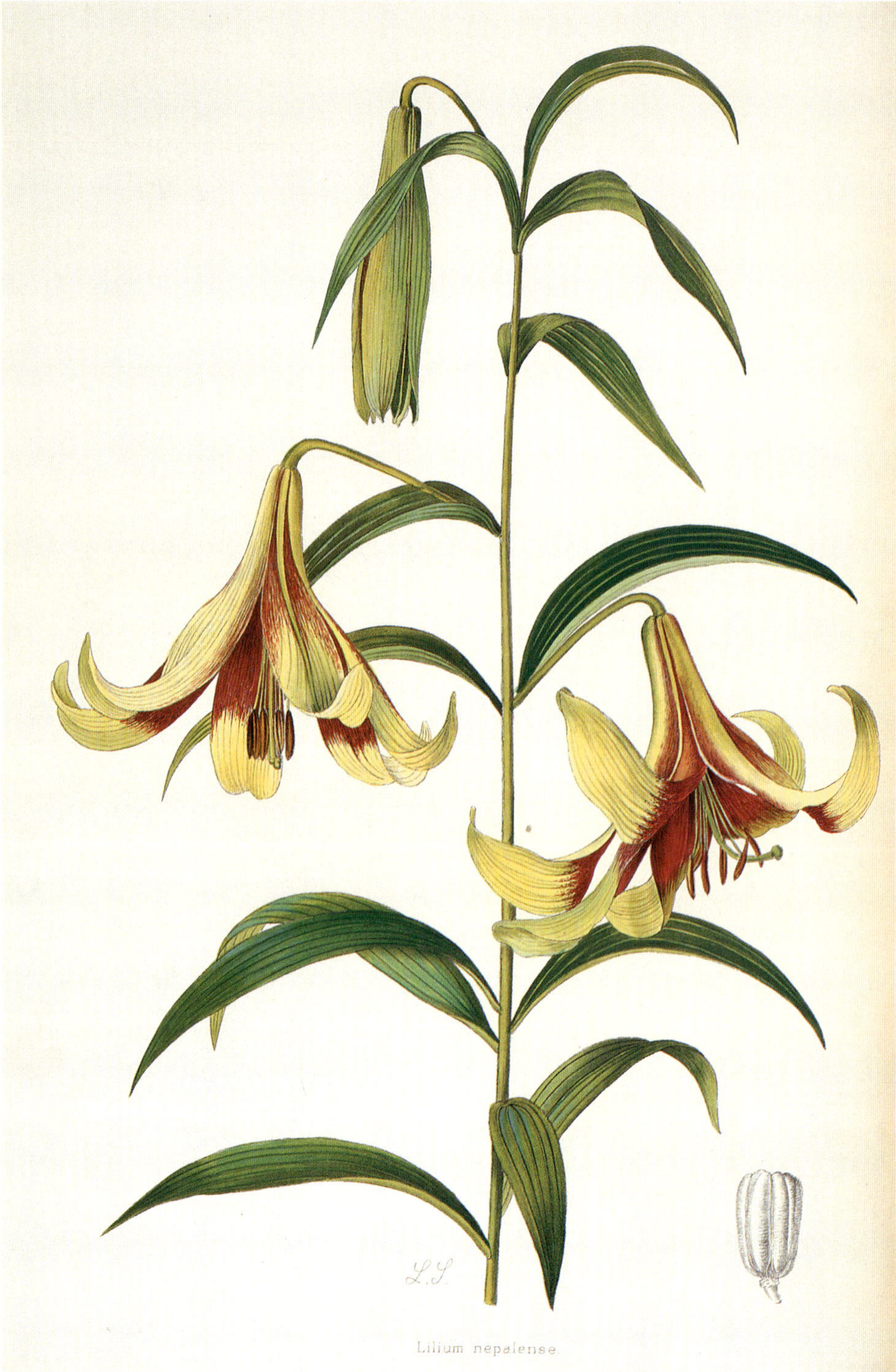

Plate 28 *Lilium nepalense*. (From H. J. Elwes (1933), *Monograph of Genus Lilium: Supplement by A. Grove and A. D. Cotton,* London.) Lithograph by Lilian Snelling. Although this lily had been discovered by Wallich's plant collectors in Nepal, it was not seen by British gardeners until it was exhibited in 1888 by Hugh Low and Company at the Royal Horticultural Society, where it received a first-class certificate.

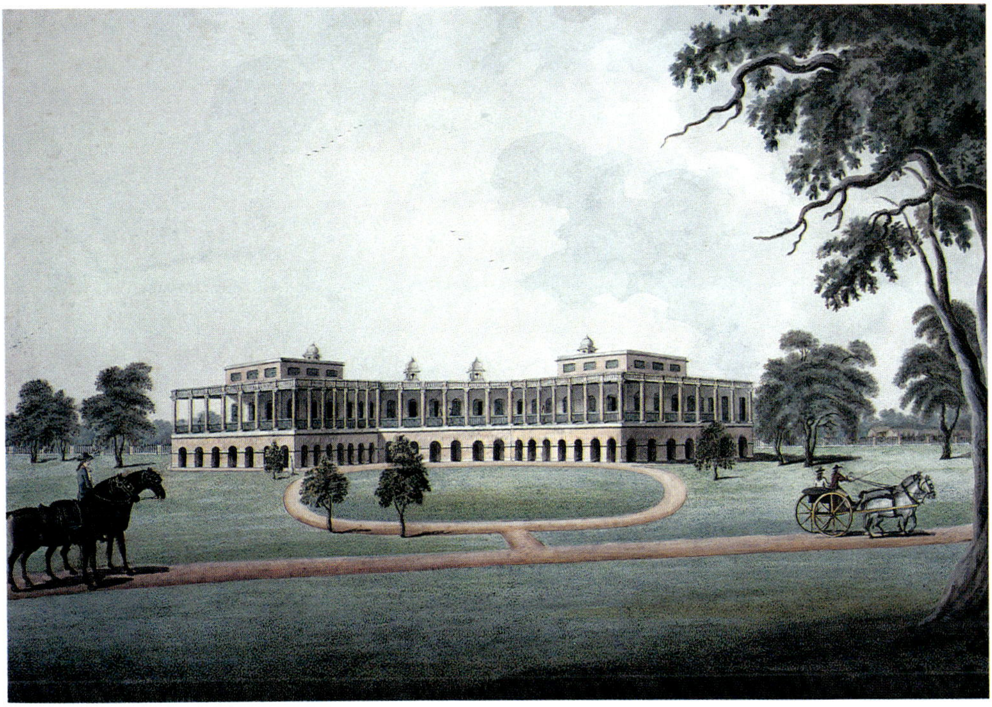

(a)

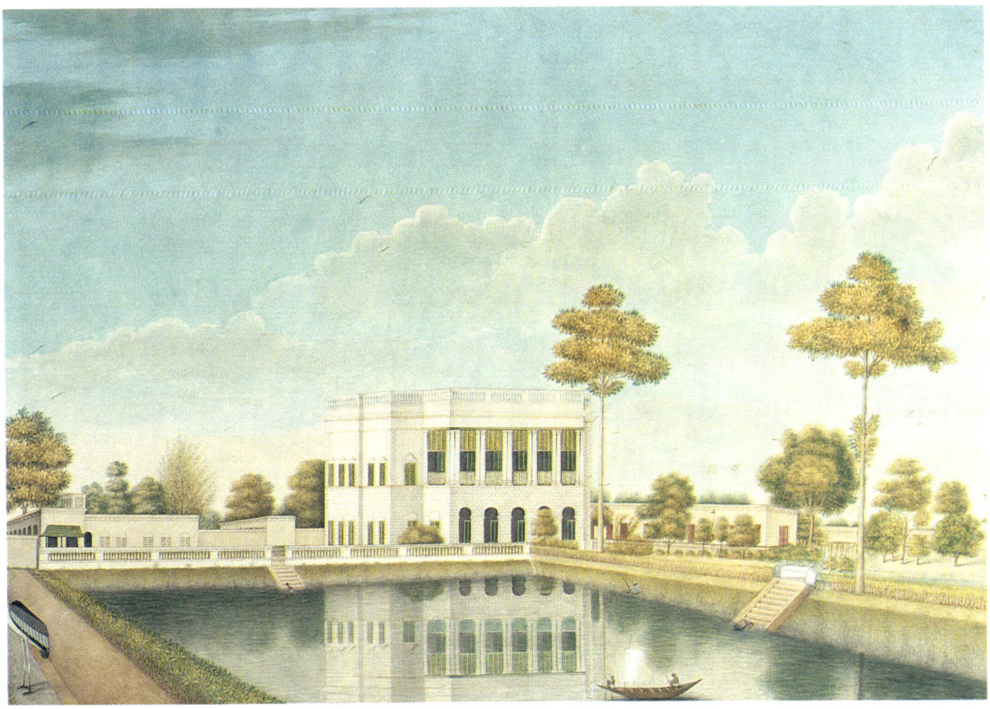

(b)

Plate 29 (a) A judge's house set in Georgian-style lawns at Dinajpur in Bengal. Watercolour by Murshidabad artist, *c.*1800. (India Office Library and Records.) (b) A Calcutta house where the tank with its essential water supplies has been made an attractive feature of the garden. Watercolour by Shaikh Muhammad Amir, *c.*1845. (India Office Library and Records.)

(a)

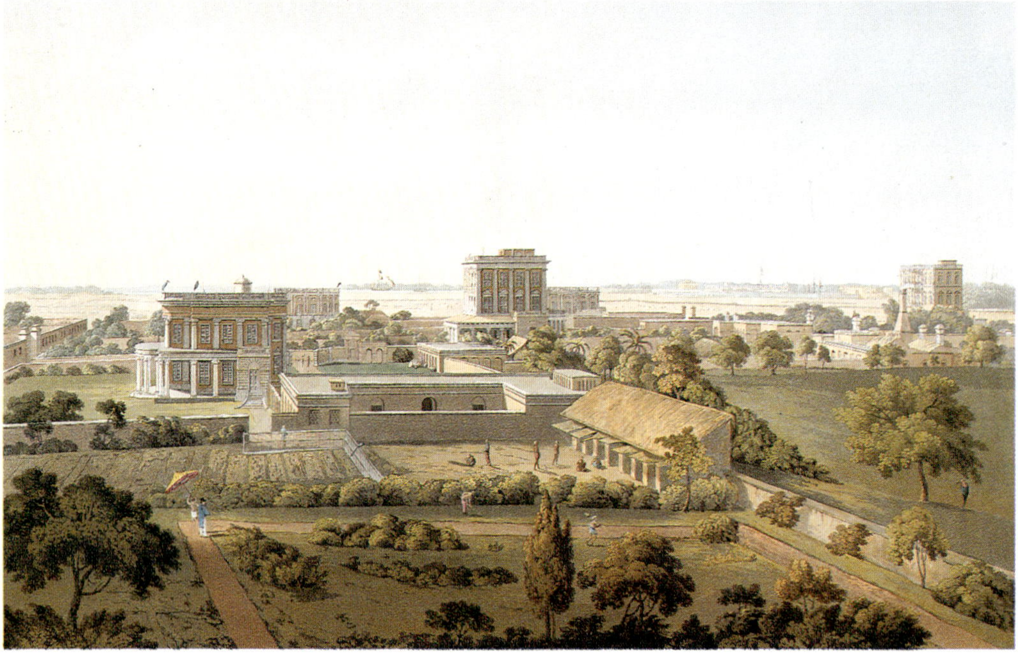

(b)

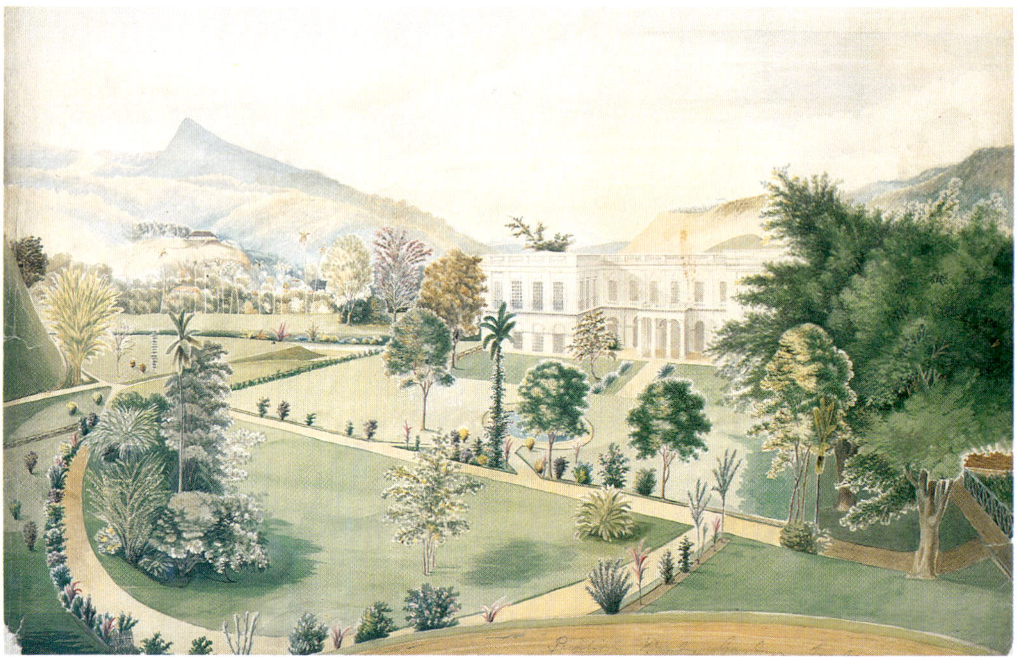

PLATE 30 (a) House and garden of Thomas Graham in Chowringhee, Calcutta. (From H. Salt (1809), *Twenty-four views in St Helena,* The Cape, India . . . , London, plate 3.) Coloured aquatint. 'The houses are placed very much like those at Kensington Gore. There is a wall and a gate for coaches, and behind the house is generally a small garden' (Thomas Babington Macaulay writing about Calcutta in 1834). (b) The garden of the Queen's Pavilion, Kandy, Ceylon. Watercolour by Patrick Lysaght, a private in the 78th Regiment of Foot, who served in Ceylon from 1826 to 1838. (India Office Library and Records.)

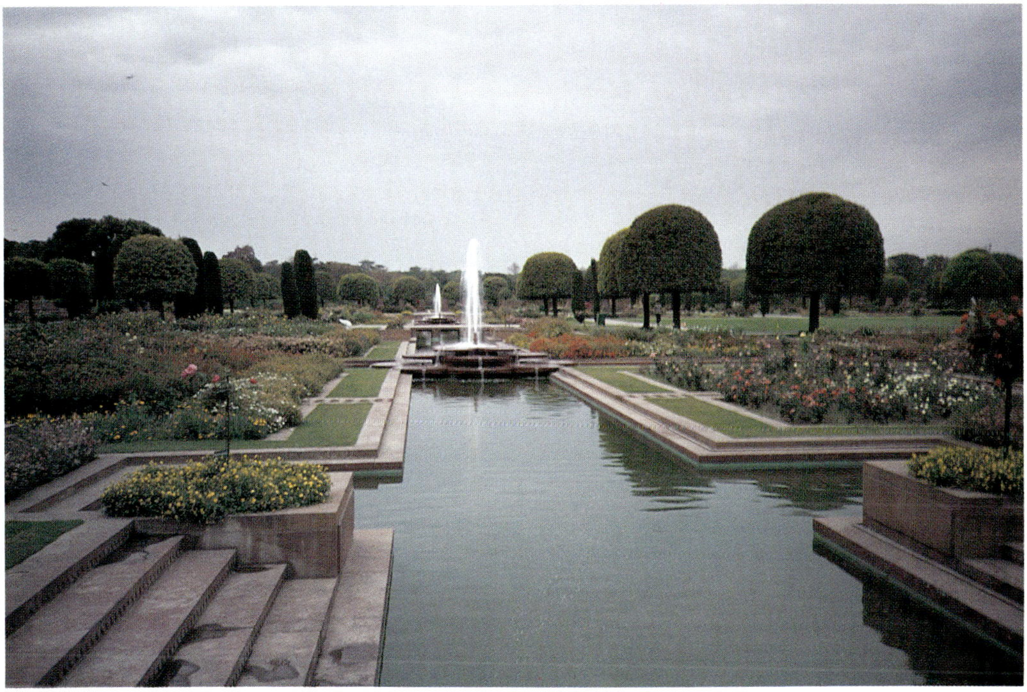

(a)

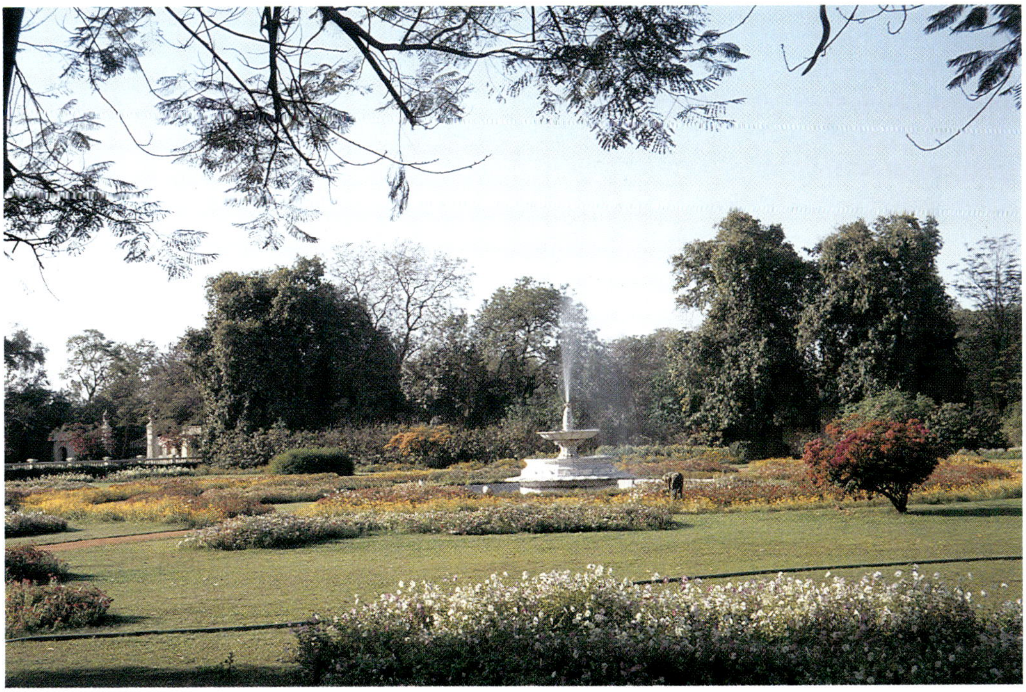

(b)

PLATE 31 (a) The Mughal-inspired garden adjoining the Viceroy's home in New Delhi (now the President of India's residence) was laid out after 1918 by Sir Edwin Lutyens. Water dominates a composition that incorporates terraces, pergolas, topiary, and English herbaceous borders and lawns. (b) The garden at the Ram Bagh Hotel, Jaipur, is clearly influenced by British garden design.

(a)

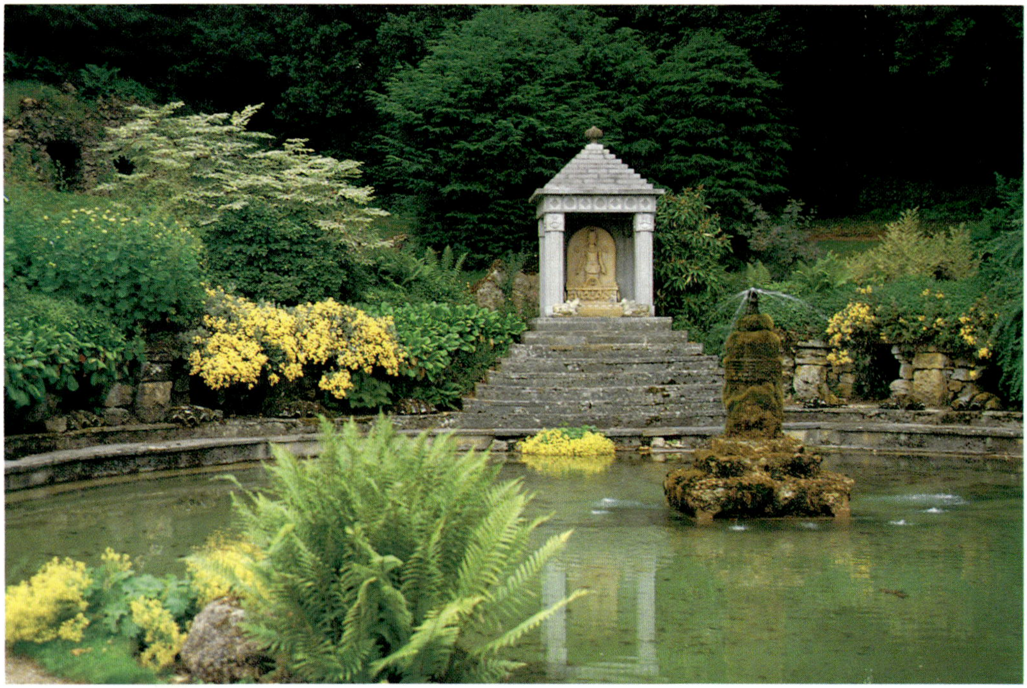

(b)

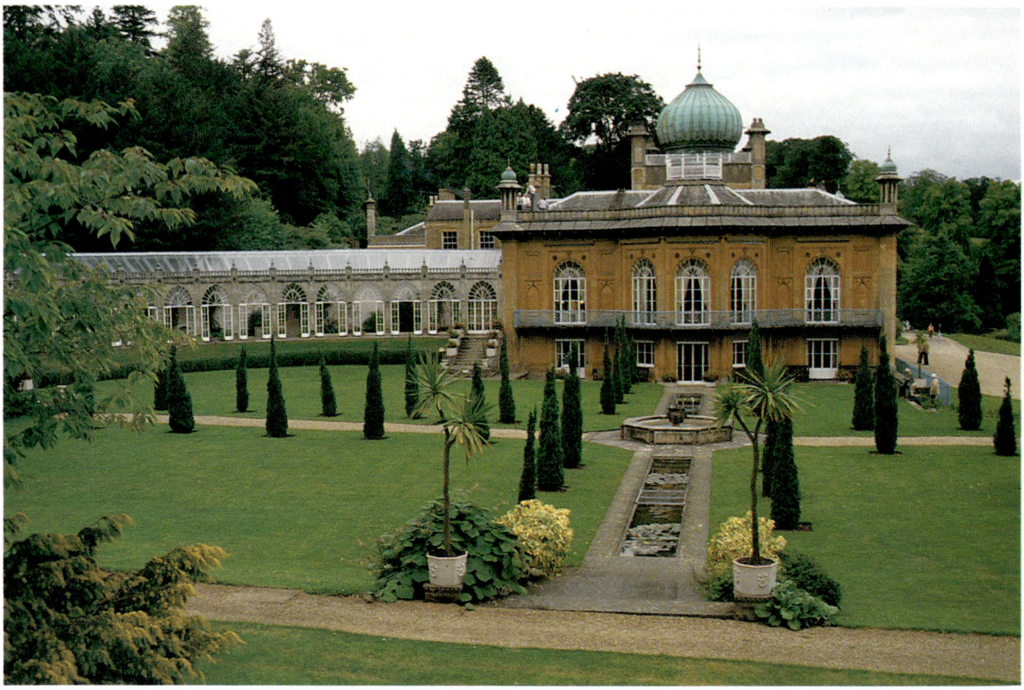

PLATE 32 Sezincote, Gloucestershire. (a) Hindu temple with a statue of the sun-god Surya. Designed by Thomas Daniell in the early nineteenth century. (b) Irish yews line the two axes of a charbagh design, constructed and planted during the 1960s and 1970s. Sezincote is the only surviving British garden of any importance with Indian architectural features.

Of the fellowship of artists in India, few were more assiduous and enthusiastic than the women who joined their husbands and fiancés in increasing numbers towards the close of the Georgian era. They socialized with their own class, supervised the household servants, gardened in a tentative fashion, and relieved the tedium of endless leisure with gossip, a discreet flirtation if the opportunity arose, or the production of pretty watercolours. Mrs Fanny Parks despised such domesticity, preferring to explore the countryside while her husband attended his office. She often rode alone in native attire, meeting Indians, improving her fluency in Hindi, developing a liking for local dishes, and becoming an authority on Hindu mythology, iconography, and ritual. Her journal, *Wanderings of a pilgrim in search of the picturesque during four-and-twenty years in the East* (1850) is the chronicle of an incurable romantic who responded impulsively to the natural world around her. She botanized, made impressions of some of the plants with lampblack mixed with linseed oil and, consulting J. C. Loudon's *Hortus Britannicus*, tried hopefully to identify them. Her collection of natural history specimens grew so large that she could with some justification and pride call it a museum. Her craving for knowledge and experience makes her journal compulsive reading; her writing communicates her enthusiasm and confirms her sharp observation. In describing the creeper, *Ipomea quamoclit*, she notes that 'the leaves, falling back, lie over one another *en masse*, spreading over a broad surface in the manner in which the feathers of the tail of a peacock spread over one another, and trail upon the ground'[14].

Like Mrs Parks, Lady Canning also travelled in the remoter regions but in the security and comfort of the entourage of her husband, the Governor-General. Thomas Thomson at the Calcutta Botanic Garden identified the plants she collected on these official tours. Her diary records a botanical foray in the south.

> *After a rainy middle of the day, we went out, carrying a great tin box, to search for ferns. A slippery path took us down a woody glen, where we gathered eight varieties, and went on & on & on down & down by zigzags into another glen, to a beautiful walk I had never seen before . . . By the side of it I saw the Hedychium growing wild, but rather of a creamy white variety, and a pretty Begonia on a great rock, for once within reach*[15].

The paintings of some of her gatherings were admired by John Ruskin who with typical hyperbole thought them 'the grandest representations of flowers he had ever seen', praising especially her 'subtle use of colour'[16]. One wonders what his reaction might have been to the fluent brush of Marianne North who travelled the world, including India and Ceylon, rapidly filling canvases with a mélange of colour—bold, emphatic, and decidedly not subtle.

The valuable contributions of the amateur naturalist were acknowledged by Joseph Hooker in introducing the native flower drawings commissioned by John Cathcart of the Bengal Civil Service.

> *Science is not yet self-supporting; it requires the countenance of amateurs no less than the severe studies of proficients to ensure its progress*[17].

Natural history has always been a field where the amateur can usefully participate through collecting and observation. 'A love of natural history', enthused Emma Roberts, 'opens endless fields of pleasurable research to those who have encouraged a taste for it'[18]. The study of nature was considered a decorous recreation for women, neither dangerous nor fatiguing; pressing plants, pinning butterflies, or cataloguing curious shells might elevate the mind as well as sharpen the perceptions. It was a popular misconception

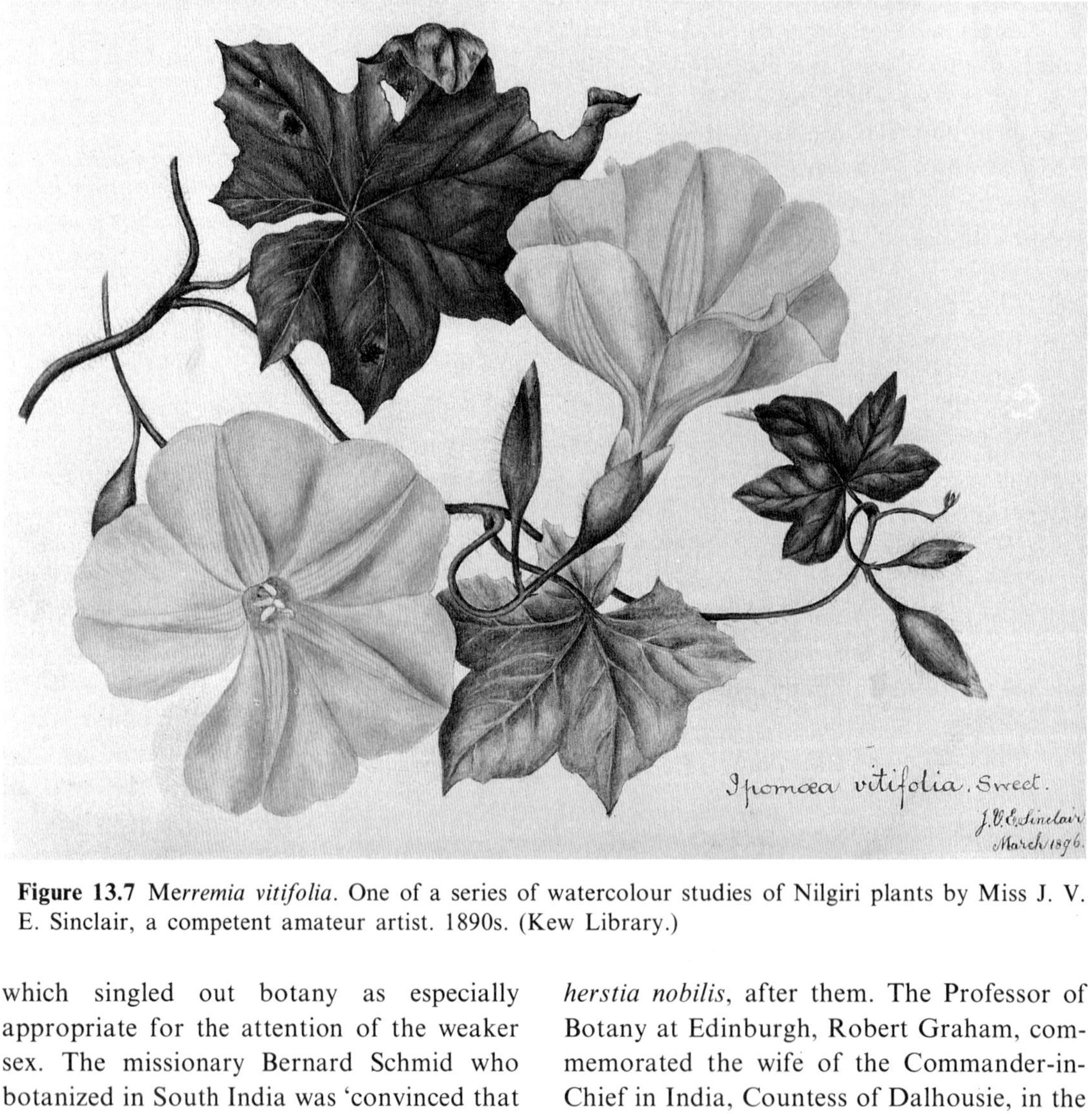

Figure 13.7 M*erremia vitifolia*. One of a series of watercolour studies of Nilgiri plants by Miss J. V. E. Sinclair, a competent amateur artist. 1890s. (Kew Library.)

which singled out botany as especially appropriate for the attention of the weaker sex. The missionary Bernard Schmid who botanized in South India was 'convinced that the study of botany is particularly suitable for ladies, and an important branch of the education of children'[19]. And it was indeed an area of study where the memsahib was prominent, with the wives of India's senior administrators leading the way. Lady Amherst, wife of the Governor-General, returned to England with a herbarium of Himalayan plants; Wallich gracefully acknowledged the contributions she and her daughter, Lady Sarah, had made to botany by naming the spectacular Burmese tree, *Amherstia nobilis*, after them. The Professor of Botany at Edinburgh, Robert Graham, commemorated the wife of the Commander-in-Chief in India, Countess of Dalhousie, in the genus *Dalhousiea*, and W. J. Hooker dedicated a volume of *Curtis's Botanical Magazine* to her. She presented her herbarium of some 1200 specimens, gathered mainly around Simla, to the Botanical Society of Edinburgh. The drawings of Indian plants by Mrs Hutton, the wife of a Calcutta merchant, were given by her daughter to Kew Gardens. Some artists suceeded in getting their flower studies published; an early and extremely rare example is Mrs James Cookson's *Flowers drawn and painted after nature in India* (1835). Mrs Margaret Brown, wife of a Major-General in the Madras Cavalry, enjoyed the privilege of dedicating her *Wild flowers of southern and western India* (1868) to the Princess of Wales.

Oriental memoirs (1813), in four substantial volumes, stands apart from much of the literary endeavour of amateur naturalists and historians. Its author, James Forbes (1749–1819), served in the East India Company in India from 1766 to 1784, never rising above the rank of a minor official. His *Memoirs*, a distillation of accumulated letters and drawings amounting to 150 volumes, examined Western India from Gujerat to Malabar—its peoples, antiquities, the fauna, and flora. This scholarly research was both relaxation and consolation.

> *The pursuit beguiled the monotony of four India voyages, cheered a solitary residence at Anjengo and Dhuboy, and softened the long period of absence from my native country: it has since mitigated the rigor of captivity, and alleviated domestic sorrow. Drawing to me had the same charm as music to the soul of harmony. In my secluded situation in Guzerat, I seemed to be blest with another sense. My friends in India were happy to enlarge my collection; the sportsman suspended his career after royal game to procure me a curiosity; the Hindoo often brought a bird or an insect for delineation, knowing it would then regain its liberty; and the brahmin supplied specimens of fruit and flowers from his sacred enclosures. India formerly was not the resort of artists; when there I had little to excite emulation, and no other instruction than a few friendly hints from Sir Archibald Campbell; who, during a short residence at Bombay in 1768, encouraged my juvenile pursuits*[20].

His isolation and self-reliance were not unusual in a country where Europeans were thinly spread and communications primitive. Even as late as the 1830s, Wallich cautioned that 'in India those who cultivate natural history are necessarily thrown on their own unaided resources'[21]. Forbes's awareness of the natural world was imbued with eastern mysticism, with a reverence for all living things, and with a sensuous regard.

> *There is, besides, a voluptuous stillness, if I may use the expression, in an Indian landscape, a serenity in the atmosphere, and a quietness in the road during a morning walk, or evening ride in the cool season, not generally known in Europe*[22].

Oriental memoirs includes nearly a hundred engravings based on Forbes's competent drawings, the best being the coloured aquatints of birds, fishes, flowers, and fruits by the botanical artist and engraver William Hooker. The five lithographs by Forbes himself are a bibliographical curiosity, being among the earliest plates printed by that process in England.

If Forbes was retiring and modest, Claude Martin (*c.*1732/5–*c.*1800), a deserter from the French forces who rose to Major-General in the Company's army, was his opposite—an adventurer who achieved fame and fortune by trading and money-lending, enjoyed the company of concubines and amassed with a connoisseur's flair a fine collection of books, manuscripts, miniatures, and about 600 flower drawings by Indian and European artists[23]. He was friendly with William Roxburgh who described some of his plants in Flora Indica.

'A more infamous or despicable character than the late General Martin never existed' was the emphatic opinion of George Annesley, Viscount Valentia (1770–1844). The purpose of his lordship's visit to India in 1803 was 'the prosecution of botanical research' and without question he was the most affluent amateur naturalist ever to explore the subcontinent. He was fêted by Lord Wellesley who reminded his officials in the three presidencies that he was to be received 'with every mark of respect and distinction suitable to his rank'. Endless receptions and dinner parties eased his itinerary through Bengal and Oudh down to the south of the peninsula. This unstinted hospitality alarmed the Court of Directors who complained that 'it was highly improper to subject the Company to the expense of Pagodas 312[24] on the occasion of the interview between his lordship and the Rajah of Tanjore'[25]. While admiring the Botanic Garden at Calcutta, Valentia regretted that 'utility seems to have been more attended to than science'[26]. He was promised a share of the nutmegs, cloves, and other plants shipped by Christopher Smith from the Dutch East Indies. Benjamin Heyne, whom he met in Mysore, presented him with seeds and with drawings by his native artist. The Governor, Frederick North, was his host in Ceylon, which Valentia pronounced 'the richest field for a botanist that I ever beheld, except the Cape of Good Hope'[27]. When he wrote an account of his travels the luxuriance of that island's vegetation still remained a vivid and agreeable recollection. Wallich named his Indomalayan genus *Anneslea* in his honour.

NOTES

1 Hodges (1793), *Travels in India*, 155.
2 Heber (1828), Vol. 2, 240.
3 J. D. Hooker (1854), Vol. 1, 144.
4 Westall to Banks, 31 January 1804. Mitchell Library, Sydney. Banks correspondence, Vol. 4.
5 F. C. Solvyns (1808), *Les Hindous*, Vol. 1, Paris, 21.
6 Ibid., (1812), Vol. 4, 9.
7 J. Fraser (1820), 133.
8 G. White (1836–8), *Views in India*, London, 44.
9 J. Fraser (1820), London, 436.
10 J. D. Hooker (1854), Vol. 1, 221.
11 T. Bacon (1837), Vol. 1, 237–9.
12 E. Roberts (1839), *East India voyager*, London.
13 *Photographic Journal*, 15 June 1860, 246.
14 F. Parks (1850), *Wanderings of a pilgrim*, Vol. 1, London, 310.
15 Allen (1986), 101–2.
16 A. Hare (1893), *Story of two noble lives*, Vol. 1, London, 478.
17 Hooker (1855), iv.
18 E. Roberts (1839), *East India voyager*, London, 122.
19 *Madras Journal of Literature and Science*, **4**, 1836, 336.
20 J. Forbes (1813), *Oriental memoirs*, Vol. 1, xi–xii.
21 N. Wallich (1830), *Plantae Asiaticae rariores*, Vol. 1, London, 4.
22 J. Forbes (1813), *Oriental memoirs*, Vol. 3, 274–5.
23 Drawings are now at Kew. *Kew Bulletin*, 1919, 207–8.
24 A pagoda was a gold or silver coin.
25 E.I.C. to Madras, 24 July 1805. I.O.L.R. E/4/895, folio 313.
26 Viscount Valentia (1809), *Voyages and travels to India*, Vol. 1, 64.
27 Ibid., 270.

A CENTURY OF CHANGE

If an amateur can be defined as one whose interests and pursuits are the products of self-education indulged in as a pastime then many of the naturalists in India qualified for that epithet. The introductory remarks to the first volume of *Asiatick Researches* (1788) reminded its readers that

> *a mere man of letters, retired from the world or allotting his whole time to philosophical or literary pursuits, is a character unknown among European residents in India, where every individual is a man of business in the civil or military state and constantly occupied either in the affairs of Government, in the administration of justice, in some department of revenue or commerce, or in one of the liberal professions; very few hours, therefore, in the day or night, can be reserved for any study that has no immediate connection with business, even by those who are not habituated to mental application, and it is impossible to preserve health in Bengal without regular exercise and reasonable relaxation of mind*[1].

The Asiatic Society was formed in 1784 to afford the opportunity for orientalists, linguists, historians, and naturalists—many of them amateurs—to enjoy the stimulus of communal interests, to submit their collections to friendly scrutiny, and to benefit from constructive criticism and advice. The Grand Jury Room in Calcutta was the venue for weekly evening meetings which had become monthly by the end of 1799 and quarterly soon afterwards. In 1806 the Society moved to its own building in Park Street where it hoped eventually to house a museum and library. Nathaniel Wallich donated some botanical works but these were later transferred to the Botanic Garden as a more appropriate repository. In September 1808 two committees were formed, one for 'Natural History, Philosophy, Medicine, Improvements of the Arts, and whatever is comprehended in the general term of Physics' and the other 'for Literature, Philology, History, Antiquities, and whatever is comprehended under the general term of literature'. Needless to say, that paragon of amateurism, William Carey, was elected to both. They languished through lack of adequate support although the Physical Committee enjoyed a temporary respite for a few years after 1818. Poor communications made the Asiatic Society essentially a local organization for the Presidency of Bengal, and indeed as early as the 1820s it was unofficially known as the Asiatic Society of Bengal[2].

Bombay, far to the west, established in 1804 its own Literary Society, which later became a branch of the Royal Asiatic Society in London. Natural history which featured among its fields of research was generally neglected for subjects with greater popular appeal: history, archaeology, and literature. By 1831 Bombay had its own

Geographical Society and local enthusiasts launched the Bombay Natural History Society in 1883, happily still flourishing.

A century after its foundation, the Asiatic Society of Bengal had become, according to Sarat Chundra Mitra, a forum primarily for the study of antiquities and literature. Dismayed that the natural sciences were being spurned, he proposed a breakaway society to be called the Linnaean Society of Calcutta, dedicated to geological, zoological, and botanical research. The usual apparatus of a society would be provided: library, museum, and publications. He also envisaged a herbarium complementing the existing one at the Botanic Garden and a museum of economic botany modelled on the one in Kew Gardens[3]. This proliferation of societies was deplored by an objector who reminded Mitra that Calcutta already had an Agricultural and Horticultural Society, a Microscopical Society, a Medical Society, and a Photographic Society for those residents interested in science. Furthermore, he disputed the thesis that science had been swamped by the humanities in the meetings and papers of the Asiatic Society of Bengal, having calculated that between 1784 and 1883, one-third of all published papers related to zoology and botany. He counter-proposed with an appeal for restructuring the Society into literary and scientific sections, each with its own secretary and separate functions.

John McClelland, editor of the *Calcutta Journal of Natural History*, doubted whether any of the existing societies had the means to fulfil all their objectives. Because there were relatively few potential European members, and these were widely scattered throughout the subcontinent, representative meetings were difficult to convene. He thought a journal alone, observing reasonable standards of scholarship, could effectively unite any membership. He advocated the creation of a society conducting such a journal, and even had a name for it—the Indian Association for the Advancement of Natural Science—confident it would not pose a threat to other organizations.

> *The object of the proposed Society would be to collect a body of information regarding the natural history of the country, and to concentrate the labours of our naturalists, which have hitherto been interspersed throughout various publications, in a manner to render it difficult to refer to what really has been done*[4].

As an editor himself, it was natural that McClelland should emphasize the pivotal role of periodicals, and for most societies the proprietorship of a publication affirmed its reputation and maturity. The founder of the Asiatic Society, Sir William Jones, had contemplated an annual journal, but up to the time of his death only three volumes of *Asiatick Researches* had been published. With literary and historical papers in the ascendency, an attempt was made from volume 18 (1833) to encourage more scientific contributions by dividing *Asiatick Researches* into two sections: Literary and Physical—the latter also being available separately.

The frustrating infrequency of *Asiatick Researches* goaded the surveyor, Captain James Dowland Herbert, to initiate, at his own financial risk, *Gleanings in Science*. It was published monthly from January 1829, its original function being to reprint or summarize articles from the English press, but by the end of its first year original contributions had outnumbered matter extracted from European sources. Following his appointment as Astronomer to the King of Oudh in 1830, Herbert surrendered the editorship to James Prinsep, one of the two secretaries of the Asiatic Society. Under his control it published all the ephemeral communications submitted to the Asiatic Society, justifying the change of name in 1832 to *Journal of the Asiatic Society*. It remained the property of Prinsep until his return to England in 1838 when he transferred it to Henry Torrens. *Asiatick Researches* made its last appearance with volume 20 in 1839 and four years later the Asiatic Society adopted the *Journal* as its official publication.

The year in which *Gleanings in Science* was published—1829—also saw the first issue of the *Transactions* of the Agricultural and Horticultural Society of India. Back in England, its emergence was welcomed in an expansive review in J. C. Loudon's *Gardener's Magazine*. The Society also optimistically translated some of the earlier volumes into Bengali.

The often precarious and brief existence of periodicals in India was dependant on a small potential readership—in Bengal in the 1820s it was estimated to be no more than 2000. The *Quarterly Oriental Journal* and *Transactions of the Medical and Physical Society* both ceased publication in 1827. Such insecurity never deterred optimistic editors like Dr Corbyn who started the *India Review and Journal of Foreign Science and the Arts* in 1837. It collapsed the following year. The Madras Literary Society, affiliated like its Bombay counterpart to the Royal Asiatic Society, ventured into publishing in October 1833 with the *Journal of Literature and Science*, the title of which was prefixed by *Madras* from volume 2[5]. Robert Wight, a frequent contributor on Indian botany, was instrumental in changing the editor in 1836, and committed himself to offering his papers in the first instance to the *Madras Journal*. But he had misgivings about such an obligation.

> *When new genera or species are published in it, it may be useful to get them transferred to some of the European periodicals to prevent their being lost, or superseded by writers in better known and more widely circulating journals*[6].

Wight was also briefly a co-editor of the *Calcutta Journal of Natural History*. McClelland, its founding editor, was confident that demand existed for a periodical specifically for geologists and biologists. When he introduced it in 1841, he feared no rivalry from the *Journal of the Asiatic Society*, now much more literary in character. His aim was 'less to afford amusement than instruction', to publicize research, and to solicit contributions from prominent naturalists. William Griffith, whom he had co-opted as a co-editor in 1844, was represented by substantial articles on palms, cryptogams, and plants growing in the Botanic Garden in Calcutta. Wight wrote on the vegetation of the Nilgiri Hills and supplied notes on Indian botany while George Gardner, another co-editor, submitted 'Contributions towards a flora of Ceylon'. Despite the support of such prestigious names, it ceased publication with the thirtieth issue in 1848. The ostensible excuse offered by McClelland was his difficulty in finding time to edit it; one suspects, however, that the real reason was more likely to be the shortage of funds and subscribers.

The *Calcutta Journal of Natural History* was undoubtedly élitist, rigorously maintaining professional standards, whereas the *Journal of the Bombay Natural History Society* positively welcomed the amateur contributor. To its credit, it has never deviated from the editorial policy announced in its first issue in January 1886:

> *to interest all students of nature, ever remembering that there are many naturalists in the highest sense of the term who have not much technical knowledge of any particular branch of science as to be able to enter with interest into questions of nomenclature and the discrimination of closely allied species.*

Some naturalists chose to submit important pieces of research to British scientific journals: the *Transactions* of the Royal Societies of London and Edinburgh and the Linnean Society of London, the *Memoirs of the Wernerian Natural History Society*, the

Edinburgh Philosophical Journal, the *Proceedings* and *Transactions* of the Botanical Society of Edinburgh, and W. J. Hooker's *Journal of Botany*. Indian economic plants featured in articles published in the *Journal of the Royal Asiatic Society*.

What interpretation can be made of the disappointing fact that there were very few articles by Indians on natural history in the publications of the Asiatic Society of Bengal? When it was formed in 1784, its President, Sir William Jones, said 'whether you will enrol as members any number of learned Natives you will hereafter decide' and Indians were subsequently admitted to membership. Sarat Chundra Mitra lamented in 1890 that although his fellow countrymen had distinguished themselves in literature and in science in general, 'Natural History pursuits, as intellectual recreations, have never been popular amongst the people of India, whether of past or modern times'[7]. A similar observation had been made by Bishop Caldwell in his address to the Convocation of Madras University in 1878; his appreciation of the performance of Indian graduates in antiquarian and literary disciplines was tinged with regret that none had elected to study the country's rich fauna and flora, a point of view also taken up by E.T. Atkinson in his presidential paper to the Asiatic Society of Bengal in 1887.

> *I should be glad to see our native members take more interest in Natural Science, and thus wipe away the reproach that, perhaps with the exception of the late Babu Harimohun Mukerji, and one gentleman in Bombay [K. P. Kirtikar], there is not a single native of India, known outside its limits, for proficiency in either botany or zoology*[8].

Indian academics earnestly contend that it was the paucity of educational facilities and also colonial proscriptions that deprived students of the advantages of western culture, especially in science and technology. This criticism is not without substance, but there were enlightened Europeans who promoted the cause of Indian education. In his *Observations on the state of society among the Asiatic subjects of Great Britain* (1792), Charles Grant, a prominent Calcutta merchant, alluded to the relevance of 'science and instruction in mechanics as applied to agriculture and other useful arts'. The Calcutta *Madrasa*, created in 1781 by Warren Hastings for Arabic and Persian studies later added mathematics and science to its curriculum. The renewal of the East India Company's Charter in 1813 included a clause authorizing an annual vote of one lakh of rupees for 'the revival and improvement of literature and the encouragement of the learned Natives of India and for the introduction and promotion of a knowledge of the sciences among inhabitants of the British territories in India'. Another ten years elapsed before it was implemented by the General Committee of Public Instruction. In the meanwhile, through the demands of Hindu reformers like Ram Mohun Roy, the Hindu College was established in 1817 'for the instruction of the sons of Hindus in European and Asiatic languages and sciences'.

The General Committee of Public Instruction, formed in 1823 to supervise educational expenditure and all government-aided institutions, was exhorted to consider 'the introduction of useful knowledge including the sciences and arts of Europe'. Thomas Babington Macaulay who was made President of the General Committee when he came to India in 1834, tried to resolve the dichotomy between oriental and occidental learning by espousing English as the principal language for instruction. The debate on education had been reflected in the reforms proposed in a further renewal of the Company's Charter in 1833. The recommendation that 'no native of the said territories . . . shall, by reason only of his religion, place of birth, descent, colour, or any of them, be disabled from holding any place, office or employment under the Company' could not be implemented without the provision of more schools and colleges. It was Macaulay who induced the retiring Governor-General, Lord Bentinck, to issue the historic Resolution of 7 March 1835:

The great object of the British Government ought to be the promotion of European literature and science among the natives of India'.

Western medicine was first formally taught in India at the Calcutta Medical College, founded in 1835; it was here, too, that Wallich held classes in botany. The need to disseminate European arts and sciences motivated the 1854 Education Despatch of Sir Charles Wood, President of the Board of Control. As a consequence, vernacular schools in villages were established, and authorization given for the creation of universities, modelled on the University of London, at Calcutta, Bombay, and Madras. These were opened in 1857; by 1902 there were five universities and 126 colleges in British India, an achievement not matched by any British colony at the time.

Calcutta University first taught botany in 1872 at its college at Berhampur where Dr Akshaya Kumar Dey, a graduate of the Medical College, became the first Indian in the subcontinent to lecture on the subject. The provision of books being an integral part of any educational programme, book societies were formed to produce cheap publications, the first in Calcutta in 1817, followed by the Bombay Native School Book and School Society in 1820. Textbooks on science featured amongst their output, but one questions the wisdom of including such a hack compilation as Oliver Goldsmith's *Animated nature*. The first botanical textbook in Bengali in 1869, was written by Jadunath Mukherjee, another graduate of the Medical College; seven years later George Watt's *First steps in botany* was translated into Bengali for use in preparing students for the Minor and Vernacular Scholarship examinations. A forester, Upendranath Kanjilal, wrote an account of the indigenous woody plants of Saharanpur and Dehra Dun as a textbook for students at the Imperial Forest School. J.S. Gamble, former Director of the School, introduced Kanjilal's *Forest flora of the School Circle, North West Province* (1901) as 'the first botanical work of any importance which has ever been prepared by a native of India'.

After a slow, hesitant start, frustrated by inadequate resources, Indian science graduates acquired confidence as well as western expertise; it was only a matter of time before they demanded emancipation. The August 1869 issue of the *Calcutta Journal of Medicine* carried an article by Mahendra Lal Sircar, yet another product of the Medical College, advocating a 'National Institution for the Cultivation of Sciences by the Natives of India'. In 1876 he was instrumental in forming the Indian Association for the Cultivation of Science, the first attempt by Indians themselves to organize their country's scientific research.

A museum with informative exhibits is a desirable adjunct to educational programmes and the activities of learned societies. A repository for its members' collections had always been a desideratum of the Asiatic Society of Bengal. Wallich proposed a museum in February 1814, offering not only duplicates from his own collections, but also his services.

In Natural History, Botany, Anatomy, Chemistry, Mineralogy and other branches, a collection would accumulate rapidly if once commenced; and from the first moment of its accumulation would furnish additional matter to the stock of knowledge[9].

Wallich's suggestion was adopted and he acted as Honorary Superintendent of the

Oriental Museum of the Asiatic Society for several years. In furtherance of mineral exploration, the Government of India set up a Museum of Economic Geology in Calcutta in 1835, voting to the Asiatic Society funds to manage it; in 1856 the Government moved their collections to a separate Geological Museum. Museums were now fashionable. Bombay opened one in Victoria Park in 1855 with economic products as its *raison d'être*; in 1857 Madras founded a museum with items retrieved from the Madras Exhibition of 1853. Belatedly in 1858 the Government acknowledged its responsibility to establish in Calcutta an imperial museum for artifacts and natural history specimens. Negotiations with the Asiatic Society culminated in the Indian Museum Act of 1866 which stipulated a display 'illustrative of Indian archaeology and of the several branches of Natural History'. Twelve years were to pass before the galleries of the new Indian Museum were accessible to the general public. The Government of India had been a late, even a reluctant, convert to the notion of a municipal museum, mindful of the fact that the East India Company already maintained a museum in India House in London. In 1839 the Court of Directors, in granting 250 rupees a month towards the salary of the Asiatic Society's curator and the care of its collections, stipulated that its museum as far as the Government was concerned, was to be subordinate to their India Museum in London. Collections initiated by the Government were still to be despatched to London; only duplicates were to be retained in Calcutta; entomological specimens, difficult to preserve in Calcutta's climate, were always to be consigned to the India Museum.

From its earliest days the artefacts, plants, animals, and minerals collected in a desultory manner by servants of the East India Company during their travels had formed a miscellaneous display in India House in London. When Cosimo the Third, Grand Duke of Tuscany, visited London in 1669, he was taken there to inspect its cabinet of curiosities. It was not, however, until Sir Charles Wilkins, an eminent Sanskrit scholar, proposed the formation of an oriental library and museum in 1798 that the Court of Directors gave any serious attention to the collecting and display of objects relating to South Asia. Wilkins conceived as a principal department of the museum a 'Cabinet of Natural Production', derived from animals, vegetables, and minerals. In collecting 'the plants, seeds and fruits of Asia', he recommended that priority should be given to those of interest to commerce and medicine. It needed the enthusiastic support of Warren Hastings before a hesitant Company appointed Wilkins its first librarian and curator at its headquarters in Leadenhall Street in 1801.

Initially, donations to the new museum were heterogeneous—a Babylonian stone inscribed with cuneiform writing, brass statues of Indian deities, three elephant heads 'with several detached parts intended to illustrate the natural history of these animals, so far as relates to their curious mode of dentition', Tipu Sultan's mechanical tiger, samples of indigo, tea, spices, and other commercial products. Specimens collected by Eudelin de Jonville in Ceylon and sent by the Governor, Frederick North, were perhaps the first significant natural history acquisition. But so disappointing was the general response from India that the Company in 1805 reminded the three Presidencies of their duty to support the museum. When Wilkins died in 1836 the charge of the museum passed to Thomas Horsfield, an American naturalist who had formerly served as a surgeon with the Dutch East India Company, had been encouraged by Raffles to study the natural history of the small island of Banka, and had generously deposited his East Indies collections in the India Museum (as it was now called) in 1819. Horsfield's predilection for natural history gave a new dynamism and direction to the India Museum's collecting policy. It soon shared with the Zoological Society of London the distinction of housing the most extensive collections of Asian material in London. It was not until the mid-nineteenth century that its pre-eminence in natural history was challenged by the British Museum.

Before long the volume of acquisitions presented acute storage problems, and the Court of Directors readily acceded to the request from the Linnean Society for duplicates from Royle's herbarium, adding for good measure Wallich's spirit collection of fruits and flowers in 172 bottles and W. H. Sykes's Deccan plants. In 1837 a natural history gallery was installed in the museum, soon to be filled as a result of Horsfield's acquisitive enthusiasm. He prompted the Court of Directors to admonish the Government of India in 1840 for failing to ensure that natural history collections made on their behalf were not retained by the Asiatic Society of Bengal. London ruled that official collections must be sent to the India Museum.

These instructions will apply to all public collections . . . should any such be still detained in the hands of individuals or remain deposited in the Botanic Garden of Calcutta or in the charge of the Asiatic Society[10].

In 1842 Horsfield complained that only part of William Griffith's Afghanistan collections had been received. Somehow extra room was found to exhibit Brian Hodgson's mammals and birds from Nepal and Hugh Falconer's palaeontological finds from the Siwalik Hills. Horsfield's influence can be detected in a memorandum to the Tibetan Boundary Commission in 1847 which stressed 'the importance which the Govt attach to the labours of the scientific department of this Mission' and listed 'a few points which have an immediate reference to the interests of the Museum of Natural History in this House [i.e. East India House]'[11]. Richard Strachey of the Bengal Engineers, while a member of the Tibetan Boundary Commission, dutifully despatched birds, animals and plants to the Museum. On their receipt, Horsfield reminded the Court of Directors

that the arrival of Capt Strachey's collections is highly opportune at the present period. The expected exhibition of 1851 [the Great Exhibition in Hyde Park] will bring a large concourse of strangers to London, and the Company's Museum may expect a proportionately increased number of visitors; among whom naturalists and men of science will duly appreciate the liberal patronage of the Honble Court of Directors to the Natural History of their oriental possessions[12].

While Horsfield was omnivorously extending his collections, his colleague John Forbes Royle, late Superintendent of the Saharanpur Botanic Garden, was contemplating their selective distribution. The East India Company had appointed him their Correspondent relating to the Vegetable Productions of India in 1838, responsible for developing these resources through correspondence and investigation. In 1849 the Court of Directors requested his recommendations for the disposal of accumulated botanical collections in India House. Royle favoured the retention of specimens of commercial and useful products for his proposed Museum of Economic Botany, regretting *en passant* that Kew Gardens had forestalled him by opening a similar museum the previous year. He reviewed the quantity and value of plant collections acquired since the 1830s: about 12 000 specimens collected by William Griffith in India, Afghanistan, and Malacca; about 600 species amassed in Tenasserim in Burma by Johann Wilhelm Helfer; probably 200 species in Theodore Edward Cantor's Chinese gatherings; between 400 and 500 species gathered by Dr Roth in Abyssinia; not less than 4000–5000 specimens in Hugh Falconer's Himalayan herbarium.

> *The placing of such [collections] within the reach of the scientific man of this country and of the continent would undoubtedly greatly benefit the science of botany at the same time that the mere fact of the collections having been made and still more if they were distributed would redound greatly to the credit of the East India Company*[13].

Royle had already sorted Griffith's collection of dried palms into sets destined for the British Museum, Linnean Society, Kew Gardens, Berlin, and Vienna. His Malacca plants had been sent to Sir William Hooker, John Lindley, and Professor Martius. But other administrative duties now intervened, in particular, his organizing of the Indian Court at the Great Exhibition, delayed the final disposal of these collections.

The news of this imminent dispersal naturally attracted a great deal of interest and hopeful anticipation in the botanical world. The Professor of Botany at the Royal Dublin Society, William Henry Harvey, wrote expectantly to Sir William Hooker at Kew.

> *You certainly hold out a tempting bait in promising me a good share in the distribution of the stores at the India House (if you mean the Falconer–Royle–Griffith items, not the relics of the former Wallichian)*[14].

But they remained in the cellars of India House, virtually inaccessible for over nine years. Within a month of Royle's death on 2 January 1858, Joseph Hooker obtained permission to examine them.

> *I found these collections to be of enormous bulk: some are in unopened chests; others are partially arranged & the confusion is so great that it will be difficult to determine where, or by whom many were collected—they have also suffered to such a degree from damp & the ravages of worms that it is feared many are irretrievably ruined*[15].

Collections had been mixed, the absence of labels in some instances hampered the identification of the collector, and much of Griffith's and Falconer's herbaria had been destroyed by decay and vermin. It was these two collections that initially had interested Hooker but he, perhaps rashly, offered to take the lot to Kew where, he reckoned, he would need a couple of years to identify and list them.

> *The ultimate destination of the collections, after arrangement, I leave wholly to the discretion of the Court. My sole purposes are to make their contents accessible to scientific men, to give the Honble Company's officers the credit of their labours, & to supply a key to Dr Griffith's posthumous works. There will probably be a few among the many thousand specimens duplicates that I should desire to retain for the Hookerian Herbm at Kew . . . As for the Drawings of Hardwicke, Roxburgh, Wallich, etc., I have to request that they may be entrusted to me, with the collections (as they were to Sir Joseph Banks, under somewhat similar circumstances)—not so much for the service I could render in naming & arranging them, for this would be a pleasure; but because the loan of them wd materially facilitate my labours.*

The Court of Directors readily agreed—eager to rid themselves of this encumbrance—to Hooker's overture, assented to the distribution of duplicates and even contributed £200 to the cost of mounting paper, etc. In July 1858, eleven waggon loads of chests, bundles, and packages—the contents of the various cellars—arrived at Kew Gardens. After eleven years' intermittent labour in sorting, numbering, and labelling the collections, Joseph Hooker, who had now succeeded his father as Director at Kew, announced in the Kew annual report for 1869 the despatch of the few remaining sets of specimens. He calculated, with merited self-satisfaction, that more than 380 000 specimens had been distributed among about 60 British and continental herbaria and museums. He made certain that the Botanic Garden at Calcutta received a complete set.

The East India Company, whose authority had been steadily eroded by successive legislation, and by the dominance of the Board of Control, was deprived of its residual powers after the sepoy rebellion or 'Indian Mutiny' of 1857–8. The Government of India Act of 1858 created a new government department, the India Office, to administer British India. The Secretary of State for India, Lord Stanley, was uncertain whether there was a role for the India Museum in his new department; a case perhaps, he thought, could be argued for the retention of its economic functions. Colonel William Sykes, a generous donor to the India Museum of plants, birds, and insects before his retirement from India in 1831, and now Member of Parliament for Aberdeen, was alarmed to learn of the gradual dispersal of the natural history collection to the British Museum in Bloomsbury. A temporary home for the unwanted India Museum was found in Fife House on the Embankment while protracted discussions about its dispersal were held among London museums. It moved again in 1869 to cramped accommodation on the third floor of the new India Office in Whitehall but much of the natural history, geological, and archaeological specimens remained packed in boxes in the India Office stores in Lambeth, a situation strongly deplored by P. L. Sclater in *Nature*[16]. The India Office, completely impervious to any public criticism, and determined to be rid of this incubus, negotiated the rental of a gallery at South Kensington with the Commissioners for the Exhibition of 1851. In June 1875 the India Museum, now a shadow of its former greatness, gained a temporary respite in the Eastern Galleries at South Kensington, but its demise was in sight. Its loyal supporters fought a rearguard action at question time in the House of Commons on 17 July 1879. The following month the Royal Asiatic Society registered its 'most earnest protest' against its dismemberment but the India Office were no longer prepared to continue financing it. Before the year was out, the carve-up of the collections had been agreed; the British Museum got the Amaravati and other ancient sculptures together with the zoological collections, the new South Kensington Museum (now the Victoria and Albert Museum) acquired the decorative artifacts, and Kew Gardens received 12 vanloads of economic botany material, duplicates of which were despatched to 25 institutions at home and abroad. In addition, 3359 botanical drawings were transferred to Kew to join the Roxburgh icones which Joseph Hooker had borrowed in 1858.

The India Museum had ceased to exist as an institution but parts of its collections had been rescued from obsurity and neglect, subjected to expert scrutiny, evaluated, labelled, and occasionally displayed for the first time. Hooker's haul of herbarium specimens from India House provided invaluable data for his ambitious resumé of the flora of British India.

The two Hookers, father and son, had led the way in the floristic survey of the British Empire. While Sir William was in the professorial chair at Glasgow, he surveyed the botany of the northern regions of British America in *Flora Boreali-Americana* (1829–40), and Sir Joseph's attachment to the Antarctic expedition under the command of Sir James Ross in 1839–43 yielded the *Flora Antarctica* (1844–7), *Flora Novae-Zelandiae* (1852–5) and *Flora Tasmaniae* (1855–9). But these were expensive works in a large format, profusely illustrated, intended rather for the researcher at his desk than for the botanist in the field. Sir William, resolved to continue the series, approached the Colonial Office in May 1857 with proposals for a number of colonial floras—cheap octavo volumes, easy to carry in a capacious pocket, without plates, and presented in a manner intelligible to the amateur. Accepting his suggestion that the British West Indies should be the next candidate, the German botanist, A. H. R. Grisebach of Göttingen, was commissioned to write it. When Sir William Denison, Governor-General of Australia in 1859, mooted the possibility of a scientific account of Australasia in 1859, the Colonial Office automatically turned to Hooker. 'Her Majesty's Government have under their consideration a project for collecting the materials of a national work on the astronomical features, the terrestrial physics, the botany, zoology and geology of the Colonial Possessions of the British Empire'[17]. For Hooker this was a heaven-sent opportunity to remind the Colonial Office of his colonial floras project. Denison was assured that accounts of the botany of several colonies were already in 'the hands of the accomplished and energetic Director of the Royal Gardens at Kew' and that Australia was near the head of the queue. With the floras of the West Indies, South Africa, Australia, and New Zealand settled, Sir William Hooker judged it an auspicious moment to persuade the remaining hesitant colonies to participate. His memorandum in 1863 attested their need—the economic, educational, and recreational benefits—and outlined a programme[18]. He had planned 12 floras, ranging from Hong Kong with an estimated 1000 species to British India with 12 000. The Kew herbarium had abundant material for India; he thought that George Thwaites who had been occupied for some years with a brief enumeration of Ceylon plants could be induced to write a descriptive flora of the island.

The abandonment of the *Flora Indica* (1855), which foundered after the first volume when the East India Company refused to sponsor it, did not deter its joint author, Joseph Hooker, who continued to cherish an ambition to write a complete flora. He kept the project alive in his correspondence with Indian botanists. Sir William Denison, Governor of Madras since 1861, promised his support at governmental level. On receiving Sir William Hooker's memorandum, Denison straightaway wrote to Sir Charles Wood, Secretary of State for India.

> *I have had a letter from Sir William Hooker, enclosing a memorandum on the subject of 'Colonial Flora', and pointing out the scantiness of the encouragement given by the East India Company towards the development of the natural history of this great country. Would it not be as well to redeem India from this disgrace, and to hold out encouragement to those who have laboured in the cause of natural science, and inducements to others to follow upon the same track? There is a very wide field open here, but few who attempt to cultivate it*[19].

A few months later the India Office received Hooker's plan for a ten-volume flora of British India, nominating his son, Joseph Hooker, and Thomas Thomson (Fig. 14.1) as its joint authors. The India Office capitulated, agreeing not only to subsidize the project but also to purchase a hundred copies of each volume.

With official approval, Hooker and Thomson resumed the daunting undertaking they had originally conceived in the early 1850s—the perfect partnership one would have thought: friends from university days, sharing the advancement of Indian botany as a common ideal. Discord, however, surfaced in a letter from Thomson in January 1870 refusing to participate in Hooker's proposed restructuring of the flora. He urged his co-author to omit Burma, Malaya, and Ceylon, confining the coverage to the three presidencies of Bengal, Madras, and Bombay, and to Sind, Punjab, and the Himalayas, a reasonable suggestion for lessening labour and time on the project[20]. But it would appear from Hooker's reply that his friend's procrastination, his habitual inability to produce contributions to an agreed time-table, was the basic reason for the friction.

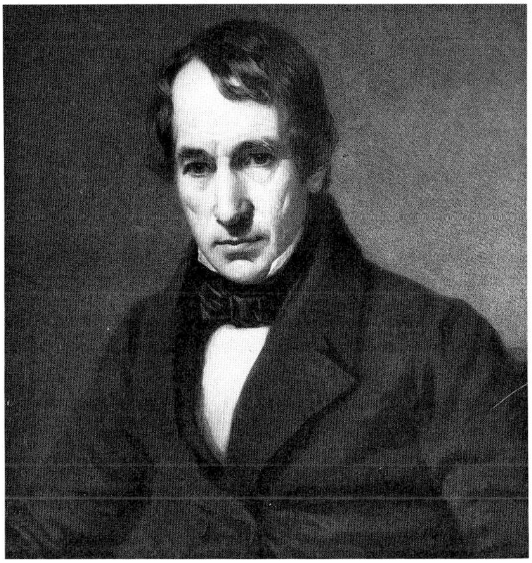

Figure 14.1 Thomas Thomson (1817–78).

Where I feel myself most to blame is in not writing to the Board years ago, & calling off the work, feeling sure as I long have, that you never wd be up to it. But you continued buoying up your own hopes at intervals, though you never could mine, & I really did not like to dispel an illusion that gave you pleasure . . . After 15 years with no published results, under our circumstances, it is clearly hopeless to expect anything from our joint action[21].

Hooker proposed tendering Thomson's ill-health, which was not entirely untrue, as a pretext for his withdrawal. He revealed that he had been tempted to quit the joint project and to proceed with his original intention of publishing a series of *praecursores* on the Indian flora[22].

I can only say in conclusion that in the matter of taking up the original Flora Indica & abandoning the Precursores for the said second Flora of India, I have acted in deference to your wishes, with loyalty to you & against my own judgment[23].

Their collaboration was at an end. Hooker notified the India Office that ill-health had forced Thomson's resignation, and that he intended enlisting the assistance of other botanists to prepare descriptions of taxa.

Hooker had pinned his hopes on Thomas Anderson, the Superintendent of the Royal Botanic Garden at Calcutta. 'In two years you might make a good stroke into the Flora Indica which Thomson will never do a stroke of'[24]. Unfortunately, Anderson died after only two years' retirement. Over the quarter of a century that the work slowly progressed to completion, Hooker enjoyed the collaboration of many distinguished botanists, but he was most indebted to Charles Baron Clarke (1832–1906), originally a teacher of mathematics with botany as a hobby. Clarke had served in the Educational Department in Bengal from 1866; his duties as an Inspector of Schools had enabled him to collect during official tours which took him all over the State; in his vacations he had botanized in the Khasia Hills; he had also visited Sikkim from time to time. When Anderson was invalided out of the Indian service in 1868, he assumed charge of the Royal Botanic Garden until 1871. He generously presented his collections to Kew where he was attached in 1879 for three years at the expense of the Government of India to assist Hooker in the *Flora of British India*.

Progress was slower than Hooker had anticipated: collections still coming to Kew had to be examined; he had deferred to the wishes of the botanical fraternity who wanted him to cite authorities for Indian plants given in every Indian publication; but above everything else, he was overwhelmed by the massive synonymy 'due to the double & even triple naming of Indian plants by authors competent & incompetent, working in England, the Continent, & India, without access to named collections, or concert with one another'[25]. With retirement from the directorship of Kew in 1885, he immersed himself in Indian botany. 'I enjoy my freedom from harness more than ever, though so much poorer', he told the American botanist, Asa Gray. His considerable Indian correspondence ranged far beyond enquiries directly pertinent to his *Flora*. He was dismayed by the lack of critical observation in fieldwork, a theme that runs like a thread through his letters to J. F. Duthie, Director of the Saharanpur Botanic Garden. 'What we want now is rather observation than specimens, especially of the smaller flowered plants, and above all of Orchidaceae'[26]. In another letter: 'Griffith set the example of what could be done by the co-ordination of observing with collecting'[27]. The lament continued: 'The rage for collecting has cold-shouldered observations in the field. Old Roxburgh stands alone'[28]. He suffered the fear that afflicts many authors—the dread of major errors and omissions. He confided to Duthie his distress regarding 'a hideous blunder in the Fl. Brit. India which fell under my eyes accidentally two days ago, and was nearly followed by a stroke of paralysis'[29]. He survived such traumas and the unremitting pressure of editorial routine and deadlines to meet; in his eighty-first year he had the relief and satisfaction of seeing the seventh and final volume rolling off the press.

It was a remarkable achievement but its accomplishment was only possible through a compromise between the ideal and the attainable, and no-one was more aware of its imperfections than Hooker himself. In another letter to Duthie, he writes

That work is a hurried sweeping up of nearly a century of undigested materials, and is in no sense a Flora like Bentham's Australian. It had to be carried out in a reasonable time, and except myself and Clarke none of my coadjutors was really well up in Indian botany, or authorities, or works, or climate, or geography. It is merely a crude guide to the extent and variety of the native vegetation of India[30].

Was this harsh self-criticism merited? It was a courageous attempt to produce for the very first time a comprehensive Indian flora which conscientiously used the vast resources of public and private herbaria. Faced with the enormous plant population of the subcontinent, it was perhaps masochistic to include that of Ceylon, Burma, and the Malay Peninsula as well. R. H. Beddome's thorough monographs on Indian ferns justified the editorial decision to exclude pteridophytes from the *Flora*. C. B. Clarke in his 1874 reprint of Roxburgh's *Flora Indica*, questioned whether 'non-professional botanists can use a book containing 15 000 species'. The *Flora of British India* which actually describes about 14 000 species is essentially a manual for the professional botanist, but it was also, as Hooker saw it, an inducement to others to produce monographs of genera and especially local floras which would serve the needs of the amateur. Now requiring revision, especially in nomenclature, it nevertheless remains a classic among tropical floras. The Government of India publicly acknowledged Hooker's services to Indian botany by elevating him to Knight Grand Commander of the Order of the Star of India.

Henry Trimen's death in 1896 left his *Handbook to the flora of Ceylon* unfinished with volume 3. Sir Joseph Hooker unhesitatingly agreed to a request from the Government of Ceylon to complete it, calculating it would take no more than a year and a half to two years. After all he had assisted Thwaites with his *Enumeratio plantarum Zeylanicae* and, furthermore, many of Ceylon's plants had already been described in the *Flora of British India*. This proved to be a rather optimistic prediction; the fourth volume did not appear until 1898 followed by the concluding volume in 1900.

Defiantly ignoring the infirmities of old age, he still managed, but progressively less frequently, to visit the Kew herbarium, and when that eventually became impossible specimens were brought to his home at Sunningdale. Balsaminaceae had always been one of his favourite plant families and he was working on Asiatic balsams to within a few days of his death (Fig. 14.2).

What may be regarded as the distillation of all his Indian researches is the essay on Indian botany he was asked to contribute to the new edition of the *Imperial gazetteer of India* (1907). He had been allocated 20 printed pages and three months in which to write it; in the event it took him six months when the length of the article had more than doubled. He subjected himself to the exacting task of a close analysis of the geographical distribution of flowering plants and ferns in his *Flora* and related works. He distinguished the main elements in the flora as Malayan, European-Oriental, African, Tibetan-Siberian, and Sino-Japanese. He came to no conclusions about 'whether India is richer in number of genera and species than any other area on the globe of equal dimensions'. It was not only a masterly summary but it was his apologia as well, an appropriate coda to a productive half-century during which his Indian interests were never neglected. His affection for the country remained constant. 'I do love Indian botany', he told A. T. Gage in 1906. 'I long to see another Griffith.'[31]

In the preface to his *Icones plantarum Indiae Orientalis* (1840), Robert Wight attributed the dearth of botanists in India to a lack of local floras, a deficiency which Hooker, too, had recognized. With the exclusion of periodical articles, one of the earliest attempts at a localized flora was compiled by the Deputy Post-Master General at Bombay, John Graham (1805–39). When he died suddenly, his *Catalogue of plants growing in Bombay and its vicinity* (1839), sponsored by the Agri-Horticultural Society of Western India, was completed by Joseph Nimmo. A straightforward listing without descriptions, its usefulness lies in its recording of localities. Doubtless N. A. Dalzell and A. Gibson consulted it for their *Bombay flora* (1861). John Lindsay Stewart (*c.*1832–*c.*1873), a forest conservator in the Punjab, enumerated plants of economic value in *Punjab plants* (1869). His *Forest flora of North-West and Central India* (1874) was completed by Dietrich Brandis. Foresters were encouraged to compile such floras for official use by their colleagues. William Alexander Talbot (1847–1917), stationed in the Bombay Presidency, was a model employee in that respect, the author of a *Systematic list of the trees, shrubs and woody climbers of the Bombay Presidency* (1894) and two volumes of the *Forest flora of the Bombay Presidency and Sind* (1909–11).

With the availability of Hooker's *Flora of British India* (1872–97) there was a burgeoning of local floras, encouraged by the Botanical Survey of India. In the course of his dealings with the Government of India about T. Cooke's proposed flora of the Bombay Presidency, David Prain, Director of the Botanic Survey, defined the style and mode of presentation of local floras. They should be

> *works compact in form and concise in style, not overloaded with critical disquisition or burdened with too much detail, but given for definite areas, in language as simple as is compatible with scientific accuracy, the means of identifying with certainty those plants that are known to occur within their limits. To be of any value these works must be in the truest sense of the word original works, not the mere compilations of untrained amateurs, but the careful production of conscientious experts personally familiar*

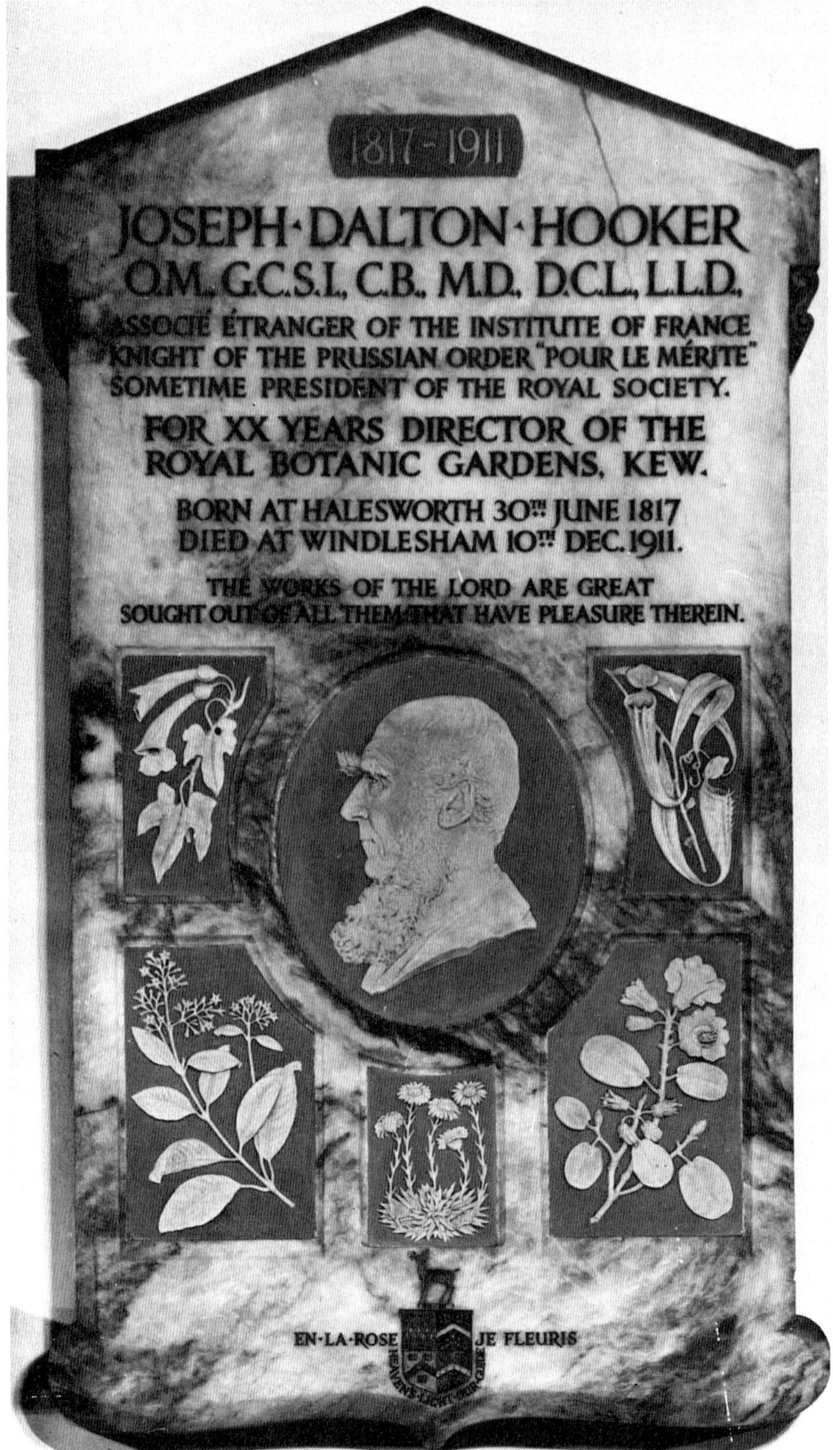

Figure 14.2 Wedgwood memorial plaque to Sir Joseph Hooker in St Anne's Church, Kew Green. The plants, which symbolise events in his long career, include *Cinchona calisaya* (bottom left) and *Rhododendron thomsonii* (bottom right).

with the vegetation of the areas whose Floras they are to describe[32].

Theodore Cooke (1836–1910), briefly Director of the Botanical Survey of Western India, was no amateur. His *Flora of the Presidency of Bombay* (1901–8) proclaims his botanical expertise. John Firminger Duthie was another professional, a stickler for taxonomic niceties. On leaving his post as Superintendent of Saharanpur Botanic Garden in 1903, he spent much of the next 19 years of his retirement on his *Flora of the Upper Gangetic Plain* (1903–29). In 1909 David Prain, now Director at Kew, proposed to the India Office that Cooke, having completed his Bombay flora, might be employed on a similar publication for Madras. With Cooke's illness and death the following year, a replacement was found in J. S. Gamble who, as former Conservator of Forests in the Madras Presidency, was eminently well qualified for the task. The first volume of the *Flora of the Presidency of Madras* was published in 1915 and Gamble continued working on it in the Kew herbarium until his death in 1925; another forest officer, C. E. C. Fischer, completed the third and final volume in 1936.

Sir Joseph Hooker's seven-volume *Flora of British India* was obviously never intended as a field manual. What Sir Henry Collett (1836–1901) wanted was a compact and portable book to identify plants in the neighbourhood of Simla. Sir Henry was in the best tradition of the informed and accomplished amateur; active service in Afghanistan, the eastern frontier and Indo-China was commemorated by the plants he collected. In retirement after 1893, he consulted the resources of the herbarium and library at Kew for his *Flora Simlensis* (1902). Although there has always been an important role for the amateur botanist in India, as universities and colleges were progressively staffed by scientists, and qualified foresters were recruited to the Forest Service, the preparation of floras eventually became the responsibility of the professional.

NOTES

1 *Asiatick Researches*, **1**, 1788, iii.
2 It was formally renamed the Asiatic Society of Bengal in 1851.
3 *Calcutta Review*, **92**, 1891, 196–203, 408–10.
4 *Calcutta Journal of Natural History*, **1**, 1841, 8–14.
5 It was suspended during 1867–77 and ceased publication in 1894.
6 3 August 1836. Hooker's *Journal of Botany*, **3**, 1841, 192.
7 *Calcutta Review*, **91**, 1890, 159.
8 Ibid., 165.
9 *Centenary review of Asiatic Society of Bengal from 1784 to 1883*, 33.
10 E.I.C. to Government of Bengal, 16 September 1840. I.O.L.R. E/4/763, folios 1155–72.
11 E.I.C. Finance and Home Committee, 1847. I.O.L.R. L/F/2/113.
12 E.I.C. Finance and Home Committee, 27 January 1851. I.O.L.R. L/F/2/140.
13 Royle to Court of Directors, 6 February 1849. I.O.L.R. Home/Misc/787.
14 Harvey to W. Hooker, 2 February 1849. Kew Archives. Hooker letters, Vol. 27, folio 237.
15 J. Hooker to E.I.C., 5 February 1858. Kew Archives. Herbarium presentations to 1900, Vol. 1, folios 249–51.
16 *Nature*, 23 February 1871, 328–9.
17 Colonial Office to W. Hooker, 5 November 1859. *Kew Bulletin*, 1905, 13.
18 *Kew Bulletin*, 1905, 20–2.
19 Denison to Wood, 12 July 1863 (W.T. Denison (1870), *Varieties of Vice-regal life*, Vol. 2, London 268–9).
20 28 January 1870. Kew Archives. J. D. Hooker, Flora of British India, folio 42.
21 J. D. Hooker to Thomson, 30 July 1870. Ibid., folio 43.
22 J. D. Hooker and T. Thomson (1858–61), Praecursores ad Floram Indicum: being sketches of the natural families of Indian plants, with remarks on their distribution, structure, and affinities, *Journal of Linnean Society Botany*, **2**, 1858, 1–29, 54–103, 163–80; **4**, 1860, 106–57; **5** 1861, 128–81.

[23] J. D. Hooker to Thomson, 2 April 1870. Kew Archives. J. D. Hooker, Flora of British India, folio 47.

[24] J. D. Hooker to Anderson, 30 May 1868. Huxley (1918), Vol. 2, 16.

[25] J. D. Hooker to India Office, 17 May 1885. Kew Archives. India Miscellaneous, 1878–1928, folio 212.

[26] J. D. Hooker to Duthie, 9 September 1890. Huxley (1918), Vol. 2, 281.

[27] J. D. Hooker to Duthie, 7 January 1896. Huxley (1918), Vol. 2, 282.

[28] J. D. Hooker to Duthie, 16 February 1896. Huxley (1918), Vol. 2, 283.

[29] J. D. Hooker to Duthie, 16 August 1896. Huxley (1918), Vol. 2, 289.

[30] J. D. Hooker to Duthie, 24 January 1901. Huxley (1918), Vol. 2, 387.

[31] J. D. Hooker to Gage, 12 July 1906. Huxley (1918), Vol. 2, 399.

[32] Prain to Government of India, 19 December 1899. Kew Archives. India. Bombay, etc., 1885–1928, folio 69.

THE SEARCH FOR USEFUL PLANTS

In a memorandum on the resources of British India submitted to the Government of India in 1894, George Watt, its leading economic botanist, postulates that in so far as cultivated plants are concerned, India had 'borrowed far more than she has given'[1]. Critics who not infrequently accuse European powers, and Britain in particular, of exploiting India's natural resources tend to overlook or to minimize their contributions to the country's prosperity. The introduction of plants of commercial value has diversified its economy and changed the eating habits of the nation. Indeed all countries have benefited from the migration of the world's economic and ornamental plants which has taken place through several millennia. Basic food crops like wheat and barley were brought by early colonizers from Western Asia, the cradle of Western civilization, to India. Traces of these cereals found in the remains of Mohenjo-Daro and Harappa in the Indus Valley civilization suggest links with ancient Mesopotamia. Indian sugarcane noted in the *Artharveda* (before 800 BC) had reached the eastern Mediterranean some time after AD 636 by way of Iran. Rice cultivation had spread from the Far East across India to Mesopotamia by the seventh century BC. Very likely the crops of China and Japan reached South Asia through the valleys of Assam or the sea route around the Malay Archipelago. The orange, believed to have originated in China, is mentioned for the first time in Indian literature in the *Caraka Samhita*, a medical treatise of AD 100 or earlier. Another well-known citrus fruit common in the tropics, the sour lime (*Citrus aurantiifolia*), is a native of India. *Tamarindus indica*, cultivated in many parts of India, may have been brought to India through the agency of Arab traders who are also credited with transporting the African baobab (*Adansonia digitata*) to the west coast, probably during Mughal times.

The New World was a veritable cornucopia of useful plants. Ships carried a bountiful harvest from America to other outposts of the Portuguese empire: the cashew nut (*Anacardium occidentale*) which Acosta saw growing in the gardens of Cochin; Cassava (*Manihot esculenta*), now a staple food crop in Kerala; and the pineapple (*Ananas comosus*), a favourite dessert of Akbar, was cultivated in the Mughal gardens of Agra. Edward Terry, in India from 1616 to 1619, mentions the potato but perhaps the sweet potato (*Ipomoea batatas*), from Central America, was meant. *Agave* species still perform the function of hedges in Mysore. Their abundance was noticed by Lord Valentia: 'it is in such profusion that it is hardly possible to suppose it could have been introduced from America.'[2] The Portuguese may have added the pepper (*Capsicum annuum*) and the custard-apple (*Annona squamosa*) to the repertoire of Indian culinary skills. Papaya or papaw (*Carica papaya*) came circuitously, according to Linschoten, from the Spanish Indies via the Philippines and Malacca to

India. India is now one of the world's major producers of maize (*Zea mays*) which originated from Mexico. Tobacco (*Nicotiana tabacum*) was not welcomed by the Emperor Jahangir, whose prohibition merely delayed addiction to it. The transmission of Central and South American plants did not cease with the decline of Portuguese power. Seeds collected in South America in 1817 by John Green, Comptroller of His Majesty's Customs in the Bahamas, were among many small gifts from overseas to the Calcutta Botanic Garden. The botanical gardens in Jamaica and St Vincent in the West Indies frequently exchanged material with Calcutta. Several thousand trees of West Indian mahogany (*Swietenia mahogoni*) had adapted to the Indian environment according to the *Hortus Bengalensis* (1814), which also recorded another 60 or more West Indian plants in cultivation in the Calcutta Botanic Garden. *Cinchona* and *Hevea* from South America revitalized Ceylon's ailing economy.

Sir George Yonge, Governor of the Cape of Good Hope, solicited Indian plants from Roxburgh for his new botanical garden. The Calcutta Botanic Garden assumed the role of an horticultural entrepôt. From Calcutta Sir Abraham Hume in England received his *Crinum bracteatum*, a native of Mauritius.

This wholesale introduction and naturalization of exotic plants has been a mixed blessing. Prickly pear (*Opuntia* species) from South America quickly spread, becoming a problematic weed. In 1825 the Military Board of Fort St George in Madras attempted its eradication from river banks and private gardens, 'it being a cover for every species of vermin and a receptacle for all kinds of filth'[3]. Sheh Singh, ruler of the Punjab, fined any of his subjects who grew it. Posing a far more serious threat, such adventive flora has upset the delicate balance of the ecosystem.

Sir Joseph Banks encouraged the international traffic in plants, genuinely believing that India as well as Britain would be a beneficiary. He alluded to the attractions of mutual gain in his exhortation to the East India Company to develop their Indian cotton trade.

> *A colony like this, blessed with advantages of soil, climate & population so eminently above its mother country, seems by nature, intended for the purpose of supplying her fabrics with raw materials; & it must be allowed that a colony yielding that kind of tribute binds itself to the mother country by the strongest & most indissoluble of human ties, that of common interest & mutual advantage*[4].

He accused the Company of commercial timidity, of a lack of entrepreneurial skills, and of an insistence on the 'most rigid economy' in their handling of any commodity. Climate and cheap labour made India well suited for the cultivation of tea; and indigo, coffee, cocoa, vanilla, cochineal, cotton, and sugar were attractive commercial ventures. So keen was Banks to promote this botanical and horticultural enterprise that he magnanimously offered the Court of Directors the benefit of his unsolicited advice. He was not alone in wanting to invest in the economic potential of India's resources. In 1795 the President of the Board of Agriculture, Sir John Sinclair, requested the Company to send him Indian plants and animals likely to be of value to British agriculture. And in India there were Company servants keen to investigate or to promote the commercial prospects of the country's unexplored flora: Sir William Jones as President of the Asiatic Society, Robert Kyd at the newly-formed Botanic Garden, and in South India, James Anderson, Patrick Russell, and William Roxburgh.

After his posting to Samalkot in the Northern Circars in 1781, Roxburgh, being persuaded that the combination of climate and soil fertility were conducive to growing pepper, recommended its cultivation to the

Governor of Madras. The profitable spice market which included pepper had led to the formation of the East India Company. Java and Sumatra were the chief sources of supply until Dutch competition forced the British out of South East Asia. The Court of Directors in 1682 had viewed the ruthless tactics of the Dutch with alarm, fearing an intensification of their belligerency.

> *If the present misunderstanding between the two nations should ferment to an open war, it would be thought by the vulgar but a war for pepper which they think to be [a] slight thing, because each family spends a little of it. But at the bottom it will prove a war for the dominion of the British as well as the British seas, because if ever . . . [the Dutch] come to be sole masters of that commodity, as they already are of nutmegs, mace, cloves and cinamon, the sole profit of that one commodity pepper being of general use, will be more to them, than all the rest and in probability sufficient to defray the constant charge of a great navy in Europe*[5].

Black pepper (*Piper nigrum*), growing wild in the forests of Malabar and Travancore, provided an alternative source of supply for the Company which was thus able to thwart Dutch efforts to monopolize the pepper trade. But by the beginning of the nineteenth century Malabar pepper commanded only eight per cent of the market whereas Sumatra contributed more than fifty per cent. Sir Thomas Munro, who inspected the Kanara pepper gardens in 1800, confirmed the decline.

> *I was now entering a country which had been long famous for the best pepper in India—an article which had been the grand object of most of the early voyages to the coast of Malabar; but there was not a single plant of it within many miles*[6].

It was Roxburgh's intention to resuscitate this languishing trade. When he heard that black pepper grew wild in the hills north of Samalkot, he despatched natives to bring back specimens and, in November 1786, reported that they were growing well in his experimental nursery. Stock was difficult to replenish since lowland natives were reluctant to explore the hills. He bribed the young Rajah of Rampa with two penknives and a looking glass in exchange for some plants and seized the offer of pepper from Malabar and Sumatra in order to diversify his trial cultivation. With 4000 pepper plants in cultivation in 1787, the Court of Directors instructed the Madras Government to render Roxburgh all the assistance in their power. By the following year he had transferred 9500 plants from the nursery to three plantations and, finding them so easy to propagate, was confidently predicting he would increase the quantity to 100 000 within a year. In 1789 he had between 40 000 and 50 000 pepper vines on about 50 acres, justifying his boast in 1790 that he had now 'brought this species of culture to such a pitch of perfection as must convince every person, that the pepper-vine will thrive as well on this coast as in any part of India'[7]. Although Roxburgh raised many healthy plants, they failed to fruit due partly, according to Heyne who succeeded him at Samalkot, to inappropriate methods of cultivation.

In his pepper plantations Roxburgh also had coffee bushes, sugar-cane, mulberry trees, breadfruit, some opuntias from James Anderson, and a few cinnamon. Since the Court of Directors had always coveted the cinnamon trade, the infant Botanic Garden in Calcutta was encouraged to experiment with its cultivation in various regions of Bengal. 'In the cultivation of the cinnamon tree in particular we foresee a great source of wealth to the Company and population and opulence to the Provinces under your administration.'[8] When Colonel Kyd's endeavours in Calcutta and Roxburgh's in Samalkot proved ineffectual, the Company reluctantly concluded that 'the cultivation of cinnamon cannot succeed in any place beside the island of Ceylon, nor indeed in all the parts of that island'[9].

After his transfer to Calcutta Roxburgh continued his predecessor's attempts to naturalize cinnamon and nutmeg as well as clove. In July 1797 he announced that over 800 nutmeg plants, collected in the Spice Islands by his nurseryman Christopher Smith, were being grown in the Botanic Garden and that another 200 had been distributed throughout the country for trials. Francis Buchanan conjectured that the only part of India where the nutmeg might succeed was at Cape Comorin in the far south. Buchanan, however, was sent in 1798 to Chittagong and Tippera in Bengal to find suitable sites for the spices of the East Indies. When Chittagong proved a failure, Roxburgh sent Christopher Smith and a cargo of nutmegs to the Prince of Wales Island (Penang) where they flourished until the nursery was abolished.

Although spices were considered a valuable commodity, Roxburgh was expected to direct his attention to other economic plants—fibres, for instance. The British Government, anxious to minimize the dependence of the Royal Navy on supplies of Russian hemp for its ships' rigging, asked the Company to report on the availability of alternative sources in India. It was known that Indians were using between 40 and 50 fibrous plants. Roxburgh had already sent India House specimens and drawings of *Sansevieria zeylanica*, a plant whose tough fibres had provided bowstrings for the armies of ancient India. The Court of Directors now instructed him to comment on the cultivation of hemp and flax in India. In 1793 London received a drawing and description of flax from Roxburgh at Samalkot and a report from its Resident at Dacca on the local cultivation of both fibres. Within a year of his installation as Superintendent at Calcutta, Roxburgh submitted a memorandum on the botany and cultivation of hemp and flax in Bengal[10], and established trial plots in the neighbourhood. From time immemorial Indians had grown hemp (*Cannabis sativa*) as a narcotic but had discarded the stem, the source of its fibre. So much importance did the Company attach to a successful outcome that they sent out an expert with European seed and when he died, immediately replaced him by another. Roxburgh in a letter to Banks peevishly questioned the necessity for these separate trials but persevered with his own experiments, examining a range of fibrous plants—Sunn hemp, jute, *Hibiscus* species, and *Agave americana*—twisting them into twine and testing their strength, with and without tanning and tarring. London was impatient for results.

> *European hemp has increased in price since the year 1792 from £23.10s to £61 per ton which is the present price. As Russia has almost the exclusive supply of this article, it is obvious that she may raise the price of it at her will. Such a precarious dependence on a foreign power for an article on which our political existence so much depends, is certainly too hazardous to be relied on, provided we can remedy the evil by obtaining a supply from our own possessions*[11].

Two hundred acres were appropriated for the cultivation of *Cannabis sativa*, more seed came from England, and Sir Joseph Banks nominated six knowledgeable men to go to India in 1801 on a three-year engagement. After such a determined effort, Roxburgh's report to Banks in December 1801 was disappointing; the seed had failed to germinate and, furthermore, he was convinced that it would never grow anywhere in the vicinity of Calcutta, although he held out hopes of its succeeding elsewhere in the country. In 1802 the Office of Trade in London presented Banks with a thick file of papers on hemp for his comments. Sir Joseph was surprisingly optimistic: some of the North American colonies were prepared to offer grants of land and premiums to settlers willing to raise and process hemp; in India trials were under way with Sunn hemp, considered by many to be almost as strong as the best quality hemp from Russia. Samples of this hemp were already being rigorously tested in England to establish its suitability for making cables, cordage, canvas, and linen for sheets and shirts.

Sunn or San hemp (*Crotalaria juncea*) is mentioned in ancient Sanskrit texts; the *Hortus Malabaricus* illustrated it; Anton Hove saw it growing to nearly ten feet near Surat; Roxburgh referred to its strength and durability in his *Plants of the coast of Coromandel.* Substantial losses had been sustained by the Company in its efforts to get it accepted as a substitute for Russian hemp in England. Rope-makers completely rejected it for maritime purposes. Robert Wissett, Clerk to the Committee of Warehouses of the East India Company, published a digest of all the available information on hemp cultivation in 1804[12].

Roxburgh published the results of his own researches on the comparative strengths of hemp and other vegetable fibres in the *Transaction of the Society of Arts*[13]. He had drawn the attention of the Court of Directors in 1795 to the potential value of a fibre 'called jute by the natives' but Buchanan, apprehensive that any production of jute might jeopardize current trials with Sunn hemp, advised against its extensive use. Roxburgh recommended it as a substitute for flax and eventually in 1820 a factory at Abingdon in Oxfordshire began using it in carpets and cloth. It was still suspect in the opinion of many manufacturers of linen goods and the declaration, 'warranted free from Indian jute', frequently endorsed their products. When, however, Dutch planters started shipping their coffee crop from the East Indies in jute rather than flax bags in 1838, and the Crimean war cut off supplies of Russian flax, the jute industry prospered, ultimately monopolizing the packing and wrapping trade.

In 1803 Roxburgh received from Bencoolen in Sumatra four plants from which he propagated several thousand seedlings of Rhea (*Boehmeria nivea*), reputed to be the source of the fibre used to make China-grass. Three bales of the fibre sent to London by Buchanan in 1811 were tested and found comparable to European hemp. Its drawback was the labour-intensive operation of separating the fibres by hand. London, having been alerted to the superior qualities of Manila hemp (*Musa textilis*) ordered Roxburgh to obtain samples from the Philippines. He busied himself selecting, importing, and growing fibrous plants, meticulously comparing their strengths, both wet and dry, but never found time to consolidate and publish his findings; this was only done shortly after his death by a friend[14].

The escalation in the price of imported Russian hemp intensified all this research on fibrous plants. Royle ran trial experiments with Himalayan fibres at his garden at Saharanpur, and all three Presidencies were prevailed upon to persevere with the cultivation of cordage plants. The success of these efforts can be gauged from trade statistics for the import of hemp into Britain: in 1847, 544 844 cwt came from Russia and only 185 788 cwt from India; by 1851 India had almost matched Russian supplies (672 342 cwt) with 590 923 cwt. Still there was no letting up in the quest for new or better-quality fibres. Sisal fibre plants (*Agave sisalana* and others) were transported from the Americas to the Royal Botanic Garden at Calcutta in the early 1890s through the agency of Kew Gardens which had now assumed the imperial mantle of Sir Joseph Banks.

Banks never once wavered in his conviction that the prosperity of Britain and its colonies could be enhanced by the bold and imaginative utilization of the world's natural resources. He never appeared daunted by the problems of distance and logistics. He was involved in the project to transport breadfruit trees from the Pacific to the West Indies as a cheap source of food for slave labour; he counselled the East India Company to acquire tea plants from China for commercial plantations in India; he schemed to transfer Indian cotton to the West Indies.

Cotton manufacture in India can be traced back at least as far as the Indus Valley civilization; the wide distribution of *Gossypium* species throughout the subcontinent has meant that for centuries cotton provided the basic material for clothes and curtains and even for the housings of elephants and horses. Banks was involved in a clandestine operation to collect cotton seed in India for experimental cultivation in the West Indies. One can only assume the operation was conducted in secrecy so as not to alarm the East India Company which would not have welcomed any competition. Anton Pantaleon Hove (*fl.* 1780s–1820s), a man with 'a daring character', according to Banks, was chosen as conspirator. A Pole by birth, he already had practical experience collecting on the west coast of Africa. The ostensible purpose of his mission to India was to collect novelties for the royal gardens at Kew, and the East India Company were requested to provide him with letters of introduction. He was working as a mere labourer at Kew Gardens when he received an official letter, confirming his appointment. Attached to it was a confidential document informing him that

> *the real object of your mission is to procure for the West Indies seeds of the finer sorts of cotton with which the Ahmood Country [Amod in the Broach District, Gujarat State] where you are ordered to reside abounds, to make yourself master of the manner in which it is cultivated, & transmit from time to time your observations upon the soil & mode of cultivation, you are continually to keep that in view as your main object, & consider the collections for His Majesty's garden as secondary to it*[15].

If he were successful he was promised the superintendency of the projected cotton plantation in the West Indies.

When Hove arrived in Bombay in September 1787 he was politely received as a servant of the King but accorded scant co-operation and no financial assistance from the Company's officials. As he was asked to swear on the Bible that he had no confidential papers in his possession, one wonders whether the East India Company had seen through the subterfuge. He inspected the cotton plantations at Broach, north of Surat, and never once did bad roads, deplorable weather, illness, and several robberies deter him from his assignment which Banks admiringly declared 'he affected in a masterly manner'. The hostility of the Company was maintained right up to his departure in February 1789. He was refused a passage on the last East Indianman of the season, allegedly because the sight of his plant cases cluttering the deck might offend the late Governor who was a passenger. Banks read his report but significantly no copy was forwarded to the Court of Directors[16].

Hove's activities may have spurred India House into drawing the attention of their Government in India in 1788 to the country's cotton, requesting that they gave 'every encouragement to its growth and improvement in general, but particularly to the species manufactured into the finer sorts of thread in use for the superior goods of the Dacca fabric'[17]. Banks, also curious about Dacca cotton which was used in the manufacture of delicate muslins, begged information and specimens from his friend, Sir William Jones in Calcutta. In 1794 the Company invested in machines for separating cotton from seeds and other impurities. Cotton seed was imported from the French island of Réunion, Persia, and the United States. By the beginning of the nineteenth century, the cotton plantations of eastern America were exporting some 20 000 000 lbs of cotton to Great

Britain, and the Company was anxious to obtain the expertise of the planters as well as the superior American varieties of Upland and Georgian cotton. In 1813 an American technician with saw-gins arrived at Tinnevelly in the Madras Presidency. Urged by the Manchester Chamber of Commerce in 1838 to expand their Indian cotton trade, the Court of Directors believing that India could produce cotton 'in quantity and quality to compete with the cotton from North America'[18], engaged ten experienced Mississippi cotton planters in 1840. These experts who brought with them ploughs, hoes, gins, and presses were shared among the three Presidencies. The trials which investigated the cultivation of cotton and its subsequent cleaning had the ultimate objective of making the British market independent of America. But they were short-lived; only two years after their arrival the three planters allocated to the Bombay Presidency resigned. The improvements that were achieved never threatened American dominance. The editor of the *Washington Union* was openly smug about his country's superiority.

> *The English experiment has signally failed. It was made under the most favourable auspices in different parts of India. It has succeeded in none of them*[19].

The first cotton mill which replaced the traditional handloom industry was built in Bombay in 1853 and within ten years Bombay and Ahmedabad became major centres for the production of cotton fabrics. Trials with new varieties continued during the 1860s. Peruvian cotton with its long staple was a favourite with C. R. Markham who had brought Peruvian cinchona to India. By the turn of the century well over a hundred cotton mills had been established but India had now become the primary market for cotton goods manufactured in Britain.

Cotton, presumably from India, was a very late arrival in China where for several thousand years the silkworm had been bred and its cocoon of thin but incredibly durable filament converted into skeins of silk for clothes and ornamental hangings and for export across the trade routes of Central Asia. Silk, albeit inferior to the Chinese product, had been made in northern India, Assam, and Bengal from ancient times. Traders of the East India Company had been drawn to Bengal by its silk but it was the Dutch who first established a presence there. They were employing 700 to 800 natives making silk at Cossimbazar when François Bernier passed through Bengal in the mid-seventeenth century. The crude methods they used in reeling silk from cocoons, producing variable thicknesses in strands, initially discouraged the Company from expanding their trade in this commodity. In 1757 the Court of Directors engaged Richard Wilder to improve the quality of Bengal raw silk. Before his death four years later, Wilder had introduced better methods for winding silk. Reelers and mechanics from Italy and France brought continental practices to India in 1770. The Company now seriously viewed the production of raw silk as a potential money-spinner. Silkworms and mulberry trees were imported from China. Farmers were offered rent reductions as an inducement to plant mulberry trees. So successful was this campaign that by the 1780s Bengal silk had eliminated all competition in the English market except from China and Italy. William Roxburgh, impressed by the ease with which the mulberry grew on the Coromandel coast, proposed to the Madras Presidency that 'poor, helpless women and children might be employed feeding and taking care of the [silk] worms, which would render the work a charitable, humane undertaking, independent of its expected advantages'[20]. Sufficiency of edible leaves for the silkworms relied on the skilled management of the mulberry trees. Roxburgh's Botanic Garden in Calcutta distributed the white mulberry (*Morus alba*) to discover whether they provided more nutritious food than the Indian species, *M. indica. M. alba*, grown extensively in Europe, grew well in the temperate regions of North India.

James Anderson (1738–1809), Physician-General at Madras, was a man of great compassion but also the personification of reckless optimism bordering on fanaticism. One of his colleagues, clearly in awe of him, wrote that 'his application is intense, sparing neither fatigue of body or mind, & a perseverance which no difficulties can check'[21]. He planted European apple trees, supported the introduction of American cotton, saw a prosperous future for sugar-cane and coffee, hunted the cochineal insect, and harassed the Government with schemes for enlarging the silk industry within the Presidency. During the 1770s he formed a plantation of *Morus alba*, obtained from Hyderabad, in his large garden in the suburbs of Madras. With 5000 healthy trees and a reel he had invented for winding the thread from the silkworms, he had little difficulty in convincing the Board of Revenue of the feasibility of sericulture in the south. He generously offered cuttings for planting on small plots of land. He cheekily proposed converting the garden attached to the Assembly Room into a mulberry plantation. He even suggested appropriating the Female Orphan Asylum.

> *Could the Governors and Governesses therefore of this Asylum be persuaded to forego the idea of educating ladies-maids, you might consider this charity as a sufficient establishment for the filature of Madras and its environs, conformable to the practice in Italy; and the young women would thus acquire the possession of an art that would be sufficient to preserve them independent through life*[22].

Like Roxburgh, he was adept in dressing up a commercial transaction as a social benefit. But his confident predictions foundered and he eventually antagonized officials with his persistence and impatience. Local conditions were frequently unsuitable and few native farmers could be persuaded to participate despite the bribe of a handsome pair of gold ear-rings. After an expenditure of £20 000, the Court of Directors abandoned the project and in 1798 the Madras plantations were up for sale. With the entry of France into the silk trade, India had another formidable rival and the demand for her silk continued to decline to the point where she was no longer a serious competitor.

James Anderson's devotion to sericulture in no way diminished his enthusiastic support for other economic products—cochineal, for instance, received the same whole-hearted sponsorship. This scarlet colouring substance, used for dyeing cotton, silk, and wool, is derived from the dried bodies of a scale insect (*Dactylopius* species), reduced to a powder. Indigenous to Central America, Mexico, and Peru, it feeds on prickly pear (*Opuntia* species) (Fig. 15.1); the problem to be resolved by Indian naturalists was matching the right species of insect with the host *Opuntia*. Its solution was hindered by the determination of the Spanish colonies, jealous of their trade monopoly, to withhold information about the production of cochineal. With encouragement from the French Government, the naturalist Thierry de Menonville had succeeded in smuggling out some cochineal insects from Mexico in 1777 and revealed the secret of their culture in *Traité de la culture du Nopal, et de l'éducation de la Cochinelle* (1787). The Court of Directors were presented with a copy of this book by Sir Joseph Banks, who also obligingly summarized all that was known about cochineal. It transpired that two sorts of insects were involved—*Grana fina*, the source of superior cochineal, and *Grana sylvestris*, a much inferior product.

Meanwhile, James Anderson was certain he had made a discovery of vital importance.

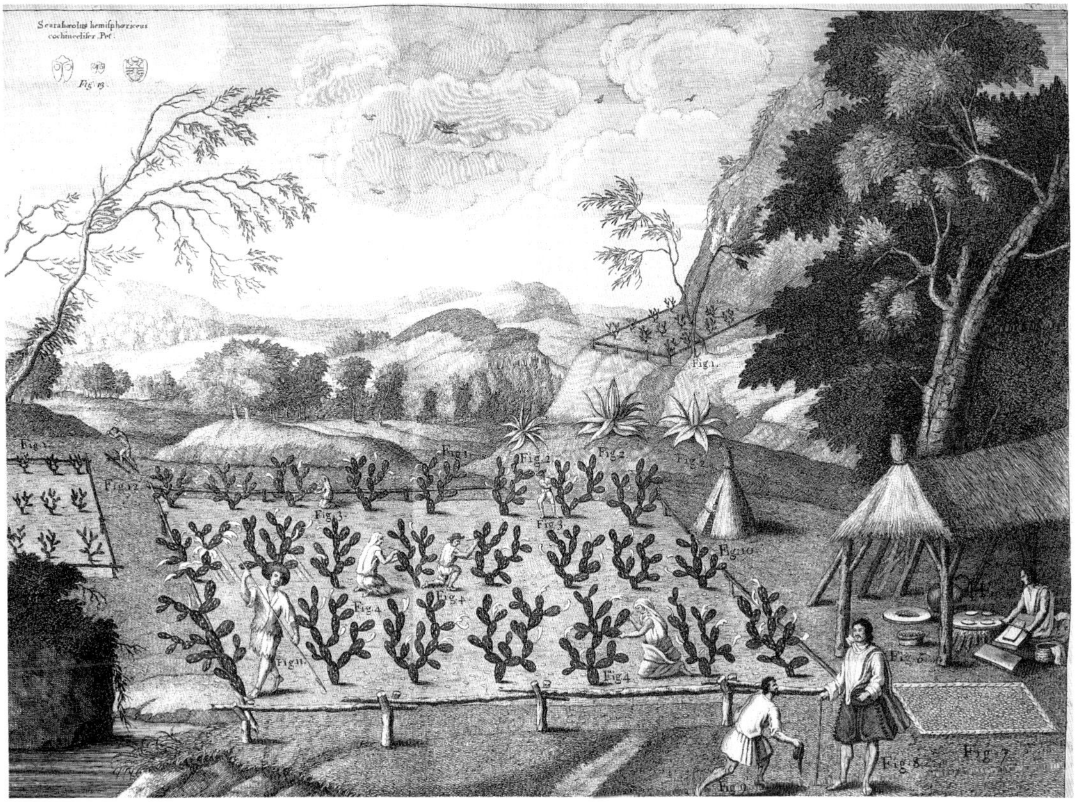

Figure 15.1 Mexican Indians tending cochineal insects on opuntias. (From Sir H. Sloane (1707), *A voyage to the islands Medera, Barbados, Nieves, S. Christophers and Jamaica* . . . , Vol. 2, London, plate 9.) Engraving.

An insect he had found in grass appeared to correspond with the descriptions of the cochineal insect in authoritative works by Leeuwenhoek, Reaumur, Linnaeus, Sir Hans Sloane, and the impeccable French encyclopaedia. By softening a number of these insects in a solution of spirits of wine and water he obtained a colour, so he claimed, similar to that of the Mexican cochineal. Without delay he planted out a plot of his garden with the grass that harboured the insect and about a thousand opuntias, and in the meanwhile sent specimens of the insect to Sir Joseph Banks[23]. The Madras Government who had encouraged Anderson's experiment assured the Court of Directors that he merited their support. A chemical analysis of these specimens indicated that they were commercially worthless but Anderson, who had been commended for his efforts, was given a garden at Marmelon in 1788. His nephew, Dr Andrew Berry, was appointed Superintendent of this garden, designated 'The Honourable Company's Nopalry'[24]. The insects were housed in a small brick building and the garden sheltered from the wind by earthen banks topped with a hedge of *Euphorbia*. William Roxburgh grew opuntias at Samalkot and had natives searching the hills for cacti while in Bengal Colonel Kyd was sanguine about finding a 'cactus with red juices'.

As opuntias were widely distributed in India and especially in Bengal, it was assumed by Anderson and other naturalists that they were native to the country. They did not know that

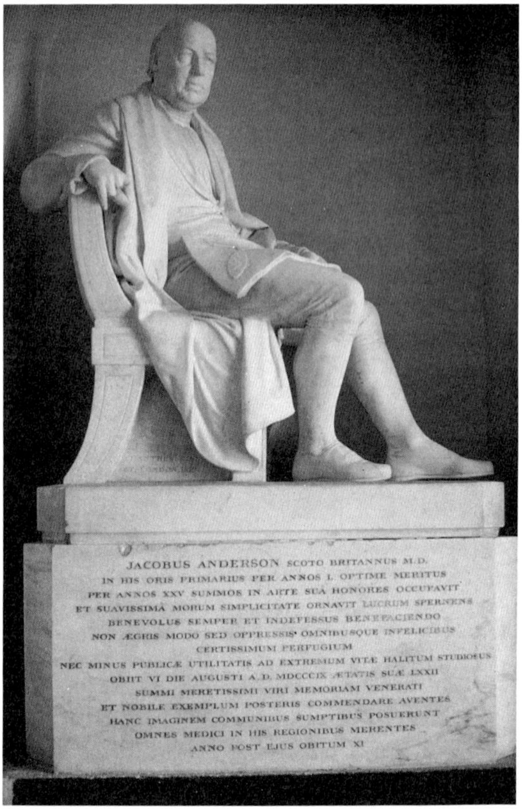

Figure 15.2 James Anderson (1738–1809). His statue in St George's Cathedral, Madras, with an herbarium specimen of an opuntia on his lap.

this invasive genus had been introduced from the New World, probably by the Portuguese although for what purpose is a matter for speculation. Since opuntias were also growing in other outposts of the Portuguese and Spanish empire, it has been suggested they were planted as a source of anti-scorbutic medicine. Anderson had no doubt about their therapeutic value which he promoted in a pamphlet, *A journal of the establishment of Nopal and Tuna for the prevention or cure of scurvy* (1808). British warships stationed in Indian waters used opuntias as an anti-scorbutic. As it remained edible long after supplies of vegetables had rotted away, Anderson also recommended it as an item of a sailor's diet.

Anderson grew opuntias shipped in from the Philippines and China; the few plants which came from Kew produced a progeny of several thousand within three years—confirmation it would appear of Sir Joseph Banks's conviction that the soil and climate of South India were ideally suited for a cochineal industry. The Calcutta Botanic Garden, too, found it delightfully easy to grow opuntias but Colonel Kyd doubted whether the peasants could ever be persuaded to rear the insects.

Following a resolution of the Court of Directors in March 1788, the captains of their East Indiamen were instructed to collect a few cochineal insects whenever they called at Rio de Janeiro. The Directors may have been persuaded that this was an easy assignment from the report of one of their surgeons who claimed to have seen the insects thickly covering plants in the Governor's neglected garden in Rio. 'Any person had access to this garden without attendants, and I cannot have the least doubt but that I might have carried away as many of the plants and insects as I pleased without giving offence to anybody'[25]. Unfortunately, the covert operation of one of the captains was disclosed by the Governor of Madras to Anderson who imprudently revealed the plan in a pamphlet which he widely distributed in 1789, even thoughtlessly sending a copy to a botanist in the employ of the Spanish at Manila. This indiscretion alarmed Banks who feared the botanist would alert the Spanish authorities to the Court's intentions. Anderson, blissfully oblivious of his blunder, busied himself with printing instructions for the care of cochineal insects during sea voyages; he and Kyd also proposed the cultivation of opuntias at St Helena where the insect could be bred for the nopalries in India.

In March 1790 Banks advised the East India Company to offer a reward of £1000 for the clandestine transfer of *Grana fina* insects to London where they could be nurtured in heated greenhouses and then despatched to India. So anxious were the Court of Directors to procure this elusive insect that they allocated £2000 for the entire operation, inviting Banks to engage reliable collaborators as smugglers. In order to establish the precise difference between *Grana fina* and *Grana sylvestris* insects and also to acquaint himself with breeding techniques, Banks asked Dr Hugh Gillan, sailing with the Macartney embassy to China in October 1792, to send him some cochineal insects when the ship docked at Rio de Janeiro. Three opuntias, crawling with insects, reached Banks's residence at Spring Grove in the following February. He placed them in a heated house with some opuntias from Kew Gardens and waited and watched. For a while the insect population increased spectacularly but the opuntias eventually died and the multitude of insects feeding on them never grew larger than 'a small turnip seed'. Disappointed, but not deterred, Banks wrote to James Bartlet in Honduras in January 1793 for specimens of *Grana fina* and the opuntias on which they fed. As an inducement he offered Bartlet a fee of £1000 with the prospect of an even greater sum if he would be willing to take them to India and instruct the natives in their culture. Bartlet never responded to this invitation. The deterioration in international relations, requiring ships to sail in convoy, made the cochineal operation more hazardous, and Banks was reluctant to support a venture 'more likely to end in the loss of liberty or even of life, than the final attainment of the object of pursuit'[26]. He deemed it prudent to postpone any further action until more propitious times.

In June 1795 Captain Neilson of HM 74th Regiment arrived in Calcutta with a rather sick opuntia to which were clinging a few cochineal insects, all that had survived of the plants he had collected in Rio de Janeiro. Before this rapidly dwindling number of insects completely disappeared they were transferred to another host plant, the well-established *Opuntia monacantha*. Within weeks Neilson was able to report that they had colonized most of the opuntias in the Botanic Garden's nopalry for which Christopher Smith took the credit. 'I flatter myself I can manage the cochineal insect now equal to any man in South America.'[27] Unfortunately, this colony turned out to be the inferior *Grana sylvestris* and there were conflicting opinions about marketing them profitably.

The cochineal programme was in danger of foundering. The botanists in India were still confused by the different species of opuntia and insects and Banks was vacillating in his support. Anderson's nopalry, which had expanded into a botanical garden, was closed by the Governor of Madras, and its more valuable stock of economic plants transferred to Bangalore. Banks made one last spirited effort. Robert Sproat to whom Banks had been introduced in 1794 had made an abortive attempt to send opuntias from Guatemala. Banks, with the backing of the Court of Directors, made Sproat, now in Honduras, a handsome offer: £500 on the safe receipt in London of the host plant of the *Grana fina* insect; £1000 for healthy specimens of the insect; and an additional £1000 if both the plants and insects were safely transported to India and successfully raised. Furthermore, if these terms were unacceptable, Banks was prepared to renegotiate them with the Court of Directors[28]. The operation does not appear to have been executed for in 1811 the Company's original reward of £2000 was still waiting to be claimed by any successful collector.

Four species of opuntia were naturalized in India: *O. monacantha* (an early Portuguese introduction), *O. stricta, O. elatior*, and *O. ficus-indica*. The cochineal insect is represented by three species: *Dactylopius coccus*, the *Grana fina* of commerce, and *D. indicus* and *D. tomentosus*, the source of the inferior cochineal, *Grana sylvestris*. Attempts in India to establish *D. coccus* upon imported specimens of its true food plant, *Opuntia coccinellifera*, failed. The trials and experiments which persisted over half a century left a legacy of opuntias rampant across the subcontinent. In the South convict labour was used in an attempt to eradicate them. The East Indian Railway Company seriously considered substituting opuntia hedges for the wire fencing skirting their tracks. *Dactylopius indicus* accidentally provided some biological control by feeding exclusively on *Opuntia monacantha*, practically wiping it out.

As late as 1859 a John Blackwood informed the India Office he was leaving Peru for India, confident he could breed the true cochineal insect there. Other isolated attempts were made from time to time but eventually technological developments provided substitute dyes. Now cochineal is used mainly as a colour ingredient in confectionery and drugs.

Competing for the attention of the East India Company was another natural dye, indigo, which held out the promise of perhaps a more profitable return on investments. This fastest of blue dyes was known to the ancient world: occasionally it coloured the embalming cloths of Egyptian mummies; Pliny called it 'a product of India'; *Periplus of the Erythraean Sea* disclosed that it was exported from a port on the Indus. Marco Polo, da Orta, and Acosta were curious about its manufacture, and the Portuguese controlled its import to Europe where it was initially opposed by woad growers who saw it as a threat to their livelihood. The dye became one of the major exports of European trade at Surat but English planters in the West Indies, Carolina, and Georgia were soon to offer serious competition with their crops of Guatemalan indigo. Indigo imports from India consequently slumped until about the middle of the eighteenth century when the West Indies abandoned the dye for the more lucrative sugar and coffee markets. The Company, now encouraged to revitalize its former trade in the commodity, brought in European indigo planters from the West Indies and opened its first indigo factory in Bengal about 1778 or 1779. Kyd grew several species of *Indigofera*, the source of the blue dye, in his Botanic Garden in Calcutta. In 1789 an English translation of J. F. C. de Cossigny's *Essai sur la culture et la fabrication de l'indigo* (1779) was published in Calcutta. Roxburgh thought that a plant he had discovered in South India and subsequently named *Nerium tinctorium* (now *Wrightia tinctoria*) might yield a dye comparable to that obtained from *Indigofera* (Fig. 15.3). Despite discouraging losses, the Company persevered in its promotion of the indigo industry and having eventually succeeded in establishing India as one of the world's major suppliers, decided in 1802 to withdraw their subsidies and grants to indigo planters. They encouraged the participation of Indian farmers as well as Europeans for whom they hoped indigo manufacture would be seen as 'a mode of remitting their fortunes to Europe which would be legal, advantageous and adequate'[29].

The legality of the way in which the industry was managed by the number of planters became a contentious issue. Adventurers and opportunists were lured to the prosperous indigo plantations of Bengal and Bihar; many planters, however, were Anglo-Indians, natives of the country. In Indian society the planters ranked just a little above shop-keepers although some lived in well- furnished homes, and sent their sons to the best schools in the Presidency and their daughters to the Loreto Convent at Darjeeling. Some owned their estates, others held them on long leases. Those who lived a bachelor existence in isolated bungalows in the interior were noted for their hospitality to travellers and, in times of famine, as generous donors to relief schemes. The unstable state of the world market in indigo with fluctuating prices made fortunes for some planters but ruined many. A sudden plague of insects could destroy both crops and future prospects. This precarious

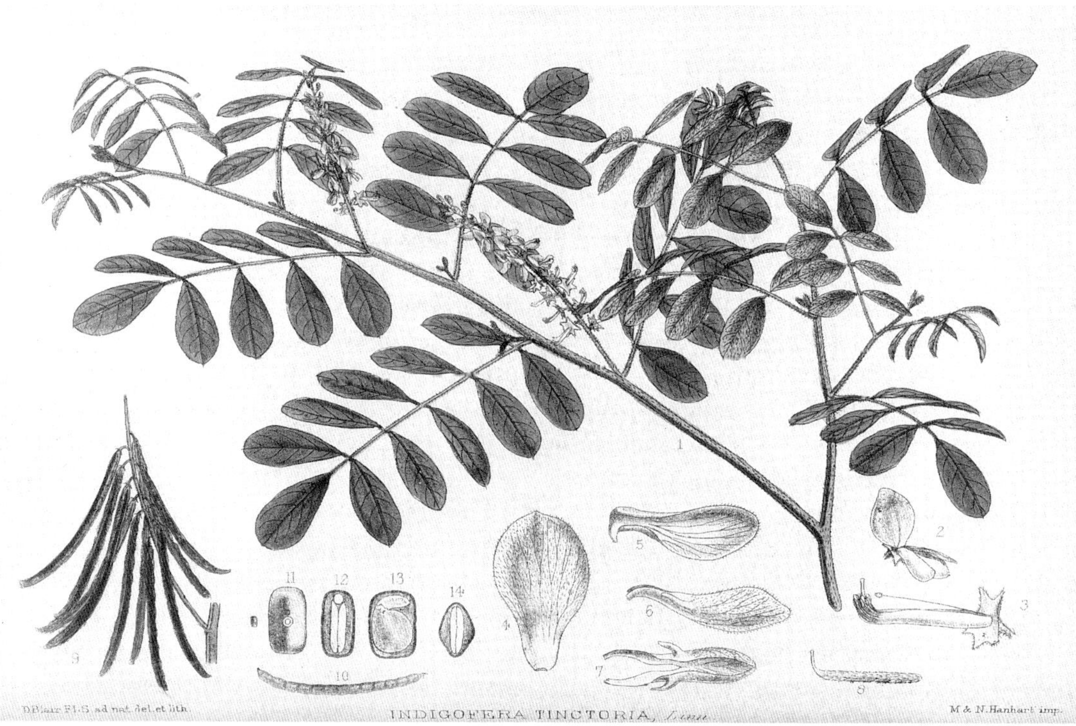

Figure 15.3 *Indigofera tinctoria.* (From R. Bentley and H. Trimen (1880), *Medicinal plants*, London, plate 72.) Drawn and lithographed by D. Blair.

and competitive nature of the industry doubtless fostered the callous behaviour of some planters towards peasant farmers who were compelled to grow a stipulated quantity of indigo at fixed prices. The contracts were so manipulated that the peasants were always in debt and, consequently, permanently in the power of the indigo factory. They were intimidated by 'bludgeon men' engaged to enforce the planters' claims despite a Government circular in July 1810 condemning this lawlessness. The Company's Judicial Department in 1828 deplored the leniency of some magistrates towards offenders and demanded an investigation of all planters—whether they were licensed, the size of their estates, and a summary of all the cases held in civil and criminal courts since 1810. By 1837, 3639 peasants had been prosecuted by indigo planters for refusing to fulfil their agreements, but of this number only 354 were convicted, the rest were either acquitted or the case against them dropped. Thomas Babington Macaulay who had been summoned to India to modify India public law, condemned this 'partial slavery' of the peasants.

John Beames, a district officer at Champaran in Bihar, a centre of indigo cultivation, was a just and impartial administrator, who challenged the planters' despotism. He issued a summons against a notorious tyrant called Baldwin who had been incensed by a ryot or peasant refusing to grow indigo instead of more profitable crops.

So Baldwin sent men with ploughs, who ploughed up every inch of ground around the ryot's cottage, sowed it with some coarse, cheap kind of pulse, put a fence of thorns around it, and so made the man and his family prisoners in their own house. He further gave the man to understand that if he set foot on the sown land he would be prosecuted for trespass and sent to jail. The wretched man and his family trembled and endured for two days, but when their food supplies were exhausted and they could not go so far as their well to draw water for household purposes, despair gave them courage. Although there were men put to watch them, the ryot managed to slip out at night and hide till the morning when he came into my court and told his tale[30].

In 1860 a Bengali, Dina Bandhu Mitra, wrote a play *Nil Durpan* ('Mirror of indigo') which attacked such exploitation. A missionary named Long who translated it into English was fined and imprisoned for libelling the planters. Villagers, no longer prepared to be submissive, formed militias to defy the planters' thugs and an ugly situation was defused only by the repeal in 1868 of the Indigo Contracts Act. Aniline dyes, invented during the second half of the last century, adversely affected the indigo trade which enjoyed a revival during the First and Second World Wars when there was a shortage of synthetic dyestuffs.

Some indigo factories which closed down were later converted to sugar production. For many years the East India Company showed little interest in Indian sugar-cane (*Saccharum officinarum*). Arab traders first brought the sugar-cane to southern Europe. By the fifteenth century the Portuguese and the Spanish had sugar growing in Madeira and the Canaries; the Dutch, British, and French established profitable sugar plantations in the West Indies. The high price and poor quality of West Indian sugar in the eighteenth century prompted a petition to the East India Company to consider the importation of Indian sugar. The first sugar factory in India opened in Bengal in 1784. Roxburgh, who submitted a report on sugar in 1792, stressed the importance of lower labour costs in India, and some sample sugar he sent to London was highly acceptable to the British market. So successful was his propagation of Chinese sugar-cane that he was able to distribute thousands of plants to cultivators in Bengal, Bihar, and Madras.

This increased production led to the imposition of discriminatory duties on Indian imports to protect the West Indian trade. After the emancipation of slaves in the West Indies in 1834 and parity in duty on sugar in 1836, a number of West Indian planters emigrated to India. Towards the mid-century India supplied about a quarter of Britain's sugar imports, but by the 1880s sugar-beet had overtaken sugar-cane in the world market.

The Americas were indirectly indebted to India for cane sugar; among the crops that came to India from the New World was tobacco, one of the most accommodating of all plants, easily grown in a range of climates. The Portuguese who grew it in their settlements on the Persian Gulf introduced it to India. In defiance of Jahangir's ban, the habit spread fast. Some addicts chose to inhale through a hookah, the Burmese used a sort of elongated cigarette-holder, and the poor made do with a rolled leaf which the Tamils called a 'shuruttu'. Most villages had a patch of tobacco. In 1786 Robert Kyd sent the Court of Directors a bale of what he considered to be good quality Indian tobacco for their comments. It was not, however, until 1829 that the Company had Maryland and Virginia tobacco seed sent to India for trial cultivation. The Agricultural and Horticultural Society of India which was involved in the project, sent samples of the processed American tobacco to London where dealers pronounced it 'the best sample of Indian tobacco they had ever seen'. This must have been an exceptional consignment because Indian tobacco was usually condemned for being either mouldy or so dry as to be fit only for snuff.

The Agricultural and Horticultural Society of India (sometimes abbreviated to the Agri-

Horticultural Society of India) participated not only in tobacco trials but also investigated or actively encouraged the cultivation of sugar, cotton, coffee, cereals, fibre and forage crops, cochineal, European vegetables, and ornamental plants. For a while after 1829 it was an important collaborator in the Government's programme for the development of economic crops. Back in 1805 Lord Wellesley, before his recall to England, had promulgated his intention of establishing an experimental farm of 300 acres at Barrackpore where European practices in animal husbandry and the cultivation of food crops could be adapted to the requirements of Indian climate, soils and labour. Probably remembering the Governor-General's abortive scheme, William Carey at Serampore discussed with his friends the formation of a society devoted to agriculture and horticulture. With their encouragement he issued in April 1820 a prospectus in which he formulated both a policy and a strategy for such a society which would recruit Indians as well as Europeans as members. Its primary objectives would embrace the best methods of agricultural and horticultural practice, the introduction of new and potentially useful plants, the latest equipment, improved animal husbandry, and the reclamation of waste land. With 32 replies pledging support, he convened a meeting at the Calcutta Town Hall on 14 September 1820. Only seven people, including two Indians, attended, but notwithstanding this disappointing response the Agricultural and Horticultural Society of India was launched with William Leycester as President and Carey as Secretary. The Society's first Patron, the Governor-General, signalled his support by an annual grant of 1000 rupees for prizes. The Court of Directors in London gave its cautious approval.

Its first task was a fact-finding questionnaire on the current state of agriculture and gardening within the Presidency. In 1827 the Government leased land at Alipore to the Society which was also given a plot of ground at Akra for the trial cultivation of cotton, sugar-cane, and tobacco. Horticultural crops were imported: grafted fruit trees from Liverpool, seeds from Europe, the Cape and New South Wales, oranges from Seville and Mozambique, lychees from Canton, and guavas from Manila. The Curator of the Liverpool Botanical Garden was asked to nominate a suitable person to be the Society's head gardener. In 1836, two years after the Society had relinquished its two small gardens, it was allocated part of the Calcutta Botanic Garden, subsequently enlarged, which it occupied until 1866. In 1872 it acquired 25 acres adjacent to the Lieutenant-Governor's residence at Belvedere; paths were made, flower-beds displayed, an orchard planned, a conservatory built, and the central portion reserved for the still popular horticultural shows. The Society had promoted shows from its earliest days as a means of improving the quality of vegetables and fruits. Gold and silver medals were awarded for the best beets, cabbages, cauliflowers, peas, turnips, and other entries.

Potatoes were always a popular exhibit. In 1827 the Society awarded a Muslim competitor a silver medal for potatoes raised from Cape stock. From about the late eighteenth century the virtues of potatoes were extolled with all the fervour of a crusade. Bishop Heber approvingly reported that potatoes 'are spoken of as the best thing which the country has ever received from its European masters'[31]. This New World root crop had apparently reached India by the early seventeenth century if one is to believe Edward Terry's account of their being served at an Indian banquet in 1617. Other travellers confirm that they were certainly in cultivation in European settlements by the end of the century. From the 1750s they were grown throughout Bengal primarily for the European market; peasants still preferred their traditional rice. The Government were anxious to remove this prejudice which hindered the substitution of alternative nutritious food in times of famine. The antipathy of brahmins to root crops made them resist potatoes, but Benjamin Heyne on his tour from Bangalore to Trichinopoly in 1802 found that peasants readily ate them in preference to yams and sweet potatoes. He had no difficulty in persuading them to grow them in their fields and gardens. George Bogle *en route* to Tibet in 1774 was not so lucky with the potatoes he diligently planted in Bhutan at the behest of Warren Hastings; they had completely disappeared when the next mission entered the country in 1783. But gradually the potato was accepted. In Bombay's Salsette Island, it vied with the onion for supremacy; bumper harvests and monster potatoes were common in the Nilgiris; the slopes around Simla were stripped of their trees to provide ground for potatoes; Sherpas managed to grow them at high altitudes in Nepal; and the prodigious crops in the Khasia Hills supplied the markets of Calcutta. A Captain Lowe, who clearly enjoyed nothing better than a tasty potato, applauded the Calcutta produce.

> *Thanks to the exertions of the Horticultural Society, both the size and flavour of fruits and vegetables have much improved, and I have seldom even in the Land of Pratees seen that delicious vegetable finer than you now get it in Calcutta*[32].

A large part of the Indian population was indifferent not only to the benefits of the common potato but also to the broader aims of the numerous agricultural and horticultural societies. Their support of such societies was disappointing. In 1829 when the Calcutta Society had 97 members, only 11 were Indians; ten years later the membership had grown to 460, but Indians were still a minority—just a mere 29 and few of them attended any meetings. Native market gardeners were generously supplied with seed and they won prizes at shows, but Carey's optimism about their wholehearted participation was never realized. The original intention of translating the Society's *Transactions* (which started in 1829) into at least two vernacular languages was abandoned after the poor response to the Bengali version of the first few volumes.

The Calcutta Society despite its title was never really national in its activities so regional societies were formed in the larger cities. Bombay had its own Society in 1830 with a garden in the picturesque valley on the east side of Bombay island near the village of Sewree. Its programme and that of Madras, formed in 1835, were modelled on the parent Society in Calcutta. Madras was fortunate to obtain the services of the energetic and capable Robert Wight as its first Secretary but, according to the *Calcutta Review*, it lay in a state of somnolency by 1854. The Lahore Society quickly languished after its formation in 1851. The *Journal* of the Calcutta Society which had superseded the *Transactions* in 1840, had reduced its frequency to an annual publication by the 1880s due, no doubt, to competition from specialist periodicals such as the *Indian Tea Gazette*, *Indian Forester*, and *Indian Agriculturist*.

The commercial management of the indigenous flora and the introduction of new species and cultivated varieties to India had depended for years on directives from the Company in London, local initiative by Governors, the participation of agricultural and horticultural societies, and the enthusiasm of individuals. The absence of any coordinated effort and the need for an overall strategy prompted the former Superintendent of the Botanic Garden at Saharanpur, John Forbes Royle, to support a motion of the Vice-President of the Royal Asiatic Society in London at its General Meeting in March 1836 to form a Committee of Agriculture and Commerce. The aims of this Committee were to identify those Indian imports considered to be of inferior quality and those products for which there was or could be a demand in the home market; to encourage the exchange of plants between India and other tropical countries; and to rationalize and make available statistics relating to Indian agriculture. As its conscientious Secretary, Royle conducted the Committee's deliberations on such matters as Indian lichens as a dyeing agent, wheat and barley, varieties of cotton, oil and cordage plants of the Deccan, and the 'vegetable butter' of northern India. It debated reports received from the agricultural and horticultural societies in India and successfully organized the despatch of some cochineal insects and their host opuntia to Calcutta. It ceased to function in 1840 on the appointment of Royle as head of a new department created by the East India Company at its headquarters in Leadenhall Street, that of Correspondent relating to the Vegetable Productions of India.

A despatch from the Governor-General in August 1838 and consultations with John Lindley and Royle had brought the Court of Directors to approving this new post. Lord Auckland, who had requested his officers serving in the North Western Provinces to collect seed for London, suggested that the Company took advantage of speedier voyages by steam ships to broaden the exchange of plants between India and Britain[33]. The Directors' reply assured his Lordship that

> *we have resolved on gradually furnishing you with the means of carrying on extensively, experiments for naturalizing in India useful and desirable plants indigenous in other countries*[34].

They enclosed a letter from John Lindley, Professor of Botany at University College, London and Assistant Secretary of the Horticultural Society of London to whom they had sent for examination a consignment of seeds sent from India by steamer and overland[35]. Lindley in his reply proposed a selection of plants from other parts of the world that might be suitable for India and ended on a confident note:

> *from the great extent of the British possessions in India, and the infinite modifications and contributions of soil and climate to be found within them, there can be no doubt whatever that almost every production of every climate, except the Arctic, may be so completely naturalized that where they are of any importance as objects of cultivation, they may be brought to all the perfection of which thay are susceptible in other countries, provided proper skill and care are shown in the selection of their situations.*

The Governor-General sent copies of the Court of Director's despatch and Lindley's letter to the Agricultural and Horticultural Society of India which promptly convened a committee to recommend plants suitable for introduction to India. The papers and correspondence of this committee were published by the Government in 1841[35].

Royle's *Illustrations of the botany . . . of the Himalayan mountains* (1833–40) and his *On the cultivation of tea in the northern provinces of India* (1834) established his credentials as an authority on plant introduction into South Asia. His *Essay on the productive resources of India* (1840) surveyed the cultivation of economic crops in an historical context and concluded that frequent disappointments and failures were often attributable to unsound methods and inadequate scientific knowledge. This *Essay* was intended as a prelude to a more ambitious work, announced in a printed prospectus in 1843, which also provided him with an opportunity to appeal for information. His *Commercial products of India* proposed to treat commodities in homogeneous groups such as cordage plants, sugar-yielding plants, gums, resins, dyeing substances, medicinal plants and so forth. In 1849, surrounded by a vast accumulation of data, he was waiting impatiently for the Company's permission to publish it. The project was delayed and, in the event, never came to fruition because Royle became the Company's representative on the body considering the forthcoming Industrial Exhibition, leaving him no time for his book. In August 1849 Royle gave the Directors his first thoughts on the items the Company could exhibit. He believed that India's animal, vegetable, and mineral resources would be best displayed in categories of potential commercial use—the arrangement he had in mind for his *Commercial products of India.* His list of raw products and manufactured articles to be obtained from India was enclosed in the Court's despatch of 7 January 1850, and was published in the official gazettes of the three Presidencies. The East India Company was a major contributor to the Great Exhibition of 1851; its Chairman was one of the 24 Commissioners; Royle served on the Vegetable Kingdom Committee and was one of the editors of the massive catalogue. The Indian section, which *The Times* for 2 October voted 'one of the most complete, splendid and interesting collections in Hyde Park', received numerous medals and commendations. The Paris Exposition Universelle in 1855 presented another international forum for Royle to display the wealth of India. He convinced the Court of Directors that the Paris exhibits could form the nucleus of a new Museum of Raw and Manufactured Products of India to complement the existing Natural History Museum and Library at India House. He was still sorting and listing the mountain of material from Paris when he died on 2 January 1858, just two months short of the new Museum being opened to the public. His successor was John Forbes Watson who filled the newly created post of Reporter on the Products of India.

NOTES

1 *Kew Bulletin*, 1894, 317.

2 Viscount Valentia (1809), *Voyages and travels to India* Vol. 1, 405.

3 Public letter to E.I.C., 29 July 1825. I.O.L.R. F/4/904, item 25650.

4 Banks to E.I.C., 27 December 1788. B.M. (N.H.) D.T.C., Vol. 6, folio 109.

5 Bastin (1960), 7.

6 J. Bradshaw (1894) *Sir Thomas Munro and the British settlement of the Madras Presidency*, Oxford, 97.

7 J. F. Royle (1840), 55.

8 E.I.C. to Government of Bengal, 31 July 1787. I.O.L.R. E/4/632, folios 50–6.

9 I.E.C. to Government of Bengal, 30 July 1806. I.O.L.R. E/4/660, folios 657–8.

10 Roxburgh (1794). A botanical history of the hemp and flax plants of Bengal with the mode of cultivation and manufacture. I.O.L.R. H/375, folios 243–54.

11 E.I.C. to Governor-General, 10 October 1800. *Transactions of the Society of Arts*, **33**, 1815, 160–2.

12 R. Wissett (1804), *On the cultivation and preparation of hemp; as also, of an article, produced in various parts of India, called Sunn, which with proper encouragement, may be introduced as a substitute for many uses to which hemp is at present extensively applied.* London. Revised edition in 1808.

13 *Transactions of the Society of Arts*, **22**, 1804, 363–96; **24**, 1806, 143–56.

14 Anon. (1815), *Observations of the late Dr William Roxburgh . . . on the various specimens of fibrous vegetables, the produce of India, which may prove valuable substitutes for hemp and flax, on some future day, in Europe*, London.

15 Banks to Hove, 7 January 1787. B.M.(N.H.) D.T.C., Vol. 5, folios 124–7.

16 A. Gibson (ed.) (1855), *Tours for scientific and economical research, made in Guzerat, Kattiawar, and the Conkuns, in 1787–88, by Dr Hove*, Bombay.

17 J. F. Royle (1840), 81.

18 E.I.C. to Govt. of India, 15 March 1839. I.O.L.R. E/4/758, folios 1089–1108.

[19] J. F. Royle (1851), 573.
[20] Dalrymple (1808), Vol. 1, 17.
[21] W. Petre to Banks, 29 May 1787. B.M.(N.H.) D.T.C. Vol. 5, folios 167–70.
[22] H. D. Love (1913), *Vestiges of old Madras*, Vol. 3, London, 406–7.
[23] Anderson to Banks, 3 December 1786. J. Anderson (1788), 1.
[24] A nopalry was a plantation of opuntias.
[25] W. Scott to R. Kyd, reported in a letter from Kyd to Nathaniel Smith, 23 March 1791. B.M.(N.H.) Banks letters.
[26] Banks to E.I.C., 18 August 1796. B.M. (N.H.) Banks letters.
[27] Smith to Banks, 17 September 1795. B.M. (N.H.) Banks letters.
[28] Banks to Sproat, 27 August 1802. B.M.(N.H.) D.T.C. Vol. 13, folios 235–8.
[29] Watt (1908), 668.
[30] J. Beames (1984), *Memoirs of a Bengal civilian*, London 174.
[31] Heber (1828), Vol. 1, 13.
[32] A. C. Lowe (1894), *The diary of an officer of the 16th (Queen's) Lancers, 1822–1840*, Calcutta, 415.
[33] Government of Bengal to E.I.C. 16 August 1838. I.O.L.R. E/4/163.
[34] E.I.C. to Government of Bengal, 13 February 1839. I.O.L.R. E/4/758.
[35] J. Lindley to E.I.C., 16 January 1839. *Transactions, Agricultural and Horticultural Society of India*, **7**, 1840, 87–93.
[36] Spry (1841).

GREEN MEDICINE

Both the Dutch and British East India Companies endeavoured to make their overseas settlements less dependent on supplies of drugs from Europe by substituting cheaper and fresher local products. Royle, asked by the Medical Board of Bengal to report on native medicines, purchased a representative selection in the bazaars, and listed them in the *Journal of the Asiatic Society of Bengal*[1], giving vernacular and botanical names and indicating those parts of plants used as drugs. The first use of the phrase, 'bazaar medicines', has been attributed to Royle who had a great respect for Indian works on materia medica 'which were far advanced and embraced an extensive range of subjects before any progress had been made in Europe'[2]. While Superintendent at Saharanpur Botanic Garden he established a garden at Mussoorie for medicinal plants including indigenous species used by native doctors. As Professor of Materia Medica and Therapeutics at King's College London, he paid tribute to the efficacy of traditional Hindu medicine in his introductory lecture to students[3].

Garcia da Orta and his contemporary, Christobal Acosta, studied Indian drug plants. Van Reede incorporated their putative cures in *Hortus Malabaricus*. Being medical men, Koenig and Roxburgh were inevitably drawn to them, persistently interrogating native doctors. Sir William Jones in his presidential address to the Asiatic Society of Bengal in 1793 regretted that 'the numerous works in Sanscrit on medical substances, and chiefly on plants, have never been inspected, or never at least understood, by any European attached to the study of nature . . . nor (what is far worse) [have any Europeans been] able to find accounts of their tried virtues in the writings of Indian physicians'[4]. Buchanan, however, questioned the benefits to be derived from translating these Sanskrit texts because of 'the terrible confusion and uncertainty' in their naming of plants and substances.

> *What is wanted is a native physician to prescribe and a European to give an account of the symptoms of the disease and of the effects of the practice. Such substances as have been actually employed, where the practice seems to have been successful, are then to be more carefully ascertained. If they are the produce of the vicinity, this may be done by sending them to the botanical garden for investigation, or if they have come from a distance, by accurately describing them and endeavouring to trace them to the country where they are produced*[5].

One of Buchanan's friends, John Fleming of the Indian Medical Service, wrote the first published digest of existing data on Indian drugs, primarily for European doctors newly arrived in the country. A *Catalogue of Indian medicinal plants and drugs with their names in the Hindustani and Sanscrit languages*

(1810)[6] is presented in two parts: the indigenous flora and introduced plants. Whitelaw Ainslie devoted nearly 20 years to identifying bazaar drugs which were also described in British materia medica. His *Materia medica of Hindoostan, and artisan's and agriculturist's nomenclature*, published by the Government of Madras in 1813[7], cites plant names in several Indian languages as well as English, Arabic, Persian, and Latin. Because it lacked illustrations, Robert Wight castigated it as 'little better than a monument of abortive labour'; both Fleming and Ainslie, he thought, were absolutely useless for identification purposes.

In 1820, implementing a recommendation of the Apothecary-General to the Company Wallich began trials at the Botanic Garden with native and foreign medicinal plants. He doubtless received guidance and encouragement from the Medical and Physical Society of Calcutta, formed in 1823. Two years later physicians in the South enjoyed a corporate identity through the establishment of the Medical Society of Madras. Their strict adherence to western pharmaceutical practices exasperated a contributor to the *Journal of Asiatic Society of Bengal*, who argued that

> *it is a desideratum to know how the natives have treated the subject of medicaments—what of good their books contain—what of error. Our medical practice pays, perhaps, too little attention to vegetable remedies, of which the orientals possess an infinite variety, many inert but many active, and many also quite unknown to Europeans*[8].

This neglect was rectified by William O'Shaughnessy's *Bengal dispensary* (1842) and *Bengal pharmacopoeia* (1844), which highlighted the remedies of native doctors and so prepared the way for the *Pharmacopoeia of India* (1868). Edward John Waring of the Madras Medical Service edited the deliberations of an official committee which met from 1865 to 1868 at the behest of the Secretary of State for India. As the author of *Notes on some of the principal indigenous tonics of India* (1858) and *Bazaar medicines and common medicinal plants of India* (1860), he was well qualified for the task. The *Pharmacopoeia of India*, modelled on the *British pharmacopoeia*, sought to bring 'to the notice of the profession in India those indigenous drugs which European experience has proved to possess value as medicinal agents, and which may be employed as efficient substitutes for imported articles'[9].

No local substitute could be found for Peruvian bark or cinchona, used in the treatment of malaria, which in mid-nineteenth-century India killed or disabled about 4 000 000 people a year. Roxburgh believed he had found an alternative in Indian Red-wood, *Soymida febrifuga*, whose bark possessed the bitter and astringent properties of cinchona. Army surgeons used it and it found a place in contemporary pharmacopoeias. Roxburgh also thought he had discovered the elusive genus in a plant he named *Cinchona excelsa*, which was subsequently identified as belonging to another related genus, *Hymenodictyon*. Until cinchona was grown in India, doctors relied on other accredited febrifuges such as the bitter roots of *Thalictrum foliolosum*, the Kut-kulega nut (*Caesalpinia bonduc*), and several species of Himalayan *Aconitum*.

The discovery of cinchona, a South American genus, is a whimsical anecdote in plant legend. The story that its powdered bark saved the life of the Countess of Chinchon, wife of the Viceroy of Spanish Peru, in 1638 is probably apocryphal. But what is certain is that it was being exported by the Society of Jesus to Europe by the mid-seventeenth century. Through this association it was also known as 'Jesuits' bark' and 'Cardinal's bark'. De la Condamine was an astronomer on a French expedition to South America in 1735 to measure an arc of an equatorial meridian in order to determine the shape of the earth. He and his botanical colleague,

Joseph de Jussieu, visited the forests of Loja in Ecuador, but the samples of cinchona they collected for shipment to Europe were lost in the Amazon. Back in France, de la Condamine published a description of quinquina ('bark of bark') in 1738 and in 1742 Linnaeus named it *Cinchona* in memory of that Spanish countess. Until the 1770s only Linnaeus's *Cinchona officinalis* was exported from Peru.

Over 60 species and varieties of cinchona[10] have been found in a crescent that sweeps for some 1500 miles through Venezuela, Colombia, Ecuador, Peru, and Bolivia. They grow in the temperate regions of dense forests, in high valleys and on mountain slopes up to 9000 feet, distinctive with glossy lanceolate leaves veined with crimson. Only a few are clinically useful: for example *Cinchona calisaya* and *C. ledgeriana*, both commercially known as yellow bark, *C. succirubra*, or red bark, and *C. officinalis*, crown or pale bark. The extraction of the alkaloid quinine by two French chemists in 1820 made medical history. Cinchona became a coveted plant for nations like Britain and the Netherlands with colonies in the tropics where malaria was endemic. Sir Joseph Banks had included it in his inventory of desirable plants which he submitted to the East India Company. When George Govan petitioned the Governor-General in 1816 for a botanical garden at Saharanpar he recommended the cultivation of cinchona there. Royle, Govan's successor at Saharanpur, echoed this request in 1835 and in his *Illustrations of the botany of Himalayan Mountains* had identified the Nilgiris, Sylhet, and Chittagong as feasible sites. He was supported by Professor Robert Chirston at Edinburgh alarmed by the rapid diminution of cinchonas in South America precipitated by increased demand in Europe. A century earlier de la Condamine had been aware of the devastation of the Loja forests, and the explorer Don Antonio de Ulloa had warned the Spanish authorities that the neglect of a replanting programme might lead to the extinction of cinchonas. Presumably when the expeditions of Mutis, Ruiz, and Pavon located other cinchona regions, thus relieving the pressure on Loja, the Spanish no longer thought any conservation policy to be necessary. When Humboldt was in the Andes in 1799, he calculated that some 800 to 900 trees were being recklessly destroyed every year. The Bolivian Government's limitation on the amount of bark that could be cut or exported in 1834 was influenced rather by the drastic fall in the price of bark which had been exported in enormous quanitites than by any concern for conservation.

When the French Government sent a scientific expedition under Count de Castelnau to South America in 1845, it instructed Weddell, its naturalist, to include cinchona in his researches. Hugh Algernon Weddell (1819–77), of English extraction, was an *aide naturaliste* at the Jardin des Plantes in Paris; on his return to France he published the first full account of cinchona, its botany and the marketing of the bark in *Histoire naturelles des quinquinas* (1849). The heedless felling by native cutters or *cascarilleros* appalled him:

> *the mode of collecting this valuable product appears to be always under the control of the half savages by whom it is performed; and if some efficient means be not discovered of counteracting the ruinous and wasteful method adopted, our descendants will inevitably have to regret the entire or at least partial extinction of the different varieties of cinchona*[11].

From the seeds of *Cinchona calisaya* which he collected in Bolivia came the first living cinchonas seen in Europe. He advocated growing cinchonas in the French colonies and some *calisayas* were sent to Algiers. But it was the Dutch who took the first positive action to acclimatize cinchona in their overseas territories. In 1851 one of Weddell's *calisaya* seedlings reached the Leiden botanical garden where it was nurtured into a healthy sapling for the long journey to Java a few months later. Miraculously it survived the voyage, a cutting was taken from it before it died, but a large supply of plants was essential before the authorities of Java could embark on commercial propagation.

For some twenty years Dutch botanists had been pressing their Government to introduce cinchona into Java. After abortive attempts to obtain seeds through their consular service in South America, they decided to appoint a competent collector. This commission was given to a former botanist at the botanical garden at Buitenzorg in Java, Justus Carl Hasskarl—fit, splendidly resourceful, and an authority on tropical vegetation. He left the Netherlands in December 1852, charged with collecting *Cinchona calisaya* and any other potentially useful species. The state of unrest in Peru, which was on the verge of war with neighbouring Bolivia, made his task more difficult. To conceal the true purpose of his mission, he assumed the identity of José Carlos Müller, a native of Cassel now living in Amsterdam, intent on an innocuous scientific excursion through South America. He failed to persuade a local governor to obtain cinchonas for him, being offered instead the services of a Bolivian of dubious reputation. Hasskarl acquired 400 plants from him which, with his own collections, were packed closely together, their roots protected in damp moss and then wrapped in sackcloth for the difficult journey to the coast. All the time he was wary of predatory bands of soldiers. When eventually he loaded his precious cinchonas into 20 Wardian cases at the port of Callao, he was unaware so it was claimed years later by Charles Ledger, that his Indian porters had surreptitiously mixed arsenic with the earth in the cases. It was not surprising, therefore, that only 70 or so plants had survived out of a shipment of 500 when Hasskarl arrived in Java in December 1854. A grateful Government made him a Knight of the Netherlands and Lion and Commander of the Order of the Oaken Crown and, furthermore, appointed him Director of the world's first cinchona plantation. Only 167 cinchonas were growing in Java when the botanist, Franz Wilhelm Junghuhn, succeeded him two years later. By 1862, when the population had soared to over 1 300 000, the Dutch were way ahead of their English competitors.

In London, Royle persisted in reminding an obdurate Court of Directors of the manifest advantages of cinchona cultivation in British India while his friend Falconer in Calcutta enlisted support from government officials and the medical and botanical fraternity. One reputed species from Brazil, *Cinchona ferruginea* (later redesignated *Remijia ferruginea*) had recently been added to the Calcutta Botanic Garden when Falconer presented a paper in February 1852[12] to the Agricultural and Horticultural Society of India. In it he endorsed Royle's preference for the Khasia Hills, the mountains near Chittagong, Upper Assam, and the Nilgiris as the most promising regions for the trial cultivation of cinchona. The Governor-General's despatch supporting Falconer's proposal that a collector should be sent to South America was submitted by the Court of Directors to Royle for comments and an estimate of likely costs.

Royle once more parading the arguments favouring decisive action, confidently asserted 'that after the Chinese teas, no more important plants could be introduced into India'. A collector should be sent to South America, preferably an employee of the Horticultural Society of London, at that time actively interested in that continent's flora. This association with the Society might cloak his search for cinchonas[13]. Conscious that supplies of quinine for India were costing them £7000 annually, the Court approved the cultivation of cinchona but not Royle's proposal to send a collector. Instead British consuls in South America were asked to obtain plants and seeds. In December 1852 Royle attempted a small pilot scheme with six cinchonas raised from Weddell's seed. These were entrusted to Robert Fortune just about to travel overland to China on another plant collecting expedition. Five were still alive when Fortune reached Calcutta but after an initial display of vigorous growth, they began to wilt; three survivors were rushed to the hill station garden at Darjeeling but they, too, were all dead by May 1855. When the attempt to procure cinchonas through the agency of British consuls failed, the Court of Directors were tactfully reminded by the Governor-General, Lord Canning, of their commitment to the project. His despatch prompted yet another resumé of the current state of affairs by Royle who again argued most persuasively for a collector to be recruited. When the Court's assent was finally obtained, it came too late for Royle who died in January 1858 before an appointment could be made.

In September 1858 the Court of Directors held 'its last solemn assembly' and issued 'its last instructions to its servants in the East' before handing over its responsibilities for British India to a new department of State, the India Office. The sepoy rebellion had hastened reinforcements to India, and the demand for quinine had gone up to £12 000 a year with every prospect of rising further. With the need to ensure adequate and uninterrupted supplies of quinine for the troops, the India Office gave careful consideration to an approach in April 1859 from one of their junior clerks, Clements R. Markham (1830–1916) (Fig. 16.1). He offered his services as a plant collector: he had already travelled in Peru; he could identify several species of cinchona; he was fluent in Spanish and knew one of the native languages; and he was friendly with a number of influential people. He appeared an ideal candidate to

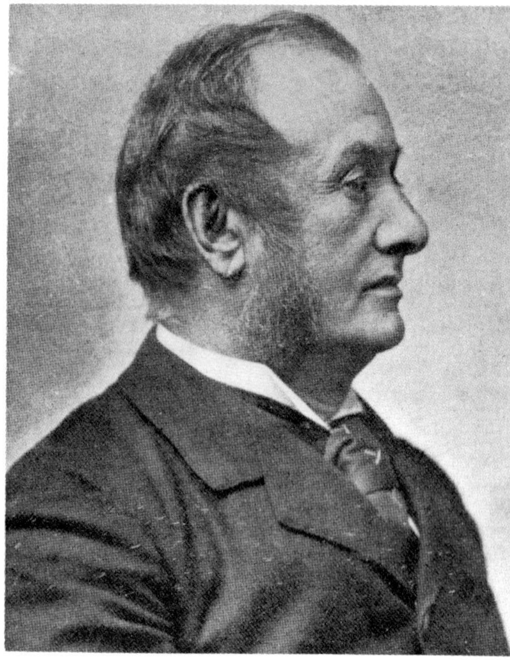

Figure 16.1 Sir Clements Robert Markham (1830–1916).

the India Office, especially as he was not demanding any increase in salary. The India Office came to a decision with a speed that would have been unthinkable in the leisurely proceedings of the Court of Directors. Within three days his appointment was approved and Markham with equal promptitude submitted his plans. Markham proposed to adopt Royle's advice to collect numerous cinchona species over as extensive an area as possible in order to maximize the chances of finding similar soils and climate in India. He therefore conceived four expeditions—two to Peru, one to Ecuador, and another to Colombia. Markham also wanted a ship to be available on the Pacific coast to get the plants to India by the quickest route. Markham and a gardener, John Weir, would seek *Cinchona calisaya* in Caravaya, the southern province of Peru, and the forests of Bolivia. *C. succuribra* would be collected by Richard Spruce and another gardener, Robert Cross, in Ecuador. And G. J. Pritchett, who was not a botanist but knew Peru well, would concentrate on the grey barks of central Peru—*C. nitida* and *C. micrantha*. Any other species would be a bonus. A tour in Colombia was deferred for the time being.

In mid-January 1860, less than ten months after he had spoken to the India Office about his plan, Markham landed at Guayaquil in Ecuador. He found Peru nervous of all foreigners. Leaving 15 Wardian cases at Guayaquil he proceeded to the Peruvian port of Islay where he deposited another 15 cases before setting out for the mountainous regions of Caravaya, believed to be the home of those cinchonas with the highest percentage of the quinine alkaloid. With four Indians he followed mule tracks across bleak plains, through dense forests, over two ranges of the Andes and penetrated the Tambopata valley beyond the furthest point reached by Weddell and Hasskarl. He had the misfortune to encounter a disaffected colonel, Don Manuel Martel, who knew about Hasskarl's visit and, doubtless aware of the purpose of Markham's presence in Caravaya, threatened to cut off the feet of anyone he caught taking cinchonas out of the country. By mid-May Markham had discreetly collected over 500 cinchonas, enough to fill the 15 Wardian cases at Islay, and having decided it was imprudent to cross into Bolivia, began the arduous return journey to the coast, his plants wrapped in mats to protect them from night frosts. He chose a circuitous route along unfrequented paths since villagers, stirred up by Martel, were determined to seize his plants. Weir acted as a decoy by taking a different route. He reached Islay with 456 budding *calisayas* which were planted into pots and firmly bedded in the soil of Wardian cases. Before they were loaded on board ship an attempt was made to kill the plants with boiling water. The authorization to cross the Pacific to India arrived after Markham had left for Southampton via Panama. When he reached England late in July 1860, just under half his consignment was alive.

Spruce had been unable to start his tour until June, incapacitated first by temporary paralysis and then further delayed by the rains. Richard Spruce (1817–93), a former schoolmaster, had been in the Amazon valley since 1849 with no official backing, botanizing, recording in his meticulously neat hand the agriculture, food, and customs of native tribes, and mapping several previously unsurveyed rivers. He accepted, with an extraordinary stoicism, solitude, disasters, illness, and threats to his life, existing frugally by selling sets of his plants to European botanists. In Peru his horses and baggage were often in danger of being appropriated by rapacious bands of troops seeking out rebels. He rested at Limon on the western slopes of Mount Chimborazo awaiting the arrival of his companion, Robert Cross, who immediately set about forming a nursery of cinchonas at this base. All the time in pain or discomfort—'to lie down quietly and die would have seemed a relief'—Spruce gathered not only the red bark but also a specimen of every plant that he thought might interest Kew. A raft of tree trunks transported the full Wardian cases to Guayaquil whence they were escorted to England by Cross just four months before the export of cinchonas was banned. Ignoring the official prohibition and thus risking imprisonment or at least a fine, Cross was back again in the autumn seeking crown barks (*C. officinalis*) in Loja. In 1863 he was sent to Colombia in search of the rare Pitaya bark (*C. pitayensis*), almost wiped out by the cascarilleros. As the seed he had collected failed to germinate, he was back in 1868, this time successfully introducing it to India. He returned to Colombia once more in 1877, bribing his reluctant Indian partners with narcotic coca. This final expedition left his health permanently impaired. G. J. Pritchett, the other collector in the team, garnered a good crop of *Cinchona nitida* and *C. micrantha* in the forests near Huanaco in central Peru.

Most of Markham's 134 *calisayas* were in a sickly condition when they reached Ootacamund in October 1860, having endured extreme heat during the overland journey from Alexandria to Suez and an unfortunate delay at Bombay. MacIvor, in charge of the Ootacamund garden, hastily made over 300 cuttings from the dying stock but not one struck root. It was a disastrous conclusion after so much effort but Markham was not entirely disheartened, remembering the 17 *calisayas* he had providentially left with Hooker at Kew. Pritchett's contribution suffered a similar fate. His grey barks were in a desperate state when they reached Bombay—some of the Wardian cases had gaping seams and their soil had dried to powder. But the project was by no means a complete failure. Pritchett's seed did germinate and Spruce's red barks which passed through the Red Sea in much cooler weather survived; 463 *succirubras* and 6 *calisayas* were passed to McIvor's care. He was able to report in June 1861 that he had 'upwards of 2000 plants in very promising condition'.

W. G. McIvor (*d.* 1876), who was praised as 'a zealous, intelligent and practical gardener', was appointed Superintendent of cinchona cultivation in the Madras Presidency in July, and it was largely due to his skilful management that cinchona in India became a profitable venture. He chose a 50 acre site in a wooded ravine that resembled as near as possible the conditions in which cinchonas thrived in South America. Unlike Junghuhn in Java who cultivated his cinchonas in shade because in their natural world they grew 'in damp forests overshadowed by trees'. McIvor experimented with partial shade and complete exposure to the sun. He increased the trees' productivity by 'mossing' and regular coppicing, methods subsequently superseded by the present practice of thinning, pruning, and uprooting. His methods, however, were criticied by Thomas Anderson, the new Superintendent of the Calcutta Botanic Garden, who had been entrusted by Hooker with carrying Spruce's *succirubra* seeds to India.

In June 1861 the Governor-General of the Netherlands Indies generously gave India 500 000 cinchona seeds, between 100 and 200 seedlings of *C. lucumifolia* and up to 100 *calisayas*. Anderson went to Java to collect this consignment for planting in the Nilgiris (Fig. 16.2). Meanwhile, McIvor had received a chemical analysis of *C. lucumifolia* from J. E. Howard, a London quinine manufacturer, who reported its quinine content to be very low. He renamed it *C. Pahudiana* in honour of the Dutch Minister for the Colonies who had sent Hasskarl to South America, but was adamant it had no commercial value other than as an attractive ornamental flower. It was the predominant species in Java's plantations and Dutch planters were now grateful to receive stock from India to replace it. But this proved to be only a temporary setback for in 1865 they purchased a pound weight of seed of a cinchona which turned out to have a very high quinine content.

The source of this seed, Charles Ledger (1818–1905), was an alpaca wool trader, resident in Peru and Bolivia since 1836. He had searched in vain for cinchona in Caravaya in 1845, and another expedition in 1850 had to be abandoned when his companion was murdered by Indians. He had met Weddell and also Hasskarl masquerading as Müller before taking a flock of alpacas to Australia where he read in a Sydney newspaper about the arrival of Markham's cinchonas in India. For some years Ledger had cherished an ambition to start a cinchona plantation in Ceylon and had no scruples about smuggling the plants out of South America. He knew how to outwit officials and suspicious Indians. The subterfuge he sometimes adopted was the role of a bird collector and general botanist. But he acquired the seeds of the cinchona which commemorate him without

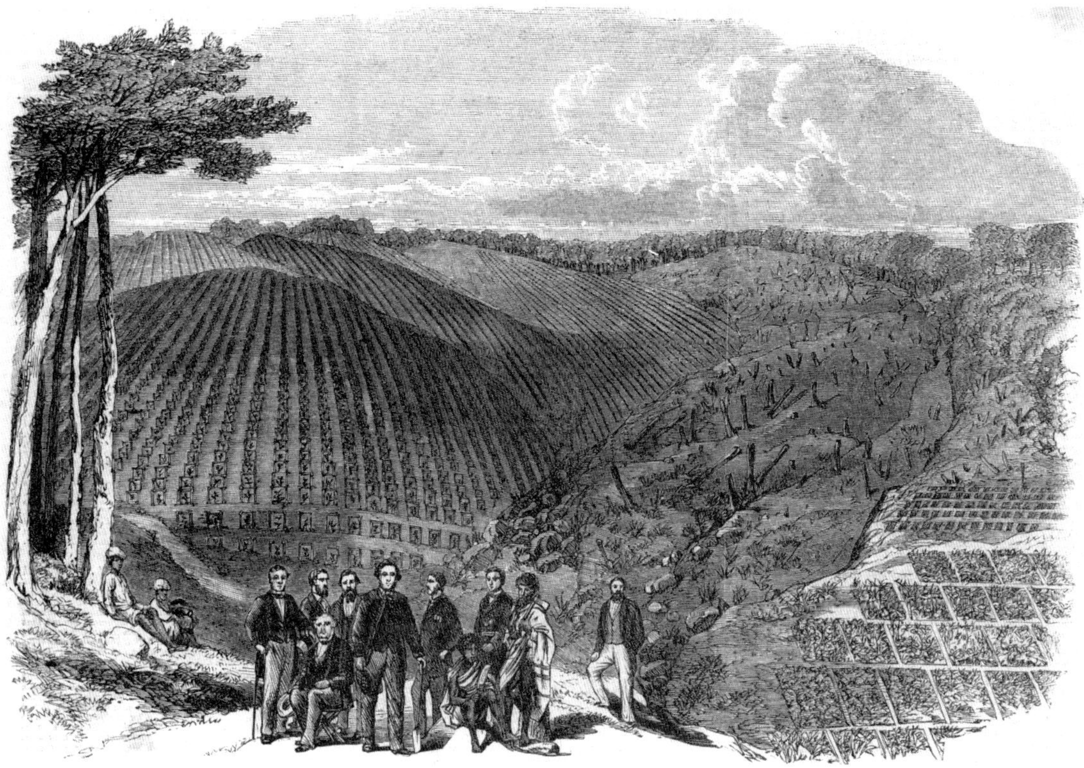

Figure 16.2 New cinchona plantations in the Nilgiri Hills, 1862. Wood engraving. Note the deforestation on the right.

the need for such a disguise. On his return from Australia in 1865 his old Indian servant brought him seed he had collected in Bolivia. Ledger, seeing it was good quality seed, sent it to his brother, George, in London for disposal. George went immediately to Kew Gardens but Sir William Hooker had just died, his son, Joseph, was away ill and the Curator was not interested. He had made several unsuccessful attempts to sell the seed before he met J. E. Howard, who referred him to the Dutch Consul-General in London. The Consul bought a pound weight for Java and an Anglo-Indian cinchona planter, J. W. B. Money, bought 14 pounds. Regretting his rashness in buying seeds of an unknown species, Money gladly exchanged them with McIvor who was about to return to India. McIvor had difficulty in establishing *Cinchona ledgeriana*—the name the Dutch gave to the new species—at Ootacamund, but a sample he forwarded to the plantations in Sikkim fared better in a more congenial climate. There was no problem with acclimatization in Java where the Dutch filled their plantations with this vigorous tree, improving its disease resistance by grafting it on *succirubra* rootstock. With the highest quinine content of all the cinchonas on sale in Amsterdam in the 1870s, it gained the best prices and confirmed Dutch supremacy in the world market. Although it is a difficult tree to grow in India, three-quarters of all the cinchonas now there are ledgerianas. 'We are indebted', wrote Howard in 1880, 'not to systematic botanists, but to the experience and practical sagacity of an Indian for our knowledge of

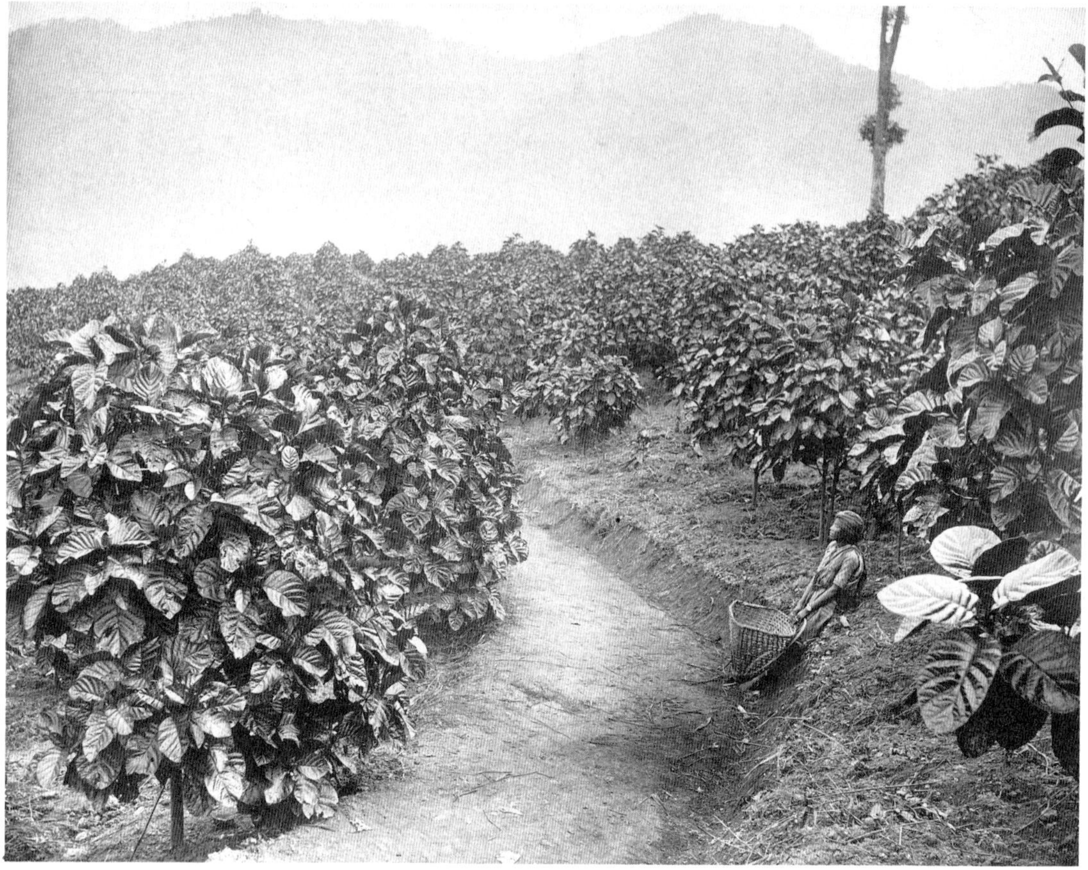

Figure 16.3 Plantation of *Cinchona succirubra* in Sikkim.

the best kind of calisaya [i.e. *ledgeriana*] bark.'[14]

The Crown Colony of Ceylon, not having contributed to the cost of the cinchona initiative, at the outset received only a token offering. Thomas Anderson brought a three foot high *calisaya* and a small cutting in 1861 and later in the year some of the seed collected by Spruce, Cross, and Pritchett, and 100 *succirubras* were donated. As Peradeniya was too hot for the cinchonas, Thwaites established a small plantation at the mountain garden at Hakgala and by September 1861 had raised 800. In 1864, with a large surplus of plants at his disposal, he tried to persuade the coffee planters to experiment with them. Although he continued to offer his cinchonas free of charge the planters manifested little enthusiasm until fungus disease ravaged their estates when they gladly turned to cinchona as an alternative cash crop.

The cultivation of cinchona was in part motivated by altruism; certainly people like Markham had always aspired to make quinine available to the poor. The annual report of the Royal Botanic Garden in Calcutta for 1892–3 announced the achievement of this goal. Quinine was being sold cheaply at post-offices in Bengal, packeted in five-grain doses with printed instructions in vernacular languages. But this facility never extended to the entire subcontinent nor endured for long in Bengal because Dutch production and business acumen soon made Amsterdam's domination of the world market unassailable. When a glut of bark depressed prices, private planters in India and Ceylon began abandoning cinchona for more profitable and stable tea.

Government plantations survived but their produce failed to satisfy the demands of India which now had to import quinine for domestic use. Just as synthetic dyes superseded cochineal so did chemical substitutes eventually replace cinchona although there is still a role for the natural drug which is cheaper to produce.

When Sir Joseph Hooker was near the end of his life, his godson asked him what commercial endeavour in which he had participated gave him the most satisfaction. Without hesitation, the old man replied: 'Quite certainly the getting of cinchona into India'. From the moment the India Office had approved the cinchona project, Sir William Hooker and his son were always consulted. In his annual report for 1860 Sir William called it 'one of the most important horticultural operations in which, as Director of this Establishment, it has been my privilege to co-operate'. A government subsidy enabled Kew to receive seeds and plants from South America, to propagate stock in a specially built greenhouse, and to despatch plants not only to India but to the West Indies, Mauritius, the Cape and Australia as well. Kew's annual reports monitored the progress of the enterprise until 1874 when Joseph Hooker, now Director, announced that 'the introduction of this febrifuge into India, Ceylon and Jamaica being now accomplished this subject no longer demands a notice in my report'.

It could be said that the most important contribution Kew made was the nomination of Richard Spruce (Fig. 16.4) and the Kew gardener, Robert Cross, both outstanding collectors. Indeed without such competent participants the whole scheme might well have foundered. As a collector himself, Markham knew only too well the demands made on stamina and endurance. When Spruce and Cross returned at last to England, both severely disabled by their service in South America, Markham tried unsuccessfully to get them an adequate civil pension. The best he could achieve was £50 a year for Spruce and a lump sum of £600 for Cross. John Weir received no recognition or reward neither did Charles Ledger, reduced to penury in his old age.

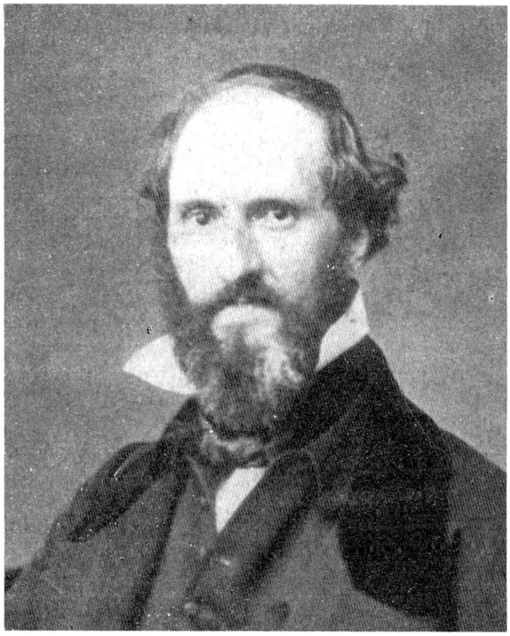

Figure 16.4 Richard Spruce (1817–93).

NOTES

[1] List of articles of materia medica obtained in the bazaars of the Western and Northern provinces of India, *Journal of Asiatic Society of Bengal*, **1**, 1832, 458–71.

[2] I.O.L.R. F/4/2186, folio 21.

[3] J. F. Royle (1837), *An essay on the antiquity of Hindoo medicine including an introductory lecture to the course of materia medica and therapeutics, delivered at King's College, London*. London.

[4] *Asiatick Researches*, **4**, 1795, 14.

[5] F. Buchanan (1936), *An account of the district of Bihar and Patna in 1811–1812*, Vol. 2, Patna 450. London.

[6] *Asiatick Researches*, **11**, 1810, 153–96.

[7] A new edition in two volumes was published in London in 1826.

[8] *Journal of Asiatic Society of Bengal*, **6**, 1837, 382–97.

[9] Preface, vi.

[10] The specific names of *Cinchona* adopted in the nineteenth century and still recognized in the trade have been retained. *C. succirubra*, however, is now *C. pubescens*, and *C. ledgeriana* and *calisaya* are varieties of *C. officinalis*.

[11] *Madras Journal of Literature and Science*, 1857, 217–18.

[12] *Journal of Agricultural and Horticultural Society of India*, **8**, 1854, 13–17.

[13] Royle to E.I.C., 27 June 1852. Return: East India (Cinchona Plant). House of Commons Paper 118, 1863, 2–3.

[14] Gramiccia (1988), 153.

Tea and Opium

Before the end of the nineteenth century, cinchona on many private plantations had been ruthlessly uprooted to make way for tea, one of the most profitable commodities handled by the East India Company. Dutch and English factors in India had soon acquired the tea-drinking habit: they served it at business meetings and on social occasions; with the addition of sugar-candy or lemon it was a cooling drink; mixed with hot spice it was a remedy for 'headache, gravel and griping in the guts'. Portuguese and Dutch merchants introduced the beverage to Europe—the precise date is unknown but by about 1610 it had reached the Netherlands through Chinese junks trading with Java. Not until 1657 did the first public sale of tea in London take place; it soon became available in the city's coffee houses and in 1660 Samuel Pepys tentatively sipped his first cup. His wife took it as a cure for 'her cold and defluxions'. Initially there was some confusion about its culinary use—some households garnished the boiled leaves with butter and salt. The correct recipe for its preparation had been presumably established when the Court of Directors decreed in July 1664 'good tea to be provided for the Company's occasions'.

In September the same year, the Company ordered a hundredweight of the best tea from Bantam in the East Indies; from 1669 tea formed part of regular shipments. Evicted from Java by the Dutch in the early 1680s, the Company turned directly to China for supplies, obtaining from the Crown a valuable concession—exclusive import rights. By the second decade of the eighteenth century, tea imports, now exceeding 1 000 000 pounds in weight, were displacing silk, for many years the most profitable merchandise in the China trade. Due largely to British addiction to tea, the Company dominated the world market with East Indiamen annually carrying about 36 000 000 pounds by 1816.

Botanists and gardeners in England were keen to examine the shrub that had so radically changed British drinking habits. James Cunningham, one of the Company's surgeons, and the first European to botanize at Amoy and on the island of Chusan during the 1690s and early 1700s, gave his collections to James Petiver, Leonard Plukenet, and Sir Hans Sloane. They included both flowering and fruiting specimens of the tea plant (*Camellia sinensis*), now in the Sloane herbarium. It is not known when a living specimen was successfully introduced but Peter Collinson reported in 1742 seeing a healthy plant in the orangery of one of the Company's directors. 'It was brought two or three years ago a present from China to his wife.'[1] John Ellis, a London merchant flatteringly described by Linnaeus as 'a bright star of natural history', raised some tea plants from seed he had obtained during the 1760s. Linnaeus himself received in 1763 some plants grown by a Swedish sea captain,

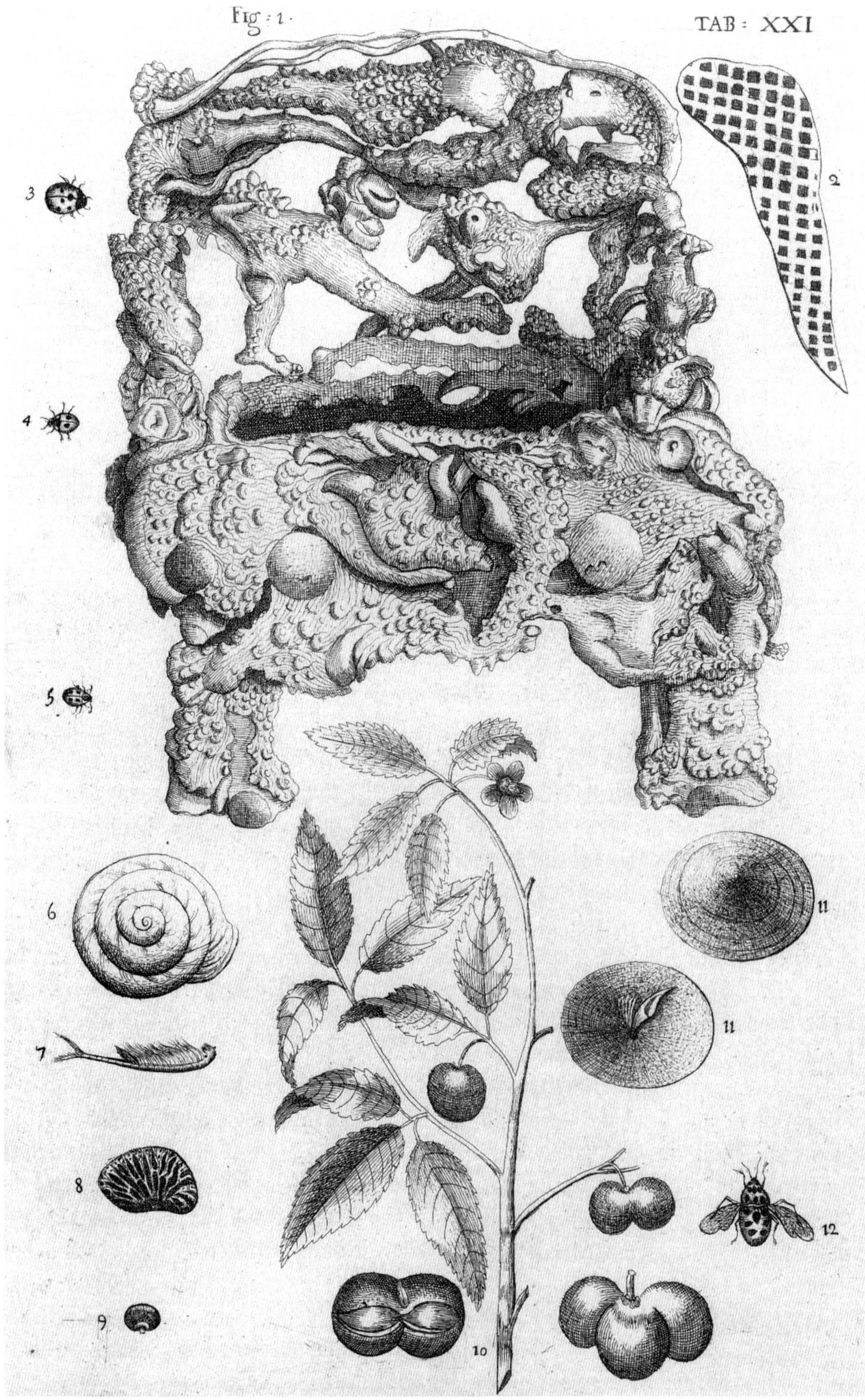

Figure 17.1 A chair (fig. 1) made out of a root of a tea-shrub and presented by the East India Company to Lord Somers who in turn gave it to the Royal Society. (From J. Petiver (1767), *Opera*, Vol. 1, London, plate 21.) Engraving.

Carl Gustaf Ekeberg, from seeds and sown when his ship left China; in 1765 Linnaeus published a dissertation, *Potus Theae*, on drinking tea. Ellis's friend, John Fothergill, flowered a five foot specimen in 1774. Two years earlier, John C. Lettsom, a London physician who was also a knowledgeable gardener, published a monograph on its history and medicinal properties—*The natural history of the tea-tree*. He forecast its successful cultivation in the colonies (it had been tried in Georgia about 1770) where cheaper labour would make it an attractive cash crop. In 1764 the Court of Directors had tried to persuade its officials at Canton to despatch some plants to the Company's factory at Fort Marlborough in Sumatra for trial acclimatization. Warren Hastings, ever aware that the Company's prosperity depended on the development of India's resources, used his authority as Governor-General to promote such commodities as indigo and cotton. Determined to encourage trade with Bhutan, he sent some tea seeds to his emissary, George Bogle, for cultivation by the Bhutanese. Robert Kyd is supposed to have had a few tea plants growing in his own garden in Calcutta in 1780. As Superintendent of the newly-formed Botanic Garden, he rated tea cultivation among his top priorities; he requested the Company's staff at Canton to send him both green and black tea plants and also some Chinese cultivators. The few plants he received never flourished.

Kyd's unsuccessful attempts to grow tea prompted Sir Joseph Banks's solution to the problem[2]. Since tea plantations in China were situated in hilly and mountainous regions, the first step should be to find a similar environment in India. Banks favoured the Rohilkhand region just south of the Himalayan range. He thought that Tibet might be persuaded to supply both plants and Chinese labour. Earlier in 1788 Banks's advice on the tea trade had been solicited by Lord Hawkesbury, President of the Board of Trade.

Will it not be possible by proper premiums and wise instructions to encourage the growth of the tea plant, and the manufacture of its leaves in some part of the British dominions in the East and West Indies, so as to be supplied from thence with a part of the tea consumed in this country, and not with the whole as at present from China[3].

Banks convinced by all the literature he had studied that tea was an inhabitant of temperate regions, indicated vaguely an area between Bengal and Bhutan as ideal for plantations. Now recognized as an authority on the subject, he was invited in November 1788 by the Deputy-Chairman of the Court of Directors to elaborate his views on a viable tea industry in India. Within a month the Company received Banks's response on a product which he rated 'an article of the greatest national importance'.

Having referred briefly to the tea-growing regions of China, Banks specified Rangpur, Cooch Bihar, and the northern part of Bihar as offering climatic conditions similar to those parts of China where black tea was grown. Green tea would fare best in cooler Bhutan whose inhabitants might be persuaded to undertake its cultivation. He attributed the poor quality of the tea grown on the Chinese island of Hainan to unsuitable soil, and wondered whether some of the growers there might be attracted to better conditions in India, bringing with them their shrubs and equipment. They could be settled on 20 acres at the Botanic Garden in Calcutta where Indians could be trained in the cultivation and manufacture of tea and subsequently sent to the proposed sites near the mountains. The success of Banks's plan depended upon the participation of skilled Chinese labour. He warned that to

intrust the execution of this business to the Company's supercargoes at Canton would be imprudent. They have an interest in its failure; & by their influence on the spot could no doubt if they chose it prevent the success not only of the present only but of all future attempts of the same nature.

The management of the operation must remain firmly in the control of the Court of Directors, concluded Banks, assuring the Vice-Chairman of his personal interest:

if my advice or assistance in preparing the details of the plan should be thought necessary, I shall at all times be ready to come forward, convinced that the object is of real importance, not to the Company only, but the country at large[4].

When Kyd in Calcutta had considered these proposals, he reported the current state of affairs at great length, marshalling new data and annotating his text with judicial skill[5]. Ostensibly this exercise was to

enable Sir Joseph Banks form a further view of the subject and with the enlightened grasp of mind which he possesses, to chalk out a final and permanent plan for bringing it to as successful an issue, as the means, resources and situation of the British Government in India will admit.

Yet as one reads his submission, one senses he was on the defensive, unsure of what was being discussed in London, determined to be heard before all decisions were taken out of his hands.

I know not from what part of the advices transmitted by the Board, it has been understood that the attempt of cultivating the tea plant has proved unsuccessful; certain it is, that such plants as have been received from China have thriven very well, altho' apparently in the most unsuitable soil and climate of all the Company's possessions.

He complained bitterly yet again about the Canton factory, their reluctance to send tea plants and native cultivators although they readily supplied individuals in India with Chinese labour. He quoted approvingly from Abbé Grosier's account of tea in China, using that author's observations as a guide to possible sites in India. He was hopeful that tea plants could be routed through Tibet and Bhutan. This verbose report was tactfully deferential to Banks whose only note of dissent when he read it was to dismiss Abbé Grosier as an unreliable authority.

I most readily admit that gentlemen who are on the spot must be able to point out the local situations most likely from the combined circumstances of climate, soil and inhabitants to suit the culture of it than anyone sitting in his chamber here can possibly do and that the reasons given against the adoption of the Rohilcund country which I originally proposed are fully satisfactory[6].

For whatever reason the tea project stagnated until 1793—the year in which Kyd died and was replaced by Roxburgh—when the former Governor of Madras, Lord Macartney, on his embassy to China to improve trade relations was asked, among his many commissions, to investigate the tea industry in China. Macartney's journal for 21 November included the following note:

In crossing into Kiangsi we passed through tea plantations and were allowed by the Viceroy to take up several tea-plants in a growing state with large balls of earth adhering to them, which plants I flatter myself I shall be able to transmit to Bengal[7].

Tea was certainly among the plants sent to Bengal by Macartney but whether they survived the sea voyage is not known. In 1798 Banks offered to send Roxburgh a hundred plants of both the green and black varieties, propagated from stock in England, 'cheaper than a similar supply could be procurd from China'. The blundering bureaucracy of the East India Company irritated Banks, himself very decisive, and in a letter to Roxburgh stated

I have some years ago corresponded on the subject [of tea] with the Court of Directors, but I am tird of promoting projects with a fluctuating body who are sure to be changed by the time I have convinced the first set of the propriety of any measure recommended to them[8].

London may have been lethargic but fortunately some of its officials in India kept the tea issue alive by comment and exhortation. George Govan in his seminal report of 1816 on the formation of a botanic garden at Saharanpur strongly advocated its introduction.

I can with difficulty bring myself to believe what nevertheless appears to be the case that the tea plant, which has occasionally been preserved in a state of miserable and feeble existence in the Calcutta Botanic Garden, should never have been tried in the climate of Saharanpore and Hurdwar, where the same latitude and elevation of soil can be obtained under which it only comes to perfection in China . . . The probabilities of success in this alone . . . I should imagine would of themselves justify the expenditure of a small sum in an attempt from which consequences so important might be expected[9].

When Royle, Govan's successor at Saharanpur, made a similar recommendation in 1827 to Lord Amherst, he was unaware that the Himalayas had already been identified as a suitable location for experimental plantations. He tried his powers of persuasion on the next Governor-General, Lord Bentinck, when he visited Saharanpur in 1831. Royle expanded his thesis in his *Illustrations of the botany . . . of the Himalayan mountains* (1834, pp. 107–27), and judged the project so important that he had the relevant pages reprinted as a pamphlet. Two years earlier Wallich, at the request of the Chairman of the Board of Control, in a paper to a Committee of the House of Commons on the Affairs of India had proposed tea plantations in Kumaon, Garwhal, and Sirmur.

There now existed both in England and India a considerable body of opinion in favour of the project, but what perhaps precipitated action by the Company was the devastating consequence of the renewal of its Charter. The first encroachment on its monopoly of eastern trade had occurred with the renewal of the Charter in 1813 when British merchants were for the first time allowed to compete in the Indian market. This major concession failed to satisfy these merchants who continued to agitate for access to the China trade which for so long had been the exclusive perquisite of the Company. Since 1699 when the Chinese government allowed the Company to establish a factory in the walled seaport of Canton at the mouth of the Yellow River, this privilege had been jealously guarded. No British subject could disembark at Canton without the Company's authority and British ships could trade there only under its licence. This exclusive right was lost when the Bill renewing the Company's Charter received Royal Assent on 28 August 1833. The China trade including tea now became open to all. Less than five months after this Bill had been passed, Lord Bentinck set about forming a committee to look into the growing of tea in India. He was persuaded that this was the right course of action after having read Wallich's submission to the House of Commons Committee and also a memorandum from a Mr Walker who argued convincingly that since Japan, Burma, and Brazil had successfully grown tea, Britain should emulate their example. He recommended Nepal and regions 'where the camellia and other plants of a character similar to the tea plant are indigenous'[10].

On 24 January 1834 Bentinck announced the formation of a committee charged to submit 'a plan for the accomplishment of the introduction of tea culture in India, and for the superintendence of its execution'[11]. The only serious difficulty he anticipated was admission to China to obtain information on methods of cultivation and manufacture, but he recommended circuitous means of gleaning that vital information by consulting Chinese agents in Penang and Singapore. George James Gordon whom Bentinck proposed sending on this investigative mission was also appointed Secretary of the new Committee of Tea Culture which comprised eleven British and two Indian members. In the event, Gordon departed not to Malaya but hopefully to China in 1834, and during his absence Wallich took over the Secretary's duties. The Committee issued a circular to local officials in March postulating optimum conditions regarding climate, soil, and topography and inviting views on recommended sites. The first response came within days, an enthusiastic letter from Hugh Falconer at Saharanpur in favour of the Himalayas. Since his case was cogently put, the Committee unanimously accepted the Himalayas as 'entirely suitable for the projected culture'. It was, however, the reply from Captain Francis Jenkins on the North East Frontier in Assam that really aroused their interest. He claimed that

> *Camellias are found in every part of this hill country, and within our jurisdiction in the Singpho district of Beesa, a coarse variety of tea plant is, as I am informed, undoubtedly indigenous*[12].

This was quickly followed by a letter from Lieutenant Charlton, an officer serving under him, who revealed that about three years earlier he had sent three or four young tea shrubs, dug up at Beesa, to John Tytler in Calcutta. What happened to them is a mystery and Wallich, much later, was insistent that they never reached the Botanic Garden. In November 1834 Charlton sent seeds and leaves of the Assamese plant via Jenkins to Calcutta.

> *The leaves you could have had before, but I was anxious to make them into something like tea, the best test that the tree is not a Camellia, as Dr Wallich imagines*[13].

With fruiting material as well as foliage to examine, Wallich was able to identify the Assamese shrub as genuine tea. It was a jubilant Tea Committee that informed Lord Bentinck on 24 December 1834

> *that the tea shrub is beyond all doubt indigenous in Upper Assam . . . We have no hesitation in declaring this discovery . . . to be by far the most important and valuable that has ever been made on matters connected with the agricultural or commercial resources of this empire*[14].

Yet this discovery could have been made years earlier if Wallich had pursued with greater pertinacity the evidence available to him. In 1815 a Colonel Latter had reported that the Singpho tribes in Assam harvested a species of wild tea which they either ate with oil and garlic or made into a beverage. In 1823 Robert Bruce trading in what was then Burmese Assam observed tea trees growing in the hills. He made an arrangement with a local chief for a delivery of tea plants and seeds on his next visit, but in 1824 the Anglo-Burmese war broke out and it was his brother, Charles Alexander Bruce, commanding a flotilla of gun-boats, who received the consignment from the Singpho chief. He gave them to Captain David Scott, a commissioner in Assam, who planted some in his own garden and sent the rest in 1826 to James Kyd, son of Lieutenant Colonel Kyd. To Wallich's share he also added some leaves.

> *The enclosed leaves are I think to be those of the wild tea plant. I have not been able to get the flowers or seeds but I have some plants that I hope will survive & plenty more are procurable. The Shanns & Burmuhs & a China man that is here say that it is the tea plant perhaps the species Mr Gardner sent down from Nepal*[15].

With only leaves as a diagnostic feature, Wallich was cautious, applying a provisional identification—'*Camellia? Scottiana*'. Had he asked for the flower and fruit to be collected in order to establish whether this was the true tea plant, there would have been no urgent need for the Secretary of the Tea Committee to go to China.

Gordon failed to reach the tea districts of Fukien and did most of his collecting around Amoy. Parcels of his seeds reached Calcutta in January 1835, the bulk going to the Himalayas and Assam with a token offering to the Madras Presidency. The seeds either failed to germinate or the seedlings perished. Gordon, recalled by the Tea Committee after the discovery of Assamese tea, returned to China to complete his task of procuring plants producing the best black tea and also skilled labour. He was reasonably optimistic since the Dutch had transported a large quantity of plants and seeds and several hundred Chinese to their experimental plantations in Java.

The Tea Committee agreed that an official deputation should go to Assam to verify this discovery of tea. Wallich who was chosen to lead the party, which was also to examine the local flora and soils, had hoped Wight and Griffith would be his companions. In the event, John McClelland, primarily a geologist, was substituted for Wight. They left Calcutta on 29 August 1835, Wallich naturally excited by the tour.

> *In matters of pure botany and horticulture, I anticipate such a harvest as was never seen before. Think only to herborize under the shade of wild tea trees, in forests never in this world examined before*[16].

After four and a half months travelling they reached Sadiya in Upper Assam, close to the Chinese border and the tea provinces of Yunnan. With C. A. Bruce as their guide, they visited five localities where tea shrubs were found in great abundance, analysed soils, and collected seeds. The deputation, having achieved their objectives, dispersed on 21 March 1836. Griffith and McClelland submitted lengthy reports whereas Wallich's observations were confined to a few letters to the Tea Committee. The mutual hostility which had burgeoned during their expedition manifested itself in acrimony and diametrically opposed opinions. Griffith and McClelland instinctively took sides against Wallich over the merits of imported and indigenous tea plants and the ideal climate and soils for their cultivation. Wallich, perhaps remembering the healthy shrub grown from Chinese seed, that he had seen in a garden in Khatmandu in 1821, echoed the preference of Royle and Falconer for the sub-Himalayas; his two colleagues petitioned for Assam. Wallich's proposal to stop the importation of further supplies of Chinese stock was opposed by Griffith who believed that 'a wild plant is not likely to give as good produce as one that has been cultivated for centuries'. There was one small area of agreement—that the tea plant in Assam had originally come from China.

C. A. Bruce, appointed Superintendent of the Tea Forests in May 1836, had, by 1839, located 120 tracts on the hills and in the plains of Assam where tea flourished. The jungle canopy of trees and undergrowth was thinned to bring more light to the shrubs. Chinese labour, skilled in manufacturing black and green teas, was recruited. Griffith's opinion that Chinese plants were superior was accepted by the Government, a decision regretted by a later generation of planters who had inherited inferior crosses between Assamese and Chinese stock.

The Government of India resolved the conflict over suitable localities by ordering that Chinese seed should be sown in all the recommended regions—Assam, the sub-Himalayas, and South India. Falconer grew it in nurseries in Kumaon and Garwhal, at heights between 2000 and 6000 feet. When he returned to England in 1842 on sick leave, he brought with him the first samples of tea manufactured in that part of India. London brokers approved its quality, comparing its fragrance and aroma with the most delicate of Chinese teas. Falconer was followed by Jameson who increased the number of tea nurseries, confident he could persuade villagers to accept tea as another crop as it 'is most hardy, and does not require much care in cultivation'. Archibald Campbell, Joseph Hooker's companion in Sikkim, encouraged tea plantations around Darjeeling. Initially tea in South India was treated as a crop of incidental value. Apart from a few abortive attempts by individuals in the 1830s in Coorg and the Nilgiris, tea planting as a commercial venture in the south was not begun until the 1850s.

In 1838, the year in which C. A. Bruce published his *Account of the manufacture of the black tea . . . in Upper Assam*[17], a shipment of 12 boxes of Assam tea, carefully packed in soldered tin boxes, was despatched to London. The East India Company auctioned eight chests—five of Assam pekoe and three of Assam souchong—the following January. The next consignment of 95 chests of far better quality was put on public sale in March 1840, and may be said to signal the end of the dominance that tea from China had enjoyed for a century and a half. Although the Company marketed these early imports, the commercial exploitation of tea had to await the formation of the Assam Company. It had always been the firm intention of the Court of Directors after the tea trials had proved conclusively the feasibility of the venture, to refer its development to private enterprise. Businessmen impressed by the high prices obtained at London auctions, sought the Company's advice on investing in Indian tea. Shortly after the formation of the Bengal Tea Association in Calcutta in February 1839, a joint stock company—the Assam Company—was launched in London to acquire the East India Company's plantations and equipment in Assam 'for the purpose of cultivating the newly discovered tea plant in Assam'. With the exception of a few small estates, the Court of Directors transferred its plantations to the Assam Company, rent free for ten years, in March 1840. Nine years later the East India Company cut its last link with the Indian tea industry by selling its few remaining gardens to a Chinese planter.

The first sale of the Assam Company's tea took place in London in January 1842 but mismanagement and complete ignorance of tea husbandry blighted this promise of future prosperity. Hundreds of Chinese were recruited indiscriminately in the bazaars of Calcutta and Singapore in the naïve belief that 'every man with a tail was supposed to be qualified to cultivate, manipulate and prepare tea'[18]. Chronic staff shortages were exacerbated by alarming mortality rates among European and native employees. Often the most unhealthy and inaccessible sites were chosen for plantations which, being inefficiently managed, soon reverted to jungle. The whole of its capital exhausted and with £20 shares selling in Calcutta for less than one rupee, the Assam Company was almost bankrupt. The resolute action of three planters saved it from insolvency in 1847; they set about reclaiming gardens choked with weeds, introducing more effective methods of tending the tea bushes, and improving the quality of the manufactured tea. Gradually as stability and confidence returned to the tea market, a flood of applications for land grants presaged a boom in Assam's tea plantations.

While this reassuring recovery was under way in Assam, the prices Jameson at Saharanpur was commanding for his tea, induced the Government to employ Robert Fortune in 1848 to go to China to purchase the very best tea varieties to restock the Himalayan plantations. Robert Fortune (1812–80) had learned his gardening skills at the Royal Botanic Garden in Edinburgh and in the hothouse department of the Chiswick garden belonging to the Horticultural Society of London. This Society had selected him in 1843 as their collector in China, then a little more open to foreigners as a consequence of the Treaty of Nanking which terminated the first Opium War. Amongst his instructions was a directive to collect 'plants that yield different varieties of tea'. His enterprise and resourcefulness on that expedition commended him to the East India Company which engaged him in 1848 to gather green tea seeds and plants and Chinese trained in tea production. He donned Chinese dress, even shaving his head to complete the guise, in order to penetrate Huchow, still relatively inaccessible to foreigners. During the summer of 1850 his first cargo reached Calcutta—the short-lived tea seeds sprouting among mulberries in Wardian cases. He himself sailed for Calcutta with 2000 tea plants, 17 000 seedlings, and six competent tea manufacturers. His second trip during 1853 to 1856 had black tea as its objective. All his plants were restored to vigorous growth at Calcutta's Botanic Garden before being transferred to the Himalayan nurseries.

It was Fortune who finally settled the controversy about whether black and green teas came from different species. The classification and nomenclature of tea has evolved through a complicated permutation of genera, species, and varieties. Linnaeus, on the authority of Engelbert Kaempfer's *Amoenitatum exoticarum politico-physico-medicarum* (1712), which described the tea plant in Japan, established two genera—*Thea* and *Camellia*—in his *Systema naturae* (1735). In his *Species plantarum* (1753), tea was designated *Thea sinensis* but in the second edition (1762), on the assumption that green and black teas originated from different plants, he distinguished two species: *Thea viridis* (green tea) and *Thea bohea* (black tea). Eminent botanists like Sir William Hooker and experienced nurseries such as Loddiges accepted without question the validity of two species. But there were contemporary writers like J. C. Lettsom who disputed this classification. In his *Natural history of the tea-tree* (2nd edn, 1799), he maintained there was only one species, the difference between the commercial teas being the result of soil, methods of cultivation, and processing of the leaves. This observation had already been privately confirmed as early as 1768 by T. Fitzhugh, one of the Company's agents in China. Gordon went to China in 1834 believing that green and black teas were exclusive to different regions of the country. Royle was still uncertain what taxonomic position he should adopt when Jameson at Saharanpur informed him in 1844 that his Chinese assistants assured him that the two kinds of tea were simply the result of different manufacturing processes. On his first visit to China in 1843, Fortune saw for himself how the leaves were dried and treated, proving conclusively that only the manner of treatment was responsible for the commercial varieties. The genus *Thea* has since been relegated to a section within *Camellia*, the correct botanical name for the species now being *Camellia sinensis*. It embraces two varieties: the hardy variety indigenous in China, *C. sinensis* var. *sinensis*, and the more tender one found in Assam, Burma, Thailand, and South China, *C. sinensis* var. *assamica*.

During the early 1860s, speculators were attracted to Assam by the Government's offer of tracts of land on attractive terms for development as tea gardens. Patches of jungle were cleared, planted with seedlings, then promptly sold for a profit, a transaction that was repeated, each time at inflated prices. It seemed an infallible recipe for immediate wealth. Misled by the widely-held belief that anyone could manage a plantation, men abandoned their professions—army and naval officers, engineers, policemen, even a few deputy and assistant commissioners—all infected by 'tea mania'. Confident young men came from England, ignorant not only of tea planting, but also of the customs and languages of the labourers they would be supervising. People invested their savings in the rash of new companies formed in Calcutta. Tea gardens were expanded without adequate labour which was both scarce and expensive; production maintained to the detriment of quality was reflected in a depreciation of market prices. The inevitable collapse came about the mid-1860s: the value of tea gardens and shares plunged, panic selling ensued and plantations were recklessly abandoned. Companies went into liquidation and many people lost their livelihood or their investments. Only the Assam and Jorehaut Tea Companies escaped unscathed. The Government, alarmed by the turn of events, appointed a commission in 1868 to enquire into the state of the tea industry.

Gradually during the 1870s confidence was restored and with it a prosperity that owed a great deal to those planters who were prepared to spend a solitary and spartan existence in isolated bungalows, dosing themselves regularly with quinine to ward off malaria. The crisis had taught the industry a salutary lesson: tea gardens henceforth were competently and profitably managed. Exports which were only 232 000 lbs in 1852, rose above 76 000 000 lbs in 1886 and for the first time in 1887 tea imports from India and Ceylon exceeded those from China. Ceylon, twice almost ruined, first by the coffee disease and then by the fall in the world price of cinchona, was saved by its tea plantations, established in 1883. Tea was one of the most profitable undertakings of British India and C. A. Bruce deserved the gold medal awarded him by the Society of Arts in London for his discovery[19]. This presentation was challenged by Captain Charlton who claimed that honour for himself. Wallich who had to mediate, tried to be fair to both candidates: Bruce undoubtedly first collected it, he declared, but it was Charlton who supplied the evidence to make a positive identification. It is doubtful that Charlton was placated by the award of the medal of the Agricultural and Horticultural Society in Calcutta.

China, enjoying a more or less self-sufficient rural economy and with little need for European goods, insisted that its tea and silk exports should be paid for in bullion, preferably silver. In 1801 the East India Company spent £3.6 million on Chinese tea and subsidized the imbalance in trade through the illicit sale of Indian opium to China. The cultivation of the opium poppy (*Papaver somniferum*) (Fig. 17.2) had been brought to India by the Arabs in the sixth century. The Moghul emperor, Babur, drank opium in rose-water; his son, Humayun, and his great grandson, Jahangir, were addicts. Their subjects treated it both as a medicine and an aphrodisiac. Rajput soldiers went into battle drugged with the narcotic.

The principal poppy growing regions were the Ganges valley, Rajputana, and central India. Dutch merchants who had bartered opium for Malabar pepper eventually wrested this lucrative drug trade from the Portuguese, becoming the main exporter of the Bihar poppy crop. Their virtual monopoly passed to the East India Company after the battle of Plassey in 1757 had established British dominance in Bengal. Company servants at Patna in the neighbouring state of Bihar supplemented their salaries with private transactions in opium until 1772 when they were expected to sell opium to the Company at less than market rates. This source of friction and the adulteration of opium prompted Warren Hastings to assume control on behalf of the Company of all the opium produced in Bihar, Orissa, and Bengal. It was the last produce of commercial significance to be converted from a private into a public monopoly. Opium contracts,

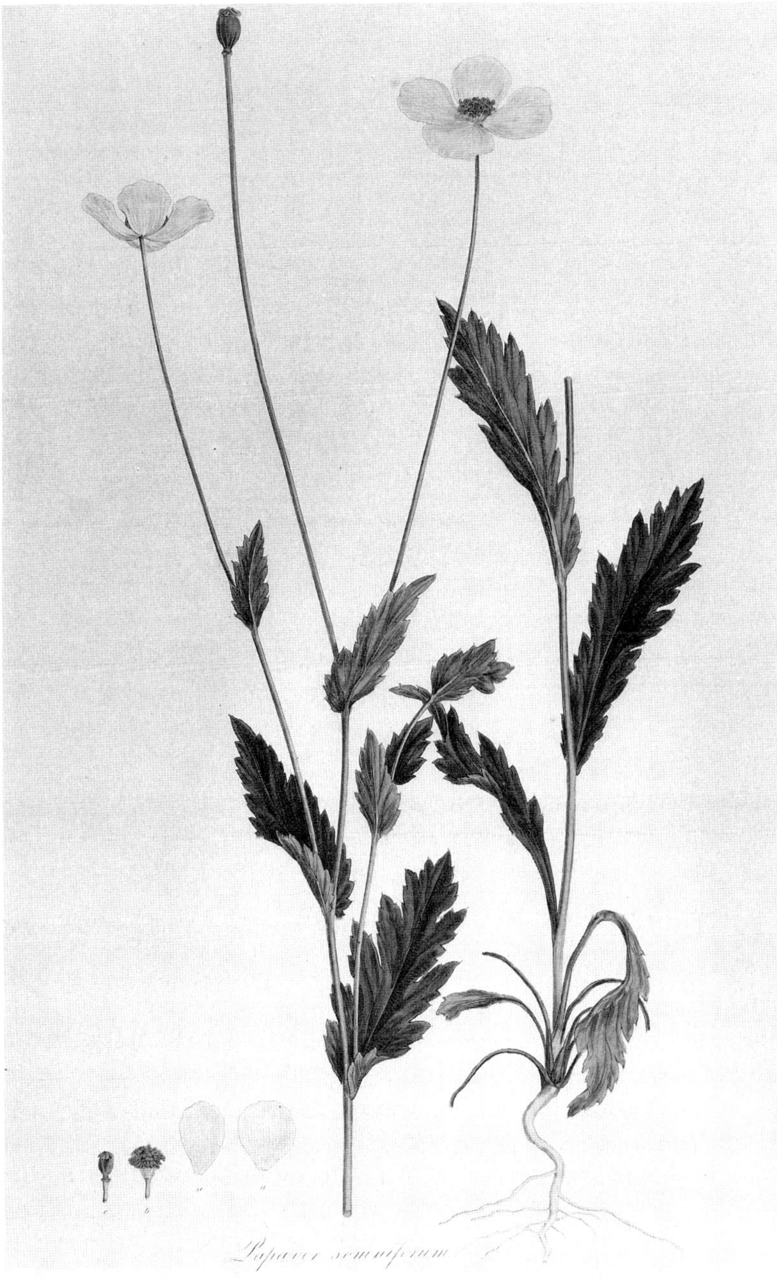

Figure 17.2 Opium poppy, *Papaver somniferum*. (*From J. Sibthorp and Sir J. Edward Smith (1825), Flora Graeca*, Vol. 5, London, plate 491.) Handcoloured engraving of a drawing by Ferdinand Bauer.

tenable for four years, were sold by public auction until abuses inherent in this system led to their abandonment in 1797 and the substitution of Government agencies. The Government jealously protected its complete monopoly of production and sale; unauthorized cultivation of the opium poppy was prohibited and any Europeans detected in illicit opium dealings risked being sent back to England. During Hastings's administration official opium transactions ranked third in importance in the revenue of British India. In 1826 the Finance Committee of the Company estimated a potential annual income of a million pounds from opium sales. Some of it was sold in India, a modest trade was conducted

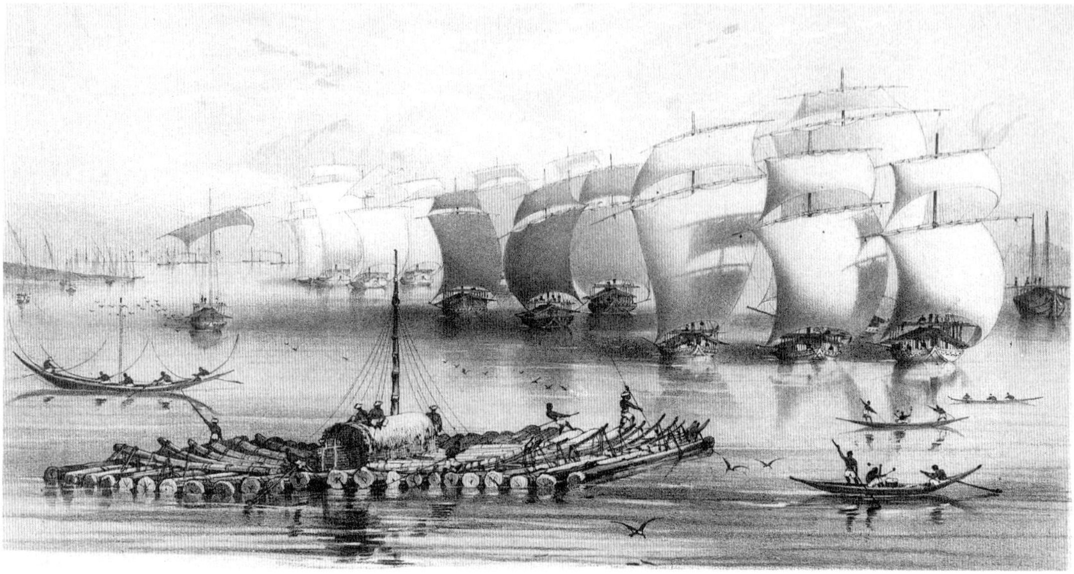

Figure 17.3 The opium fleet approaching Calcutta. The raft floated down river from Nepal's forests was destined to be made into packing cases for the opium trade. Lithograph of a drawing by W. S. Sherwill.

with Malaya and the Dutch East Indies, but the bulk of it went to China (ig. 17.3).

The Chinese had been cultivating the opium poppy from about the fifteenth century, originally for culinary and medicinal purposes; its use as a narcotic was forbidden by a 1729 edict but this and subsequent prohibitions failed to suppress the practice. While China was an obvious market for India's surplus opium, the Court of Directors expressed their displeasure on learning that opium had been taken on a British frigate and sloop to China in the 1760s.

As opium is prohibited by the Chinese Government to be imported in that kingdom, and the introducing it there can only be in a clandestine manner we have too much reason to apprehend very bad consequences may result to the Company by the embarrassment of their affairs with the said Government[20].

The Directors' concern on that occasion may have been sincere but the profits to be made soon banished any such qualms. In July 1782 the *Nonsuch* sailed for China with 1601 chests of Patna opium. Warren Hastings cautioned 'that the *Nonsuch* will enter the river at China as an armed ship, and will not be reported as bearing a cargo of opium, that being a contraband trade'. By 1785 the annual production of Indian opium averaged about 3500 chests (Fig. 17.4), and the Court of Directors acknowledged that it made a significant contribution to the country's export trade.

Having already given our opinion that this article is undoubtedly a proper object of encouragement for foreign commerce and remaining of the opinion contained in our letter to you of 12 July 1782 that the Company cannot on any account engage directly in an

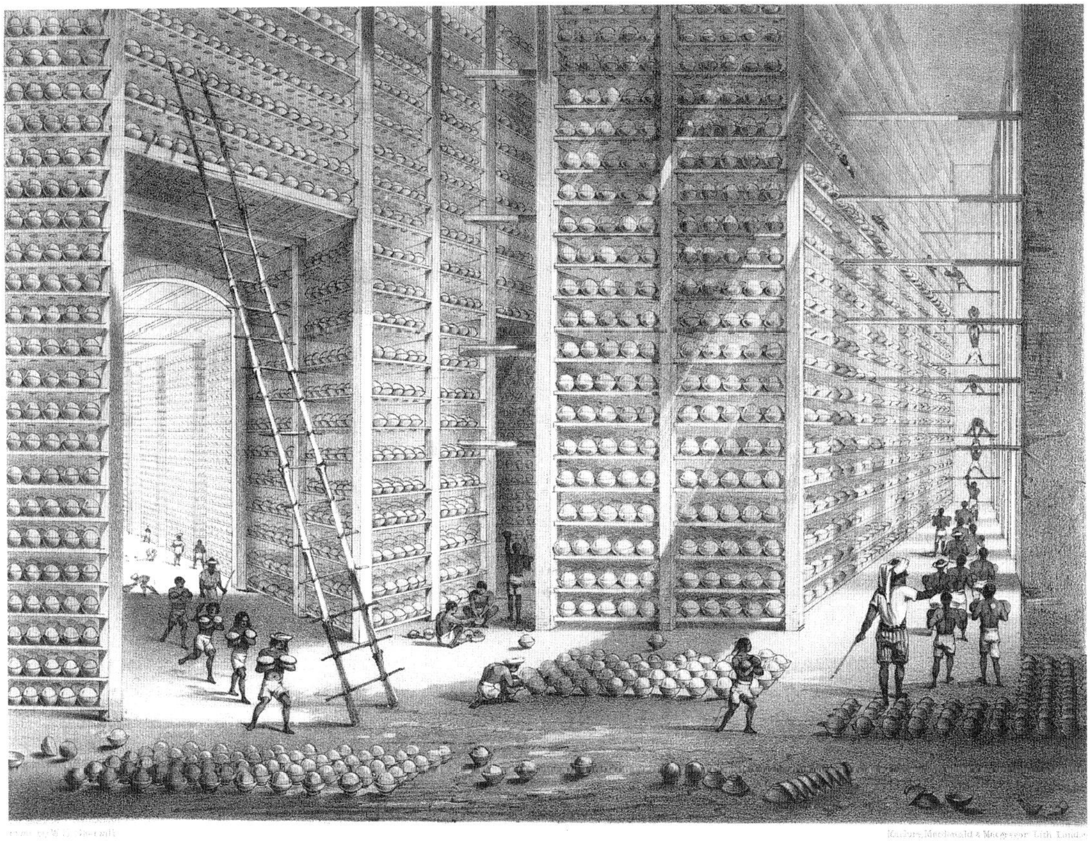

Figure 17.4 Balls of opium in a factory at Patna awaiting despatch to Calcutta for shipment to China. Lithograph of a drawing by W. S. Sherwill.

illicit traffick in it to China, where it is prohibited, it remains to be considered in what manner, consistently with those opinions, this produce of our province can be most beneficially converted to the purposes of foreign commerce and particularly of a supply to our Treasury in Canton[21].

A neat solution was to leave the trade to private merchants using non-Company vessels and to deny blandly any knowledge of the traffic. Nevertheless, the Company was still in control since no ship at that time could trade with China without its licence. In 1800 when exports had risen to 4000 chests the Chinese promulgated yet another edict, this time banning the import of opium on any ship. This was circumvented by wholesale smuggling with the connivance of corrupt Chinese officials. Other nationalities also wanted a share of the market, foremost among them American tea merchants. The Chinese Government, now alarmed by an excessive number of addicts and the drain on its silver resources, in desperation seized and burnt contraband opium in Canton and arrested a few British sailors. Retribution came swiftly: 20 armed ships filled with troops sailed to China in 1839. In a patently ill-matched confrontation, the Chinese were forced to capitulate in 1842, made to pay compensation for the confiscated opium and to cede the first of the treaty ports including Hong Kong. In 1857, the year before the outbreak of the second Opium War, India exported well in excess of 60 000 chests of opium to China. The British involvement in the opium traffic did not end until 1917, ultimately yielding to Chinese resistance and international disapproval.

NOTES

[1] R. Richardson (1835), *Extracts from the literary and scientific correspondence of Richard Richardson* (ed. D. Turner), 391.

[2] Banks to E.I.C., 25 November 1788. I.O.L.R. E/4/634, folios 41–3.

[3] Carter (1988), 271.

[4] Banks to E.I.C., 27 December 1788. B.M. (N.H.) D.T.C., Vol. 6, folios 103–4.

[5] Kyd. Remarks on the President of the Royal Society's propositions for the introduction of the tea plant into the Company's provinces. 23 September 1789. Kew archives.

[6] Banks to E.I.C., 17 January 1791. I.O.L.R. E/1/86, folio 57.

[7] Sir G. L. Staunton (1797–8), *Authentic account of Embassy . . . to Emperor of China*. London.

[8] Banks to Roxburgh, 9 August 1798. B.L. MS 33980, folio 159.

[9] *Royal Central Asian Journal*, **49** (1), 1962, 50.

[10] House of Commons. Tea cultivation (India), 1839, 10.

[11] Ibid., 5.

[12] Jenkins to Committee of Tea Culture, 7 May 1834. House of Commons. Tea cultivation (India), 1839, 34.

[13] Charlton to Jenkins, 8 November 1834. House of Commons. Tea cultivation (India), 1839, 35.

[14] House of Commons. Tea cultivation (India), 1839, 32.

[15] D. Scott, presumably to Wallich, n.d. Kew Herbarium. Letter attached to Wallich herbarium sheet no. 3668.

[16] Wallich to J.C. Loudon, 16 February 1835. *Gardeners Magazine*, **11**, 1835, 29.

[17] *Account of the manufacture of the black tea, as now practised at Suddeya, in Upper Assam, by the Chinamen sent thither for that purpose, with some observations on the culture of the plant in China, and its growth in Assam*. Calcutta.

[18] *Friend of India*, 9 September 1841.

[19] In 1822 the Royal Society of Arts offered a gold medal to the person who discovered the Chinese method of manufacturing tea.

[20] E.I.C. to Government of Bengal, 26 March 1766. I.O.L.R. E/4/618, folio 287.

[21] E.I.C. to Government of Bengal, 27 March 1787. I.O.L.R. E/4/59, item 1343, para. 205.

RUBBER, COFFEE, AND FORESTRY

No one today would dispute the immorality of Britain's defiant continuance of the opium trade in the face of official Chinese opposition. It was a sordid episode in British imperialism, yet it had its supporters apart from the principal beneficiaries, the Government of India and British and Chinese merchants. They even included some deluded Christians. The German missionary, Karl Gutzlaff, convinced himself that the 'traffic would tend ultimately to the introduction of the Gospel'. Religious tracts often accompanied each case of opium. An American missionary compared the British expeditionary force in 1839 to 'God's Axe' and hailed this intervention as 'The Divine will'. The Government of India casually refuted all charges of unethical behaviour. It argued that Chinese officials were more concerned about the drain on their bullion than about the drug's pernicious effects. The Bengal Finance Minister in 1862, defending the export of Indian opium, reassured his audience that 'as the Englishman is a drinker of beer, so is the Chinaman a smoker of opium'. Some advocates of opium went so far as to claim positive benefits. G. H. M. Batten of the Bengal Civil Service saw it as a means 'to resist fatigue, to sustain privation, and to save life'. The botanists George Christopher Birdwood and George Watt perceived no harm in its moderate use, wilfully ignoring the tendency to progressive addiction. The Inspector of Hospitals of Hong Kong in his report for 1881 gave his opinion 'that opium smoking is a luxury of a very harmless description'. He doubted whether 'the dull, sodden-witted, debilitated opium smoker' existed in great numbers. But this complacency was by no means universally shared. James Kerr in his 1773 submission on opium cultivation in Bihar confirmed that 'too frequent use emaciates the person and a languid stupification appears in his countenance'[1]. C. A. Bruce, superintending the tea plantations in Assam, saw 'this vile drug' frustrating his efforts to recruit dependable labour.

Back in Britain public opinion was also divided. The young Gladstone led a vocal opposition to a 'most infamous and atrocious trade'. Lord Palmerston supported the recommendations of a Parliamentary Committee of 1832 that it was 'inadvisable to abandon so important a revenue'. When the Society for the Suppression of the Opium Trade was founded in 1874 with Lord Shaftesbury as its first President, it had the support of the Archbishop of Canterbury and Cardinal Manning. Castigated by one opponent as 'an irresponsible party of philanthropists', this society succeeded in getting the issue raised in Parliament. A motion in 1891 that 'the system by which the Indian opium revenue is raised is morally indefensible' was lost only on a technicality. Gladstone, now Prime Minister, was more politically circumspect. In 1893 he attempted to reduce tension and delay action by the time-honoured device of appointing a Royal

Commission. This Commission, manned by his supporters, predictably 'found the evil effects of opium in India greatly exaggerated'. Moreover, not only did it decline to recommend any measures to ban or restrict the trade but it disingenuously blamed the Chinese for any addiction in their country.

The stance adopted by British officials in the opium trade was as morally reprehensible as the indigo planters' coercive measures against peasant farmers, but the outright condemnation by some historians of so-called 'plant imperialism' is debatable. The charge is that British and other Western colonial powers undermined certain trade monopolies enjoyed by Latin America by unlawfully taking their plants for commercial exploitation. Although cochineal dye production was a protected industry in Brazil, it did not deter the British from trying to establish the insect and its host plant in India. There are, however, authenticated reports of the cochineal insect being collected by British agents not by subterfuge but quite openly in Rio de Janeiro. Lucile H. Brockway in 1979[2] questioned the ostensible motivation behind the transfer of the cinchona—to save the tree from extinction in South America and to make quinine easily accessible to all who needed it in India. Markham always indignantly repudiated the imputation that his intentions were anything but humanitarian.

In obtaining plants and seeds of the valuable chinchona from South America, it would have been a source of deep regret to me if that measure had been attended by any injury to the people or the commerce of Peru, Ecuador, or Colombia . . . Hitherto they have destroyed the chinchona trees in a spirit of reckless short-sightedness, and thus done more injury to their own interests than could possibly have arisen from any commercial competition[3].

He denied that his removal of cinchona plants and seeds had been illegal. There was at that time, he said, no Peruvian law prohibiting their exportation and cited the authorization he had received in June 1860 from a Peruvian minister. Spruce had collected his *Cinchona succirubra* before the ban imposed by the Government of Ecuador. On some of his trips Robert Cross had, however, contravened Government prohibitions. Markham has been accused of manipulating the facts and there can be little doubt that the legality of the whole transaction was suspect, but to their credit the British did not keep the cinchona enterprise to themselves. They generously gave plants and seeds to the Dutch, French, and Portuguese, and to Emperor Maximilian of Mexico. But the *cause célèbre*, the episode that every schoolchild knows, is the story of the 'smuggling' of rubber plants out of Brazil. Once again Markham was a key figure.

Michele do Cuneo, who accompanied Colombus on his second voyage to America in 1493, wrote home about 'trees that give milk when cut, of which they make a kind of wax, which we used'. The first mention of rubber in print appeared in Peter Martyr's *De orbo nova* (1511) where he described Amerindians playing with bouncing balls made from the juice of a vine. Spanish colonists in seventeenth-century Mexico discovered how to waterproof textiles with the 'milk' or latex but failed to prevent its melting in the sun. When de la Condamine, a scientist attached to a French expedition to South America in 1735, travelled at night, his way was illuminated by torches made of rubber. His report in the *Mémoires* (1745) of the Académie Royale des Sciences noted this and also the bottles and boots Indians made from the sap of Caoutchouc or 'weeping wood'. The French botanist, Aublet, named the tree he found in French Guiana, *Hevea guianensis*, adapting the native name 'Hhévé'. When de la Condamine's account of Caoutchouc was published by the Académie, French boys were already playing street games with balls made from this substance. It is supposed to have acquired the name 'rubber' when the scientist, Joseph Priestley, discovered it erased pencil marks from his manuscripts. Rubber manufacturers first applied it to clothes and one of them—Charles Mackintosh—in 1823 patented his invention of waterproofing fabrics by bonding two layers together, thereby perpetuating his name in a familiar garment. When Nelson Goodyear discovered in 1839 how to make rubber tough and durable by a process of vulcanization, the rubber industry entered an era of expansion and prosperity. Solid rubber tyres attracted attention at the Great Exhibition of 1851 and soon engineers found every conceivable use for it.

South America was not slow to respond to the demands of a new market; Peru began exporting rubber in 1853, joined a decade or so later by Bolivia and Venezuela. In 1870 the Amazon basin produced about 15 000 tons of rubber. There were three commercial rubbers: *Castilla elastica* from Central America and *Manihot glaziovii* and *Hevea brasiliensis* from South America. Because of its superior strength, its high yield of latex and its ability to survive repeated tappings, *Hevea* was favoured. Richard Spruce, who resided in Brazil from 1849 until 1864, and had witnessed the frantic rubber boom in the province of Pará, was the first botanist to gather *Hevea* species on the Amazon and Rio Negro. His specimens were described by George Bentham in Hooker's *London Journal of Botany* in 1854 and by J. Müller of Aargau in *Linnaea* in 1865. Plants yielding latex were not confined to the Americas; West Africa and Mozambique had various species of *Landolphia* and *Ficus elastica* grew in Assam, Burma, and Java.

William Roxburgh became aware that there was a tree in South Asia producing latex when he received from Matthew R. Smith in Sylhet in 1810 a crudely-made basket, the inside of which was thinly lined with a substance he identified as Caoutchouc. The tribals who had made it obtained the rubber solution from gigantic trees with distinctive aerial roots which they felled or savagely mutilated in their determination to get the last drop of latex. Unfortunately, no notice was taken of Roxburgh's paper until Lieutenant Charlton confirmed in 1832 that *Ficus elastica* grew in Assam's forests (Fig. 18.1). Four years later the London Caoutchouc Company was formed to meet the expanding demand for rubber in the English market. Through the agency of Royle, whom they had employed as their botanical adviser, they offered rewards for a hundredweight of Assam rubber of acceptable quality. In 1838 they awarded 25 guineas to a Lieutenant Vetch for a small sample of Assamese rubber while the Agricultural and Horticultural Society presented him with their gold medal. Little was done to organize a viable rubber industry in Assam; trees were destroyed or over-tapped, there was no systematic replanting programme and production inevitably declined. The formation of plantations on a large scale was recommended by Dietrich Brandis, Inspector-General of Forests in India, in his observations on James Collins's *Report on the Caoutchouc of commerce*; but *Ficus elastica*, which was also tried by the Dutch in Java, could not compete with the superior *Hevea*. Collins, who had written two papers on rubber in 1869 and 1871 and was Curator of the Museum of the Pharmaceutical Society in London, was asked by Markham to report on the current state of the rubber industry. Markham later claimed that it was in 1870 that he had resolved that an operation similar to the one mounted for cinchona was needed for the survival and development of rubber[4]. He had witnessed the maltreatment of trees through slashing and over-tapping during his first visit to Peru in 1853. Collins's *Report* printed at the request of the Secretary of State for India in 1872, was a digest of all available information on the American, African, and Asian

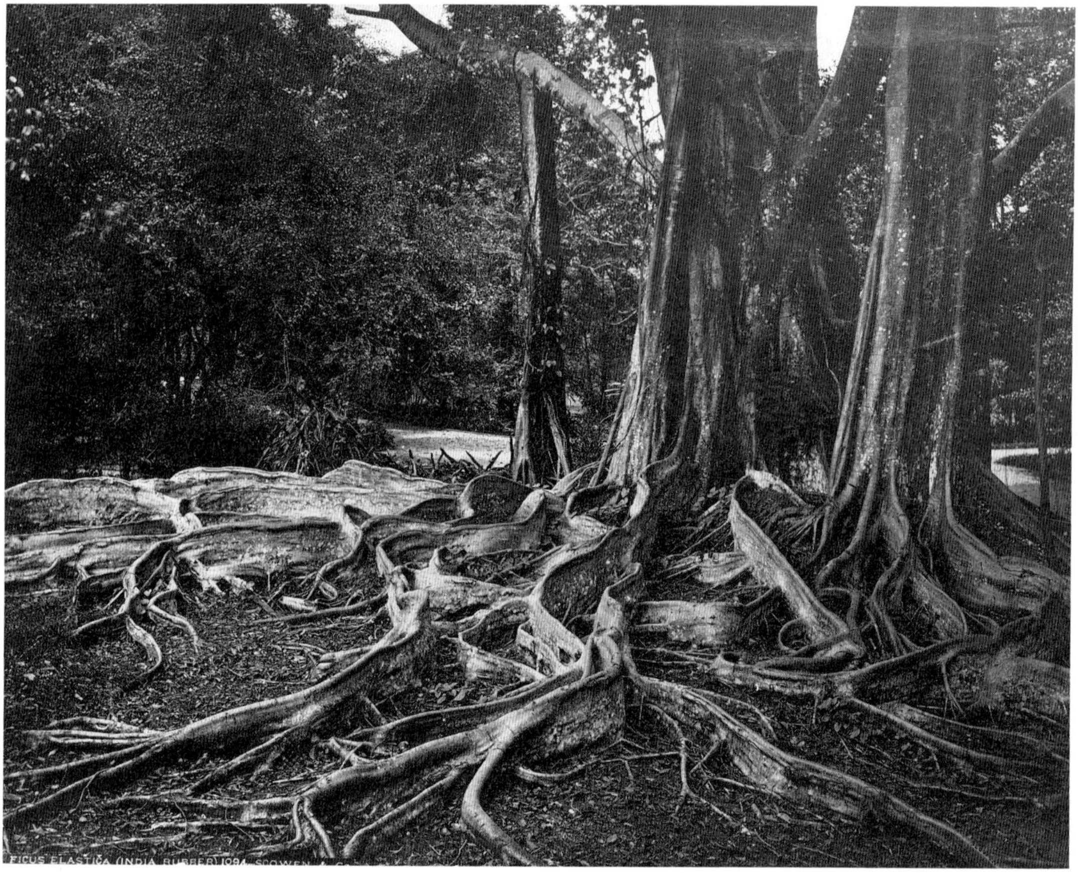

Figure 18.1 *Ficus elastica,* known as the 'snake tree' because of its sprawling roots above ground.

plants and their methods of cultivation. He favoured the introduction into India of *Hevea* and *Castilla* species; the latter, he thought, would do well in Assam, but he left the choice of location of *Hevea* to others better informed on Indian topography and climate. Brandis's memorandum which was appended identified Ceylon, Malabar, Travancore, and Burma as suitable habitats.

The India Office sent a copy of this report to Joseph Hooker at Kew in May 1873, at the same time requesting the British Consul at Pará in Brazil to obtain *Hevea* seeds for transmission to India. Hooker's opinion on the development of *Ficus elastica* plantations and on the introduction of foreign rubber plants to India was sought. He was also asked whether *Hevea* seeds should be first sown at Kew and the seedlings sent to India, confirmation that Kew was seen from the outset as central to the project[5]. At Collins's instigation, a Mr Farris arrived from Brazil in June 1873 with a package of *Hevea* seeds. Only about a dozen germinated at Kew and in September half of them were taken in a Wardian case by George King, returning to the Royal Botanic Garden in Calcutta. They succumbed in Bengal's dry, cool winter and cuttings from them which were sent to Sikkim also died. It was then decided that Brandis's advice should be followed, and future consignments should go initially to Ceylon.

More seeds and plants were urgently required if the momentum of the programme were to be maintained. It so happened that

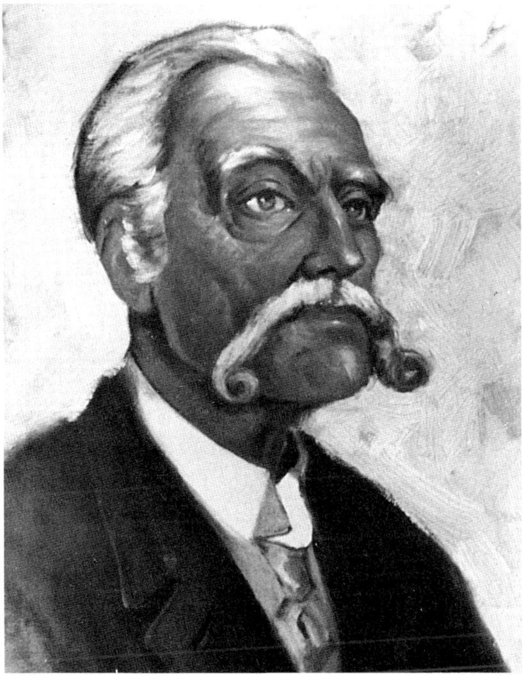

Figure 18.2 Henry Alexander Wickham (1846–1928).

an Englishman, Henry Alexander Wickham (1846–1928) (Fig. 18.2), trading in the Amazon, was known to Kew through a small gathering of plants he had donated 'of possible interest to the British Empire'; moreover his book, *Rough notes of a journey through the wilderness from Trinidad to Para* (1872), included his simplified drawing of the foliage and fruit of *Hevea brasiliensis* (Fig. 18.3). The British Consul at Pará asked Wickham whether he would accept a commission from the India Office to collect rubber seeds for India. Wickham, who was at that time developing a coffee plantation at Santarem about 500 miles in the interior, consented but because of the short viability of rubber seeds, suggested that he germinated them first and transported the seedlings to England. Protracted negotiations about his fees, finally settled at £10 per 1000 seeds (the India Office was reluctant to accept his proposal to ship plants), delayed his first consignment until 1876. He obtained his seed of *Hevea brasiliensis* in the forests covering the broad plateau dividing the Tapajos and Madeira rivers. They were packed in well ventilated crates of split *Calamus* canes, between layers of dried banana leaf. Speed was now essential if the seeds were to reach England alive. In the name of the India Office he chartered an ocean-going steamer and circumvented any bureaucratic delays by the port authorities at Pará by declaring that he had on board 'exceedingly delicate botanical specimens designated for delivery to Her Britannic Majesty's own Royal Gardens at Kew'. These formalities cleared in the friendliest manner, the ship proceeded to the open sea, 'the hatches off with the open-work crates slung up on lines fore and aft in the air and free of danger from ship's rats'[6]. This episode has been denounced as yet another example of smuggling economic plants out of an unsuspecting country. P.R. Wycherley, who has researched the relevant archives, refutes the charge. He claims that the shipment of *Hevea* seeds was not prohibited at the time and that subsequently an official statement—*Borracha no Brasil* (1913)—confirmed that Wickham's operation had had the authorization of the Brazilian authorities[7].

The cargo of 70 000 *Hevea brasiliensis* seeds, which was transferred to a goods-train at Liverpool docks, reached Kew Gardens on 14 June 1876. They were all sown the next day in the propagating house which had been made ready. 'We knew it was touch and go, because it was likely the seeds would not germinate', recalled W. Thiselton-Dyer, then Assistant Director. 'I remember well on the third day, going into the propagating house where they were planted and seeing that by good luck the seed was germinating.'[8] A goodly number of the 2700 seeds that had been potted were, within a few weeks, 15 inches high. In fact they grew so vigorously that larger cases for their shipment to Ceylon

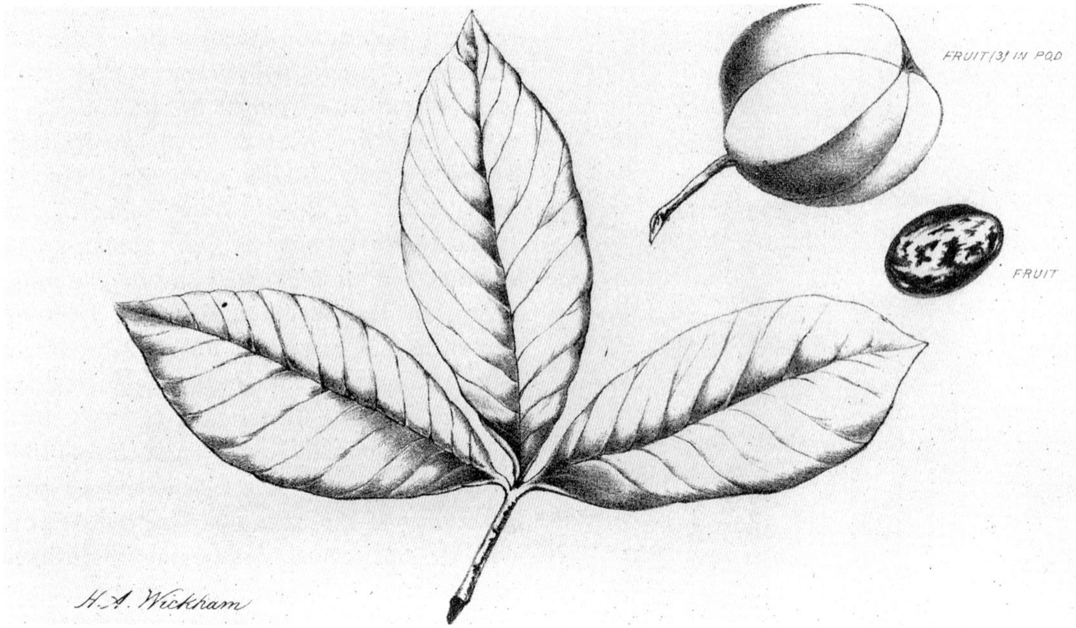

Figure 18.3 H. A. Wickham's sketch of the leaf and fruit of *Hevea brasiliensis*. (From H. A. Wickham (1822), *Rough notes of a journey through the wilderness* . . . . , London, p. 295.)

had to be made to allow for their anticipated growth during the voyage. On 12 August 1876, 38 Wardian cases of *Hevea* containing 1919 plants and a case of *Castilla* were loaded on a ship bound for Colombo, and ninety per cent of the consignment reached Ceylon in perfect health.

When negotiations with Wickham about his fees delayed the shipment of seed, Markham, at the India Office's behest, found another collector. He had no hesitation in re-employing Robert Cross who had proved his reliability in the cinchona transfer. In May 1875 Cross left for Central America seeking *Castilla elastica*. None of the 7000 or so seeds he collected germinated, but his three boxes of cuttings survived and were distributed between West Africa, Ceylon, and Java. In June 1876, almost coinciding with the arrival of Wickham's seeds, he sailed direct to Pará for *Hevea*. Five months later he returned with just over a thousand *Hevea brasiliensis* seedlings bedded in decayed leaves mixed with wood ash plus a bonus—42 *Manihot glaziovii* cuttings. Four hundred of the *Hevea* were retained at Kew and the remainder presented to Bull's nursery in Chelsea, but of the entire delivery scarcely three per cent were saved. About a hundred of the Kew plants subsequently went to Ceylon.

King at Calcutta and Hooker at Kew had agreed that Ceylon was the most suitable region in South Asia for the propagation and distribution of rubber to India and Burma. A new garden at Henaratgoda, just above sea-level, was opened in 1876 to receive the rubber introduction. Wickham's seedlings, first planted in bamboo pots at Peradeniya, were transferred to Henaratgoda the following year. When Trimen succeeded Thwaites in 1880 only 300 slender rubber trees remained of the original planting; the rest had been distributed or lost. The first Ceylon-grown seeds were garnered in 1881; in June 1882 Trimen presented Kew with a sample of rubber from *Manihot glaziovii*, the first produced in the island. At the outset, planters in Ceylon through ignorance of cultivation and tapping procedures and the low prices in the world's rubber market were cautious about investing in a new cash crop. Their indifference caused Trimen to complain to Kew in 1887: 'We do not know what to do with *Hevea* seed—no one here wants it'[9]. John Dunlop's invention of the pneumatic tyre in 1888 and the growth of the car tyre industry produced a world boom which even the tea planters in Ceylon could not resist. By 1903 over 3000 acres were planted with *Hevea* and briefly Ceylon became the world's third largest supplier.

But Ceylon's significance in the story of rubber was as a distributor rather that as a producer. The recipients of its seeds and plants included India, Burma, the Andaman Islands, Penang, Singapore, North Borneo, Java, Fiji, Queensland, Jamaica, and the German East Africa Company. Singapore was to be the most important destination, the birthplace of the Malayan rubber industry. The first batch of *Hevea* to Singapore, sent by Kew in 1876, died in transit. Kew had chosen Singapore in order to give the rubber seedlings 'an additional chance of being established in the East'[10]. Another attempt in 1877 with 22 plants succeeded. In 1888 when Ceylon sent Singapore 11 500 seeds from which 8000 plants were raised, Henry Nicholas Ridley (1855–1956) arrived as the new Director of the Botanic Garden. His method of tapping the rubber trees, considerably increasing the latex yield, superseded that used in Brazil and is still employed today. Ten years earlier Thiselton-Dyer had confidently predicted the successful outcome of the rubber transfer.

> *It appears therefore that while upwards of 2000 plants are safely established in Ceylon, smaller parcels are also growing in Burma, Calcutta, Mauritius and Singapore ... The plant is now therefore to be regarded as definitely established in the East Indies*[11].

Retaining only a residual stock, Kew's participation was now officially over. All future requests for *Hevea* were to be met by Ceylon. With the Japanese occupation of South East Asia, British and Dutch domination of the rubber market came to an end, and with the advent of synthetics so did the supremacy of natural rubber.

Of all the crops tried and developed in Ceylon, it was 'king coffee' that briefly underpinned the island's economy and, when fungus disease killed the plants, ruined many growers. *Coffea arabica*, notwithstanding its specific name, is probably a native of Ethiopia and was domesticated by the Arabs who introduced Indians to the fragrant bean. When Edward Terry was in Western India from 1616 to 1619, he noticed that natives who abstained from alcohol for religious reasons, indulged in 'a liquor more healthfull than pleasant . . . : a blacke seed boyled in water, which doth little alter the taste of the water'[12]. By the early eighteenth century coffee was being cultivated on the Malabar coast and in Ceylon. In 1699 the Dutch had shipped some coffee plants from Malabar to Java. Two trees were sent to the botanical garden in Amsterdam in 1706 and their seeds provided the initial stock for the coffee plantations of the Dutch and French possessions in the New World. The Dutch coffee plantations in Ceylon were far more profitable than those in Java and stimulated competitive action by British India. Under the authority of the Governor of Bombay, Charles Whitehill left for Mocha in the Yemen in 1761 to obtain coffee plants. When these died on the voyage, several more attempts were made by the Bombay Presidency to acquire saplings planted 'in boxes of mould taken from the soil in which the trees grow'. In 1786 Robert Kyd in Calcutta was appealing for 'some plants of the best species [of coffee] from Mocca, with a native of that country to show the method of cultivating this valuable article'[13]. At his suggestion, a small vessel was fitted out for the sole purpose of bringing back to Bengal 'the best species of coffee, only obtainable from Arabia[14]'. In 1798 Murdock Brown, overseer of the Company's 200 acre plantation at Randaterra in Malabar, applied to the Governor of Bombay for permission to grow pepper, cotton, and coffee. C.W. Young, a former coffee planter in the East Indies and Malacca, was leased land in the Madras Presidency in 1805 to cultivate coffee at his own risk.

Roxburgh, who had successfully grown coffee on his experimental plots at Samalkot in the South, discovered that coffee preferred shade in the Bengal teak plantations. The Calcutta Botanic Garden propagated thousands of young plants for cultivation by somewhat unco-operative native farmers[15]. Wallich, anxious to redress this neglect of coffee within the Presidency, readily agreed to advise a Calcutta merchant, George James Gordon, in trials in 1823, 'to afford satisfactory evidence as to the expediency of engaging extensively in coffee plantations in Bengal'[16]. Some businesses as well as individuals experimented with coffee. The Agricultural and Horticultural Society encouraged its cultivation in Assam. In South India coffee diversified the crops of Mysore and the Nilgiri Hills. Following instructions from the Governor of Ceylon, Alexander Moon established the first British coffee plantation on the island in 1824. The availability of cheap land and the removal in 1835 of a differential duty favouring West Indian coffee, precipitated a boom—colonial officials, Company servants in India, and even the clergy purchased land or invested their savings, and speculators from England arrived by every ship. The removal in 1847 of duties that protected the island's product from the competitive coffee of Brazil and Java, alarmed many planters who hastily abandoned their plantations. The fluctuating fortunes of coffee in the world market governed Ceylon's economic progress for half a century: hills were stripped of their forest cover to satisfy an insatiable demand for land, new roads were constructed to the plantations, and migrant labour from South India swelled the population. During the 1870s when coffee leaf rust devastated a quarter of a million acres of plantations, a desperate situation was averted only through the timely substitution of cinchona and then tea. The Government of Ceylon faced with the folly of dependence on a monoculture, followed Sir Joseph Hooker's advice by appointing a commission in 1881 to recommend a diversification of crops. Kew sent some Wardian cases filled with cacoa (*Theobroma cacoa*) from the West Indies; mahogany, teak, tobacco, vanilla, and cinchona were other crops specifically recommended by Hooker and his deputy, Thiselton-Dyer.

Coffee planters in South India cleared virgin forest by felling and burning. One of them years later recalled that 'burning a clearing of 150 or 200 acres in extent was usually a grand sight, the huge columns of smoke and flame rising to an immense height, and sometimes destroying the surrounding standing forest to a considerable extent'[17]. Yet when he saw the Ghats gradually being denuded of their vegetation by this wanton destruction and fertile soil washed down to the sea, he wished the Government of Madras had more firmly limited the encroachment of new plantations. The tea planter's axe rang in the primeval forest in the Himalayan foothills. India's forest cover had been disappearing over several millenia as a pastoral society sought land for their animals and crops. The ancient epic, the *Mahabharata*, tells of the burning by man of the great Khundava forest which covered much of the country between the Ganges and Jumna rivers. The traditional cycle of *kumri* cultivation—the felling of woodland, planting seed in the ash, and after two seasons repeating the operation elsewhere—was still practised by villagers during the Company's administration. At first British officials took no restraining action, judging the seemingly inexhaustible forests as a hindrance to agricultural development.

For centuries Indian teak, sandalwood, satin-wood, ebony, and blackwood or rosewood had been exported to Arabia and Persia, and the East India Company naturally wanted a share of this valuable trade. In 1799 the Court of Directors requested samples and commercial data regarding woods suitable for finely-crafted furniture and mathematical instruments, but they were really concerned about timber for maritime and military purposes or for public works. As early as 1796, through a Company surgeon's initiative, a timber syndicate had been formed in Malabar. An acute shortage of English oak for ship-building stirred the Court of Directors to enquire about the teak resources of Malabar. A Plantation Committee, convened by Warren Hastings, busied itself with the practical aspects of forest management. When an investigation by a Forest Committee in the Bombay Presidency revealed that the supply of mature teak trees was less than had been estimated, the Bombay Government took charge of the forests. In November 1806 a police officer, Captain Watson, was appointed the first Conservator of Forests in India. Work had already begun in the Bombay docks on the experimental construction with indigenous teak of a ship of the line and a frigate. The teak monopoly Watson established in Malabar and Travancore was resisted by independent timber merchants and Watson's conservatorship was eventually abolished in 1822.

The strength, lightness and durability of teak made it versatile timber (Fig. 18.4). It is often stated that the demand for teak had led to the formation of the Botanic Garden at Calcutta, a statement that is not supported by any archival evidence. Colonel Kyd, the founder of the garden, casually mentioned teak among a host of plants he hoped to grow. Nevertheless, in 1797 Roxburgh had several thousand teak trees there and, moreover, had liberally distributed plants and seeds within Bengal and Bihar. The vast teak resources of Burma passed to British control following the country's annexation. When Tenasserim and Martaban provinces were ceded in 1826, Wallich sent there to report upon the forests, judged them superior to any in British India. Remembering the devastation wreaked by timber merchants in the Sal forests of the North West Provinces, he was acutely sensitive to the consequences of uncontrolled felling. But his code of conservation was ignored by speculators who were leased tracts of the forests. The devastation appalled Johann Wilhelm Helfer who carried out another examination of the Tenasserim forests in 1837, but his earnest recommendation that properly managed plantations should be created was ignored, the spoliation continued, and Falconer's visit in 1849 confirmed a belated recognition of the disastrous results of the license system. When in 1852, on the conclusion of the second Burmese War, the province of Pegu was annexed, the Government, now anxious to avoid what had happened in Tenasserim, claimed the ownership of the forests, appointing John McClelland as superintendent. McClelland's report which condemned private enterprise's exploitation, prepared the way for Lord Dalhousie's historic memorandum of 1855 which defined for the first time an official policy for the management of government forests.

To some extent regional conservation was already operating in the subcontinent. During the 1830s the Rajah of Travancore, following the advice of a Scottish forester, stipulated that of every ten teak trees fit for felling, two were to be retained for seed, and that ten young trees replaced every one cut down. A conservation programme had been practised in Travancore at least as early as

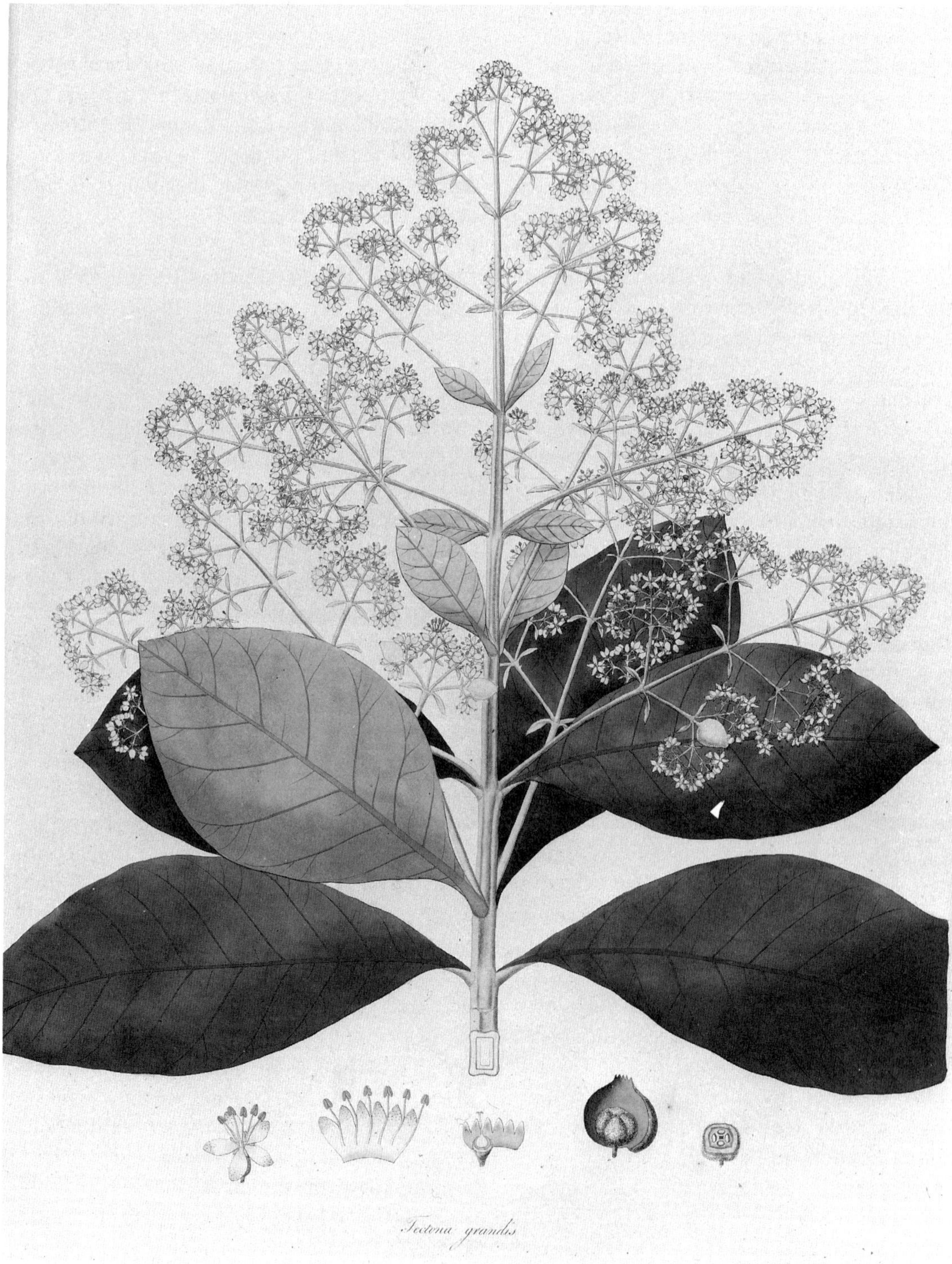

Figure 18.4 Teak, *Tectona grandis.* (From W. Roxburgh (1795) *Plants of the coast of Coromandel*, Vol. 1, London, plate 6.) Hand-coloured engraving. 'For ship building the teke is reckoned superior to any other sort of wood, being light, strong, and very durable, either in or out of the water' (Roxburgh).

1817. The Superintendent of the Dapoori garden, Alexander Gibson, also acted as 'Interim Conservator' until he was gazetted as Conservator to the Presidency of Bombay in 1847. Foremost amongst his duties was the maintenance of timber supplies for the Bombay dockyard. His attempts to control the collection and sale of timber and firewood inevitably brought him into conflict with long-established communal rights. His unremitting crusade against *kumri* cultivation eventually achieved its banning within the Presidency. In the Madras Presidency a few enlightened individuals demonstrated in a small way a feasible conservation practice. H. Conolly, an administrator in that part of Malabar devastated 'by a selfish and short-sighted policy', started an experimental teak plantation at Nilambur in 1843, the first to be managed successfully in southern India. When James Michael was placed in charge of the Anamalai forests in 1848, he devised a fire protection system which was adopted throughout India.

When Hugh Francis Clarke Cleghorn (1820–95) was posted to Mysore as a civil surgeon, it was not long before he took up the cause of forest conservation with evangelical fervour. He persuaded the annual meeting of the British Association for the Advancement of Science at Edinburgh in 1850 to appoint a committee to consider the destruction of tropical forests. Cleghorn, as its Secretary, submitted its deliberations to the British Association's meeting in Ipswich the following year. It surveyed the state of forests in South India, the Himalayas and Burma, Malabar, and Sind. It recommended controlled felling, replanting and the provision of nurseries. It deplored the loss of plants of potential value to science through the complete destruction of forests, and expressed anxiety about the effects of forest clearance on local water supplies. But officials in India seemed more concerned about the levels of teak supplies than the consequences of forest clearance on soil erosion and the silting of rivers. In December 1856, four months after he had submitted a report on conservation to the Government of Madras, Cleghorn was made Conservator of Forests in the Presidency. His discreet and cautious extension of the forestry practices at Anamalai throughout the Presidency met with little opposition, and in 1861 he had the satisfaction of achieving the official abolition of *kumri* cultivation.

With the acquisition of the forests of Pegu in Burma, Lord Dalhousie had no intention of letting them be exploited by timber merchants. His memorandum of 3 August 1855, which has been described as the 'Charter of the Indian forests', ordained that Pegu's teak resources were state property, to be cropped with due regard to proper maintenance. When McClelland, whose report had influenced Dalhousie, resigned, he was replaced in January 1856 by Dietrich Brandis (1824–1907) a German botanist with an interest in forestry. His management for the next eight years of the forests of Tenasserim and Martaban established his reputation as an outstanding forest conservator. Brandis and Cleghorn were the founders of scientific forest management in Burma and India respectively; they were appointed 'Commissioners of Forests' in 1862 to advise the Government on forestry matters, and in 1864 Brandis became the first Inspector-General of Forests. The first Indian Forests Act, passed in 1865, empowered local authorities to promulgate rules for the better management of forests. In the absence of suitably qualified British foresters, Brandis recruited his staff from forestry schools in Germany and France. The first of this continental intake, Wilhelm Schlich and Berthold Ribbentrop, eventually attained the office of Inspector-General of Forests. The first Forest School in the British Empire was inaugurated at Dehra Dun in 1878 for the training of Indian personnel. Schlich left India in

1885 to form a School of Forestry at the Royal Indian Engineering College at Cooper's Hill near London. As more qualified staff were appointed, forest conservancy gradually extended to the whole of British India. On returning to Europe, Brandis spend eight years in Kew and Bonn, his home city, preparing his great work, *Indian trees* (1906, five times reprinted), published a few months before his death.

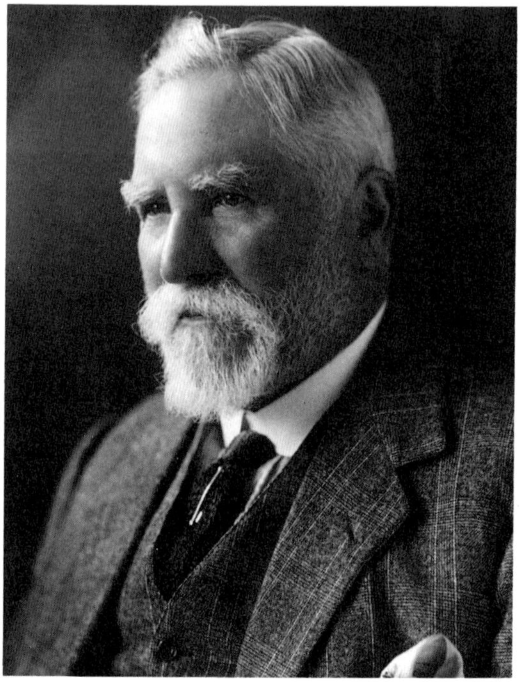

Figure 18.5 Sir George Watt (1851–1930).

(Sir) George Watt (1851–1930) (Fig.18.5) obtained much information from forest conservators during the compilation of his encyclopaedic *Dictionary of the economic products of India* (1889–96). The first book to attempt a comprehensive survey of India's economic products was Royle's *Essay on the productive resources of India* (1840). As Correspondent relating to the Vegetable Productions of India at India House in London, Royle tirelessly lectured and wrote on India's potential wealth in natural resources. *On the culture and commerce of cotton in India and elsewhere* (1851) and *The fibrous plants of India* (1855) were among his more substantial contributions. His pioneering efforts found a response in India from Colonel Heber Drury of the Madras Light Infantry with his *Useful plants of India* (1850) and from the Professor of Botany at Bombay, George Birdwood, with a *Catalogue of the economic products of the Presidency of Bombay* (1862). In 1874 a museum on the economic products of Bengal opened in Calcutta with a display of cereals, vegetable oils, jute, silk, indigo, cloth, and models of implements and machines. It received little support or attention from the commercial and manufacturing community but George Watt retrieved some of the samples for the Calcutta International Exhibition of 1883–4. Watt had been employed in the Bengal Education Department before he began organizing exhibitions. His catalogue on *Economic products of India*, prepared for the 1883–4 Calcutta exhibition, was the precursor to his *magnum opus*. Two years after the Government of India made him Reporter on Economic Products in 1887, the first volume of his *Dictionary of the economic products of India* appeared. It is a remarkable distillation of information gleaned from reports, books, experts and colleagues. It concentrates on plant products—animals and minerals receive scant attention—summarizing their botany, cultivation, and commercial applications, and is especially valuable for their regional vernacular names in the various languages of India. For the historian the work is a treasure trove of facts extracted from classical authors, accounts of early travellers to India, botanical texts and periodical literature. In just over 5000 pages Watt consolidated all the relevant data he could find on India's useful plants, indigenous and introduced, that were tested, accepted or rejected; it is a history of achievement and failure, intended both as a reference work and as an inspiration for the new breed of economic botanists in the approaching twentieth century[18].

NOTES

[1] I.O.L.R. Eur MSS E11, folios 9–12.
[2] Brockway (1979).
[3] Markham (1880), 266–7.
[4] H. A. Wickham said it was J. D. Hooker who had approached Markham about the project. Wickham (1908), 47.
[5] India Office to J. D. Hooker, 7 May 1873. Kew Archives. India Office: Caoutchouc, Vol. 1, folios 2–3.
[6] Wickham (1908), 54.
[7] Wycherley (1968), 1–11, 127–37.
[8] *Kew Bulletin*, 1912, 65.
[9] 26 August 1887. Kew Archives, India Office: Caoutchouc, Vol. 1, folio 119.
[10] Thiselton-Dyer to India Office, 12 August 1876. Kew Archives. India Office: Caoutchouc, Vol. 1, folio 31.
[11] Thiselton-Dyer to India Office, 17 April 1878. Kew Archives. India Office: Caoutchouc, Vol. 1, folio 463.
[12] Foster (1921), 300.
[13] Kyd to Governor-General, 1 June 1786. B.M. (N.H.) D.T.C., Vol. 9, folio 64.
[14] Government of Bengal to E.I.C., 25 June 1793. I.O.L.R. E/4/640, folio 119.
[15] Roxburgh to Government of Bengal, 20 Fe-lbruary 1802. I.O.L.R. F/4/134, item 2427, folio 33.
[16] Wallich to Government of Bengal, 7 February 1822. B.L. MSS Add. 9869, folio 224.
[17] *Journal of Horticulture and Cottage Gardener*, 12 May 1887, 373.
[18] An abridgement—*The commercial products of India*—was published in 1908.

Gardening in India

The hymns that constitute the Rig Veda required flowers to be used in religious ceremonies and through these ancient Hindu texts we discover that gardens have enjoyed a long tradition in India. Garland-makers grew flowers for festivals and most households tried to cultivate a few blooms to offer in prayer and ritual. Trees with sacred associations such as Kadam (*Neolamarckia cadamba*) and Champaka (*Michelia champaca*) are still familiar sentinels in temple gardens. Public parks were laid out for recreation, usually on the outskirts of towns and frequently near rivers. Asoka, the greatest emperor of the Mauryan empire and a Buddhist convert, encouraged the planting of trees along highways as well as in parks and gardens. His pillar in Delhi, one of a number erected throughout the empire proclaiming royal edicts, announced 'that everywhere wholesome vegetables, roots and fruit trees shall be cultivated, and that on the roads wells shall be dug and trees planted to give enjoyment for both men and animals'[1].

The historian Aelian, writing in Hadrian's Rome, gives a glimpse of India's royal gardens:

> *among cultivated plants there are some to which the King's servants attend with special care, for there are shady groves and pasture-grounds planted with trees, and branches of trees which the art of the woodsman has deftly interwoven. And these very trees, from the unusual benignity of the climate, are ever in bloom, and, untouched by age, never shed their leaves; and while some are native to the soil, others are with circumspect care brought from other parts, and with their beauty enhance the charms of the landscape*[2].

During the Gupta era (AD 4th–6th centuries), sometimes referred to as India's classical age, gardens were an important facet of Hindu culture. Vatsyayana's *Kama Sutra* which treats of love as a refinement of erotic pleasure, provides an insight into the manners and morality of contemporary social life. The services that a dutiful wife rendered to her husband included the care of the garden.

> *She should surround the house with a garden. In the garden she should plant beds of green vegetables, bunches of sugar cane, and clumps of the fig-tree, mustard, parsley, and fennel. Clusters of various flowers such as the jasmine, the yellow amaranth, the wild jasmine, the china rose and others should likewise be planted. She should also have seats and arbours made in the garden, in the middle of which a well, tank or pool should be dug*[3].

The first Muslim incursion took place in AD 711 when Sind was annexed. But Islamic culture made little impact on the Hindu way of life until the thirteenth century when the Punjab, Bengal, and the Gangetic Plains were dominated by the Delhi Sultanate. One of its rulers, Firuz Shah Tughlaq, is credited with having planted some 1200 gardens around Delhi and restoring 30 more. If one can believe historians, he must have been obsessional about gardens. He created, so we are told, another 80 near Salaura and 44 in Chitor. Since he grew grapes in all of them, it is not unreasonable to suppose that many of them were no more than vineyards.

His dedication to horticulture was shared by Babur, the first Mughal emperor. He had designed gardens in Kabul, the base for his invasion of India in 1525. Repelled by the absence of water and the lack of symmetry in Hindu gardens, he created his Indian gardens in the Persian style of a square divided into four by the crossed arms of a water-course or channel, known as a charbagh or four-fold garden. Water, the elixir of life and an essential component of all Persian and Mughal gardens, flowed through shallow channels into ornamental pools linking the various elements of the garden, contributing to the mood of tranquility and contemplation. Fruit trees and cypresses, symbolic of life and death respectively, provided shade and contours, and bulbs and annuals were freely planted in parterres or scattered in lawns—all confined within high boundary walls. A Mughal garden was an affirmation of Muslim concern for order and pleasure in geometry. The burial place in Delhi of Humayun, Babur's son and the next emperor, established the pattern of the Mughal tomb: an imposing mausoleum in the centre of an enclosed charbagh.

Jahangir had been on the throne almost three years when the English merchant, William Finch, landed at Surat in August 1608. It took him nearly 20 days to travel the 400 miles between Lahore and Agra to see the emperor. By imperial command an unbroken parade of trees lined the highway linking these two great cities. The planting of trees for shelter and shade had always been considered an act of piety and charity in the East. According to Marco Polo, Kublai Khan ordered all the main roads in China to be distinguished by avenues of trees for the convenience of travellers. Asoka, in India, was assiduous in observing a similar tradition. Akbar began the planting of an imperial highway which his son, Jahahgir, conscientiously maintained. A symbol of Mughal confidence and authority, this king's highway impressed all early European visitors who saw it—'one of the great woorkes and woonders of the world', marvelled Sir Thomas Roe, England's first ambassador[4]. The 'Long Walk', as it was generally called, was likened to a 'delicate walke' by Edward Terry as it progressed across flat, fertile plains, escorted by fruit trees, the ubiquitous banyan and peepul and flowering trees like the Flame of the Forest (*Butea monosperma*). From Peter Mundy we learn that the space between trees was 'about 8 or 9 ordinairie stepps, and the rancks from side to side about 40'[5]. And Tavernier reports that where there were gaps small whitened stones were erected 'so that the letter carriers can distinguish the road on dark and rainy nights'[6].

Usually the first glimpse that European venturers had of Indian gardens was at Surat, then the principal port on the west coast. Edward Terry, chaplain to Sir Thomas Roe, was charmed by 'fruitfull trees and delightful flowers, to which Nature daily lends such a supply that they seeme never to fade'[7], and also the pervasive sound of water. The fountains, pools and garden pavilions in the walled town of Toda in Jaipur State seduced Sir Thomas Roe: 'I must confesse a banished Englishman might have been content to dwell there'[8]. The Parisian Jean de Thévenot had wandered through North Africa and the Middle East as far as Persia before reaching Surat in 1666, where he visited the garden belonging to Aurangzeb's sister. In his eyes its straight walks and channels of water flowing from a central pavilion could not match the splendour of French formal gardens. 'There is nothing to be seen of our arbours, borders of flowers, nor of the exactness of their compartments and far less of their waterworks'[9]. When John Splinter Stavorinus saw the same gardens a century later, most of the ornamental trees had been cut down and only tamarinds flanking the four main walks remained. After three weeks in Surat, de Thévenot left for Ahmedabad—the 'handsomest town in Hindustan, perhaps in the world', vouched Roe—noted for its enclosed gardens, havens of greenery. The emperor's garden outside the town had long straight walks of trees which reminded de Thévenot of the Cours de la Reine in Paris while the surrounding countryside could be viewed from a raised terrace. Whilst the gardens at Bhagnagar or Golkonda in Hyderabad lacked Ahmedebad's abundance of flower beds and water features, de Thévenot rated the palace gardens the best in India. They were laid out compartmentally, one area being so thick with palms that 'the sun can hardly pierce through them'[10]; elsewhere flower beds ran parallel with gravelled paths; a terrace walk 'two hundred paces long and fifty broad' overlooked a pool with an octagonal pavilion. The weight of large trees upon the arched terrace usually astonished European travellers.

Not every observer responded so appreciatively. John Fryer, who reached India about six years after de Thévenot's departure, tended to be dismissive. 'Though these people delight much in gardens, yet are they but rude, compared to ours of Europe'[11]. The four quarters of those gardens he visited were thick with trees or planted with potatoes, yams, brinjal (eggplant), or gourds; flowers were often confined to dianthus, hibiscus, and jasmine; 'roses would grow here if they would but cultivate them'. One suspects that he did not see Mughal gardens when they were at their best in spring and early summer. Peter Mundy was impressed by the variety of vegetation in Agra's Mughal gardens: oranges, mulberries, mangos, coconuts, figs, apples, roses, marigolds, poppies 'and divers other sortes of faire flowers which wee knowe not in our parts'[12].

Ralph Fitch's Agra—'a very great citie and populous, built with stone, having faire and large streetes'—was intermittently the capital for several Mughal emperors. It was there that Babur designed his first garden on the banks of the River Jumna and that Akbar built a massive fort; it is where Jahangir's father-in-law lies in an elegant tomb just as perfectly conceived as the resting-place of Shah Jahan's queen beyond the bend of the river. The Taj Mahal is the best loved of all the Mughal tomb gardens which Edward Terry fancifully compared to 'Elysian fields such as the poets dreamed of, wherein their soules might take their repose'[13]. François Bernier had the privilege of viewing a pristine Taj, just 20 years or so after its completion in 1642. 'Nothing offends the eye', wrote Bernier: the main terrace 'wide enough to admit six coaches abreast'—'several garden walks covered with trees and many parterres full of flowers'—'stones on which you walk, cut and figured in various forms'[14] and, of course, the tomb itself, placed at one end of the garden contrary to convention, serene in its proportions, symmetry and simplicity. When the artist William Hodges was there during the 1780s, the fountains still played and disciplined parterres still resisted the peripheral sprawling mass of vegetation. It became a favourite resort for Agra's cantonments in the evenings while a regimental band played on the terrace. The Taj, Babur's Rambagh garden and Akbar's tomb at nearby Sikandra (Fig. 19.1), all maintained out of state funds, came under the control of the Superintendent of the Botanic Garden at Saharanpur in 1839. Hugh Falconer saw the Taj functioning as a public park, its thick canopy of trees offering welcome shade dur-

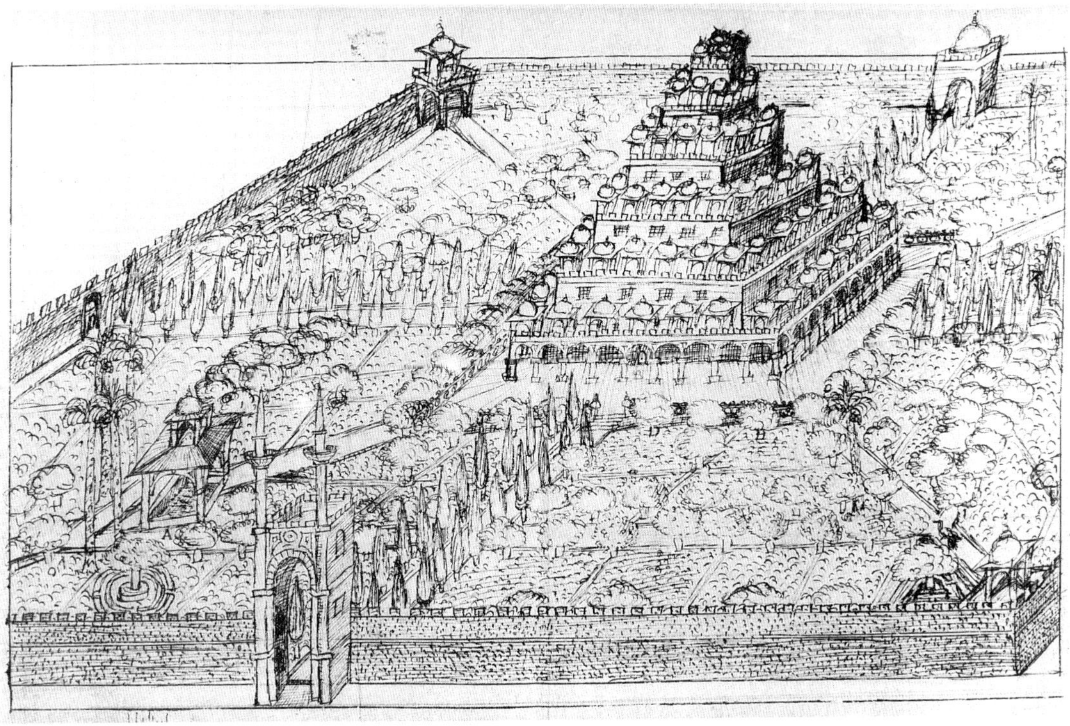

Figure 19.1 A sketch about 1632 by Peter Mundy of Akbar's tomb and garden at Sikandra. (From *The travels of Peter Mundy . . . , Vol. 2: Travels in Asia, 1628–1634*, London, 1914.) The unusual luxuriance of the vegetation and the tomb itself call into doubt the accuracy of the drawing.

ing the hot season, but he was, at the same time, sensitive to the delicate relationship between the garden and the tomb itself. His report to the Government of the North West Provinces makes depressing reading[15]. The maintenance of the gardens at the Taj and Rambagh was the responsibility of a Mr Kaine, the Superintendent of an experimental penitentiary at Sikandra. All three gardens grew fruit for local consumption. At the Rambagh the grounds were 'altogether excessively crowded both with fruit and ornamental trees . . . [many] have been left to grow where they have sprouted . . . most of them are scraggy, worn out and neglected'. While tactfully commending Kaine's efforts to improve these gardens with an inadequate and untrained labour force, Falconer deplored their status which made them 'subservient to the ends of public utility'—a ready source of fruit trees, shrubs, and flowers for local residents. The uninhibited luxuriance of shrubs and trees smothered in creepers at the Taj excited Edward Lear. 'Poinsettias were in huge crimson masses, and the purple flowered bougainvillaea runs up the cypress tree'[16]. Marianne North shared his pleasure in this chromatic confusion. 'The great lilac masses of colour often ran up into the cypress-trees, and the dark shade of the latter made the flowers shine out all the more brightly.'[17] Sir Edwin Arnold, not an artist but a poet, was in tune with their vision.

It is such an orderly wilderness of rich vegetation as could only be had in Asia, broad flags of banana belting the dark tangle of banyan and bamboo, with the white pavements gleaming crosswise through the verdure[18].

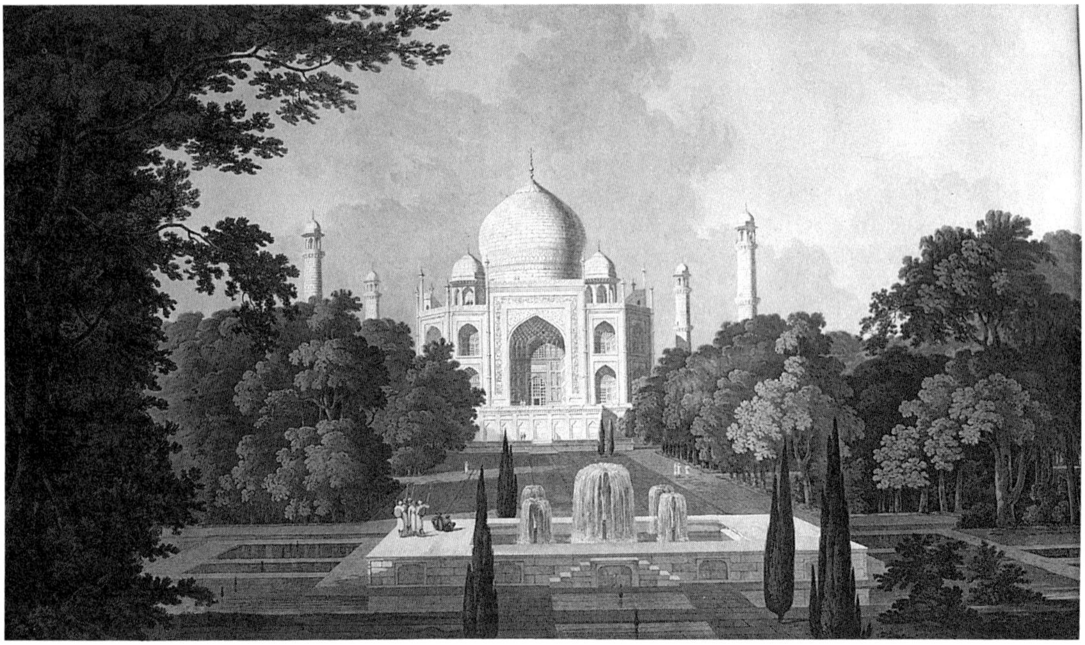

Figure 19.2 Taj Mahal at Agra in 1789. (From T. and W. Daniell (1801), *Views of the Taje Mahel . . .*, London.) Coloured aquatint. This thick screen of trees was severely thinned during the nineteenth century.

When Alexander B. Westland took charge of the Taj gardens in November 1891, its rampant disorder offended his sense of neatness and authenticity. He recommended replacing the central beds of shrubs with turf to reveal the stone geometrical scrolls. The process of reducing the number of trees would continue and there were to be fewer rose-beds; 'luxuriance [must] be restrained and . . . no stiff nor continuous screen be allowed to impede the view of the tomb'[19]. The Commissioner, while endorsing Westland's thinning-out policy, cautioned restraint (Fig. 19.2).

It is not, in my opinion, aesthetically correct that the Taj, the whole Taj, and nothing but the Taj should strike the eye from every quarter. The builders' idea, I believe, to have been that this and similar palaces and tombs should be visible through long vistas of heavy and narrowly divided foliage, so that height, distance, and the charm of the unknown might be communicated to them. If you can see them fully exposed at once to view, with only a series of devious gravel walks and multiform beds in the intervening space, there is nothing left to the imagination; and the spectator soon begins to believe that there is little difference between this and a thousand other flower gardens in all parts of the world.

If any reliance can be placed on Peter Mundy's rough sketch in 1632 of Akbar's tomb at Sikandra, depicting a dense mass of foliage almost obliterating the charbagh layout, then the Commissioner's thesis is amply substantiated. With the decline of Moghul power, monuments such as Sikandra inevitably suffered neglect. No water spurted from its fountains and the channels were dry when William Hodges walked in its grounds with a 'melancholy pensiveness'. A half century later, that is during the 1830s, Lieutenant

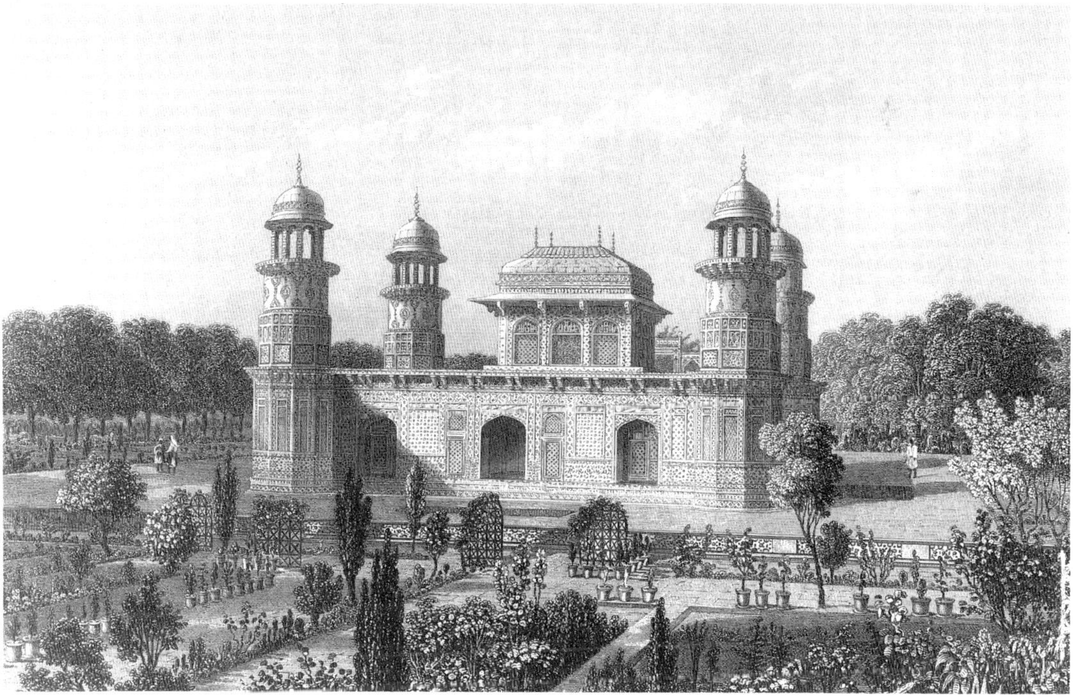

Figure 19.3 The Mughal garden surrounding I'timad-ud-Daula's tomb at Agra was filled with Victorian planting and potted plants. Wood engraving.

Thomas Bacon noted that 'the gardens about the place are now a complete jungle, and both the walls and gateways are falling fast to decay'[20]. If the gardens had deteriorated to such an extent during the rule of the Mahrattas, it becomes less reprehensible that the British should have chosen Sikandra for 'the prosecution of agricultural experiments' in 1838. Falconer reported that

> *The old trees are chiefly pines and fine tamarinds which are more or less thickly scattered over the garden in detached trees. Hardly a vestige besides remains of the trees, shrubberies and flower beds with which it was formerly ornamented*[21].

Fourteen native gardeners and 90 prisoners from the local jail tended the nurseries, the forest and ornamental trees, and assisted in cotton and senna trials. One of the squares having been laid out in the 'modern style', Falconer vaguely contemplated planting the entire garden with ornamentals. Unfortunately, during Holi, Diwali, and other festivals, the garden was regularly stripped by Indians of all its flowers.

Agra was too small and the streets too narrow for Shah Jahan who moved his capital to Delhi where the foundations of a new city, Shahjahanabad, were laid in 1638. The emperor entered the Red Fort, the only part of the complex to be completed, ten years later. The unifying theme of the palace within the Fort was water which flowed through the terrace of marble pavilions and fed the canals, cascades, and pools in the gardens. Bernier wrote

> *Nearly every chamber has its reservoir of running water at the door, on every side are gardens, delightful alleys, shady retreats, streams, fountains, grottoes, deep excavations that afford shelter from the sun by day, lofty divans and terraces on which to sleep coolly at night*[22].

The royal apartments overlooked the two principal gardens: the Hayat Baksh (the Life-giving Garden) and the Mahtab Bagh (the Moonlight Garden). Water sprayed from nearly 200 jets in a square tank and its four radiating channels; embowered fruit trees provided essential shade. The Life-giving Garden was planted with red and purple flowers, symbolic of life, whereas pale colours were chosen for the flowers in the Moonlight Garden. Oranges and other fruit trees, a rash of rose-bushes and a few jonquils in bloom were all that remained when Bishop Heber saw it in 1824. Its decline was complete when Bacon was there in the 1830s.

> *. . . the buildings are overgrown with weeds and rank mosses, the tanks are stagnant and corrupt, the channels are broken up and choked, the fountains are silent, the beautiful marble baths are filled with dried leaves and rubbish, and the grottoes are for the most part half buried*[23].

After the sepoy rebellion, when Bahadur Shah, the last Mughal emperor, was exiled to Burma, the Red Fort was occupied by British troops. The Moonlight Garden and half of the Life-giving Garden disappeared under barracks and a parade ground.

Sir Edwin Lutyens, one of the architects of New Delhi, studied the surviving Mughal gardens at Delhi, Agra, and in Kashmir in order to fulfil the wish of the Viceroy, Lord Hardinge, and his wife who wanted Mughal features in their new Viceregal garden. In 1918 Lutyens submitted a design of a formal water garden of imperial grandeur: broad canals, gentle cascades, reflecting pools, columnar jets, and fountains of tiered lotus leaves. Trim lawns and drifts of flowers in herbaceous borders that would have delighted Gertrude Jekyll softened the geometrical severity of the Mughal concept in an inspired fusion of eastern and western gardening tastes.

Lutyens provided a spectacular finale to a tradition that reached back to the early European diplomats and traders who created gardens for fresh food and for relaxation. The navigable rivers and natural harbours of Goa made it a strategic base for the Portuguese who, once they had fortified it, planned a European city with churches, colleges, civic buildings and houses 'surrounded with groves of palm-trees and delightful gardens'[24]. In 1636 Peter Mundy was entertained in one of these houses situated 'in a most delicious shady grove on the side of a hill replenished with multitudes of tall, spreading and allwaies flourishing fruite trees, pleasant springs, walkes, and a curious tancke lying under a rocke (where the water issues), quite overshadowed with trees'[25]. When the British took possession of the island of Bombay from the Portuguese in 1664 they found a house inadequately defended by four brass guns with 'a delicate garden, voiced to be the pleasantest in *India*, intended rather for wanton dalliance, love's artillery, than to make resistance against an invading force'[26]. John Fryer regretted the destruction of its 'walks which before were covered with Nature's verdant awning, and lightly pressed by soft delights . . . now open to the sun, and loaded with hardy cannon'. Could this garden of 'terrestrial happiness' have been the home of the Portuguese physician, Garcia da Orta, the first European to examine India's flora with a botanist's curiosity? When that shadowy figure, John Burnell, was there about 1710–11, only a forlorn summerhouse of coconut palms stood in a patch of ground quite devoid of trees and flowers with 'a small kitchen ground at one corner which had hardly had greens enough to compose one salad'[27].

The impermanence of a garden, the ease with which neglect will cause it to disappear, reverting to scrub, is sadly confirmed by the loss of so many European creations in India. The outline of formal Dutch colonial gardens, their flower beds and paths lined with brick, furnished archaeological evidence of their former existence long after they had been abandoned. In the 1860s a visitor investigating the ruins of the Dutch Station at Sadras (40 miles south of Madras), captured by the British in 1765, could still discern—despite invasive opuntias—the straight walks, the water channels and the regular disposition of the parterres in the Governor's garden. European gardening styles in India were usually a reflection of national taste. Fryer mentions a Dutch garden 'with arbors and beds after the *Europe* mode'[28]. All contemporary accounts of the gardens attached to Dutch stations suggest they were well maintained—'very spatious, well kept, with tarrass walks, and full [of] lettice and good herbage'[29] was the approving comment on a Dutch establishment in Bengal (Fig. 19.4). The Dutch commander of the factory at Bhagnagar boasted to Abbé Carre in the 1670s that 'one million'[30] was spent on bringing water about 12 miles by underground channels to irrigate the kitchen garden and for household use. A kitchen garden was necessary for salads and vegetables, 'the seed of which comes from Japan, the Dutch, desiring to have all kinds of herbs and pulses in their gardens'[31]. John Burnell admired the garden of the Danish factory at Gondalpara in Bengal with its abundance of vegetables and flowers, a shady walk and a fishpond, in striking contrast to the run-down English factory at nearby Gholgat 'in which nothing grows but weeds'[32].

Gabriel Dellon was impressed by 'gardens surrounded with fine arbours, bearing grapes

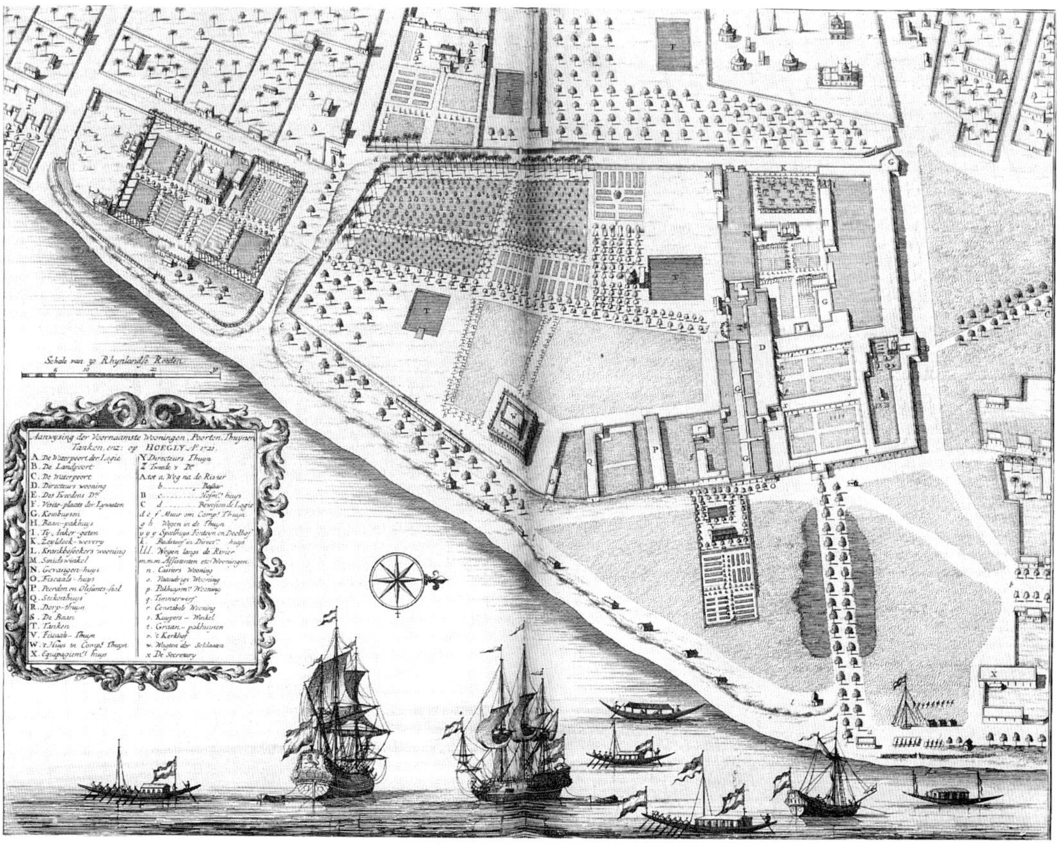

Figure 19.4 Plan of the Dutch station or factory on the River Hooghly in Bengal with gardens and avenues of trees. (From F. Valentijn (1724–6), *Oud en Nieuw Oost-Indien*, Amsterdam.)

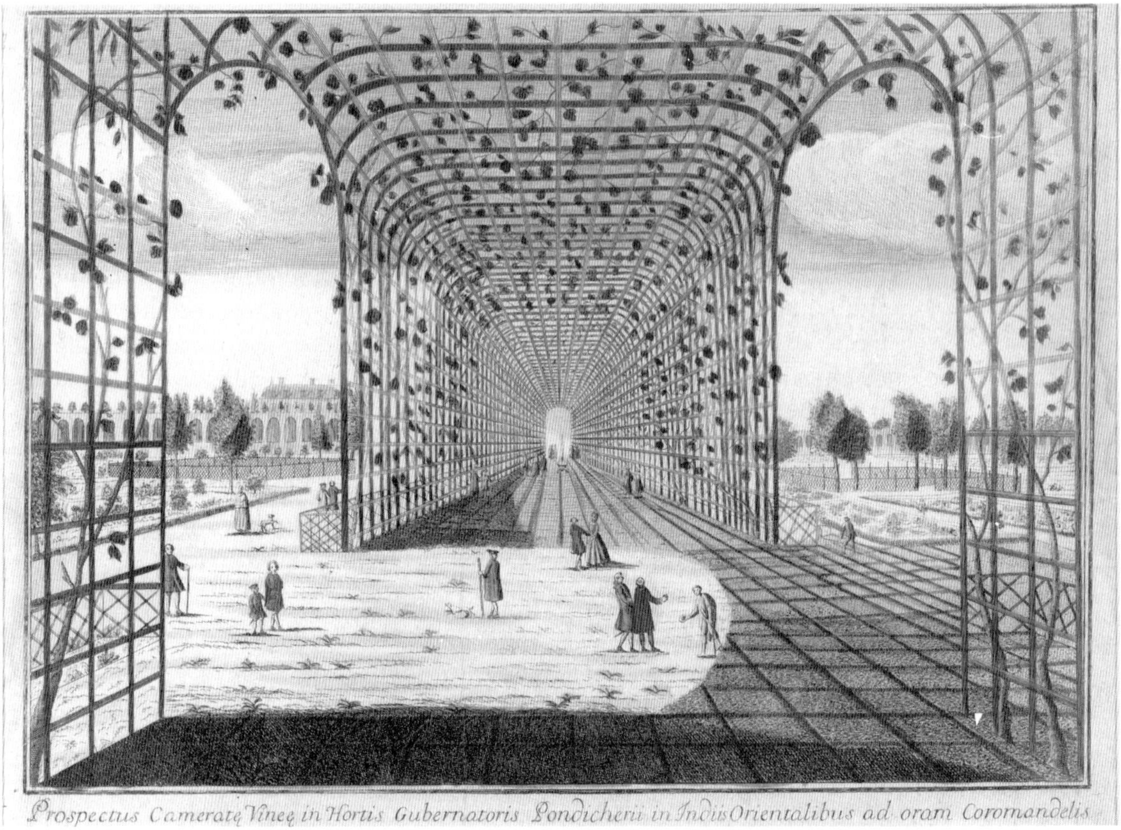

Figure 19.5 A vine on a trellis arcade in the Governor's garden in the French settlement at Pondicherry. Late eighteenth century. Hand-coloured engraving.

twice a year'[33] at Surat. Pineapples and custard-apples, introduced from the Americas by the Portuguese, could always be found, when in season, on European tables, and mangos, especially those grown in Goa, were a favourite fruit, eaten raw or made into a chutney which was even exported to Britain. Peter Mundy's description of the East India Company's garden at Surat suggests either the adaptation of a native garden or of one imitating Indian design.

We have also a garden which for its bignes is the neatest and costlyest in all the countrey hereaboute; being neare 4 square, having 4 verie faire longe walkes round about, all covered over with vynes supported with tymber, very curiously contrived. It hath 4 other allies which go from the midle of the longe walkes into the midle of the garden, where stands a chowtree, or prettie roome, covered overhead to sitt and passe the tyme. Before it stands a little tancke to wash in tyme of heate and rayne[34].

When the Company moved its operational base from Surat to Bombay, the Governor, Gerald Aungier, set about recreating the comfortable and relaxed surroundings he and his colleagues had enjoyed at Surat when they retired to their gardens in the evening with a bottle of wine. Young citrus trees were planted in ground near the Fort but seed imported from England failed to

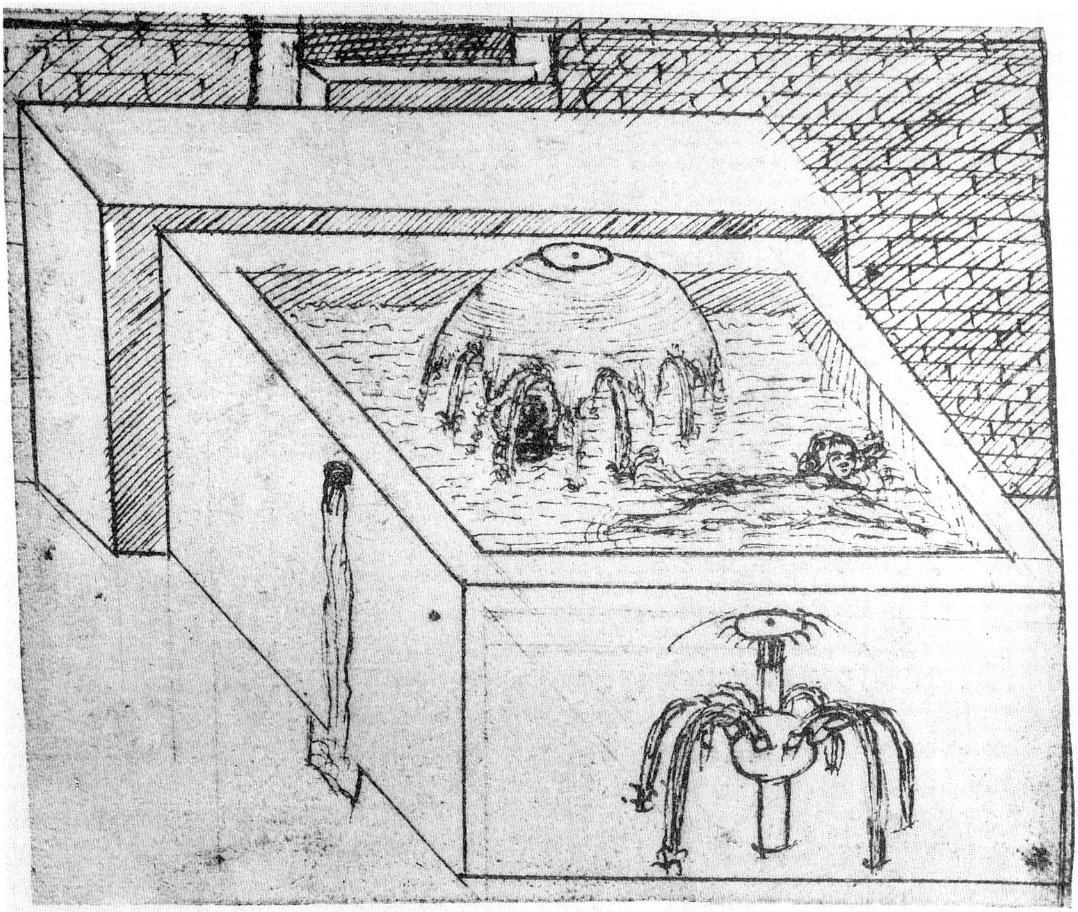

Figure 19.6 Sketch by Peter Mundy of a small tank in the East India Company's garden at Surat with a fountain showing the details of its mechanism. (From *The travels of Peter Mundy . . . , Vol. 2: Travels in Asia, 1628–1634*, London, 1914.)

germinate. They were more successful with grapes and water-melons and some were even sent to Surat.

When Calcutta was little more than a precarious settlement, Madras was already a prosperous town, laid out on a grid pattern of streets, sheltered by the protective presence of Fort St George. Houses brilliant with a rendering of white chunam plaster stood in spacious grounds 'where [there were] besides gourds of all sorts for stews and pottage, herbs for sallad, and some few flowers as Jassamin, for beauty and delight'[35]; palms, bananas, mangos, guavas, and other trees offered fruit or shade. Governor Thomas Pitt's map of Madras (*c.*1710) (Fig. 19.7) reveals a generous scattering of private gardens, the orderly arrangement of plots in the Company's new garden and, in the bend of the river outside the town, the Governor's avenue of trees or Great Walk (Fig. 19.8).

As the countryside was made safe for excursions, Europeans built garden houses outside town boundaries as an escape from the noise, smells and dust of streets, as a week-end retreat or, if not too distant, as a place for supper with friends. Standing in their own grounds, they were one or two storeys high, frequently classical in style with colonnades and wide verandahs. They populated St Thomas' Mount some miles inland from Madras, their lawns sloped down to the River Hooghly near Calcutta, and they jostled for the best view on Bombay's Malabar Hill. At first the Company attempted to control 'the folly and vanity of merchants in having the parade of country houses and gardens'.[36] For a time the Madras Government restricted its grant of land for building garden houses to officials above the grade of writer. The Court of Directors in 1758 cautioned Calcutta's authorities to 'be not lavish of the ground in building and gardening; allow only what may be convenient'[37]. These measures seem to have been completely ineffectual since there was a brisk business in the building, buying, and selling of garden houses. The *Calcutta Gazette* for 15 April 1784, for example, announced the letting of a garden of 'about 50 beegahs'[38], laid out in pleasant walks, and with the choicest and rarest fruit trees in India; a smaller garden up for auction a few months later had over 500 fruit trees.

The country residences of governors and senior officials in the civil and military establishments reflected the status of their occupants. Seventy-five acres of parkland with banyans, peepul, palms, tamarinds, and a lake were the setting for the Governor's garden house at Madras. George Pigot spent an agreeable bachelor existence there from 1755 to 1763, supervising his estate with the help of a European gardener and showing guests his menagerie. Although the house was enlarged by his successor, it was still considered too small for official receptions by Lord Clive, son of Robert Clive, when he became Governor in 1799. As a riposte to the splendours of the new palace of the Nawab of Arcot, he commissioned John Goldingham, Madras's Civil Engineer, to redesign the garden house. Goldingham almost doubled its size, added deep-set verandahs on three sides, colonnades on both storeys and a separate banqueting hall perched high on a terraced podium. His plans for the grounds included a grand processional avenue of trees, a mound, and a sunken garden. The enormous expense of this ambitious rebuilding alarmed the Company's shareholders in London.

It by no means appears to us essential to the well-being of our Government in India that the pomp, magnificence and ostentation of the Native Government should be adopted by the former; the expense to which such a system would naturally lead must prove highly injurious to our commercial interests[39].

The Court of Directors showed their displeasure by recalling Clive in 1803.

The magnificence of the Governors' residences at Madras and Calcutta was not replicated at Parel, the country retreat of the Governors of the Bombay Presidency since 1720. Extended in instalments, it lacked their cohesion and symmetry but it had the mandatory banqueting hall with a ballroom above it. The only feature that seems to have caught the attention of Bishop Heber in 1825 in its 'moderate-sized, old-fashioned garden' was a willow purported to be a scion of the tree near Napoleon's grave at St Helena. In 1829 the house replaced the Governor's official residence in the Fort. Emma Roberts, ten years later, enthused about the grounds, so 'perfectly Italian, with their fountains and cypress trees'[40]. At the end of the flower garden a broad terrace overlooked a large tank or pool. Roses grew in wanton profusion. Trees flanked a straight avenue a mile long, and elsewhere were grouped informally, a blending of native and introduced species. Lord Canning, on his way to Calcutta as the new Governor-General, stayed briefly at Parel in January 1856. His wife, charmed by the place, her first introduction to India, wondered whether Calcutta would match it.

From my window I see groves and groves of cocoa and palm overlapping round-headed

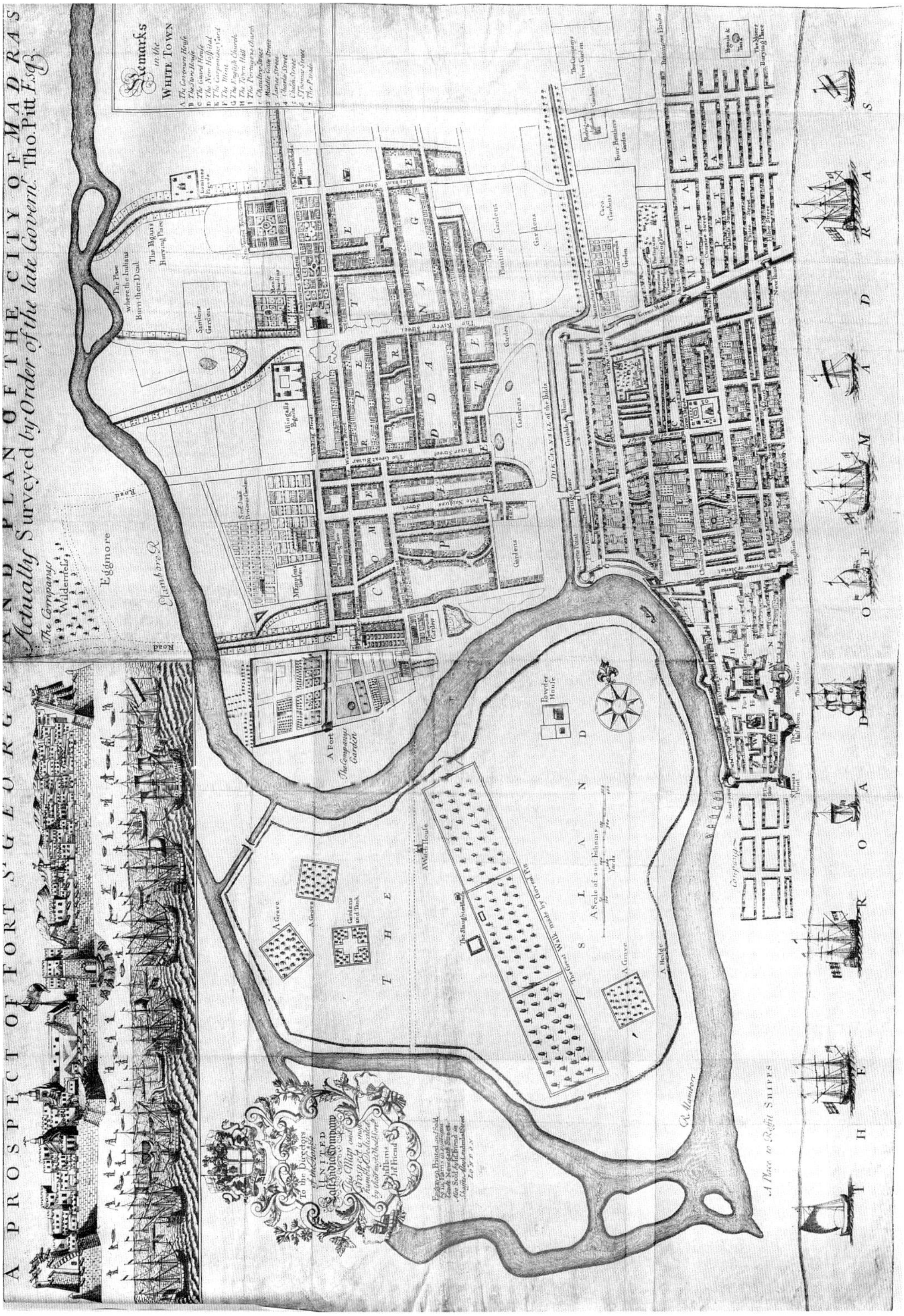

Figure 19.7 Plan of Fort St George and Madras, *c*.1710, well-provided with public andivate parks.

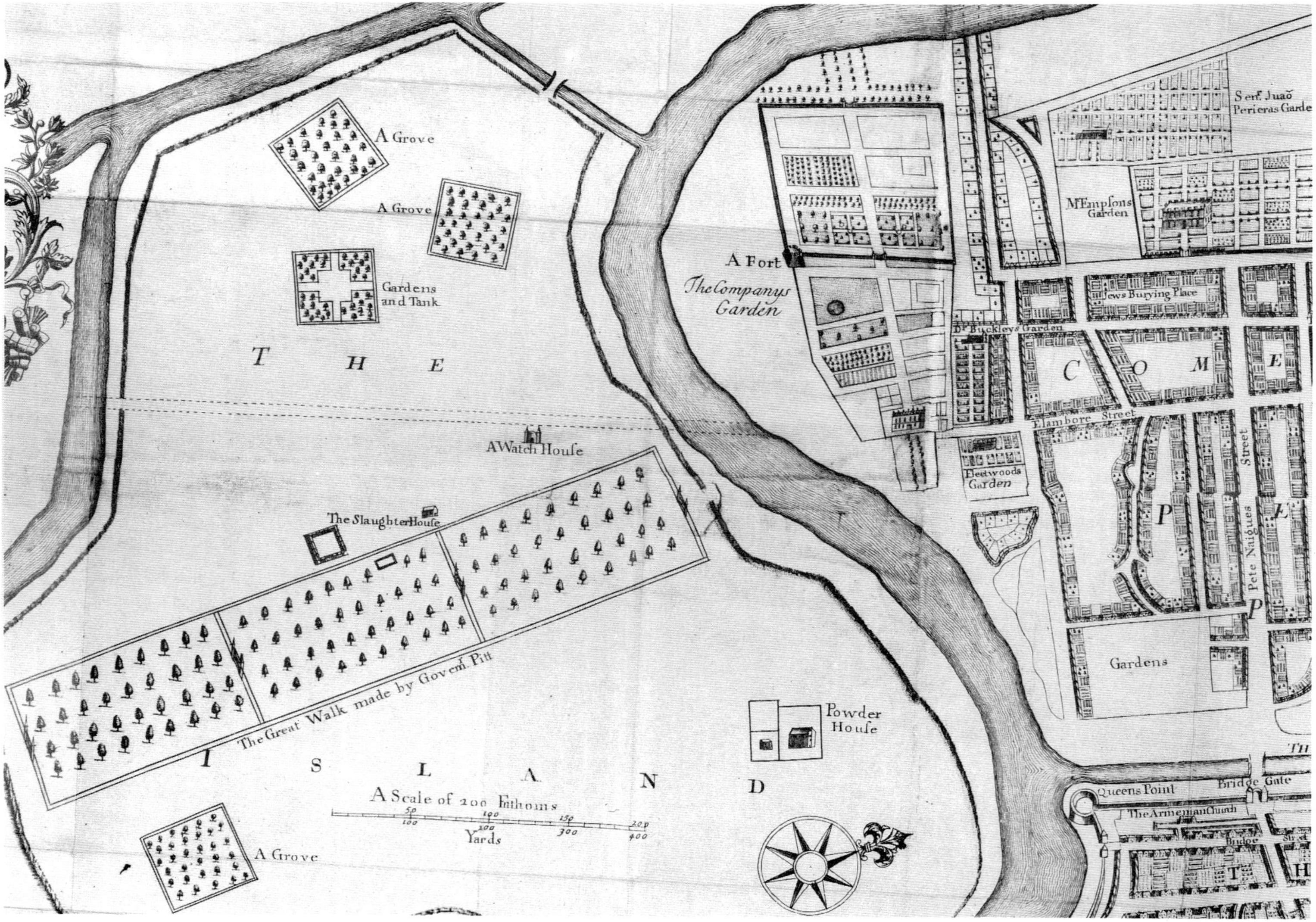

Figure 19.8 Enlargement of part of the plan (see Fig. 19.7) showing Governor Pitt's Great Walk, which was about 900 yards long, and also the Company's garden on the other side of the river.

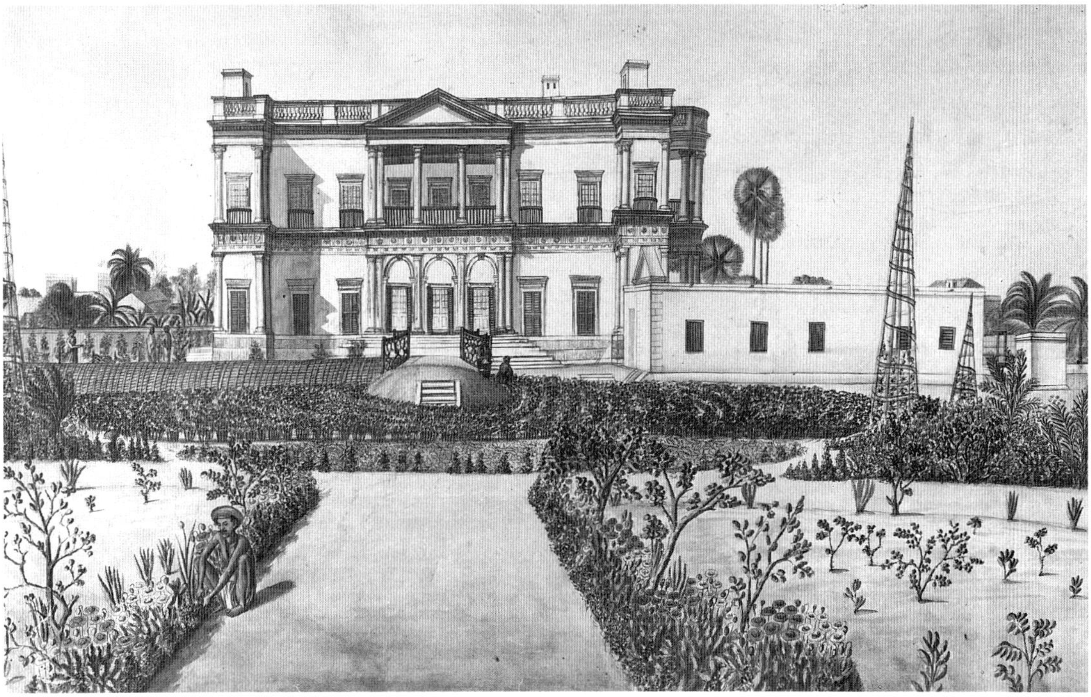

Figure 19.9 House and garden, probably at Chapra near Patna. Watercolour, *c.*1796. Note the curious frame-work for climbing plants. (India Office Library and Records.)

> *trees; then burnt-up ground; mango-trees in flower exactly like Spanish chestnuts; tamarinds in the style of acacias, but much thicker; peepul-trees, higher a good deal than the rest, with trembling leaves—very green, and pinkish stems, like white poplar; a teak tree with large leaves and sort of bunches of berries. Little green parrots fly about, and settle in the tamarind tree, looking like green peas, and hawks and odd crows—grey and black—flutter about. Very slightly clothed peasants fill up the picture, and sometimes beautifully draped figures in white*[41].

The house was relinquished as an official residence after Lady Fergusson died there in 1883. The present Plague Research Laboratory was transferred there in 1899.

In the 1860s the Governor of Bombay decided to transfer the Government House at Dapoori to a site nearer to Poona. Ganeshkhind, as the new building was called, was low and sprawling, declaring its presence with a flag tower, 80 feet high. The house incorporated a winter garden for exercise during the rains. Its garden in the charge of a former Kew gardener, George Marshall Woodrow, was voted 'the prettiest I saw in India' by William Mudd, Curator of the Cambridge Botanical Garden and a member of the Prince of Wales's Indian entourage in 1876. Cascades of climbers on buildings and walls 'drew out acclamations of rapture and surprise from more than one of the Royal Party'[42]. Now it is the University of Poona and the campus still maintains the dazzling display of colour that so impressed Mudd.

After Lord Wellesley's extravagance and recall, his successors practised a cautious economy at the new Government House in Calcutta. Simple lawns sufficed although in 1818 Lord Hastings dared to import gravel from England for the paths. It was usually left to wives and other women of the Viceregal household to indulge any horticultural flair. Lady Amherst organized flower-beds which her successor, Lady Bentinck, convinced that 'flowers were very unwholesome', instantly removed. Emily and Fanny Eden, the sisters of their bachelor brother, Lord Auckland, dallied with a little gardening and added two small fishponds, but it was Lady Mayo who instituted a programme of systematic planting. Earl Lytton, who was responsible for an ornamental pond crossed by a rustic bridge, raised a mound. By the 1880s a mature thicket of palms, bamboos, and shrubs provided privacy and, at the same time, hid the untidy huts of some 30 to 40 native gardeners. But it was the Governor-General's country residence at Barrackpore that received unstinted devotion and the full impact of innovative design.

All guests invariably succumbed to the charms of Barrackpore adjacent to the placid Hooghly river with its gently undulating parkland invoking nostalgic remembrances of home; 'if I had not from time to time seen a tall coco-tree towering above all the rest, I could have fancied myself on the banks of the Thames instead of the Ganges,'[43] sighed Bishop Heber. The river was also evocative of the Thames for Lady Canning who thought of Syon and for Lady Dufferin who recalled Cliveden. It was Wellesley who had the jungle cleared, swamps drained, ponds dug, and the flat ground shaped into shallow declivities to create a Georgian landscape. Even the wished-for 'distant view of a steeple' was eventually granted when the church in Danish Serampore on the opposite bank obligingly acquired a spire. The arches of his unfinished house fortuitously added a picturesque ruin which stood until Lord Hastings had it demolished. The road between Barrackpore and Calcutta lined with young trees 12 to 15 feet apart—banyan, pepul, mango, teak, etc.—was opened the very week Wellesley ceased to be Governor-General. Lord Minto who assumed office in 1807 was appreciative of his predecessor's efforts. 'The grounds are a mixture of park and pleasure-grounds. They are laid out with the greatest judgment and taste, and their extent is considerable.'[44] The Corinthian temple commemorating officers who died in Mauritius and Java which Lord Minto added is now flanked by statues of imperial worthies discreetly removed after Independence in 1947 from the open space or Maidan in Calcutta. Lady Hastings built a large conservatory on the site of the shell of Wellesley's house to grow British plants—the intention was to shut out the heat, a rather naïve concept since it was certain to have the opposite effect. This needless expenditure irritated J. C. Loudon, the editor of the *Gardener's Magazine*:

> *After similar instances of extravagance and folly, it is not greatly to be wondered at that the East India Company have in a great measure ceased to patronise botanic gardens*[45].

Emily Eden, agreeably surprised by the lushness of the grass ('much greener than the grass near London in summer'[46]), and her sister Fanny converted an old thatched bungalow in the flower garden into a seed and plant house. Lord Ellenborough laid out a broad terrace walk and Lady Canning contributed a raised walk from the house to the landing stage. She also made a terrace, erected a balustrade around the small Italian garden, and removed the shrubs around the great banyan tree. Barrackpore crops up frequently in her correspondence. She tells Queen Victoria that

> *The Park is carefully planted with round headed trees to look as English as possible more so than I approve, and I am glad when bamboos & cocoanuts & palms have crept in. All the apparent chestnut, elms & ash prove quite unknown on nearer view. The luxuriant growth in the jungly ground outside, of daz-*

zling green during the rains, is more beautiful than I can describe & I always think of the Palm House at Kew which gives a faint idea of it[47].

Nothing in her role as Vicereine pleased her more than the days she spent at Barrackpore, reshaping the garden, planting more exotic species, or painting its flowers. When she died of fever in November 1861 she was buried in the park near a favourite spot overlooking the river. Alterations were made by later occupants: Lady Ripon planted a bamboo avenue and a rose garden, Lord Curzon enlarged the grounds by taking in some of the foreshore, and at some stage a melancholy Gothic-style ruin appeared. It was undoubtedly the best-known and the best-loved of the many English gardens laid out in India.

The gardening fraternity in India—those who enjoyed gardens as well as those who created or worked in them—was drawn from the whole spectrum of society. Officers on active service collected seeds for Warren Hastings's garden at Alipore where he tried, not always successfully, to grow plants from Britain. Mountstuart Grant Duff found relief from the stresses of gubernatorial duties by fixing plant labels to the trees and shrubs at Government House, Madras. The garden of the Chief Justice of the Supreme Court of Calcutta, Sir Lawrence Peel, on the banks of the Hooghly made a great impression on Joseph Hooker who dubbed it 'the Chatsworth of India'. When James Forbes, an official in the Company's service, was posted to Broach, north of Surat in the Bombay Presidency, he purchased a small house with six acres of ground in a nearby village. There he made a garden 'as much as possible after the English taste'[48], creating shade with mango, tamarind and 'burr' trees, and making it distinctive with flowers from China and the distant regions of India. His constant regret was his inability to produce a smooth, green lawn. The missionary, William Carey, used his botanical connections to make his plot at Serampore a place of pilgrimage for all garden-lovers in Bengal.

The owner of possibly the best-stocked private garden in British India, Carey, like all obsessional gardeners, was always greedy for more acquisitions. He cajoled one of his friends in England.

Were you to give a penny a day to a boy to gather seeds of cowslips, violets, daisies, crowfoots, etc., after they have done flowering, you might fill me a box every quarter of a year; and surely some neighbours would send a few snow drops, crocuses, etc., and other trifles. All your weeds, even nettles and thistles are taken the greatest care of by me here; . . . indeed, though you cannot buy a little cabbage seed here under about £2.2s., yet, I have never been able to extort an ounce or a quarter of kidney beans from all the friends in England[49].

General traders in Calcutta, anticipating a profitable side-line, speculated in the importation of seeds. Mr Oakes of the Europe, China, and India Warehouse, announced in the *Calcutta Gazette* for 11 September 1794 a stock of fresh garden and flower seeds, 'warranted having been tried in a garden near Calcutta, and all have succeeded'. They were all the familiar European vegetables and flowers, deserving 'the attention of the curious', and included some of the favourites of the cottage garden—stock, carnation, marigold, larkspur, mignonette, honeysuckle, polyanthus, poppy, and Sweet William. A competitor in the same issue of the *Calcutta Gazette* offered a wide choice of vegetable seed from Portugal. H. C. Broeger's 'horticultural establishment' at Edgar advertised in March 1822 eleven sorts of European fruit trees and several thousand annual and perennial European flowers. On 13 June, having just received a consignment of seed from Liverpool, Broeger requested his subscribers 'to intimate if they will partake of this batch, or wait for the Cape supply'. The exchange of seeds and plants between Britain and India enjoyed official backing after Lord Auckland ordered in August 1838 that officers on duty in the North West Provinces were to collect seeds, bulbs, and roots for despatch to Britain. The Horticultural Society of London, several nurseries, and individuals reciprocated with plants suitable for India.

The climate and garden pests, the scarcity of native gardeners with horticultural skills and the nomadic life of many officials all conspired to frustrate any dedicated gardener. The gardens of small and modest residences tended to be rather functional with few imaginative touches: some paths covered with brick dust or gravel intersecting a lawn of sorts, sporadic clumps of shrubs, flower-beds, and serried ranks of flower pots (Fig. 19.10). The author of one gardening manual recommended the simplest parterre since the average native gardener was, in his opinion, incapable of maintaining anything elaborate. The house where Thomas Babington Macaulay stayed in Calcutta had a typical town garden: 'a fine sheet of turf with a gravel walk round it, and flower beds scattered over it . . . It is so well shaded that you may walk there till ten o'clock in the morning.'[50] Arched trellis-work often got out of control in Madras gardens while garden sheds in Bombay were frequently converted into ferneries. In the absence of a European cemetery at Dinapore near Patna, the deceased were buried in their gardens which then acquired unusual garden ornaments—commemorative stone urns and obelisks.

It was often the memsahib who supervised the garden, inspecting it during the cool hours of morning and evening and for longer periods in the cold season. The perennial challenge of a lawn was usually solved by

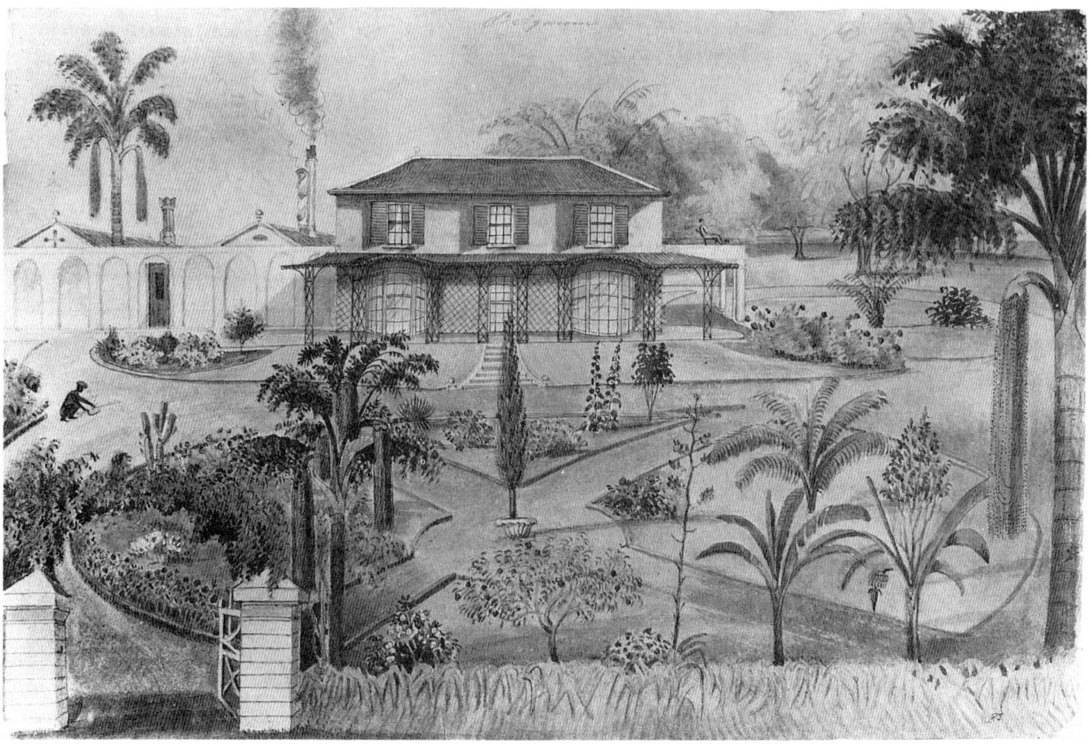

Figure 19.10 John Brownrigg Bellasis's watercolour sketch of his home, *c.*1850s. Island lawns and flower-beds separated by paths were favoured by Indian gardeners.

doob grass (*Cynodon dactylon*), a plant of trailing habit, and so compact that it effectively smothered all weeds. The horticultural calendar consisted of English annuals during the cold months, petunias, verbenas, phloxes, and the like in the hot weather, followed by balsams and reliable zinnias during the monsoon. Trial and error established a repertoire of dependable plants: stocks, camellias, and balsams were vigorous growers; petunias adapted well to any situation; verbenas were prolific self-seeders. Ixias, jasmine, lantana, chrysanthemum, hibiscus, rose, hoya, and petunia survived the heat of the plains but crocus, snowdrop, hyacinth, anemone, ranunculus, tulip, dahlia, auricula, and geranium responded only to the cooler climate of the hills. Roses from the time of the first Mughal emperor, and possibly earlier, are still everybody's favourite: they bring colour in May to Kashmiri gardens and are striking sentinels to the entrances of Delhi's hotels in December.

India's garden flora is as cosmopolitan as that in any other part of the world. Favourites among native trees are Flame of the Forest (*Butea monosperma*), Champaka (*Michelia champaca*) memorable for its heavy fragrance, Indian laburnum (*Cassia fistula*) with cascading golden clusters, the Coral tree (*Erythrina variegata*) with its starkly red inflorescence on bare branches, the Glory lily or Tiger's claws (*Gloriosa superba*), a vigorous climber, and, of course, the rhododendron. Some introduced plants, like the Royal Gold Mohur (*Delonix regia*) from Madagascar or Frangipani (*Plumeria*) from Central and South America, adapted so well to their new environment that they have become indispensable adornments of large estates, public parks, and even the few acres around unpretentious bungalows.

The universal bungalow which spread like a contagious rash through the British colonies from Canada across the African continent to remote Australasia had its birthplace in Bengal. Originally it was no more than a single-storey hut with a thatched roof and a verandah, home for officials and planters. Gradually these bungalows evolved into substantial residences of brick and plaster, endowed with the distinction of classical columns, ornamental balustrades, elaborately carved gable-ends, fretted barge-boards, and Venetian shutters, any residual stark lines being softened by enveloping bougainvillaea. The bungalow compound, whether a miniscule apron of ground or a more generous allocation of space, was an invitation to horticultural ingenuity. A lawn was usually attempted with varying degrees of success, roses were predictable with the less innovative, the porch screened with shrubs and perhaps a few trees, and trellised verandahs festooned with climbers. Potted plants such as begonias, crotons, ferns, fuchsias, pelargoniums, hoyas, and salvias competed for space on the verandah. Impenetrable opuntias or *Agave americana* formed ideal boundary hedges. The greatest concentration of bungalows was always to be found in cantonments, normally sited on the periphery of towns, the base for the local garrison. Laid out on a grid pattern of broad roads, the military was separated from the civil lines and the officers' quarters segregated from the barracks. Each bungalow was set within its neat square of turf, colour was provided by communal flower beds, and any recreational area planted with trees. Soldiers were given tools, flower and vegetable seeds, and the lure of prizes to encourage them to cultivate cantonment plots. Soldiers' gardens were seen by a benign authority as a means of healthy relaxation, a distraction from the canteen and an additional source of fresh vegetables. G. Marshall Woodrow wrote his *Hints on gardening in India* (5th edn., 1889) specifically for the guidance of British troops.

Bungalows were ideal for hill stations lodged on the lower reaches of the Himalay-

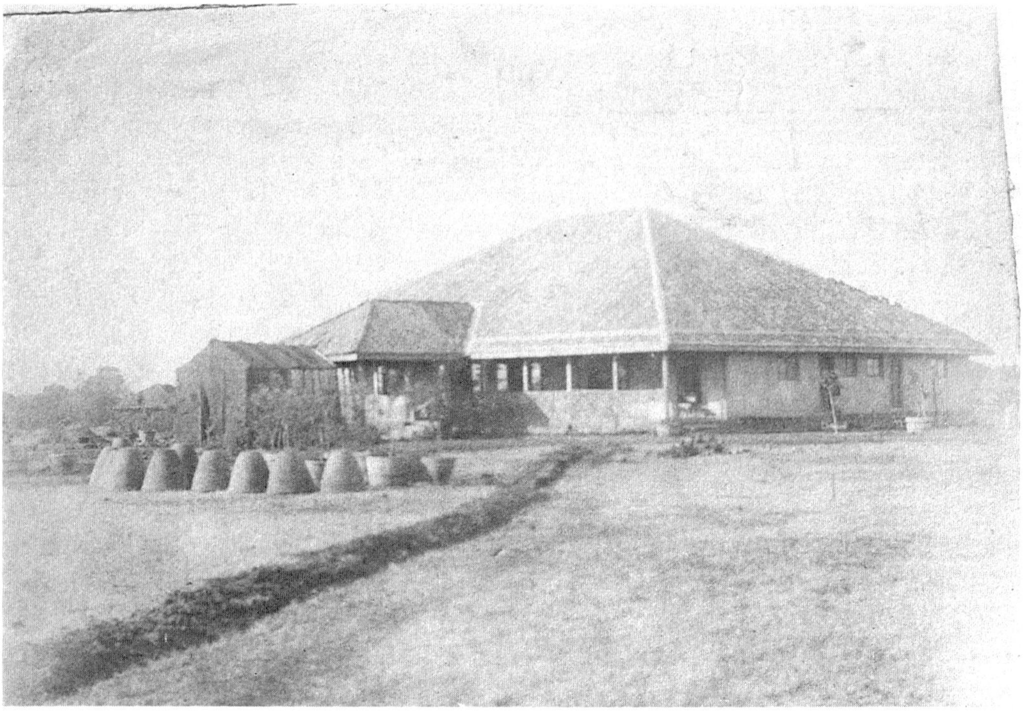

Figure 19.11 A bare bungalow compound in 1884.

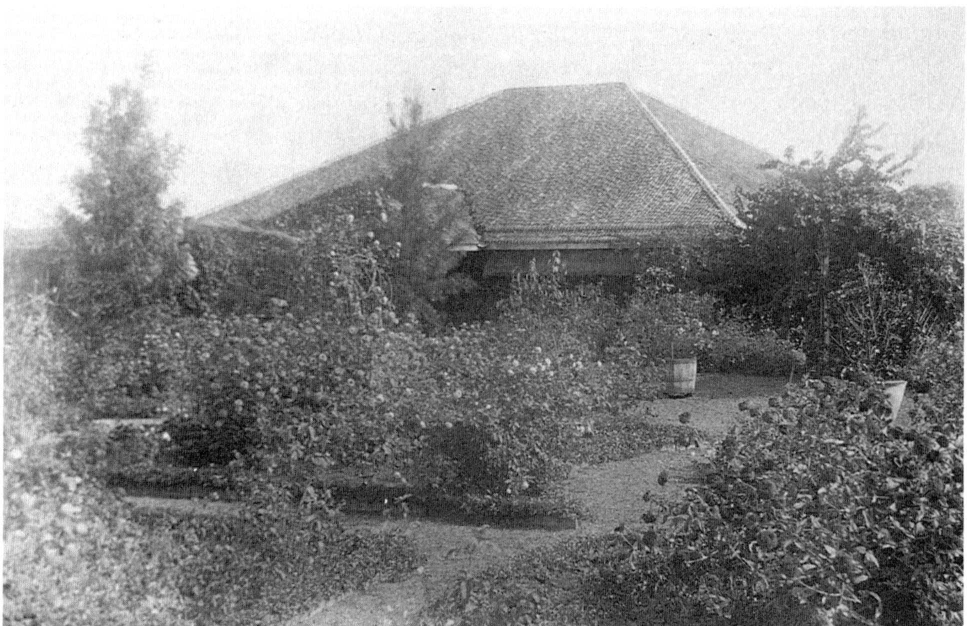

Figure 19.12 The same bungalow (See Fig. 19.11) and its garden four years later.

as, in the Nilgiris in the South, or on any other eminence that offered a cool retreat from the heat of the plains. They attempted to evoke in a domesticated landscape the ambience of spas and seaside resorts, centres of recuperation and relaxation. The sites and sizes of houses and bungalows and their compounds were regulated by social status, the Governor and senior officials always enjoying the best views. The names of the dwellings were wistfully nostalgic—Apple Cottage, Grasmere Lodge, Holly Bank, Sunnybank, Tudor Hall, etc.—and the familiar flowers of an English garden—campanulas, carnations, convolvuli, fuchsias, pansies, primulas, roses, salvias, sweet peas, wallflowers, etc.—completed the illusion of home.

Every year the Governor-General and his staff, with their clerks and their files, migrated to Simla perched 7000 feet up on a ridge in the lower Himalayas. Since many bungalows were deprived of gardens, so steep are the hills and so scarce the land for building, flowers were pampered in pots and hanging baskets. The fortunate few who could claim a patch of ground grew potatoes and cabbages and recklessly dug up bulbs in the local forests to complement their English seeds. There were evidently enough gardens in Simla to provide a living for H. Bwye in the 1870s as a gardener, seedsman, designer, and plant collector. Doubtless one of his best customers would have been Torrentium House, notable for its flowering shrubs and fruit trees with a pond encircled by weeping willows. Experimental planting of European vegetables and fruits was carried out at Mashrobra a few miles out of Simla, and some of its ornamentals were presented to the Viceregal Lodge. The first Viceroy to occupy the new Viceregal Lodge was Lord Dufferin in 1888; his successor, Lord Lansdowne, planted the grounds with the assistance of Parsons, his gardener. Lord Curzon's contribution was more lawns, flower-beds and a rose pergola; stone terraces commemorated Lord Minto's residence there.

No matter how hard Simla's residents tried to fabricate an English setting, the distant Himalayas constantly reminded them they were still in Asia. Ootacamund in the Nilgiris, on the other hand, was convincingly evocative of pastoral England or the Scottish hills. It recalled for Earl Lytton 'Hertfordshire lanes, Devonshire downs, Westmorland lakes, Scotch trout streams'. Macaulay compared its scenery to 'the fine works of the present English landscape gardeners. It is exactly a Wentworth Park or a Bradgate Park as large as Devonshire'. 'Malvern at the fairest season,' was the approving verdict of Stephen Lushington, Governor of Madras. Edward Lear perversely perceived a resemblance to Leatherhead. It was John Sullivan, the Collector at Coimbatore, who persuaded the Government of Madras that Ootacamund was well-placed for a sanatorium. In 1821 he started a garden there with apple and peach trees, strawberries, and European vegetables, and a few years later built a bungalow for his family. In 1822 he was allowed to engage as gardener, John Johnstone, recently arrived from the Cape. Many of the early plant introductions from Europe and the Cape to this hill station are credited to Sullivan, and the lake in a snug valley was his conception (Fig. 19.13). In 1829 the Government of Madras recommended to London the emigration of small farmers and 'mechanics' to the Nilgiris.

Ootacamund's communal endeavour in 1845 to grow vegetables and flowers at a reasonable cost led to the formation of a horticultural society two years later when subscriptions were invited for a public garden. With the Governor of Madras's support, William Graham McIvor was enticed from Kew Gardens in 1848 to take over a garden of some six acres, formerly managed by a corporal in a light infantry regiment. He came with a handsome donation of seeds

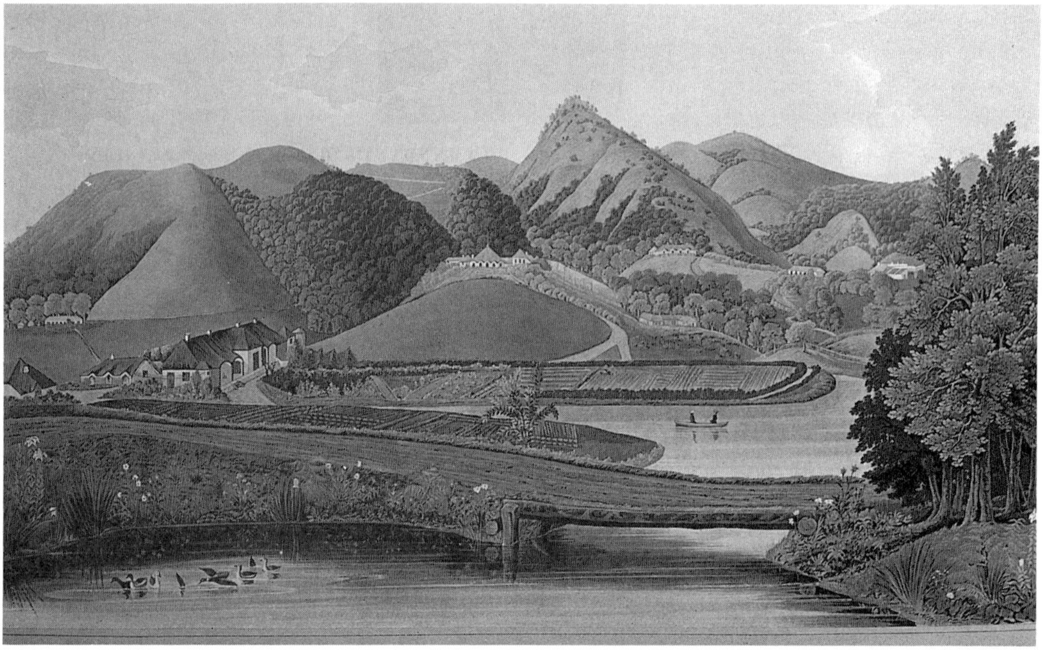

Figure 19.13 The lake at Ootacamund in the Nilgiri Hills, conceived by John Sullivan, the hill station's first European resident. (From R. Barron (1837), *Views in India, chiefly among the Neelgherry Hills*, London.) Coloured aquatint.

from Kew—British and North American shrubs and many Australian species, especially acacias. By filling in a ravine and swamp he enlarged the garden, redesigning it with a pond, terraces, flower beds bordered with ivy, walks overarched with trellis-work, a scattering of urns crammed with flamboyant blooms, clumps of rhododendrons, hedges of heliotrope and fuchsias, and an arboretum of exotics (Fig. 19.14).

Another hill station that encapsulated the feeling of the English countryside was Gulmarg in Kashmir. The traveller, G. T. Vigne, thought that all it needed to make it a splendid English park was a noble house and a herd of deer. Unlike Ootacamund, everything here was on a miniature scale, just a smooth flowering meadow in a valley only about two miles long by half a mile wide. Areas of encircling forest were thinned to scenic clumps of trees. It was a favourite retreat for 'second colonels and other superfluous officers' where 'they are lively as schoolboys, playing cricket and catching butterflies'. It sat high in the long, narrow Vale of Kashmir, renowned for temperate summers, fertile soil, and abundant water. Akbar who had annexed the valley created the first Mughal garden there, Nasim Bagh, which his son, Jahangir improved and his grandson, Shah Jahan, planted thickly with chenar trees (*Platanus orientalis*). It became a summer resort for the Mughal emperors and their courtiers who lined the Dal Lake and the River Jhelum with their gardens. It was 'a garden of eternal spring' for Jahangir. For François Bernier who was there in 1665 it presented 'the appearance of a fertile and highly cultivated garden'. The gardens were complemented by the dramatic backdrop of mountains and a foreground of water, the traditional Mughal walls having been modified to achieve a fusion of garden

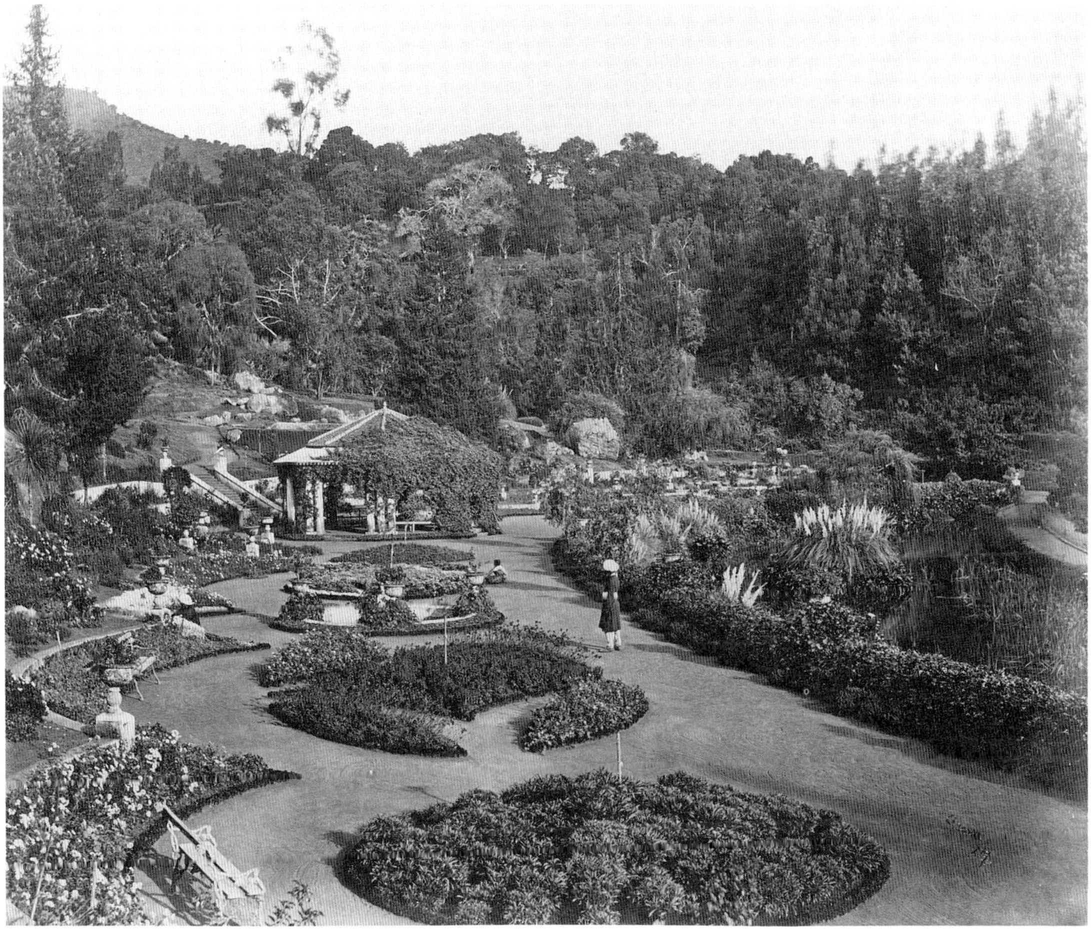

Figure 19.14 Victorian bedding in the public gardens at Ootacamund.

and natural landscape. The majestic chenar trees, spectacular in their gold and russet autumnal tints, helped to relate the gardens to the vast panorama around them. Bernier assessed Jahangir's own garden, Shalamar Bagh, the most beautiful of all. Gone, however, are its pergolas, admired by Baron Hügel in 1835. Some 30 miles away to the south is Achabal, a favourite with Jahangir's consort, Nur Jahan. Even today with inferior Kashmiri buildings replacing the Moghul pavilions, it is still a magical spot: sunlight filtering through the branches of chenar trees and everywhere the sound of water from cascades and fountains. The extravagant use of water and its adroit manipulation impressed Bernier.

What principally constitutes the beauty of this place is a fountain, whose waters disperse themselves into a hundred canals round the house, which is by no means unseemly, and throughout the gardens ... Jets-d'eau in various forms and fish-ponds are in great number, and there is a lofty cascade which in its fall takes the form and colour of a large sheet, thirty or forty paces in length, producing the finest effect imaginable; especially at night, when innumerable lamps, fixed in parts of the wall adapted for that purpose, are lighted under this sheet of water[51].

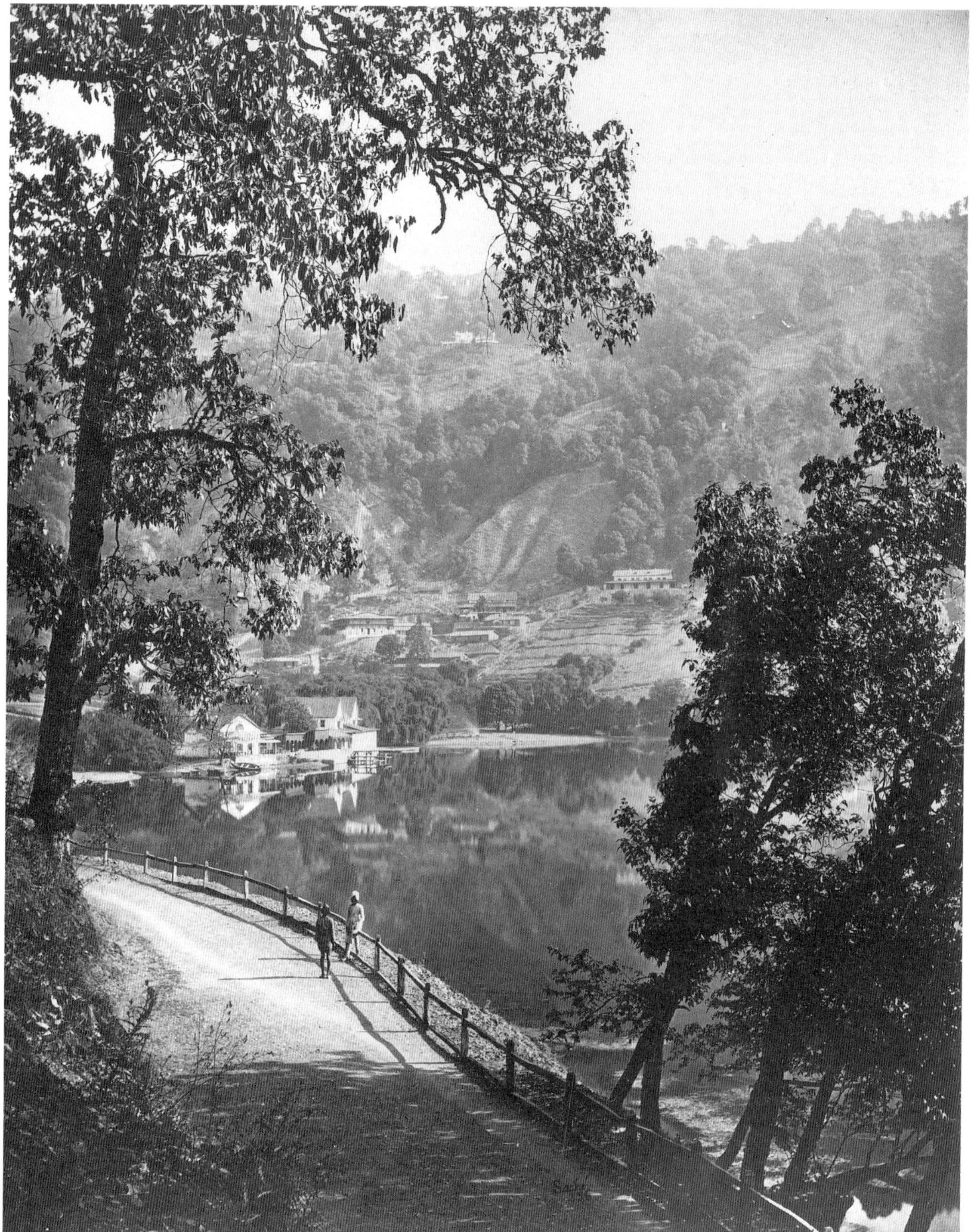

Figure 19.15 Naini Tal, the summer resort of the government of the North-West Provinces, situated on the shores of a thickly wooded lake.

It was not the Mughal gardens but the small rafts of vegetables floating on the Dal Lake at Srinagar that intrigued William Moorcroft. A shortage of land inspired local Kashmiris to construct floating gardens of matted reeds, some 20 yards long and a yard or two wide, supporting a layer of soil and decayed vegetable matter. Anchored to the shallow lake bottom by poles thrust through them, they grew cucumbers, melons, pumpkins, tomatoes, and a variety of root crops. The custom persists today, and it is not uncommon for these gardens to break adrift or to be stolen during the night.

Faced with the Maharajah of Kashmir's refusal to sell them land to build holiday bungalows, the British also took to the water, living in well-furnished houseboats moored around the Dal and Nagin Lakes. Consequently, the only notable British garden in Kashmir belonged to the British Residency, celebrated for its wide lawns shaded by chenars and willows, and immaculate flower-beds.

When the British could not escape to Kashmir or to the hill stations, they took their recreation in public parks, sometimes little more than an open space of worn grass, wide paths, forlorn flowers, a few trees, and a dismal bandstand, all encircled by forbidding iron railings. There were, of course, exceptions like the Eden Gardens in Calcutta, said to have been laid out about 1840 by Fanny and Emily Eden. There walks wind through flowering trees and shrubs, boats can be hired on a serpentine lake in which is reflected the pyramidal outline of a Burmese pagoda—'a sort of Cockney-Calcutta Kensington Gardens', wrote Edward Lear. The oldest park in Bombay is Victoria Gardens, planned in 1861 when it was filled with the produce of the garden of the Agri-Horticultural Society of Western India which managed it until the Society was disbanded in 1873 and it was handed over to the municipal authority. Few public parks can compete with the Lal Bagh in Bangalore whose magnificent trees are a reminder of its earlier function as a royal garden and a horticultural station. Sim's Park at Coonoor in the Nilgiris, established in 1874, vied with neighbouring Ootacamund's public gardens; it had meandering paths, lawns, rockeries, a small lake, and creeper-hung pergolas. Indian princes sponsored public parks which were usually designed by British gardeners and officials: former Kew gardener G. M. Woodrow was largely responsible for the park in Baroda, and the Residency surgeon at Jaipur, F. W. A. Defabeck, planned the city's 'Ram Newas' public garden. When John Beames was a district officer at Cuttack in 1874, he took it upon himself to provide a public garden and, at the same time, to begin a tree-planting programme for roads in the neighbourhood. Opened by the Vicereine, Lady Willingdon, in 1936, the attractive Lodi Gardens in Delhi take their name from the sombre tombs of the Lodi dynasty. The Brindavan Gardens in Mysore with terraces, fountains, and running water, originally laid out in 1932 and subsequently developed, is a rather garish recreation on an impressive scale of some of the features of a Mughal garden.

Although not their primary function, botanical gardens were and still are considered by many who frequent them as public parks, places for leisure as well as education, for horticultural expertise as well as botanical display. Sir Joseph Banks indignantly repudiated Colonel Kyd's proposal that the new Botanic Garden at Calcutta might also be 'a not inelegant retreat for such of the members of administration whose indisposition may preclude from the duties of their station'. There could be no compromise for Banks: the Botanic Garden was 'intended solely for the promotion of public utility & science'[52]. But Roxburgh, Wallich, and certainly Sir George King, aware that the garden they supervised had this dual role, skilfully avoided any serious conflict of interest. Lieutenant G. F. White said approvingly of the Botanic Garden at Saharanpur that it combined 'all the advantages of a highly established pleasure-ground with the interest of the nursery'[53].

Many British residents found Hindu gardens an acquired taste. The artist William Hodges was repelled by their overpowering perfume. Francis Buchanan deplored the unimaginative allocation of plots to a single plant—roses, for example, or pomegranates—in the gardens in the South. R. W. Danvers detected 'a want of freshness' in all the gardens he viewed. Straight walks—a legacy of Mughal precepts—and unpruned fruit trees offended the English taste for meandering paths and an aesthetic juxtaposition of trees and shrubs.

In their selection of flowers, considerations of beauty have no influence with natives. For them, the prime recommendation of a plant is that its flowers have a sweet smell, the second that they are of gaudy and distinct colours. Delicacy of shading, gradation of tint and grace of form, are unappreciated; and beauty of foliage and habit are still more utterly so . . . Landscape gardening is unappreciated, nay unknown among them[54].

Hindu gardens were an undisciplined mass of foliage, united by rampant creepers. Flowers were grown mainly for religious festivals and ceremonies. Wallich believed the British had a duty to inculcate an understanding of gardening principles, to wean natives from

straight and narrow rows of arecas, plants at right angles, beds of endless tagetes and a few others of the most common flowers; wildernesses of inferior sorts of plantains, and some other ordinary fruits such as the mango in its natural state, the guava, and custard apple[55].

Wealthy Indians engaged British gardeners to design their estates in the prevailing western fashion. The Gaekwar of Baroda had the benefit of William Goldring's skill in composing an Italianate garden at Makurpuri Palace (Fig. 19.16); Charles Maries reorganized the Maharaja of Durbhanga's grounds in the 'gardenesque' style before moving on to Gwalior to transform Maharaja Scindia's garden into a Victorian pastiche of lawns, parterres, shrubberies, and ferneries, all linked by a maze of paths. The gardens of those Indians who copied European styles without professional advice to control their enthusiasm sometimes became an exercise in unrestrained vulgarity. The garden of the Kaiserbagh at Lucknow was populated with marble and plaster statues of shepherds and shepherdesses, Greek deities, sundry animals and mythological creatures, even busts of Napoleon and Rousseau. A liberal coating of lamp-black emphasized hair, eyes, and feet. The excessive use of marble prompted Lord Minto to label Raja Rajendro Mullick's residence in Calcutta, the 'Marble Palace'. Its garden, too, flaunted a gallery of busts, but the *pièce de résistance* was an exuberant fountain with frolicking sea gods and naiads. The trend was contagious; even in distant Khatmandu General Mahtabur Sing's garden had unusual fountains—statues of soldiers, for instance, with water spurting from their muskets.

A note of disillusionment introduced the second edition of Thomas Firminger's *A manual of gardening for Bengal and Upper India* (1869):

Under the most favourable point of view it can hardly be said that horticulture has as yet made advancement in India. Of the natives, those of the higher class, it would seem, have never manifested much fondness for it, nor taken much interest in the pursuit; while those who follow it for a livelihood, have not found it sufficiently remunerative to devote to it more than the least possible of their time and thought.

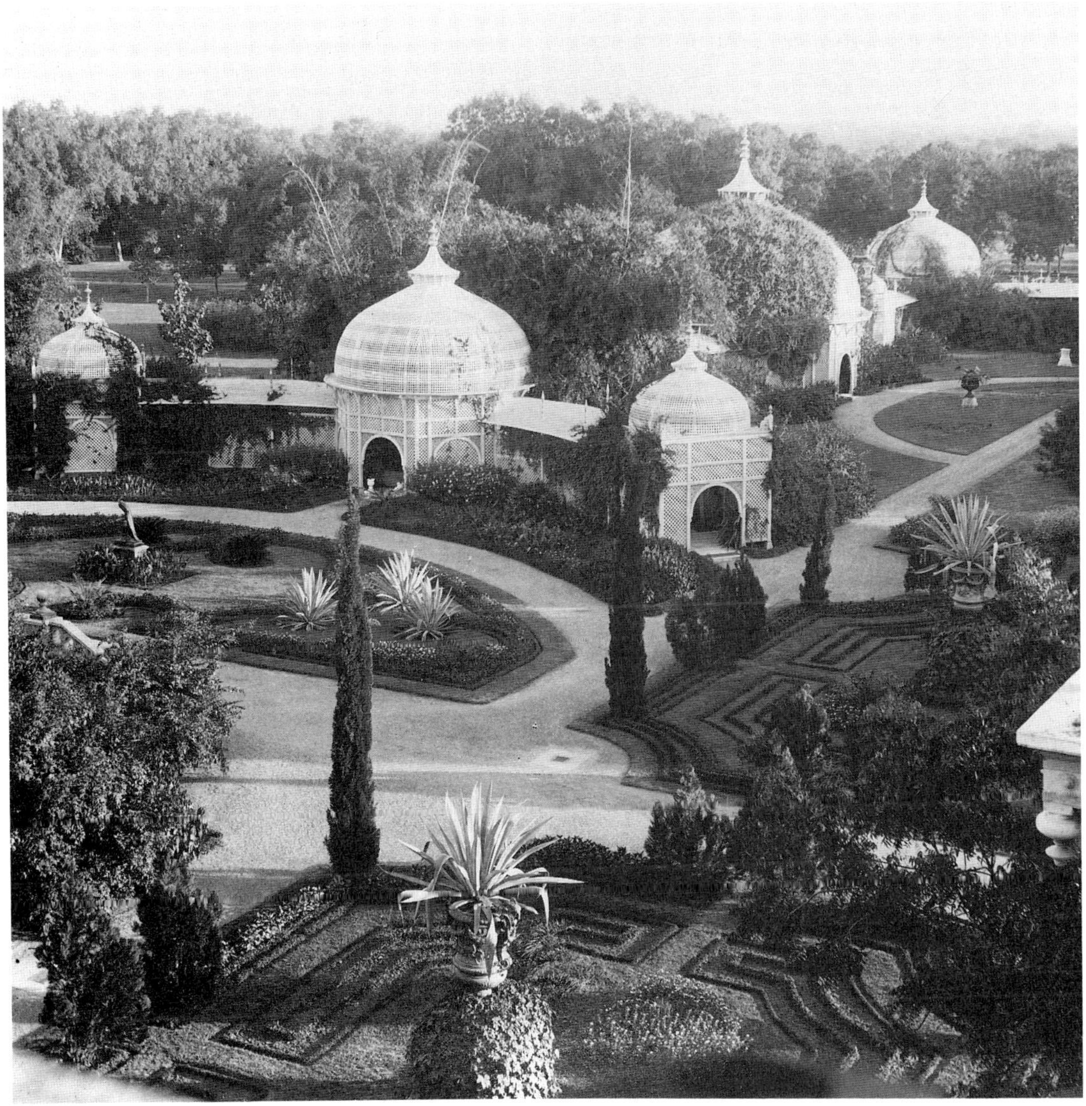

Figure 19.16 The gardens at Makurpuri Palace, Baroda, designed by William Goldring.

He had little confidence in the competence of Indian gardeners or malis: their dedication was proportionate to the amount of supervision they received. Many of them could remember the scientific names of plants with impressive ease but were capable of only the simplest horticultural tasks—clipping lawns, repotting, weeding, perhaps grafting and budding, but never pruning. Left to their own devices they would instinctively fill a garden with familiar balsams and marigolds and plants with religious associations such as Holy Basil, *Ocinum sanctum*, which they then considered a sacrilege to weed. In the retinue of servants that every British household employed, the mali was low in the domestic hierarchy, treated as little better than a common labourer. Even those homes without a garden always had a mali whose function was to supply flowers for the table, though no one would dare ask how he got them. Lady Dufferin, who considered all gardens 'damp and nasty' was resolute in her advice that

the proper and healthy thing to do is to have a gardener, but no garden—his duty being to provide you with flowers at your neighbour's expense, so that you always have as many as you possibly can want, and are spared the disagreeables incident to growing them for yourself[56].

Attempts were made to instruct malis in basic horticultural practices—the Botanic Garden at Dapoori offered such tuition as early as 1828. The Agricultural and Horticultural Society of India set up a school in Calcutta in 1847 for the sons of their employees. They were taught reading, writing, geography, and some botany by native teachers during the day, and worked in the garden under the head gardener in the morning and evening. Although only two boys out of an intake of about 60 over a period of six years took up gardening, it proved a better result than the complete failure of the school run by the Calcutta Botanic Garden. The Victoria Gardens in Bombay, which had 21 malis, six boys for manuring and watering, and four women for sweeping and weeding on their establishment, defined their duties precisely even to the days of the week on which they were to be performed. Since they were not expected to display any initiative, all their routine operations were summarized like a drill manual.

In watering, the 2nd head mallee is to stand at the nozzle of the hose, and an assistant of the 4th head mallee at the hydrant, and the boys are to hold the hose and clear the water channel as the water flows. Watering should not be continued after 9 am, not be commenced earlier in the afternoon than 3 pm. The water channels should not be carried to the base of the trunk of a tree, but should be traced round the shadow of the tree as it falls at noon[57].

Today's malis who tend with exemplary flair Lutyens's gardens at the President of India's official residence or the grounds of luxury hotels have progressed far beyond such elementary guidance.

The enthusiasm and initiative of horticultural societies, regional and local, brought a confident expertise to the art of gardening in India. They imported plants and seeds, maintained demonstration plots and held regular horticultural shows. For years the Agricultural and Horticultural Society of India offered in vain a premium for the best gardening manual. The first in the field was G. T. Frederic Speede's *Indian hand book of gardening, containing directions for the management of the kitchen and flower garden in India* published in Calcutta in 1840. A second edition which appeared two years later was followed by Speede's *The new Indian gardener* in 1848. This was eventually superseded by Thomas A. C. Firminger's *Manual of gardening for Bengal and Upper India* in 1863 which became the standard authority on horticulture in northern India, reaching a sixth edition in 1918. G. Marshall Woodrow's *Hints on gardening in India* which also enjoyed the distinction of six editions, had been intended originally for soldiers' cantonment gardens. Woodrow adapted garden lay-outs published in the English horticultural press. A classic on tropical gardening based on the author's experience at Peradeniya is H. F. Macmillan's *Tropical planting and gardening with special reference to Ceylon* (1910).

It is surprising that none of the botanical gardens in India, always the mentors of the local gardening community, ever considered the compilation of a practical manual. The usefulness of J. O. Voigt's *Hortus suburbanus Calcuttensis* (1845) is strictly limited to informing the amateur gardener of the range of plants successfully grown in the Calcutta Botanic Garden and at William Carey's garden at Serampore. Nevertheless, Wallich saw his Botanic Garden as adequately fulfilling a role in promoting gardening in the Bengal Presidency.

> *The art of gardening which not many years ago was in its infancy in Bengal has by the beneficial influence of the garden in question [i.e. Calcutta Botanic Garden] attracted general attention, and forms at present an object of pursuit among all natives of respectability, there being few among them, who have not improved their nominal gardens of betle nuts and the like into real ones of utility as well as ornament*[58].

Sir Joseph Banks and the directors of the Royal Botanic Gardens at Kew from Sir William Hooker onwards advised either the East India Company or its successor, the India Office, on the maintenance and development of India's botanical gardens. In 1870 there were about 50 Kew-trained gardeners working in the botanical gardens, the gardens of agri-horticultural societies, in public parks, on cotton, tea and cinchona plantations or on princely estates. This supply of qualified gardeners continued up to the Second World War; after independence, India and Pakistan established their own schools of horticulture.

NOTES

1 J. Fergusson (1873), *Tree and serpent worship*, London, 254.
2 McCrindle (1901), 141.
3 *Kama Sutra* (trans. Sir Richard Burton and F. F. Arbuthnot), 1963, 197.
4 W. Foster (1899), *The embassy of Sir Thomas Roe to the court of the Great Mogul, 1615–1619*, Vol. 2, London 537.
5 Sir R. C. Temple (1914), *The travels of Peter Mundy in Europe and Asia, 1608–1667*, Vol. 2, London 83.
6 Tavernier (1925), Vol. **2**, 233–4.
7 Foster (1921), 303.
8 W. Foster (1899), *The embassy of Sir Thomas Roe to the court of the Great Mogul, 1615–1619*, Vol. 2, London, 361.
9 S. Sen (1949), 36.
10 Ibid., 134.
11 J. Fryer (1909), *A new account of East India . . .*, Vol. 1, 263.
12 Sir R. C. Temple (1914), The travels of Peter Mundy in Europe and Asia, 1608–1667, Vol. 2, London, 215.
13 Foster (1921), 316.
14 Bernier (1934), 295–7.
15 2 April 1841. I.O.L.R. F/4/1957.
16 D. Carroll (1972), *The Taj Mahal*, New York, 149.
17 North (1892), Vol. 1, 343.
18 E. Arnold (1886), *India revisited.*
19 *Kew Bulletin*, 1892, 293–6.
20 Bacon (1837), Vol. 2, 369.
21 2 April 1841. I.O.L.R. F/4/1957.
22 Bernier (1934), 267.
23 Bacon (1837), Vol. 2.
24 Grey (1892), Vol. 1, 154.
25 Sir R. C. Temple (1919), *The travels of Peter Mundy in Europe and Asia, 1608–1667*, Vol. 3, Part 1, London, 56.
26 Fryer (1909), *A new account of East India . . .*, Vol. 1, 165.
27 G. Tindall (1982), *City of gold*, London 87.
28 Fryer (1909), *A new account of East India . . .*, Vol. 1, 289.
29 Sir R. C. Temple (1911), *The diaries of Steynsham Master*, Vol. 2, 40–1.
30 Currency not specified, but presumably livres or francs were intended.
31 Tavernier (1925), Vol. 1, 109.
32 J. Burnell (1933), *Bombay in the days of Queen Anne*, London, 152.
33 Dellon (1698), *A voyage to the East-Indies*, London 38.
34 Sir R. C. Temple (1914), *The travels of Peter Mundy in Europe and Asia, 1608–1667*, Vol. 2, London 25–6.
35 Fryer (1909), Vol. 1, 110.
36 *Calcutta Review*, 1919, 234.
37 J. Long (1869), *Selections from unpublished records of Government*, Calcutta, 171.
38 A measurement of land varying from one-third to two-thirds of an acre.
39 Bence-Jones (1973), 30.
40 E. Roberts (1841), *Notes of an overland journey through France and Egypt to Bombay*, London, 278.

[41] A. J. Hare (1893), *The story of two noble lives*, Vol. 2, London 16–17.
[42] *Gardeners' Chronicle*, 19 August 1876, 231.
[43] Heber (1828), Vol. 3, 229.
[44] Countess of Minto (1846), *Lord Minto in India*, London, 31–2.
[45] *Gardener's Magazine*, **10**, 1834, 516.
[46] E Eden (1872), *Letters from India*, London, 94.
[47] Allen (1986), 41.
[48] J. Forbes (1813), *Oriental memoirs*, Vol. 2, 239.
[49] *Bulletin Botanical Survey of India*, **3**, (1), 1961, 6.
[50] T. Pinney (ed.), (1976), *The letters of Thomas Babington Macaulay*, Vol. 3, Cambridge, 93–4.
[51] Bernier (1934), 413.
[52] Banks to E.I.C., 17 January 1791. I.O.L.R. E/1/86, folios 57a–d.
[53] G. F. White (1836–8), *Views in India chiefly among the Himalaya Mountains*, London.
[54] *Calcutta Review*, **49**, 1869, 202.
[55] Wallich to Government of Bengal, 1 October 1836. I.O.L.R. F/4/1761.
[56] Lady Dufferin (1889), *Our Viceregal life in India*, Vol. 2, London, 227.
[57] G. Birdwood (1864), *Report on the Government Central Museum and on the Agricultural and Horticultural Society of Western India for 1863*, Bombay, 163.
[58] Wallich to Government of Bengal, 3 April 1830. I.O.L.R. F/4/1139, item 30257, folio 101.

INDIAN FLOWERS IN EUROPEAN GARDENS

Tudor England grew plants mainly from Western and Central Europe, the Mediterranean region, and the colonies in North America; in the absence of reliable facilities for keeping plants alive during long sea voyages or the means for protecting them in northern winters, comparatively few came from the tropics. Such tender exotics that did arrive were consigned to wooden huts heated by coal pans or placed in a dung bed and covered by mats, cloth, or straw. Some miraculously survived this treatment long enough to flower briefly and be included in such works as Gerard's *Herball* (1597). He was familiar with the 'Tree of sorrow' (*Nyctanthes arbor-tristis*) (Fig. 20.1), a very common Indian shrub: 'most odoriferous and sweet smelling flowers, of the colour of Saffron, which flourish and shewe themselves only in the night time, and in the day time looke withered and with a mourning cheere'[1]. Pierandrea Mattioli (1501–77) saw pepper growing in the gardens of Naples and Venice. When the Portuguese Viceroy of India, João de Castro returned home in 1548, he planted Indian trees and shrubs in his terraced garden at Pehna Verde near Sintra. Some of the Indian flowers painted by Pieter van der Borcht, the principal artist of the Antwerp printer, Christopher Plantin, may have been growing in the Leiden botanical gardens, supervised by Clusius.

In 1599 Clusius had a brick gallery built for his tender plants, which he later fitted with leaded glass windows. Botanical gardens pioneered the development of plant houses—Padua and Pisa, for example, had greenhouses to shelter their precious stock during the winter. A glazed wooden structure heated by four furnaces in 1620 housed orange trees belonging to the Elector

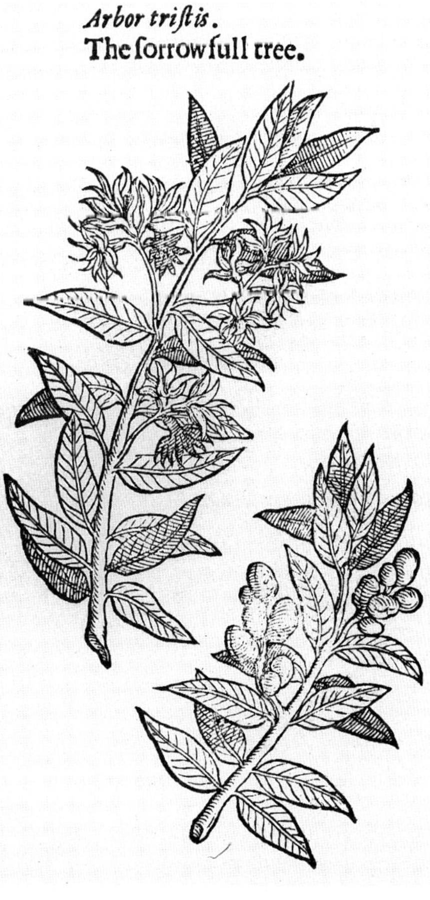

Figure 20.1 *Nyctanthes arbor-tristis.* (From J. Gerard (1597), *The herball or general historie of plantes*, London, pp. 1342–3.) Woodcut.

Palatine at Heidelberg. Orangeries, a Dutch speciality, frequently with south-facing glass windows, were heated by stoves.

The Netherlands was also a trendsetter in the introduction of exotic plants which came from their settlements at the Cape, in India, Ceylon, and the East Indies. Their introduction had no overall strategy until the Stadholder, William of Orange, requested the Dutch East India Company in 1675 to send him every year 'all sorts of animals, birds, tissues, cabinets and other curiosities'[2]. The Governor of Ceylon was asked specifically to send elephants, birds, cinnamon, pepper, and floral novelties. Officals in Ceylon were conscientious in despatching their annual consignment of seeds and bulbs, but Van Reede evidently found it necessary to remind other Dutch settlements of this obligation. In March 1691, in his capacity as Commander-General of the Western Quarters, he instructed all his governors in South Asia, Persia and the Cape

to send annually by the homeward-bound ships from Ceylon to the fatherland all kinds of seeds, bulbs, or roots of the trees, plants, herbs, flowers, etc. which each of you is able to collect in his district for a whole year . . . on leaving Europe I had received a large order from distinguished gentlemen to send annually such a consignment from the Indies, not only to satisfy the curiosity of many wise men and amateurs, but also in order that the world might enjoy and be served by the medicinal virtues which the above-mentioned wise men might discover and find out . . . You should also know that by the name of seeds we do not mean seeds of garden plants, which are tame and known in Europe, but exotic, wild and unesteemed seeds which occur and can be found without exception in the forests[3].

After William, the botanical gardens at Leiden and Amsterdam were first among the privileged beneficiaries of this influx of plants. The earliest attempts to germinate exotic seeds were made, according to Jan Commelin, by Clusius after his appointment as Praefectus Horti of the Leiden garden in 1594. Under the directorship of Arnold Syen from 1670 to 1678, the garden built up a small collection of Ceylonese plants, but it was not until his successor, Paul Hermann, erected tropical greenhouses that adequate conditions were available for this floral bounty from the Dutch colonies. During his eight years as medical officer in Ceylon, Hermann had sent Leiden many plants, including several cinnamons which died during a severe winter. He took the opportunity to bring seeds and bulbs with him when he returned to the Netherlands in 1680. Leiden's collections had more than doubled, amounting to some 3800 different kinds of plants, when Hermann published his catalogue of the garden's stock: *Horti academi Lugdano-Batavi catalogus* (1687). Under Herman Boerhaave, who succeeded Hermann, Leiden became one of the most important botanical gardens in Europe, notable for its tropical plants, but it had a rival in the younger garden in Amsterdam.

When an Hortus Medicus was formed at Amsterdam in 1682, Jan Commelin was one of the two commissioners appointed to supervise it (Fig. 20.2). On the death of the first editor of the *Hortus Malabaricus*, Arnold Syen, in 1678, Commelin had agreed to assume the editorship and his subsequent friendship with Van Reede benefited the Amsterdam garden. The banyan sent by Van Reede from Bengal, was displayed in Commelin's hothouse. This heated shelter protected many tender specimens: a *Cucumis* from Ceylon eventually flowered after three years; both the double and single forms of Paul Hermann's *Nerium indicum* flourished; and it held an impressive presentation of Araceae including several species from Ceylon. Seventy-five of the plants listed in his *Catalogus plantarum Horti Medici Amstelodamensis, pars prior* (1689) were identified from the *Hortus Malabaricus*.

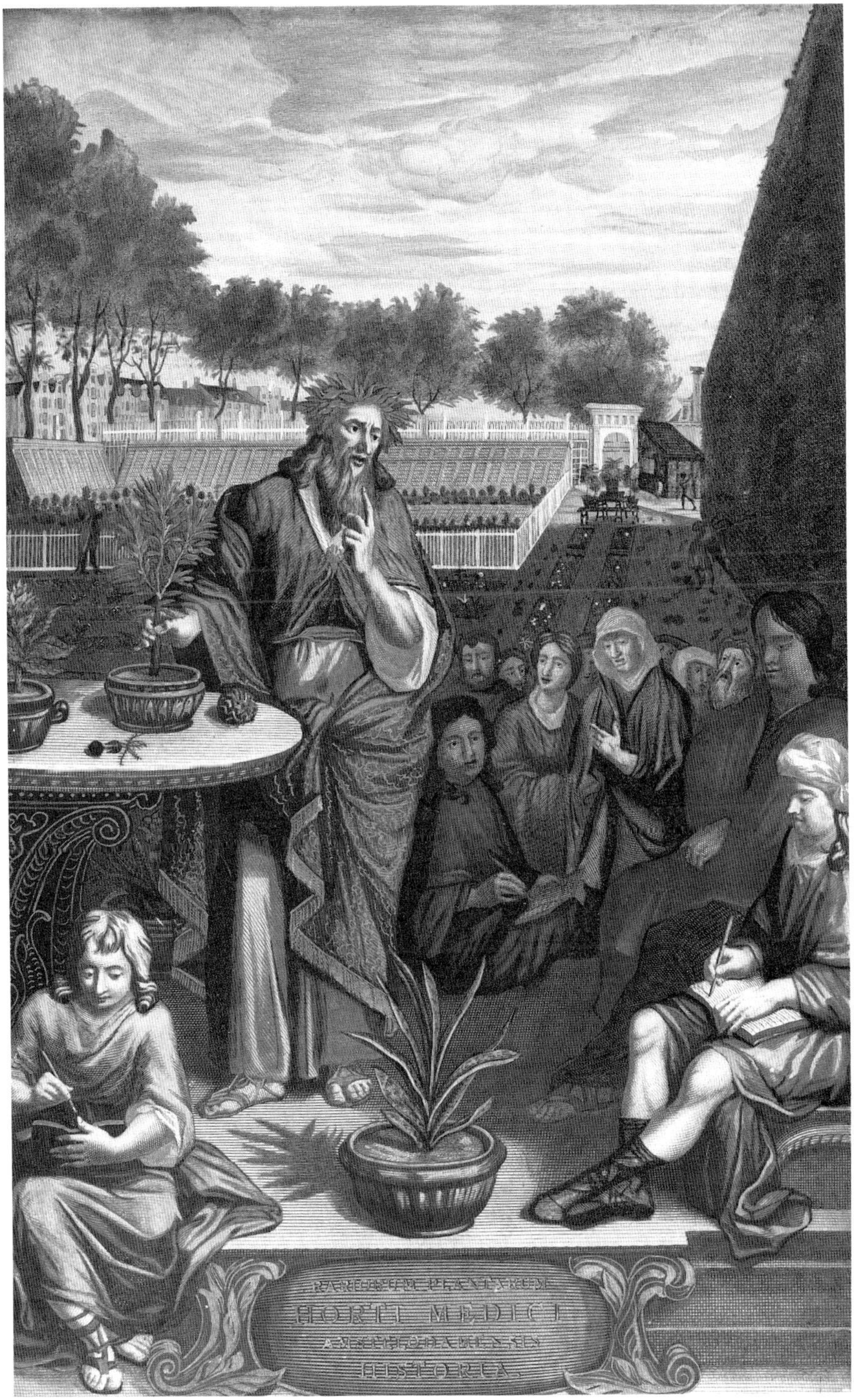

Figure 20.2 Frontispiece from J. Commelin (1701), *Horti medici Amstelodamensis rariorum tam orientalis*, Vol. 2, (Amsterdam.), showing the range of glasshouses at the Amsterdam botanical garden where exotics from the tropics were cultivated.

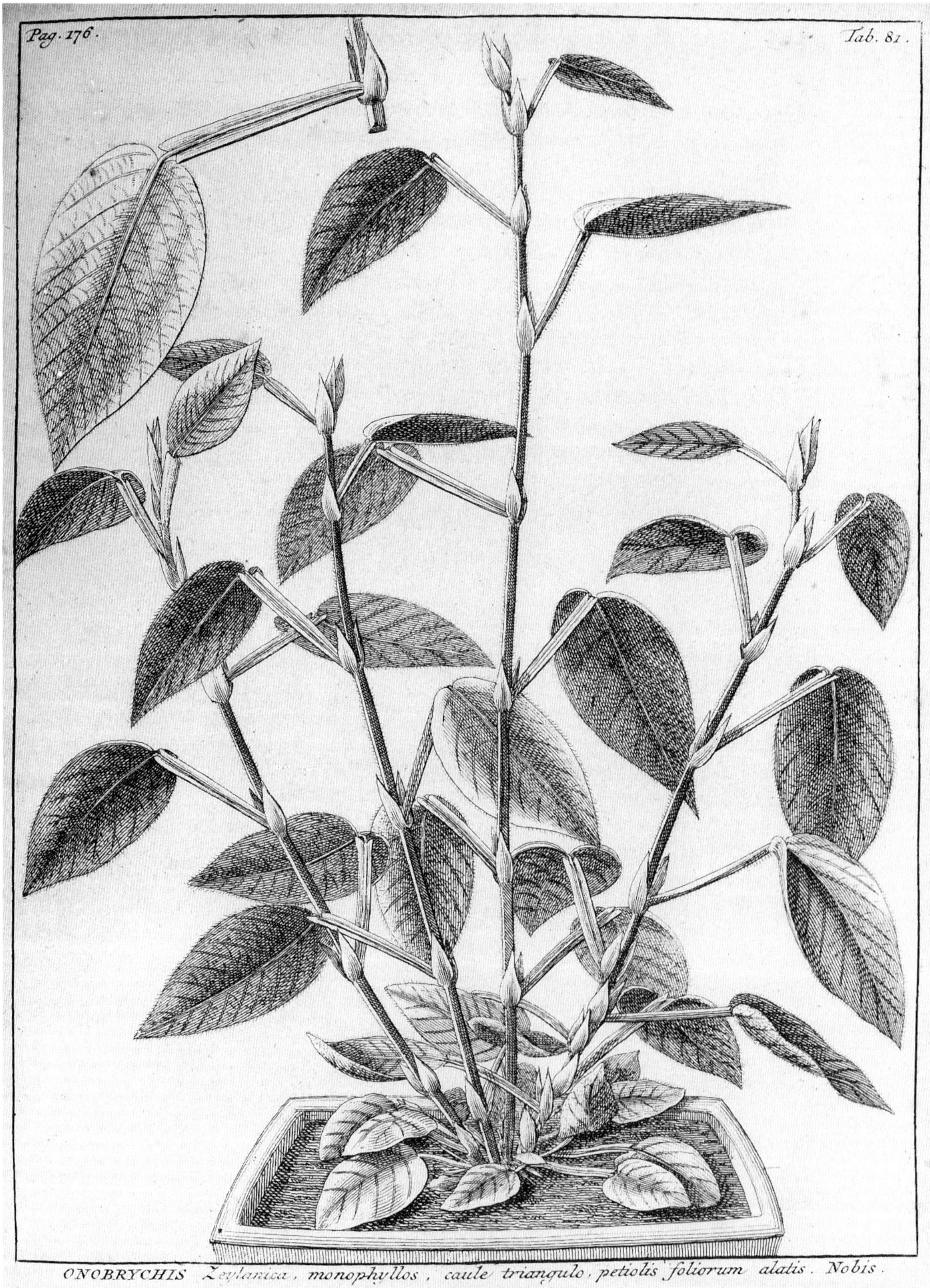

Figure 20.3 *Desmodium triquetrum*. (From J. Burman (1737), *Thesaurus zeylanicus*, Amsterdam, plate 81.) Engraving. Burman's plant was drawn from a living specimen at Amsterdam, grown from seed sent from Ceylon.

In his book *Nederlantze Hesperides* (1676) Commelin defined a hothouse as 'a lightsome room . . . which is free from all cold piercing winds and hath glass-windows to the south, that the sun shining in the winter may refresh these tender strangers'[4]. Amsterdam's hothouses were glass frames built against a brick back wall, stretching over 100 feet, just under 6 feet wide, and with underfloor heating. Gaspar Fagel, the Grand Pensionary of Holland, undoubtedly sheltered his renowned exotics at De Leeuwenhorst in glasshouses. He and Hieronymous Beverningk, another garden enthusiast, succeeded in flowering the spectacular *Gloriosa superba*. One of the finest private gardens in Holland belonged to George Clifford (1685–1765), an Amsterdam banker and a trustee of the city's Hortus Medicus. Being a director of the Dutch East India Company probably made it easier for him to acquire specimens for his glasshouses, one of which was devoted to the Asian flora. Linnaeus who catalogued the contents of Clifford's garden at Hartecamp named 95 plants with the aid of the *Hortus Malabaricus*. In his introduction to the *Hortus Cliffortianus* (1737), he made the observation that plants from South Asia, 'for example from Malabar or Ceylon, take hardly to our winters and prefer a warm greenhouse, though they thrive better below 50 degrees of heat'[5].

While the Dutch and British East India Companies were rivals in trade and territorial expansion, the botanists and gardeners of the two countries exchanged hospitality and plants. Paul Hermann was welcomed at the botanical gardens at Oxford and Chelsea; John Watts, Curator at Chelsea, William Sherard, and George London visited Dutch gardens, and Dutch fashions in florist's flowers found enthusiasts in England. The accession of William of Orange to the English throne in 1688 introduced Dutch glasshouse technology and a taste for exotics to Britain where tender plants were being lost through the lack of suitable hothouses to protect them. Moses Cook, the Earl of Essex's gardener at Cassiobury, experienced such a disappointment.

> *My Lord had thirteen sorts of strange seeds sent him, as I remember, from Goa; I never saw the like . . . I rais'd ten of the thirteen sorts, tho some of them lay almost a year in the ground; but I must tell you, I lost all my ten sorts the first Winter, but one sort, and that the second for want of a green-house*[6].

Enterprising English garden-lovers lost no time in installing hothouses; when J. Gibson made a tour of London gardens in 1691 he discovered that 22 of them including Hampton Court could boast some sort of greenhouse.

William and Mary who chose Hampton Court as a royal residence planned to replace Cardinal Wolsey's Tudor palace. Wren was appointed architect and Dutch gardeners were engaged to work on redesigning the gardens under the new Deputy Superintendent, George London. When Gaspar Fagel died in 1688 William purchased his entire stock of exotics, and the following spring an estimate was drawn up for building glasshouses at Hampton Court to display them. A Dutch carpenter built three 'glass cases', each 55 feet long, 8 feet wide at the bottom, tapering to 5 feet at the top. Their fronts were entirely of glass with a generous number of doors to facilitate the movement of plants. Each glass case had four mobile grates. Fagel's collections had been installed when Sir Christopher Hatton of Kirby Hall went to Hampton Court in April 1690.

> *. . . there is about 400 rare Indian plantes wch were never seen in England; and there is scarce any desirable Indian plant, but a specimen may be seen ther; and some very curious Indian plants are in so great perfection y^t^ it is very wonderfull and scarce credible. The stoves [i.e. hothouses] in wch they are kept are*

much better contrived and built than any other in England[7].

Mary, Duchess of Beaufort, who sent one of her servants to Hampton Court to report upon these new glass cases, was an authority on exotics which she grew at the family home at Badminton and at Beaufort House in Chelsea. Some of her Indian specimens would have come from the botanical gardens at Amsterdam and Leiden with whom she exchanged plants. Another grower of exotics was Henry Compton, Bishop of London, with over a thousand in his glasshouses at Fulham Palace, tended by George London before he moved to Hampton Court. His diary notes the receipt of seeds from Fort St George, Madras on 7 April 1687, 'Indian sedds from Mr London' on 3 May 1687, and the flowering of *Crotalaria juncea* from Malabar in 1700. Within ten years of its foundation, the reputation of the Society of Apothecaries' physic garden at Chelsea lured Paul Hermann to London in 1682; the following year its Curator visited Leiden when seeds and plants were exchanged. The diarist, John Evelyn, saw cinchonas growing in the new Chelsea conservatory warmed by 'subterranean heate . . . all vaulted with brick, so as he [i.e. Watts, the Curator] has the doors and windows open in the hardest frosts, secluding only the snow[8].

The number of hothouses or stoves, as they were generally called, increased significantly during the eighteenth century to accommodate the flow of foreign plants. Flues embedded in walls replaced inefficient iron grates which distributed heat unevenly. There were two types of stoves: dry and bark. Flues below the floor or running several times the length of the back wall heated dry stoves in which potted plants were placed on stepped tiers of shelves in order to receive the maximum amount of natural light. Bark stoves, introduced from Holland early in the eighteenth century, had a pit about three feet deep and six or seven feet wide filled with fresh tanner's bark to make a hotbed. Oak bark, used by tanners to dress leather when mixed with elm sawdust, provided after fermentation a moist heat considered essential for most tender exotics. The allocation of species to dry or bark stoves was often a matter of personal experience or experiment. The Hoxton nurseryman, John Cowell, favoured the bark stove for India's mango trees[9] whereas Philip Miller at Chelsea preferred the dry stove:

> *. . . this plant [i.e. mango] will not thrive in a hot-bed of tanners bark, so that the only way to have them succeed, is to plant them in pots filled with light kitchen-garden earth, and place them in a dry stove*[10].

James Sherard's garden at Eltham in Kent was among the earliest to make use of tanner's bark. Assisted by his brother William, who brought plants and seeds from the Amsterdam and Leiden gardens, James Sherard (1666–1738) built up a collection of rare and exotic plants which he commissioned Johann Dillenius to describe in *Hortus Elthamensis* (1732) (Fig. 20.4). His gardener, Thomas Knowlton, declared that his employer 'was ye first y[t] over designed our now new improved stoves to y[t] perfection we now have'[11]. It is reasonable to assume that a number of the Indian plants in these stoves came from William Sherard's friend, Paul Hermann.

Thomas Knowlton (1691–1781) left Eltham in 1724/5 to become the 3rd Earl of Burlington's gardener at Londesborough in Yorkshire. When a stove was erected there in 1729, Knowlton convincingly demonstrated his expertise in cultivating exotics. His South Asian plants included *Ichnocarpus frutescens, Euphorbia antiquorum*, and *Datura metel.* He affectionately referred to *Crinum zeylanicum* from Ceylon as 'an old acequaintance of mynse & is a prity tender stove plant'[12]. He enriched his collections

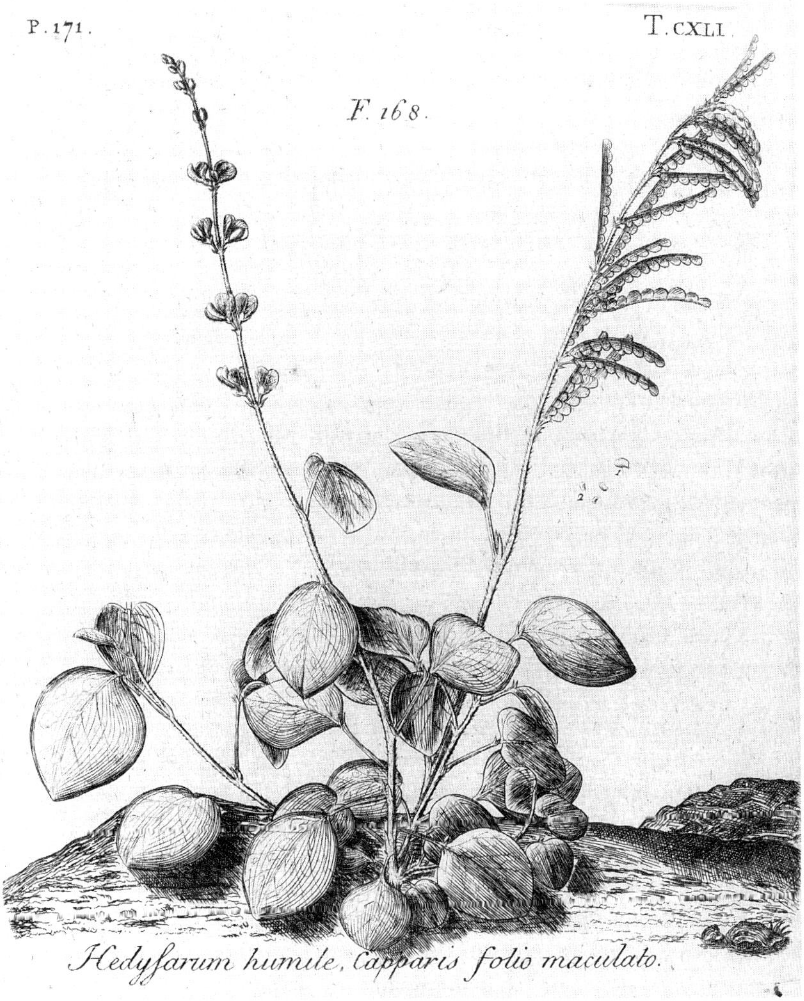

Figure 20.4 *Desmodium gangeticum*. (From J. J. Dillenius (1732), *Hortus Elthamensis*, London, plate 141.) Dillenius described, drew, and engraved the plants growing in the garden of James Sherard at Eltham in Kent. Most probably the seeds of Indian plants such as this *Desmodium* came from William Sherard's friend, Paul Hermann, at Leiden.

through a mutual exchange with other growers who must surely have included John Blackburn of Orford Hall near Warrington, reputed to have built the first hothouse in northern England. Peter Collinson claimed that Blackburn had grown the largest sago palm in the country.

Collinson (1694–1768), who exchanged mainly American species from his garden at Mill Hill for specimens from other countries including India, was at the hub of England's horticultural activities. One of his closest gardening friends, Lord Petre, had at Thorndon Hall in Essex a great stove, 30 feet high, 'the most extraordinary sight in the world', besides several smaller ones, all full of choice plants; inside a screen of floral curtains was contrived with climbers massed on trellis-work. The sago palm and *Ficus religiosa* dominated his Indian collection. On Petre's death in 1742, Collinson estimated his total collection to exceed 200 000 plants, 'mostly exotic'.

Another of Collinson's friends, John Fothergill (1712–80), filled a stove nearly 260 feet long at Upton Park near Stratford in Essex with more than 3400 exotics, 'the natives of Asia and Africa' among them. At Woodford, also in Essex, Richard Warner waited expectantly for seeds from Bengal to germinate, 'but neither he nor Miller knows to what genus they belong, he believes that they have never been in Europe before'[13]. Lord Bute's conservatory at Highcliffe, Hampshire, nearly 300 feet long, displayed its contents, 'growing and flourishing in the soil (not in pots), like an Indian grove'[14]. The library of the Royal Horticultural Society possesses a collection of 737 botanical drawings, formerly belonging to Lord Bute, entitled, 'Flora Asiatica, consisting of coloured drawings of flowering plants chiefly indigenous to Hindostan . . . partly taken from living specimens'. The models for these drawings would have been found not only at Highcliffe but also in the royal gardens at Kew where Bute advised Princess Augusta on their development (Fig. 20.5). 'From his lordship's great knowledge of botany', noted Collinson in his commonplace book, 'the gardens at Kew have been furnished with all the rare exotick trees and flowers that could be procured.'[15] Sixty of the plants listed in the catalogue of the late Princess Augusta's garden—W. Aiton's *Hortus Kewensis* (1789)—had an Indian provenance and others vaguely described as 'East Indies' may also have come from South Asia. Out of that 60, 17 had been cultivated at the Chelsea Physic Garden by Philip Miller who had no doubt presented them to Her Royal Highness.

By the mid-eighteenth century, the Physic Garden had become an important source of new plants for British gardeners, thanks to the efforts of its energetic and erudite curator. During his time there which extended over nearly half a century, Philip Miller (1691–1771) added two stoves and a greenhouse, transformed the collections, and entertained distinguished visitors such as Linnaeus, whom he informed in 1768 that he now had *Cinnamomum camphora* growing at Chelsea and was 'in hopes of getting other rare plants from India next summer'. Collinson was unstinting in his praise of his achievements.

> *He had raised the reputation of the Chelsea Garden so much that it excels all the gardens of Europe for its amazing variety of plants of all orders and classes and from all climates*[16].

He is remembered today chiefly for his *Gardeners dictionary* (1731) which went through eight editions during his lifetime, each revision incorporating the latest horticultural practices and innovations, and describing plants added to the Physic Garden. It became an indispensable work of reference for all serious gardeners who sought his sound, practical advice. He recommended that 'Arum Ceylanicum' (*Typhonium divaricatum*) should be 'kept in the bark-stove, otherwise it will not live thro' the winter in England'. He noted that *Jasminum officinale* 'grows naturally at Malabar, and in several parts of India, yet has been long inured to our climate, so as to thrive and flower extremely well, but never produces any fruit in England'. His comments are a constant reflection of floral fashions. Indian species of *Impatiens*, for instance, were not grown in England because, he said, they were considered to be 'of little beauty'.

No section of the gardening community was better placed for receiving Asian plants than the servants of the East India Company. Charles du Bois (1656–1740) used his position as the Company's Treasurer to procure plants from India for his garden at Mitcham in Surrey. Sir Joseph Banks in return for services rendered to the Court of Directors was able to sow at Spring Grove seeds received from Koenig, Kyd, Roxburgh, and Sir John Murray. When James Forbes returned to England he brought seed collected in Gujarat to Stanmore Hill in Middlesex. His garden was for him a recollection of India epitomized by an octagonal temple adorned with Hindu sculpture, and in his conservatory he

> *had the pleasure of beholding the tamarind-tree, custard-apple, and cotton plant, flourishing with ginger, turmerick, and coffee; and*

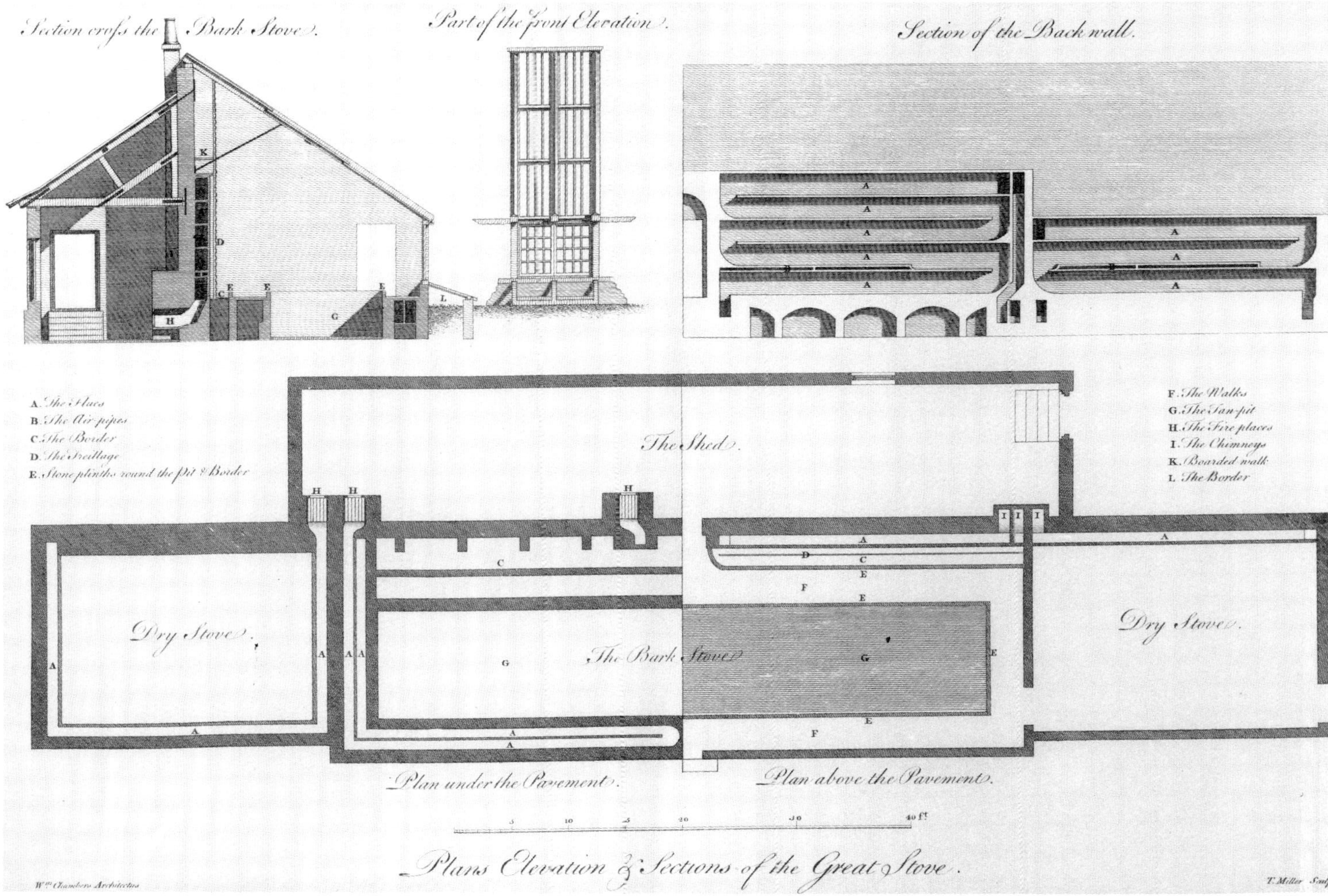

Figure 20.5 The Great Stove erected in Princess Augusta's garden at Kew by Sir William Chambers in 1761. (From W. Chambers (1763), *Plans, elevations . . . of the gardens and buildings at Kew in Surry*, London.) The central portion of this conservatory, which was 140 feet long, was located to a bark stove for growing tender exotics. It was heated by flues in the wall and under the floor.

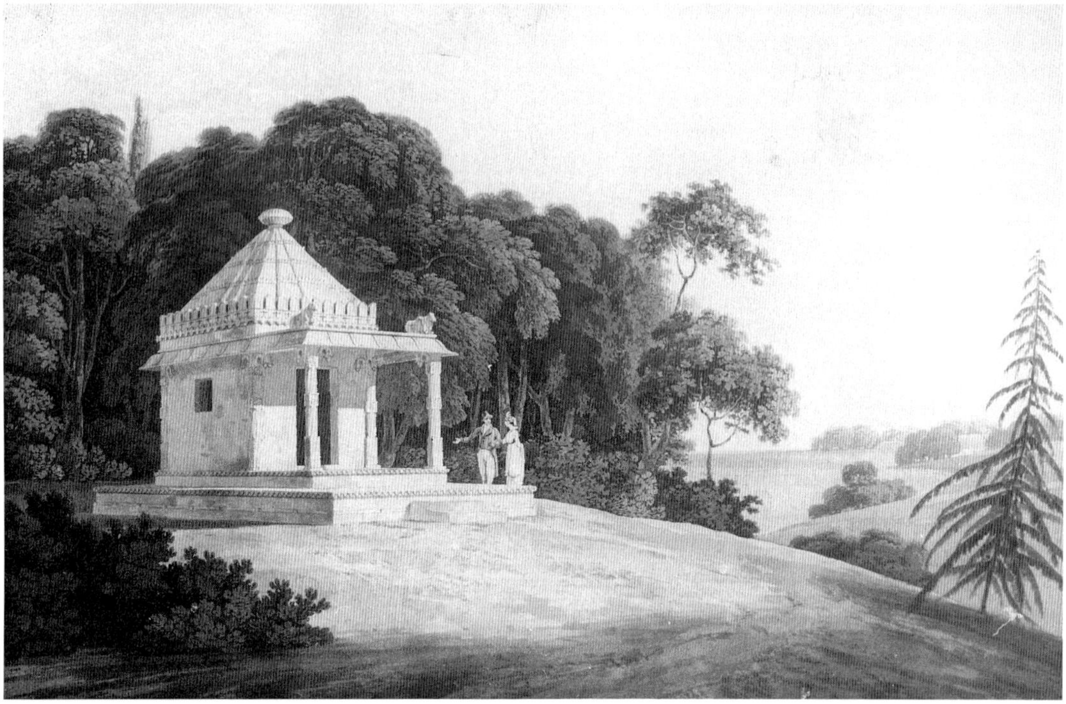

Figure 20.6 The Hindu temple designed by Thomas Daniell for Sir John Osborne at Melchet Park, Hampshire. Aquatint by William Daniell, 1802.

> *... ripe guavas ... entwined by the crimson ipomea, the lovely Mhadavi-creeper of the Hindoos; encircled by the changeable rose (hibiscus mutabilis), the fragrant mogree, attracting alhinna, and sacred tulsee*[17].

Warren Hastings escaped from the stresses of his impeachment trial to his garden at Daylesford near Chipping Norton. There he landscaped 600 acres with woods, a lake and a grotto, tried to acclimatize lychees and the Bhutan turnip and sent for seeds of a favourite custard-apple in the garden of his former villa at Alipore. Using meteorological data he had meticulously recorded in Calcutta, he designed a stove which he hoped would simulate a climate suitable for tropical fruit trees.

Hastings was a frequent guest at Melchet Park on the Wiltshire/Hampshire border, the home of Sir John Osborne, late Colonel of the Nawab of Oudh's Light Infantry Battalion. Their friendship was symbolized by a Hindu temple (Fig. 20.6) in the grounds which enshrined a Coade stone bust of Hastings emerging from a lotus flower, designed by Thomas Daniell, joint author with his nephew, William, of *Oriental scenery* (1795–1808). Daniell was given a much more demanding commission by Sir Charles Cockerell, a Calcutta nabob, who inherited Sezincote in Gloucestershire in 1798 (Fig. 20.7). Here Daniell installed a temple, a snake fountain, a bridge, as well as sundry sculptural embellishments, all in an Indian idiom, to match a new house designed by the owner's younger brother, Samuel Pepys Cockerell, that was a blend of Hindu and Mughal features—minarets, widely-projecting cornices, and a central blue-green dome. At the head of a narrow ravine with a chain of pools linked by winding paths and partly concealed by luxuriant planting is Daniell's temple to the Hindu god Surya. On the lawn below the curving arm of the orangery, a

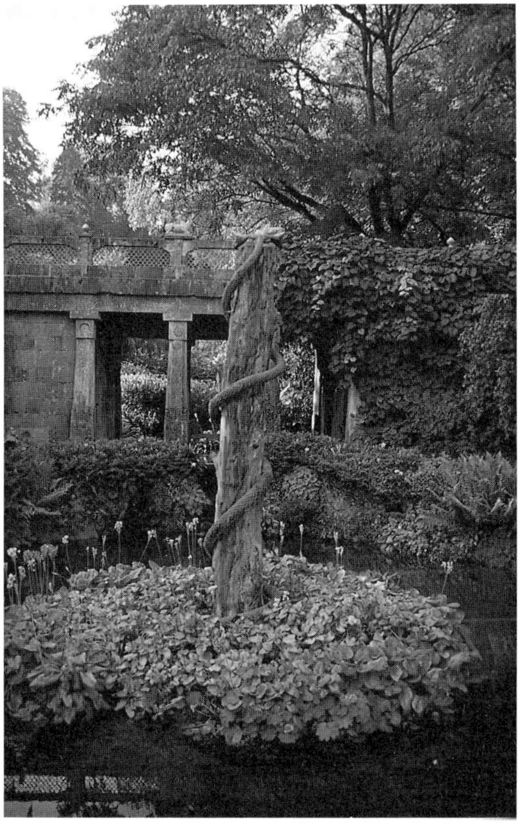

Figure 20.7 The snake fountain and bridge at Sezincote, designed by Thomas Daniell. The bridge is adorned with small Brahmin bulls.

path and a canal are crossed in a manner reminiscent of a Mughal charbagh, a recent imaginative addition by Graham Thomas and Sir Cyril and Lady Kleinwort. The landscape designer, Humphry Repton, was consulted about the general lay-out of the grounds—the lake is attributed to him but the extent of his involvement is uncertain. Repton's enthusiastic response to his introduction to Indian architecture at Sezincote influenced his designs for the Brighton Pavilion.

> *I cannot suppress my opinion that we are on the eve of some great future change in both those arts [architecture and gardening] in consequence of our having lately become acquainted with scenery and buildings in the interior of India*[18].

His optimistic forecast proved ill-founded and Sezincote has remained an isolated attempt to recreate the ambience of an Indian setting. The real impact of the Indian subcontinent lies in its contributions to the English garden flora.

The Sezincote orangery was never particularly noted for its display of tender exotics which were now relentlessly taking over hothouses. Although a devotee, even Sir Joseph Banks found himself yearning for plants

> *such as will thrive in our green houses if not in the open air; in truth our hothouses overflow at present so much with intertropical plants, that I scarce wish for additions of that kind*[19].

Repton also made the point that

> *the numerous tribe of geraniums, ericas and other exotic plants, requiring more light, have caused a very material alteration in the construction of the greenhouse . . .*[20]

Overhead natural light in hothouses was now normal; new methods of improving ventilation were being tested; the source of heat had been gradually progressing over the past two centuries from smouldering peat, free-standing grates, flued walls and flooring to steam and hot water pipes. The wrought-iron glazing bar and the repeal of the glass tax in 1845 ushered in elegant and seemingly insubstantial glasshouses. The words 'conservatory' and 'greenhouse', formerly synonymous, were now differentiated: the former referred to a building which housed exotics planted permanently in the soil or in large tubs; the greenhouse over-wintered plants in flower-pots kept on wooden stages. Gardeners, having slowly discovered that not all exotics required high temperatures, tentatively moved some from the conservatory to the cooler greenhouse, and sometimes boldly wintered out reputedly tender plants such as camellias and rhododendrons.

Any achievements in raising exotics were due as much to the skill of professional gardeners as to technical improvements in glasshouses. Gardeners like Philip Miller, Thomas Knowlton or James Gordon were deservedly renowned for their collections of tropical plants. Gordon who had trained under Lord Petre and James Sherard established his own nursery at Mile End in London in 1742. John Ellis recommended him to Linnaeus.

> *If you send him anything rare, he will make you a proper return . . . I have lately got him a curious collection of seeds from the East Indies, many of which are growing, which are quite new to us*[21].

The nursery trade which had always been active in importing new species expanded spectacularly from about the mid-eighteenth century. By 1812 London alone had over 60 nurserymen, seedsmen, and florists, some of whom specialized in the flora of the tropics. When Conrad Loddiges died in 1826, J. C. Loudon ranked his Hackney nursery as having 'the best general collection of greenhouse and hot-house exotics of any commercial garden'. Loddiges had launched the *Botanical Cabinet* in 1817, an illustrated periodical to promote the sale of his stock, one of a growing number of serial publications describing garden plants, especially new introductions. This genre of literature had been established by William Curtis, the proprietor of the Brompton Botanic Garden. The first monthly issue of the *Botanical Magazine* (later changed to the more familar *Curtis's Botanical Magazine*) appeared in February 1787, its title-page promising subscribers that 'the most ornamental foreign plants, cultivated in the open ground, the green-house and the stove, will be accurately represented in their natural colours'. For nearly two centuries *Curtis's Botanical Magazine* recorded changing tastes in hardy and tender plants, depicted until 1948 in either hand-coloured engravings or lithographs. Its first Indian contribution, *Jasminum officinale*, appeared in the December 1787 issue, and by the time the periodical was merged with the *Kew Magazine* in 1984, nearly 650 Indian plants, mostly from the Himalayas, had been figured from specimens in botanical gardens, nurseries, and private collections.

When *Curtis's Botanical Magazine* featured *Stemona tuberosa* in 1812, it acknowledged its indebtedness to the Comtesse de Vandes's stove at Bayswater in London where many 'curious and valuable tropical plants are cultivated in great perfection'. James Vere's hothouse in nearby Kensington Gore provided Indian species of *Nymphaea, Arum, Hedychium*, and *Amorphophallus*. Between 1810 and 1821 *Curtis's Botanical Magazine* presented seven Indian plants that had flowered at Wormleybury in Hertfordshire, the home of Sir Abraham and Lady Amelia Hume (Fig. 20.8). Their usual source of supply was William Roxburgh in Calcutta, grateful for their support in his efforts to secure an Indian appointment for his son William. One of the magazine's most loyal supporters was the Reverend William Herbert (1778–1847). His garden at Spofforth near Manchester, where he was rector, was filled with exotics-his stove displayed a *Gloriosa superba* and a *Crotalaria pulcherrima* raised from seed to a vigorous three-footer within just four years. The family Amaryllidaceae being his speciality, his first major contribution to *Curtis's Botanical Magazine* was *Crinum zeylanicum* (plate 2121 (1820)) for which he was both artist and author of eight pages of text.

When Nathaniel Wallich despatched 'chests and baskets of plants and parcels of seeds' to London during the period from August 1817 to August 1818, the recipients included not only individuals but also the botanical gardens at Edinburgh, Cambridge, Liverpool, and Hull. On his arrival in London in 1828, Wallich presented to the royal

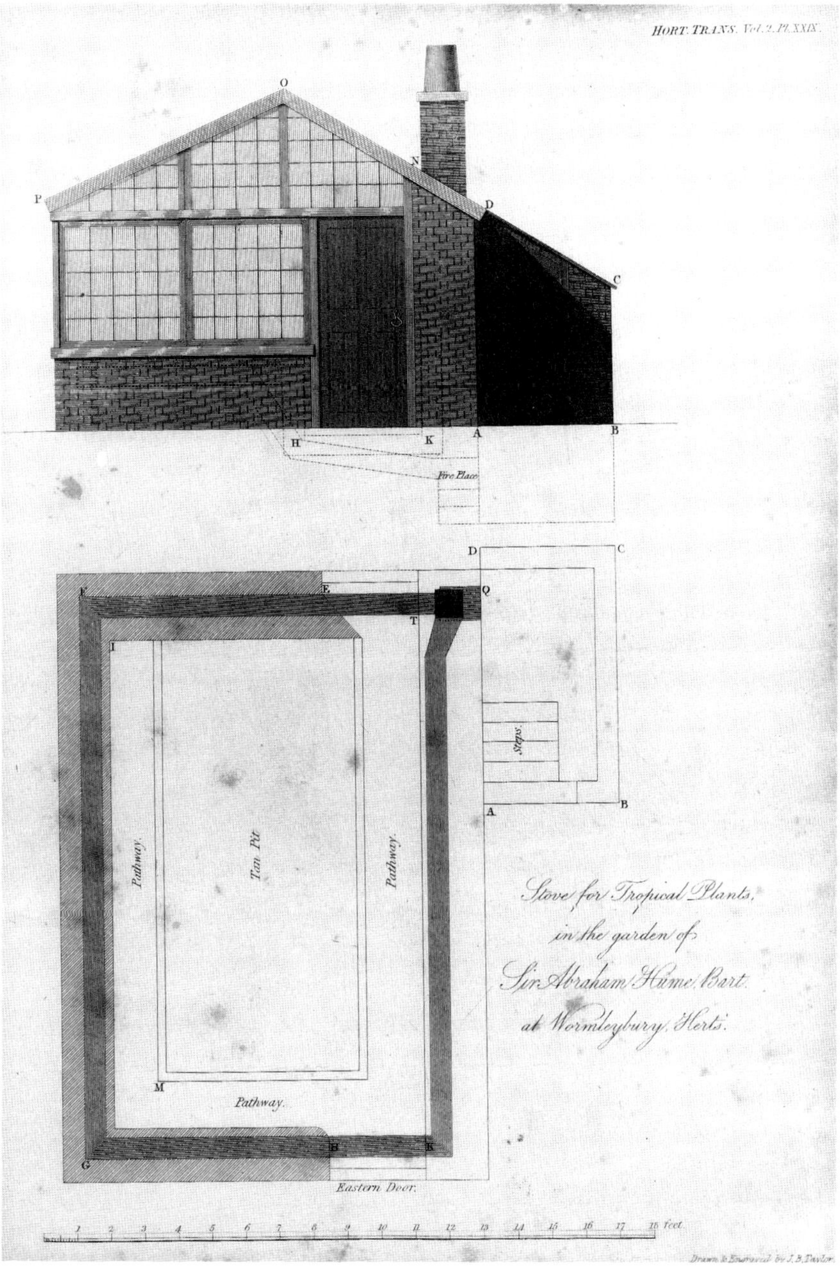

Figure 20.8 One of the stoves in Sir Abraham Hume's garden at Wormleybury in Hertfordshire. (From *Transactions Horticultural Society of London*, Vol. 2, 1817, Appendix, 4–5.) A temperature of between 65° and 70° Fahrenheit was maintained for tropical plants.

gardens at Kew a root of *Hedychium flavum* and a *Ceropegia elegans* from the Nilgiris, both of which flowered within a few years. Kew, an 'odious and useless establishment' in the opinion of Herbert, seldom featured as a source of plants in the early volumes of *Curtis's Botanical Magazine*. The botanical gardens at Edinburgh and Glasgow proved more co-operative during the early years of William Hooker's editorship which began in 1827. When Francis Buchanan retired to Scotland he busied himself naturalizing Nepalese hardy plants in his garden at Leny but, not having a hothouse, gave his Indian seeds to William McNab, Edinburgh Botanic Garden's able Curator, some of whose successes were drawn for *Curtis's Botanical Magazine*.

A key figure in the promotion and development of the botanical garden in Liverpool, founded in 1802, was William Roscoe (1753–1831), banker and historian. He recruited John Shepherd as its Curator to fill its ten acres with plants obtained from nurseries and public and private gardens, and persuaded ships' captains to bring plants to Liverpool. At his instigation, perhaps, Shepherd established a rewarding relationship with the Calcutta Botanic Garden. Roscoe's particular interest was Scitamineae (now Zingiberaceae) with many genera native to India, and both Carey and Wallich sent specimens to Liverpool and Roscoe's own garden at Allerton. The culmination of Roscoe's obsession with this group, his *Monandrian plants of the order Scitamineae* (1824–8), is magnificently illustrated with 112 hand-coloured lithographs of *Hedychium, Zingiber, Curcuma*, and *Kaempferia*. In a limited edition of 150 copies, the fifteenth and final part appeared in 1828 when Roscoe was 75 years old, regretful that he could no longer continue his research.

From 1793 to the mid-1840s, the two superintendents at Calcutta, Roxburgh and Wallich, diligently attended to the needs and requests of British gardeners. Fifty-two species of the Indian flora illustrated in *Curtis's Botanical Magazine* between 1830 and 1934 had either been collected by Wallich and his assistants or had been despatched by him to Britain. His year's residence in Nepal had fostered an abiding interest in that country's flora. He rated *Hedychium gardnerianum* which he had named after the Hon. Edward Gardner, British Resident in Khatmandu, 'the queen of the genus if not of the whole order'. Among his numerous contributions to British gardens are cotoneasters from the Himalayas—*C. frigidus, microphyllus*, and *rotundifolius*, several species of Geranium—*lambertii, nepalense*, and *wallichianum*—*Bergenia ligulata*, and *Pinus wallichiana*. Sir James Edward Smith, being a botanist rather than a gardener, usually disposed of the seeds that Wallich thrust upon him.

> *I received one of your rich packets of seeds, & after letting Mr Shepherd take his choice, I have divided the rest among several friends, particularly the worthy old Bishop of Winchester, who propagates the cinnamon tree by seeds ripened in his own hothouses*[22].

When Wallich died, one of his obituaries acknowledged that 'there was scarcely an English garden of magnitude, which was not indebted to his liberality'.

After the Duke of Devonshire had read Wallich's ecstatic account of the Burmese flowering tree, *Amherstia nobilis*, in his *Plantae Asiaticae rariores*, he resolved to have one for his great conservatory at Chatsworth. John Gibson (1815–75), a member of his gardening staff, was selected for the mission which was also to embrace orchids and other rarities. A visit to Loddiges's nursery alerted Gibson to some of the exotics he should seek in India. He collected profitably in the Khasia Hills after his arrival in Calcutta late in 1835, but as he had never reached Burma to harvest the seed of *Amherstia nobilis*, Wallich gave him two *Amherstias* from the Botanic Garden—one for the Duke, the other for the East India Company. When one of them died on the voyage, the Court of Directors generously allowed the Duke of Devonshire to keep the survivor. Though Paxton tended it carefully at Chatsworth it never flowered because, so it has been suggested, it had been placed in a tub treated with a deleterious wood preservative. The honour of flowering it fell to Mrs Louisa Lawrence, wife of the celebrated surgeon William Lawrence, at Ealing Park in Middlesex. In 1847 Lord Hardinge, the Governor-General of India, gave her a young tree which, two years later when it was 11 feet tall, produced its spectacular racemes of red flowers tipped with yellow. One pendulous inflorescence was presented to Queen Victoria, and another to the editor of *Curtis's Botanical Magazine*. She presented the tree itself to Kew in 1854 where it continued to flower until its transfer to the Palm House killed it.

Gibson went to India at the whim of a Duke, whereas Thomas Lobb (1820–94) travelled as an employee of that shrewd nurseryman, James Veitch of Exeter. Lobb had only recently returned from a plant collecting expedition in the Far East for his firm when he was despatched to India in 1848. High in the Khasia Hills he encountered J. D. Hooker and T. Thomson. 'We spent the evening and following morning most pleasantly', Hooker informed his father, 'but he would not stay even a day with us, though in no hurry—it appears odd to me—he talks very slightingly of the plants and seeds as usual and to judge by what he pays he cannot be worth 6d to Veitch and Co.'[23] Lobb's taciturnity probably fostered this unfavourable impression which was not confirmed by results. He discovered several new orchids in the Khasia Hills, Assam, and Burma, and from Nepal sent back *Berberis wallichiana* and *Lilium wallichianum* var. *neilgherrense*. When *Curtis's Botanical Magazine* figured his *Hypericum oblongifolium* (*H. hookerianum*) from Assam in 1856, it predicted enthusiastically that 'so ornamental a plant will find its way into every garden and shrubbery'.

Lobb had found a number of his plants in the Himalayas, that seemingly inexhaustible supply of hardy plants which have continued to enhance our gardens over the last hundred years or so. John Lindley, writing in his capacity as Assistant Secretary of the Horticultural Society of London, urged the Court of Directors to concentrate on the flora of what was then a relatively little-explored region.

> *It is now more desirable to limit the supplies of seeds from India to those which come from the cold provinces because in those districts there is a great abundance of species the introduction of which is an object of national importance. The similarity of the climate of Great Britain and of many parts of the Himalayas is well known*[24].

After a couple of decades into Queen Victoria's reign collectors had become a familiar sight in northern India and the Himalayan foothills. Among them were army officers like Lieutenant Colonel Edward Madden (1805–56) of the Bengal Artillery who, between 1841 and 1851, presented seeds to the botanical gardens at Glasnevin and Kew. His most memorable introduction was the giant Himalayan lily, *Cardiocrinum giganteum*, described by Wallich in 1824 as *Lilium giganteum*, which first flowered in the Reverend J. T. Boscawen's garden at Truro in Cornwall in 1851 (Fig. 20.9). Several of his introductions—the butterfly bush *Buddleia crispa* and *Abelia triflora*—had already reached Europe but his specimens would appear to be the first to get established in British gardens. *Rhododendron maddenii* commemorates him.

Himalayan trees and shrubs with the potential to be hardy in Britain were always sought. Out of well over a hundred introductions one can but serendipitously mention a few: the fir *Abies spectabilis*, the alder *Alnus nepalensis*, *Cotoneaster frigidus*, *Magnolia campbellii*, *Picea smithiana*, and *Sorbus cashmiriana*. Kew Gardens received a handful of *Cedrus deodara* seeds in 1831 from Leslie Melville who had casually brought a few cones back with him, and raised a solitary specimen which had attained a height of 32 feet by 1884. In 1841 the Commissioners of Woods and Forests in London resolved to encourage the cultivation of this majestic timber tree. The Court of Directors agreed to import 400–500 lbs of seed annually for distribution to public and private gardens.

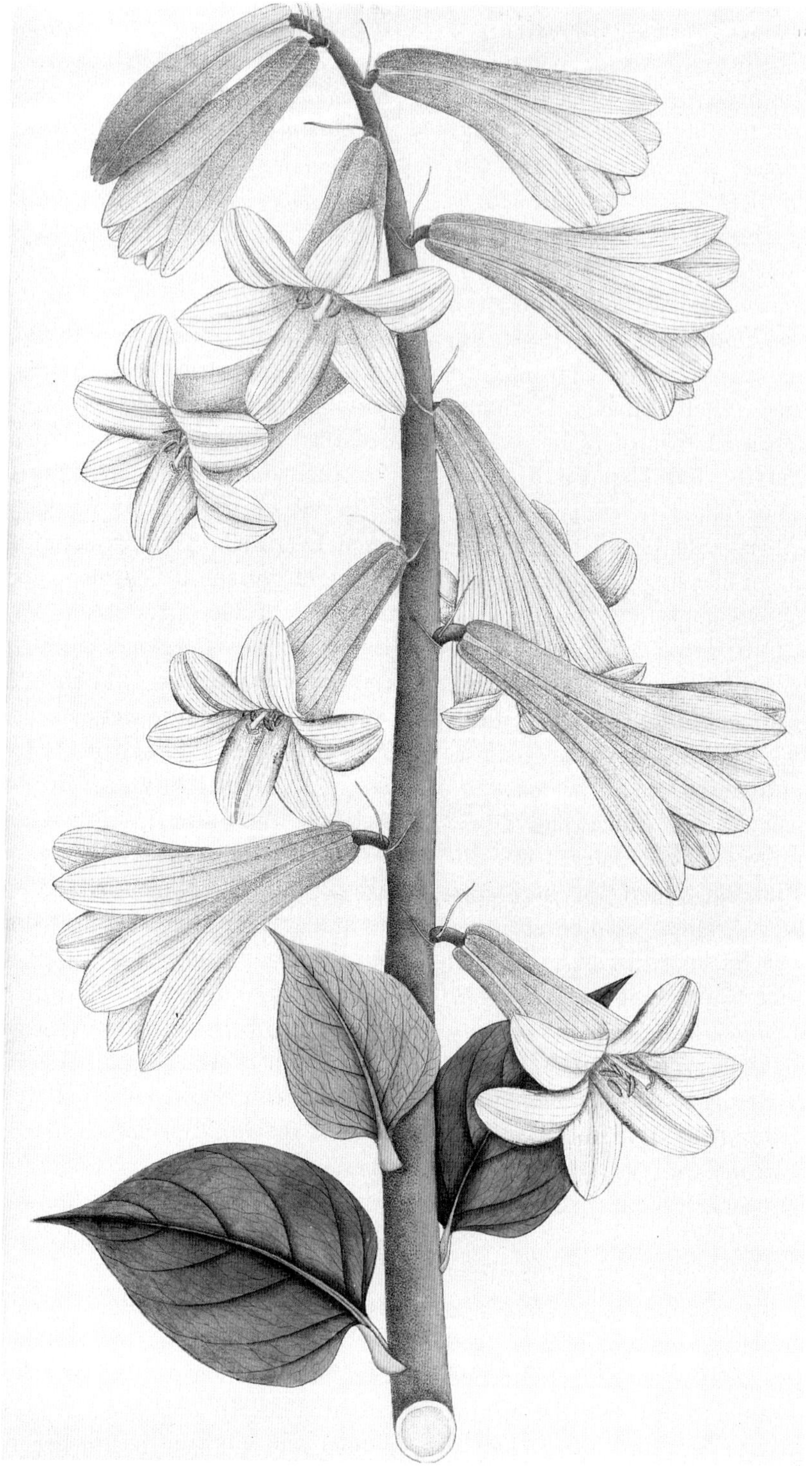

Figure 20.9 *Cardiocrinum giganteum*. (From N. Wallich (1824), *Tentamen florae Napalensis illustratae*, Calcutta, plate 12.) Lithograph. This giant lily was discovered by Wallich but Edward Madden was the first to send seed which reached Glasnevin in Dublin in 1848. The first seedling from this stock flowered at Lamorran Rectory near Truro in Cornwall.

By 1863 when Saharanpur Botanic Garden sent 2000 lbs of seed in that year alone, the supply had become so plentiful that saplings which used to sell for £5 and £6 each were now available from nurserymen at £1 per hundred!

Of all the trees and shrubs native to the Himalayas, it is the rhododendron which has had the greatest impact on British garden design—being easy to cultivate, with bold masses of colour in early spring, made it an obligatory component of all woodland gardens. The Highlands of Scotland, Surrey woods, and Cornish estates tried to emulate the scenery of the Himalayas with congregations of species and hybrids, often creating an insensitive clash of colour. Over 450 species are now in cultivation, the greatest number coming from the Sino-Himalayas. The first Indian species to be described—*Rhododendron arboreum*—was discovered by Thomas Hardwicke in Kumaon in 1796 (Fig. 20.10); Sir James Edward Smith used his notes and drawings when featuring it in *Exotic botany* in 1804. Smith succeeded in germinating at least one of Hardwicke's seeds, but the first recorded flowering occurred at The Grange, Alresford in Hampshire in 1826. Wallich who also sent seed was certain that most rhododendrons would adapt to the British climate. It was Joseph Hooker's collection of 25 species in the Sikkim-Himalayas that confirmed Wallich's optimistic prediction. *Rhododendron ciliatum*, while still only seven inches high, was the first of his discoveries to flower at Kew; *Rhododendron dalhousiae* and *R. maddenii* evoked the admiration of *Curtis's Botanical Magazine* but they were not the equal of *R. grande*; Sir William Hooker thought that none could surpass the brilliant colour of *R. fulgens*. Of all Hooker's introductions, *R. falconeri* proved to be the most hardy. When Thomas Nuttall saw the flamboyant lithographs in the first part (1849) of Joseph Hooker's *Rhododendrons of Sikkim-Himalaya*, he wasted no time in despatching his nephew, Thomas J. Booth, to India to bring back specimens for his garden at Nutgrove near Liverpool. Booth, barred from turbulent Sikkim, collected several new species in Assam, among them *R. nuttallii* which was acclaimed by *Curtis's Botanical Magazine* as 'the Prince of Rhododendrons'[25].

John Gibson's haul of plants included *Rhododendron formosum*, one of Wallich's early discoveries, but the core of his collection was some 50 species of orchids, mainly from the Khasia Hills—*Dendrobium*, *Phaius*, *Vanda*, *Coelogyne*, and *Saccolabium*—all destined for the Chatsworth hothouses. William George Spencer Cavendish (1790–1858), sixth Duke of Devonshire, an early victim of orchidomania, amassed one of the largest collections of these aristocrats of the plant world. The honour of flowering the first tropical orchid in Britain has been bestowed on Sir Charles Wager at Fulham in 1731 with a *Bletia* from the Bahamas. Difficult plants to rear, orchids remained rarities in hothouses for much of the century—in 1789 even the royal gardens at Kew could offer only 15. In 1812 Roxburgh sent *Aerides*, *Dendrobium*, and *Vanda*—the first Indian species to be seen alive in England—to Loddiges's nursery, soon to become a major orchid supplier. The newly-formed botanical garden at Liverpool impressed Sir William Hooker by its success with these seemingly temperamental plants and he had their *Dendrobium aphyllum* from Carey and *D. fimbriatum* from Wallich copied for his *Exotic flora* (1823). His son was in India at a time when the fashion for orchids dictated absurdly high prices for new or striking species such as *Vanda coerulea* which Joseph encountered at 3000 to 4000 feet in the Khasia Hills.

We collected seven men's loads of this superb plant for the Royal Gardens at Kew, but owing to unavoidable accidents and difficulties, few specimens reached England alive. A gentleman who sent his gardener

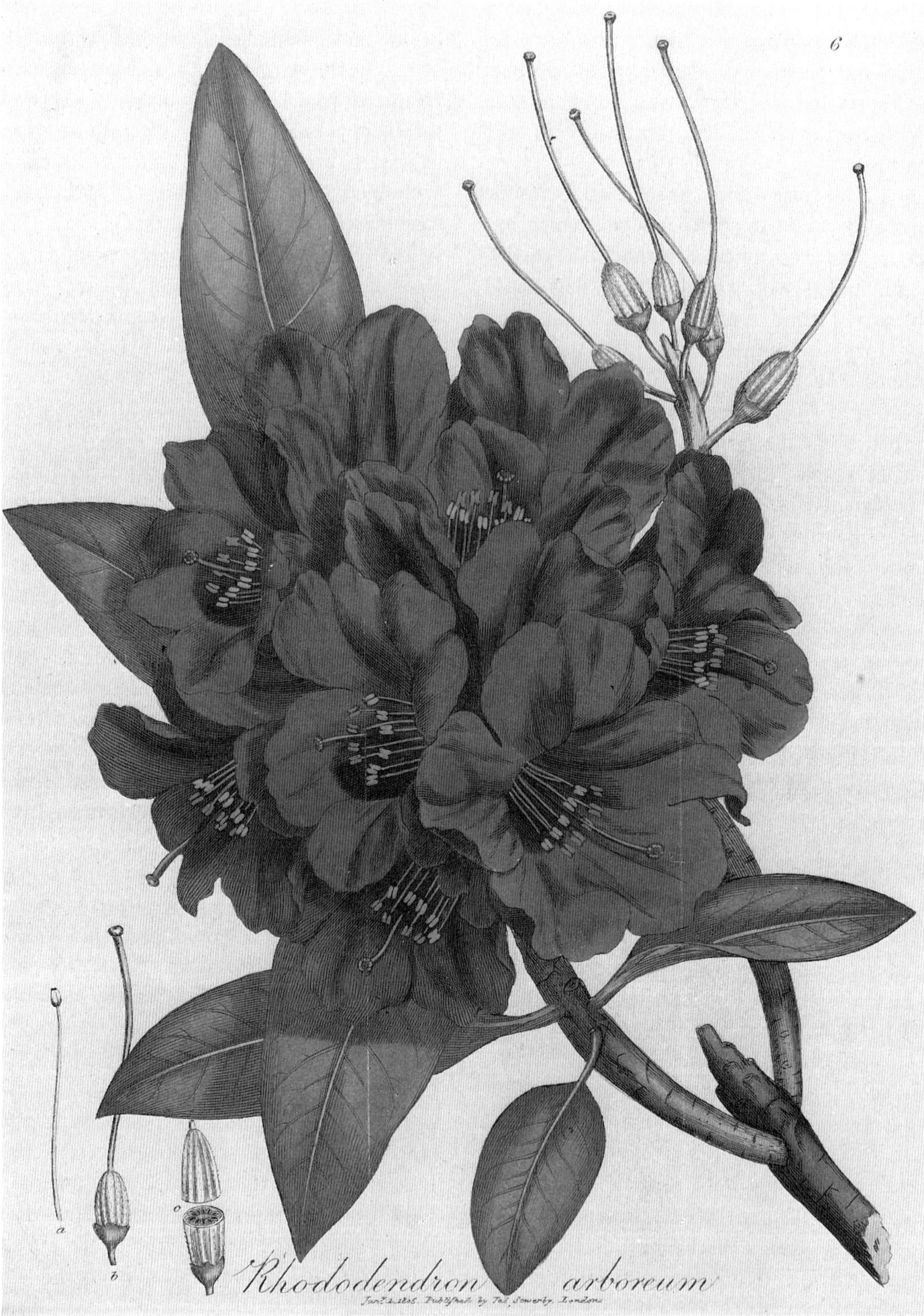

Figure 20.10 *Rhododendron arboreum.* (From Sir J. E. Smith (1804), *Exotic botany*, Vol. 2, London, plate 6.) Hand-coloured engraving. Collected by Thomas Hardwicke, of whom Smith wrote: 'It is hoped the seeds which that gentleman has liberally distributed in England will enrich collections with this noble tree'.

with us to be shown the locality, was more successful: he sent one man's load to England on commission, and though it arrived in a very poor state, it sold for 300l; the individual plants fetching prices varying from 3l to 10l. Had all arrived alive, they would have cleared 1000l. An active collector, with the facilities I possessed, might easily clear from 2000l to 3000l, in one season, by the sale of Khasia orchids[26].

Thomas Lobb, working for James Veitch, also collected this coveted *Vanda* besides new species of *Aerides*, *Coelogyne*, and *Dendrobium* gathered over an area extending from the Khasia Hills to Lower Burma.

Too often collectors laboured in vain when their prizes languished or died in inhospitable hothouses. The search for the correct treatment of orchids was very much an empirical exercise founded on inadequate data about their habitats. In the 1780s Chelsea Physic Garden was potting orchids in a composition of rotten wood and decayed leaves, then plunging them into a tan pit. Kew bedded them in fragments of half-rotten bark and applied very great heat. For a long time epiphytic orchids were believed to be parasites like mistletoe; in the 1820s Kew placed orchids on branches and pieces of wood in a loose, turfy soil, achieving an encouraging growth before heat and moisture eventually killed them. India's *Cymbidium aloifolium* grew in James Vere's tan bed but never flowered; Greenwood and Wyke, nurserymen at Kensington, were more successful: by placing their specimen on the floor of the hothouse they were rewarded with a bloom. Lady Banks flowered her *Vanda tessellata* by following her late husband's practice of suspending epiphytes in moss-filled baskets from the hothouse roof. Loddiges's orchid house had brick flues with a steamy tan bed to produce the humid atmosphere believed essential. John Lindley, having been informed by Wallich that Nepalese orchids were more abundant in the dense, dark, and humid jungle prescribed, not unreasonably, a combination of shade, high temperatures, and extreme humidity. Growers who accepted implicitly the *obiter dicta* of the country's leading orchidologist, saw their treasures bravely but briefly bloom and then die; England had become, in Joseph Hooker's words, 'the grave of tropical orchids'. Joseph Cooper, gardener to Lord Fitzwilliam at Wentworth, contrary to current practice, reduced temperature and ventilated his orchid house, and was rewarded with a pageant of colour. A similar approach was adopted by Paxton at Chatsworth where conditions, so said Lindley, were 'as mild and delightful as that of Madeira'. James Bateman at Knypersley had discovered that *Coelogyne* preferred coolness. Through trial and error, the optimum conditions for different species were discovered and basic requirements established: light, ventilation, and houses heated preferably by hot-water pipes and divided into compartments to provide variable temperatures.

Conservatories and hothouses, always expensive to maintain, had become rarities by the Edwardian era when tropical exotics yielded to the alpine flora of the Sino-Himalayas. The regimented display of bedding plants, so beloved by Victorians, was forsaken for the new hardy plants collected by Wilson, Forrest, Kingdon Ward, and the partnership of Ludlow and Sherriff. They were grown in wild and woodland gardens and alpines, cherished for their individual merit, were displayed to advantage on rockeries.

Wallich discovered the first of the Himalayan poppies—*Meconopsis napaulensis*—but it was Joseph Hooker who introduced its spikes of blue, puce, brown or white tinged with blue to woodland gardens. For H. J. Elwes who grew it in his Gloucestershire garden at Colesborne, it was 'the most beautiful herbaceous plant in the world'.

Imagine a large rosette of leaves clothed with long golden hairs, which, when covered with raindrops, glisten in the sunshine, running up into a branching spike of golden green buds covered with similar hairs, and opening from the top downwards into large poppy-like flowers, normally of a bright pale purple, whose centre is filled with a mass of golden anthers[27].

Alert to the horticultural potential of the genus, Hooker also gathered seed of *Meconopsis simplicifolia* and *M. paniculata*. The most dazzling of all the blue poppies, *M. grandis*, was found by George Watt among the ruins of some stone huts south of Kangchenjunga. For ten years after 1927 when Nepal was closed to foreigners, two local collectors—Lall Dhwoj and Professor K.N. Sharma—sent seeds of *Meconopsis regia*, *longipetiolata*, and *dhwojii* to Thomas Hay's nurseries in the Royal Parks. Nepal is the home of some dozen species of this genus now well-established in our gardens.

Primulas are also prominent in Nepal's rich flora but Western China is their principal centre; outstanding amongst them is the tall yellow *Primula sikkimensis*, one of Hooker's most popular introductions whose range extends from Nepal to Western China. When the drumhead primula, *Primula denticulata*, one of the commonest species, flowered in Veitch's Exeter nursery in 1842 it received the Banksian medal of the Horticultural Society of London. Arthur K. Bulley who made a speciality of primulas at his Ness nursery employed R. E. Cooper in Bhutan and Sikkim during 1914 and 1915. Nine of Cooper's primulas which were drawn for *Curtis's Botanical Magazine* between 1919 and 1963 included *P. eburnea* and *P. griffithii* which were later reintroduced by Ludlow and Sherriff whose best find was *P. sherriffae* in Bhutan in 1934. A. F. R. Wollaston collected *P. wollastonii* during a Mount Everest expedition in 1921 and Assam yielded *P. whitei* and *P. kingii* to Kingdon Ward.

A number of Wallich's Nepalese discoveries were never seen in British gardens until a century or so later: *Gentiana ornata* was first raised from seed by Thomas Hay in 1930; *G. depressa* flowered two years later and in 1949 Ludlow and Sherriff sent seed from Bhutan. Many of the Asian garden plants illustrated in *Curtis's Botanical Magazine* over the two centuries of its existence, were found in Nepal by Buchanan and Wallich and a succession of collectors right up to the present day but there are still species of gentians and primulas known only from herbarium specimens awaiting introduction. W. Griffith contributed herbarium material from Bhutan in 1838 but the first collector to send back seed from that country was R.E. Cooper in 1914 and 1915. Many Bhutanese plants in cultivation today come from the seven expeditions that Ludlow and Sherriff made between 1933 and 1950.

Griffith found *Arisaema griffithii* in Bhutan and it was first flowered by Elwes in 1879 who also had *Arisaema nepenthoides* and *A. utile* at Colesborne. These aroids flourish in the temperate forests of the Himalayas and in Ceylon where Thwaites collected them for Kew. Thwaites also had a fondness for balsams (*Impatiens* species) which also inhabited the subtropical regions of the Himalayas. Few of them remain in cultivation but *Impatiens glandulifera* or 'Policeman's helmet', introduced about 1839, has become naturalized in parts of Britain, colonizing the banks of streams and rivers (Fig. 20.11). How it escaped and spread so quickly puzzled Sir Joseph Hooker.

It appeared for the first time in the shrubberies of my garden near Sunningdale three or four years ago and rapidly increased, so as to become a weed, until 1898, when the long-continued drought decimated the plants, and prevented the seeding of the few that were spared. How or whence it was introduced I cannot form an idea. I have never

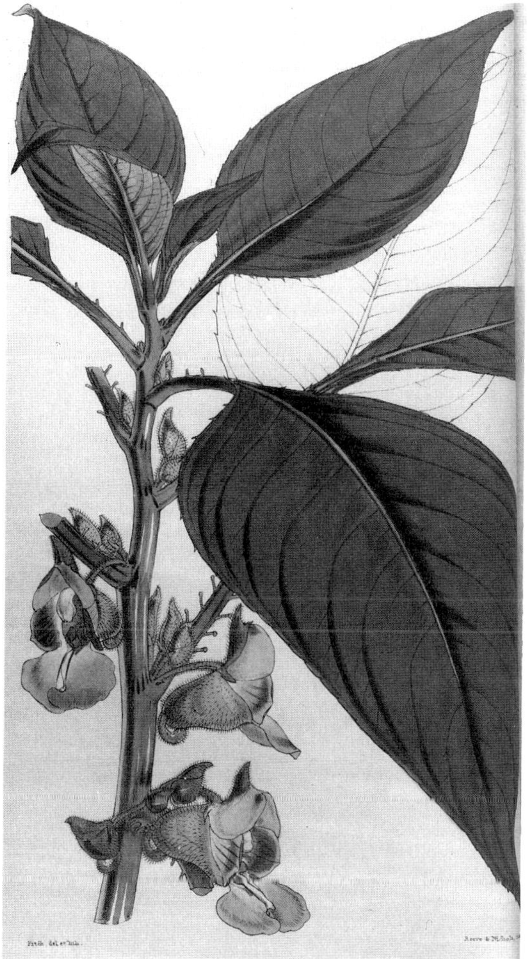

Figure 20.11 *Impatiens glandulifera*. (From *Curtis's Botanical Magazine*, 1852, plate 4623.) Drawn and lithographed by W.H. Fitch.

raised an Indian balsam in my garden, nor had I at that time received the seeds of any, and it is not in the Kew collection, or in that of other botanical gardens to which I have directed inquiries. On the other hand, I have seen it in several cottage gardens, one at Marlow being the nearest to my place (about 15 miles in a straight line). Also I saw it growing in profusion in a cottage garden near Mr Mitford's place, Batsford Park, but it was not grown at Batsford Park![28]

Unfortunately, many Indian alpines lack the robustness and adaptability of this *Impatiens*. Primulas, in particular, are susceptible to a change of habitat. Initially they appear healthy but then decrease in number, probably the victims of some virus disease. *P. klattii*, *amethystina*, and *dickieana* are just a few of the species that have lasted only a few seasons before disappearing. *Gentiana kurroo* obstinately refuses to set seed. Temperamental *Meconopsis*, too, can test the gardener's skill and patience. *Saxifraga* and *Androsace*, however, are less vulnerable to the vagaries of the British climate. Despite such frustrations and casualties, the gardens of the West would be impoverished without trees like the deodar, numerous species of rhododendrons some of which are the parents of notable hybrids, vigorous climbers like *Clematis montana*, indoor favourites like *Begonia rex*, and rock-garden plants like *Polygonum affine* and *P. vacciniifolium*.

NOTES

1 J. Gerard (1597) *Herball*, London.
2 Heniger (1986), 44.
3 Ibid., 269.
4 *The Belgick, or Nederlandish Hesperides*, 1683, 163 (a translation of Commelin's *Nederlantze Hesperides*). London.
5 *Taxon*, **17**, 1968, 677.
6 D. Jacques and A. J. van der Horst (1988), *The gardens of William and Mary*, London, 177.
7 E. M. Thompson (ed.) (1878), *Correspondence of the family of Hatton*, London, 148.
8 J. Evelyn (1685), *Diary*, London, 7 August
9 J. Cowell (1730), *The curious and profitable gardener*, London, 37.
10 P. Miller (1768), *Gardeners dictionary* (8th edn), London.
11 B. Henrey (1986), *No ordinary gardener; Thomas Knowlton, 1691–1781*, London, 64–5.
12 Ibid., 174.
13 Peter Kalm to Linnaeus, 23 May 1751. *Essex Naturalist*, **20**, 1921–3, 249.
14 R. Pulteney to J. E. Smith, 14 October 1792. *Memoir and correspondence of . . . Sir James Edward Smith*, (ed. Lady P. Smith) Vol. 1, 1832, 402. London

[15] N.G.B. James (1926), *The life of Peter Collinson*, 87. London
[16] A.P. Paterson (1980), *The Chelsea Physic Garden*, 6–7. London
[17] Forbes (1813), *Oriental memoirs*, Vol. 3, 409.
[18] H. Repton (1806), *An enquiry into the changes of taste in landscape gardening*, London, conclusion.
[19] Banks to N. T. Host, 25 February 1812. B.M. (N.H.) D.T.C., Vol. 18, folios 146–9.
[20] H. Repton (1803), *Observations on the theory and practice of landscape gardening*. London.
[21] Ellis to Linnaeus, 25 April 1758. Sir J. E. Smith (1821), *A selection of the correspondence of Linnaeus and other naturalists*, Vol. 1, London, 93.
[22] Smith to Wallich, 11 May 1819. L. S. Smith letters, Vol. 10, folio 107.
[23] Coats (1969), 154.
[24] Lindley to E.I.C., 16 January 1839. I.O.L.R. L/F/2/38.
[25] Plate 5146 (1858).
[26] J. D. Hooker (1854), Vol. 2, 321–2.
[27] H. J. Elwes (1930), *Memoirs of travel, sport and natural history*, London, 113.
[28] *Curtis's Botanical Magazine*, 1899, plate 7647.

TRANSPORTATION OF PLANTS

The global traffic in plants had more failures than successes until 1833 when Dr Nathaniel Ward invented the Wardian case. Tender exotics were especially susceptible to climatic changes on voyages of several months' duration, and for much of the seventeenth century British botanists and gardeners concentrated on hardy North American plants which had only to endure the relatively short passage across the Atlantic. We learn from Van Reede's despatch in March 1691 to his Governors, Directors, and Commanders in the Cape, Persia, India, and Ceylon that seeds from their outposts were not fit for sowing when they reached Holland, the result, possibly, of either being too old when collected or having been packed in damp conditions. His advice was that preference should always be given to 'young, hard and bone-dry seeds'. It would seem that those sent in bags were more prone to decay than boxed consignments.

James Petiver had a broadsheet printed with instructions for preserving specimens collected for his museum and herbarium (Fig. 21.1). In a pamphlet entitled *Brief instructions for making observations in all parts of the world* (1696), James Woodward assured collectors that bulbs and tubers, provided they were packed in dry moss, cotton, or sand, would survive a prolonged stay in a ship's hold. Practical measures to safeguard seeds in transit were expounded by Richard Bradley in the July 1722 issue of his *Monthly Register of Experiments and Observations in Husbandry and Gardening*, the first gardening periodical published in this country. Seeds, having been left in their cases to serve both as protection and as a means of identification, were to be stored in tightly sealed bottles which, in turn, should be placed in containers packed around with common salt as a deterrent to insects and as an insulation against heat. 'By this means, I judge, that it will be impossible for any of the seeds to be injured in their passage through different climates.' But many seeds perversely defied all ameliorative treatment. Philip Miller had no success with the seeds of the Indian mango.

> *The only method to obtain the plants in England is to plant a good quantity of the nuts in a tub of earth in the country where they grow naturally; and when the plants are grown a foot high, to ship them for England, placing a covering over them to defend them from salt water and the spray of the sea, being very careful not to let them have much wet in their passage*[1].

The Reverend Stephen Hales writing from his Teddington parish applauded Linnaeus's *contrivances for bringing seeds from India. As small ventilators are now put into all our transport ships, and into some India ships, and probably will be put into all, so it will be the better for your seeds to have a fresher, cooler air*[2].

The Swedish naturalist's solution was similar to that advocated by Bradley: seeds were to be placed in paper packets with fine sand to

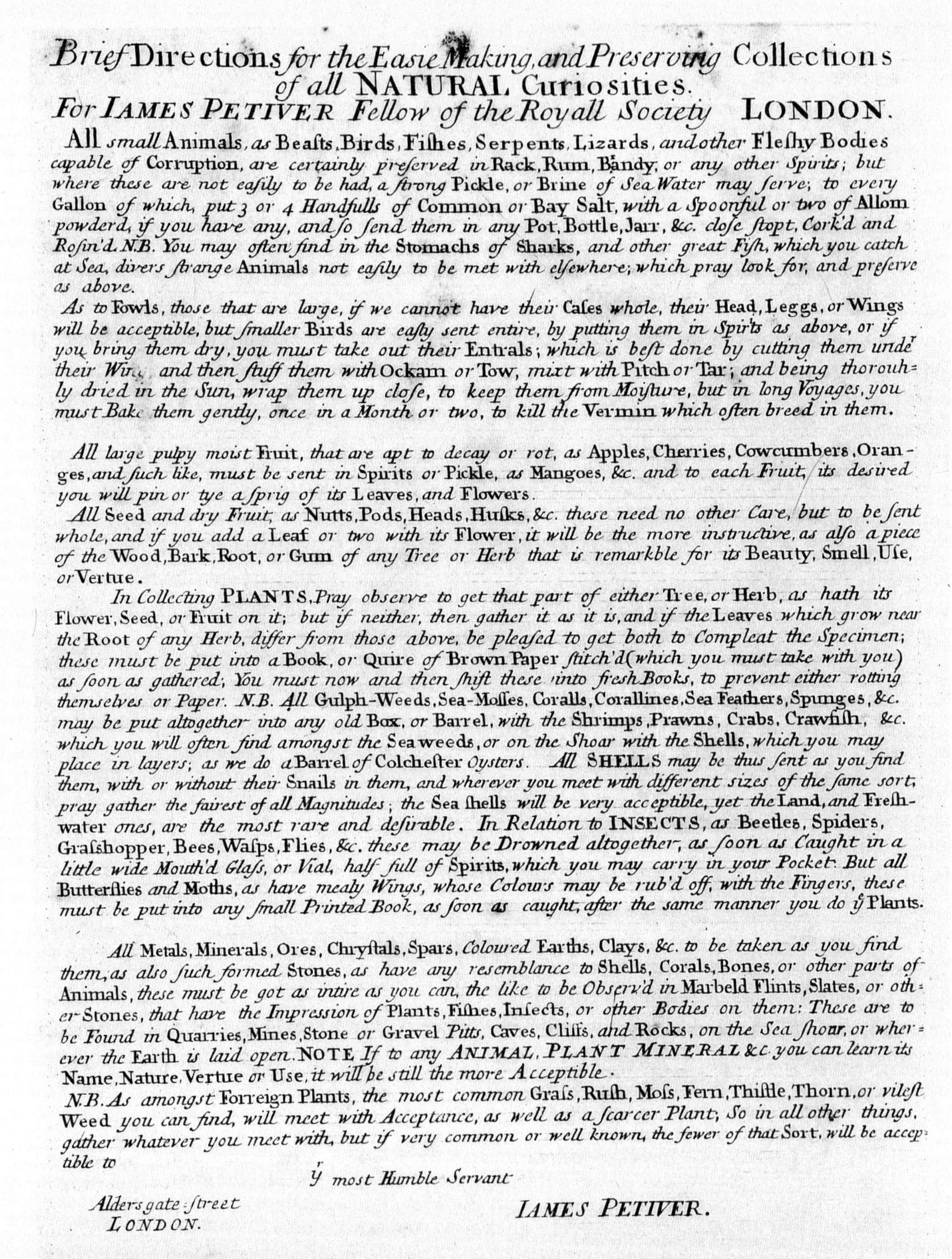

Brief Directions for the Easie Making, and Preserving Collections of all NATURAL Curiosities.
For IAMES PETIVER Fellow of the Royall Society LONDON.

All small Animals, as Beaſts, Birds, Fiſhes, Serpents, Lizards, and other Fleſhy Bodies capable of Corruption, are certainly preſerved in Rack, Rum, Brandy, or any other Spirits; but where these are not eaſily to be had, a ſtrong Pickle, or Brine of Sea Water may ſerve; to every Gallon of which, put 3 or 4 Handfulls of Common or Bay Salt, with a Spoonful or two of Allom powderd, if you have any, and ſo ſend them in any Pot, Bottle, Jarr, &c. cloſe ſtopt, Cork'd and Roſin'd. N.B. You may often find in the Stomachs of Sharks, and other great Fiſh, which you catch at Sea, divers ſtrange Animals not eaſily to be met with elſewhere; which pray look for, and preſerve as above.

As to Fowls, those that are large, if we cannot have their Caſes whole, their Head, Leggs, or Wings will be acceptible, but ſmaller Birds are eaſly sent entire, by putting them in Spirits as above, or if you bring them dry, you must take out their Entrals; which is beſt done by cutting them under their Wings, and then ſtuff them with Ockam or Tow, mixt with Pitch or Tar; and being thorouhly dried in the Sun, wrap them up cloſe, to keep them from Moiſture, but in long Voyages, you must Bake them gently, once in a Month or two, to kill the Vermin which often breed in them.

All large pulpy moist Fruit, that are apt to decay or rot, as Apples, Cherries, Cowcumbers, Oranges, and ſuch like, must be sent in Spirits or Pickle, as Mangoes, &c. and to each Fruit, its desired you will pin or tye a ſprig of its Leaves, and Flowers.

All Seed and dry Fruit, as Nutts, Pods, Heads, Huſks, &c. these need no other Care, but to be ſent whole, and if you add a Leaf or two with its Flower, it will be the more instructive, as alſo a piece of the Wood, Bark, Root, or Gum of any Tree or Herb that is remarkble for its Beauty, Smell, Uſe, or Vertue.

In Collecting PLANTS, Pray observe to get that part of either Tree, or Herb, as hath its Flower, Seed, or Fruit on it; but if neither, then gather it as it is, and if the Leaves which grow near the Root of any Herb, differ from those above, be pleaſed to get both to Compleat the Specimen; these must be put into a Book, or Quire of Brown Paper ſtitch'd (which you must take with you) as ſoon as gathered; You must now and then ſhift these into fresh Books, to prevent either rotting themselves or Paper. N.B. All Gulph-Weeds, Sea-Moſſes, Coralls, Corallines, Sea Feathers, Spunges, &c. may be put altogether into any old Box, or Barrel, with the Shrimps, Prawns, Crabs, Crawfiſh, &c. which you will often find amongst the Seaweeds, or on the Shoar with the Shells, which you may place in layers; as we do a Barrel of Colcheſter Oysters. All SHELLS may be thus ſent as you find them, with or without their Snails in them, and wherever you meet with different sizes of the ſame sort, pray gather the fairest of all Magnitudes; the Sea ſhells will be very acceptible, yet the Land, and Freſh-water ones, are the most rare and deſirable. In Relation to INSECTS, as Beetles, Spiders, Graſshopper, Bees, Waſps, Flies, &c. these may be Drowned altogether, as ſoon as Caught in a little wide Mouth'd Glaſs, or Vial, half full of Spirits, which you may carry in your Pocket. But all Butterflies and Moths, as have mealy Wings, whose Colours may be rub'd off, with the Fingers, these must be put into any ſmall Printed Book, as ſoon as caught, after the same manner you do yͤ Plants.

All Metals, Minerals, Ores, Chryſtals, Spars, Coloured Earths, Clays, &c. to be taken as you find them, as also ſuch formed Stones, as have any resemblance to Shells, Corals, Bones, or other parts of Animals, these must be got as intire as you can, the like to be Obſerv'd in Marbeld Flints, Slates, or other Stones, that have the Impression of Plants, Fiſhes, Inſects, or other Bodies on them: These are to be Found in Quarries, Mines, Stone or Gravel Pitts, Caves, Cliffs, and Rocks, on the Sea ſhoar, or wherever the Earth is laid open. NOTE If to any ANIMAL, PLANT MINERAL &c. you can learn its Name, Nature, Vertue or Use, it will be still the more Acceptible.

N.B. As amongst Forreign Plants, the most common Graſs, Ruſh, Moſs, Fern, Thiſtle, Thorn, or vileſt Weed you can find, will meet with Acceptance, as well as a ſcarcer Plant; So in all other things, gather whatever you meet with, but if very common or well known, the fewer of that Sort, will be acceptible to

yʳ most Humble Servant

Alders gate ſtreet
LONDON.

IAMES PETIVER.

Figure 21.1 James Petiver, *Brief directions for the easie making and preserving collections of all natural curiosities.* The leaflet was distributed to travellers prepared to collect specimens for Petiver.

absorb any moisture; the packets were then inserted into a glass or earthenware jar which, in turn, was put into a larger container filled with a mixture of nitre, salt, and sal amoniac to keep the seeds cool. Linnaeus applauded the researches of John Ellis, a London merchant with an obsessive interest in the transportation of seeds and plants.

> *I am glad that you take so much pains about preserving seeds, If your attempts should succeed, the whole world will be obliged to you, for our gardens will, through your means, be enriched with abundance of Indian plants, whose seeds have hitherto been imported to no purpose. Thus we may, before long, see the tea growing in Europe, as well a numerous plants besides, never yet seen among us*[3].

A healthy tea plant actually reached Sweden in 1761 only to be killed, so Linnaeus told Ellis, by mice stripping off the bark. Ellis was luckier with the chance discovery of a single tea seed in the bottom of a tin canister from China. This he succeeded in germinating and had raised to an encouraging six inches when he informed Linnaeus.

Ellis considered two methods for conveying tea seeds in his *Directions for bringing over seeds and plants, from the East-Indies and other distant countries in a state of vegetation* (1770). His preference was a protective coating of deep yellow English beeswax, but then he had a predilection for coating seeds with all manner of substances: gum arabic, rosin, and a mixture of rosin, beeswax, and pitch. His other method was more conventional: the seeds still in their cases were simply stored in canisters in a cool part of the ship. 'Tea seeds, put up in this manner, require less trouble than those that are rolled up singly, and afterwards immersed in melted w ..' Always a pragmatist, Ellis conceded that 'experience could determine which is the best method'. The seeds could also be sown in boxes or casks at the beginning of a voyage, bedded down in loamy soil with an admixture of moss for moisture retention. His *Description of the mangostan and the bread-fruit* (1775) illustrates and prescribes methods for sowing seeds and putting plants into special boxes (Fig. 21.2). But without a tolerant captain and a co-operative crew the contents of these boxes were sure to be neglected, seldom watered or given light and air, and afforded no protection against storms and sea spray. Seeds were less demanding in attention and more portable. They were stored in linen bags, bottles, and pots, even snuff boxes of tortoiseshell, horn, and silver. Hung up in a bag in a ship's cabin they benefited from circulating air, which both Miller and Banks believed to be essential.

About 1770 the Quaker physician John Coakley Lettsom compiled some guide-lines for plant collectors on behalf of a fellow Quaker, Dr John Fothergill, whose fee on one occasion for curing a sea captain of some tropical fever was 'two barrels of earth from Borneo taken from as many points as possible'. In this way Fothergill hoped to recover some adventitious seeds from the soil. Lettsom had been infected with Ellis's enthusiasm for wax-embedded seeds, but cotton saturated with beeswax, time-honoured dried sand, and Linnaeus's salt-crammed containers also met with his approval. Lettsom demanded an unreasonable and unlikely horticultural devotion from the ship's crew. He expected the captain to give up a part of the great cabin for boxes of plants, 'certainly by far the best station for them'; on deck they were to be protected from the 'minute particles of salt water with which the air is charged'; their covers were never to be removed in high winds; a regular watering was mandatory and also 'sprinkling all the leaves . . . to wash off the salt drops which cover them . . . 'In this manner, Lettsom confidently predicted that 'plants may be brought from almost any distance; many come from China every year

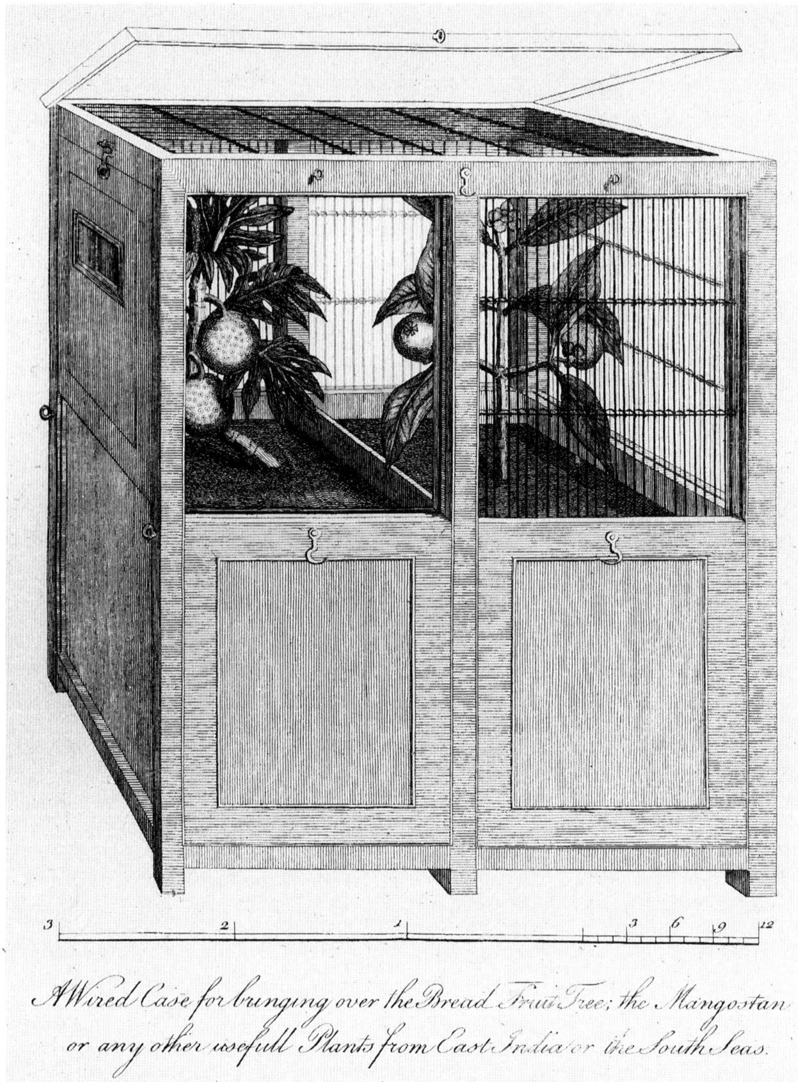

Figure 21.2 The plant box recommended by John Ellis in his *Description of the mangostan and the bread-fruit*, London, 1775. 'This box is wired on every side, and on the top; having shutters that slide up and down at pleasure, to protect the plant from the weather . . . The back pannels, instead of wire, may consist of glass, by which means the plants may receive the benefit of the sun, and at the same time be preserved from the severity of the cold, when other lights are shut up.'

in a flourishing state'[4]. Knowing that the plant cases cluttering the deck were always the first to be thrown overboard in emergencies, a grateful Horticultural Society of London presented in 1820 its medal to nine exemplary captains who had safely delivered plants from China that year (Fig. 21.3).

When Colonel Kyd in Calcutta was told in September 1786 about a royal command for Asian plants to be sent to the new botanical garden in St Vincent in the West Indies, he considered the various options available for transporting them[5]. He could not send seeds of grain and fruit in hermetically sealed bottles because he lacked the necessary pump. Few seeds had ever arrived in a viable state in the packets of oiled paper used by English nurseries. On balance he favoured the Dutch practice of placing them in coarse linen bags suspended in open wicker baskets in an airy part of the ship, convinced that ventilation was absolutely essential. Chinese gardeners successfully sent him fruit trees packed horizontally in well-ventilated boxes. He attributed the loss of an entire consignment of English fruit trees to the excessive protective packing of ferns and mosses which obstructed the free flow of air. The safe arrival of fruit trees planted in garden mould in open boxes placed on the stern balcony of the ship, persuaded him that this was the method he should adopt for the transit of plants to St Vincent. Sir William Jones, curious to learn the success rate of Kyd's despatch of seeds, wrote for information to Sir George Yonge who was involved in the developments at St Vincent.

I have tried several methods without success: a covering of wax I thought likely to succeed. But when I received the seeds of the Cabbage Tree, the Sago Tree, and the Sandalwood Tree (the two first of which were sent me from the Eastern Islands, and the last, from the

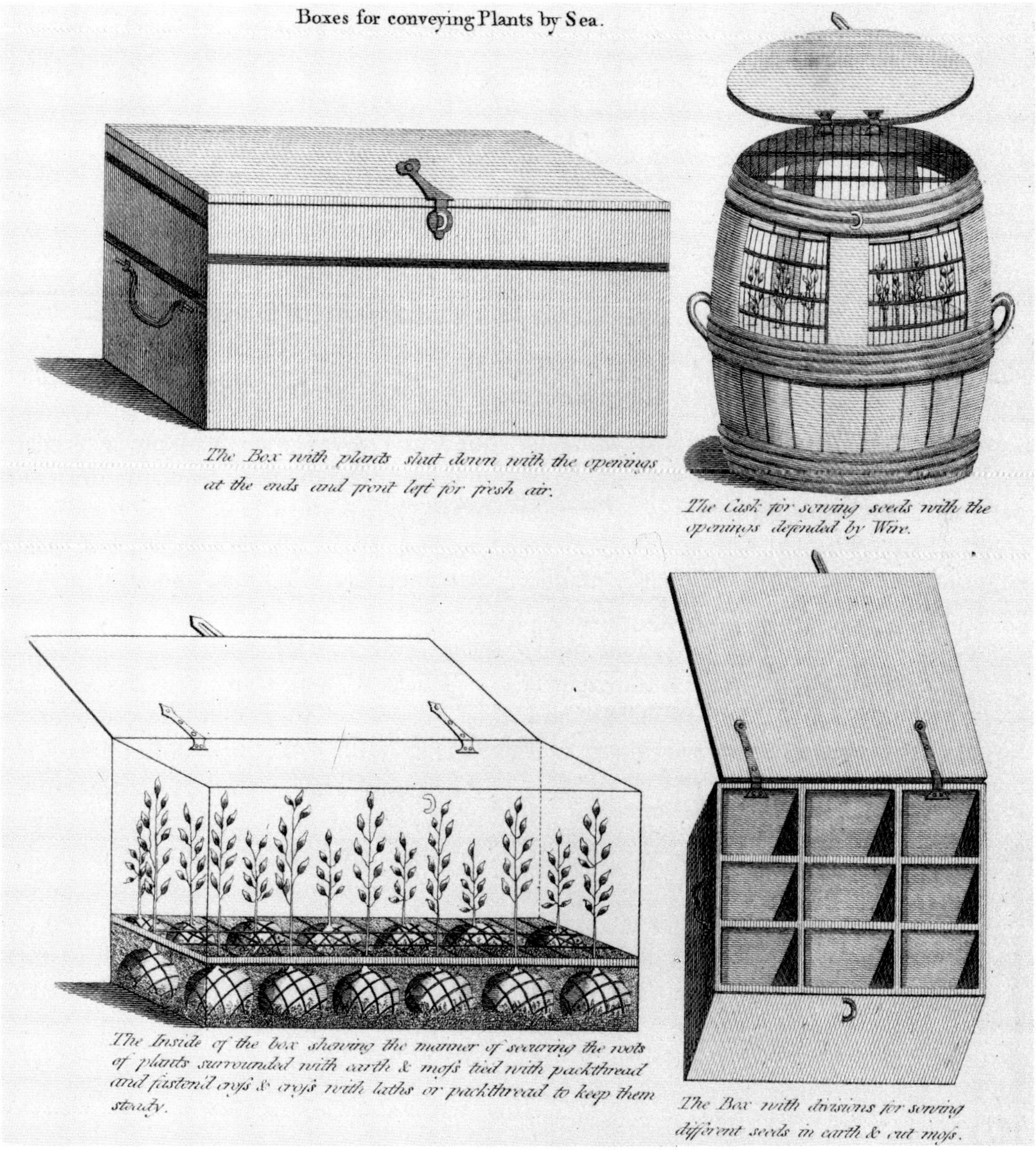

Figure 21.3 Some of the containers for transporting plants by sea which were in use during the latter part of the 18th century. (From J. C. Lettsom (1772), *The natural history of the tea-tree*, London.) This plate was copied from John Ellis's *Directions for bringing over seeds and plants* . . . London, 1770.

West) both covered with wax, I sowed them in my garden, but none of them germinated[6].

On one of his visits to Kew in 1790 to inspect a box of plants from India, Yonge found

> *nothing, but four stumps of plants! apparently dead—Aiton [the Superintendent] was of the opinion one of them might be the Mangasteen and there appearing some chance tho' small, of life, we assisted the gardner in potting them and plunging them in a fine bed of tann, where if not too late, they may live, tho' I much doubt it . . . the fate of all these has been like all others, and the cause the same viz. total neglect & inattention almost wilful of what was committed to their care [i.e. ship's crew]—something must be done to correct this, or all our labours will be in vain*[7].

After a succession of cargoes of dead and sickly plants, the Government of Bengal in January 1792 urgently requested the Court of Directors to instruct their ships' captains to take better care of them. Late that same year the Hackney nurseryman, Conrad Loddiges, experimented with the delivery to India of four boxes of common hardy plants: a deal box with apple, pear, plum, and rose trees packed in moss and stowed in the hold, not to be opened until India was reached; the remaining boxes containing shrubs and more fruit trees were potted in ordinary soil and stacked either on deck or the stern galley. When the ship reached Madras in April 1793 all the gooseberry and currant bushes were dead; peaches and nectarines were barely alive; only the pears and apples had survived and these were ceremoniously presented to the Nabob of Arcot. The lesson deduced from this experiment was that plants would only endure their time at sea and fluctuations in temperature if they had been potted a year or so before shipment and had left Europe in summer rather than in winter[8]. Since this discouraging casualty rate persisted Kew Gardens actually set up its own 'infirmary' for any survivors, and Sir Joseph Banks insisted that plants destined for him were always promptly unloaded when the ship docked and despatched without delay.

Roxburgh proposed to Banks in December 1796 that plants in need of remedial attention could convalesce at the Cape of Good Hope, to be reshipped after recovery. He had selected the Cape because he was hoping for a part-time posting there, but in the event lonely, rocky St Helena in the South Atlantic served as a transit nursery. The Dutch who had colonized it in 1645, abandoned it when they settled at the Cape in 1652, and the British promptly took possession of this strategic watering station for their ships. During the 1760s tea seeds were sometimes planted on St Helena and the seedlings potted for the last stage of the homeward voyage. Kyd suggested that trees from England could recuperate at St Helena before continuing their journey to India. The formation in 1789 of a small botanical garden on the island provided a depot for this oceanic traffic in plants, a facility much appreciated by the Horticultural Society of London which believed that many more plants to and from the East could be saved if their passage were broken at St Helena[9].

Before William Carey sent his son, Jabez, to Amboina in 1813, he briefed him thoroughly on collecting procedures: 'plant tubers and bulbs in a box so thickly as to touch one another, or hang them dry in a well-covered basket in an airy part of the ship'[10]. Seeds were to be liberally sown in the soil of boxes of plants or slipped into paper envelopes; epiphytes should be hung in baskets; strips of wood were to be nailed over all boxes of plants to keep out rats; and, above all, absolutely nothing was to be stored in the ship's hold. It was quite customary for European gardeners to save the soil in which their plants arrived for any chance seeds it might contain. When Carey emptied the soil from a bag of plants from

Lord Fitzwilliam's gardener at Wentworth, he was rewarded, some weeks later, with an emerging daisy.

> *I know not that I ever enjoyed, since leaving Europe, a simple pleasure so exquisite as the sight of this English daisy afforded me—not having seen one for upwards of thirty years, and never expecting to see one again*[11].

Carey calculated that scarcely one in ten plants reached India in a viable state, and of those that were planted perhaps a tenth might recover. Seeds still irritatingly arrived at the wrong time to sow them. He was resigned to the fact that 'little dependance can therefore be placed on supplies from England'[12].

But over the years there gradually evolved a code of practice for the care of plants in transit with Roxburgh in 1800 being the first person in India to draft *Directions for taking care of growing plants at sea*[13]. His instructions were a summary of accepted practice. Plants should be protected against salt sea spray, given light and air, watered in moderation, and their growth controlled as they passed into warmer climates; baskets with roots and succulent plants could be hung over the stern or from the projection above the wheelhouse; insect pests were to be ruthlessly eliminated; seeds were to be kept in a cool, dry place, preferably not below deck.

It has already been observed that the limited life of many seeds and an imperfect knowledge of their physiology produced conflicting recipes for their preservation. Roxburgh usually coated them with

> *a thick mucilage of gum arabic, which was then suffered to dry with the seeds incorporated therewith; in this mode the vegetative power of the seeds is well preserved, it being necessary only when they are to be sown that the mixture of gum and seeds should be put into water, which will redissolve the mucilage and leave the seeds in a state ready to be put in the earth*[14].

William Carey packed his seeds in mould. When English hemp seed was sent to India in 1803 by several ships, their commanders were ordered to hang the linen seed bag in a well-ventilated spot, secure from rain and spray and to 'pay the utmost attention to its preservation during the voyage'. So much importance did the Company attach to these Indian hemp trials that they sent further consignments variously packed to preserve their 'vegetative powers': they were mixed with wood ash, with cedar-wood shavings, with white calcined bone ash, with pepper, with sugar 'as recommended by Dr Roxburgh', with camphor, with dried sawdust, with cuttings of cotton yarn, or with saltpetre. Some were packed in paper parcels, others in linen bags. This operation was repeated several times with recipients of the seeds in India obliged to report on the success or failure of the method employed. Wallich sent seeds of *Rhododendron arboreum* and other plants bedded in boxes of coarse brown sugar, covered with a Nepalese paper made from the bark of a species of *Daphne*, then placed in soldered tin boxes, which were inserted in wooden boxes. These elaborate procedures evidently did not always work, for in 1819 he informed Sir Joseph Banks that

> *I have latterly resolved to send seeds simply put in papers, without any admixture of sand or sugar fearful lest the other methods which I adopted for some years, should not have succeeded in the generality of instances*[15].

Within a few years of his appointment as Superintendent at Calcutta, Wallich replaced Roxburgh's *Directions* with his own printed *Directions relative to the preservation of growing plants and seeds on a sea voyage*. He expanded his predecessor's instructions, adding nothing particularly significant. We learn that Calcutta's plant chests had 'movable railed frames and covers' to shelter their contents from sea-water, wind, and foul weather. A lengthy letter to Joseph Sabine, Secretary of the Horticultural Society of

London, summarizes his recommendations for transmitting seeds[16]. Seeds, always recently gathered and ripe, should leave Britain in March or April in order to arrive in India in August or September near the approaching cool season. Small quantities of seed were more likely to germinate than large packages. Having tried bedding seeds in a wide range of substances, he came to the conclusion that brown paper envelopes covered with oil cloth or just a deal box were as effective. While he was in England Wallich presented a paper to the Horticultural Society of London in July 1831 'Upon the preparation and management of plants during a voyage from India'[17]. He stipulated that only strong, healthy plants would withstand the rigours of the voyage, that plant chests must not be too large or unwieldy if the assistance of the ship's crew were to be enlisted, and that every plant should be in a square wooden pot. John Lindley had also read a paper on the same subject in November 1822[18] and in 1832 the *Journal of the Asiatic Society of Benegal* published William Hooker's guidelines[19]. There was now a corpus of experience and advice to ameliorate the excessive losses not only in plants but also in investments (Fig. 21.4). An Ootacamund resident who imported in 1829 a hundred guineas worth of plants of which only a pear tree and a currant bush survived was not an isolated case. Royle recalled that an apple tree, the residual offering of a consignment from Liverpool, represented an expenditure of £70 by the time it was planted at Mussooree.

The distance between Europe and India seemed an insuperable obstacle. A voyage around the Cape in sturdily built but incredibly slow East Indiamen could take between five and six months or even longer in time of war, and few plants would still be alive at the end of it. By 1820 small steamers were operating regular services across the English Channel and the Irish Sea. Within a few years officials in both Britain and India petitioned for trans-oceanic steam links and Calcutta merchants offered a lakh of rupees to the first British steamship to reach Bengal in less than 70 days. The challenge was taken up by the *Enterprise* which left Falmouth in 1825 reaching Calcutta in a disappointing 113 days, of which, however, only 62 days had been entirely under steam. The Cape route was temporarily abandoned when a steamship built of Indian teak in Bombay left that harbour for Suez which it reached in only a month, thus establishing the feasibility of another route—the overland route. Passengers and cargo were conveyed overland by camels, donkeys, or caravans between the ports of Suez and Alexandria, where small wooden or iron paddle steamers were waiting to complete the journey either west through the Mediterranean or east into the Indian Ocean. The Peninsular and Oriental Navigation Company had a fleet of such steamers to take passengers from Suez to Bombay, Ceylon, Madras, and Calcutta, and in 1842 inaugurated a monthly service around the Cape. Soon Calcutta, the capital of British India, was only a six weeks' voyage from London.

The advent of steamships was enthusiastically welcomed by Wallich in 1833.

> *I hail with anxious impatience the establishment of this mode of communication, by means of which, seeds and plants will have a chance of surviving the transmission, beyond anything they now possess, even under the most attractive treatment and greatest solicitude for their preservation*[20].

The manifest benefits of speedier sea communication for botanists and plant collectors were brought to the attention of the Court of Directors in Lord Auckland's despatch of 16 August 1839. When the Company received some 50 packets of seeds in 1839 via the overland route they gave them to Professor Lindley for distribution and comment. Lindley, disappointed that many of these

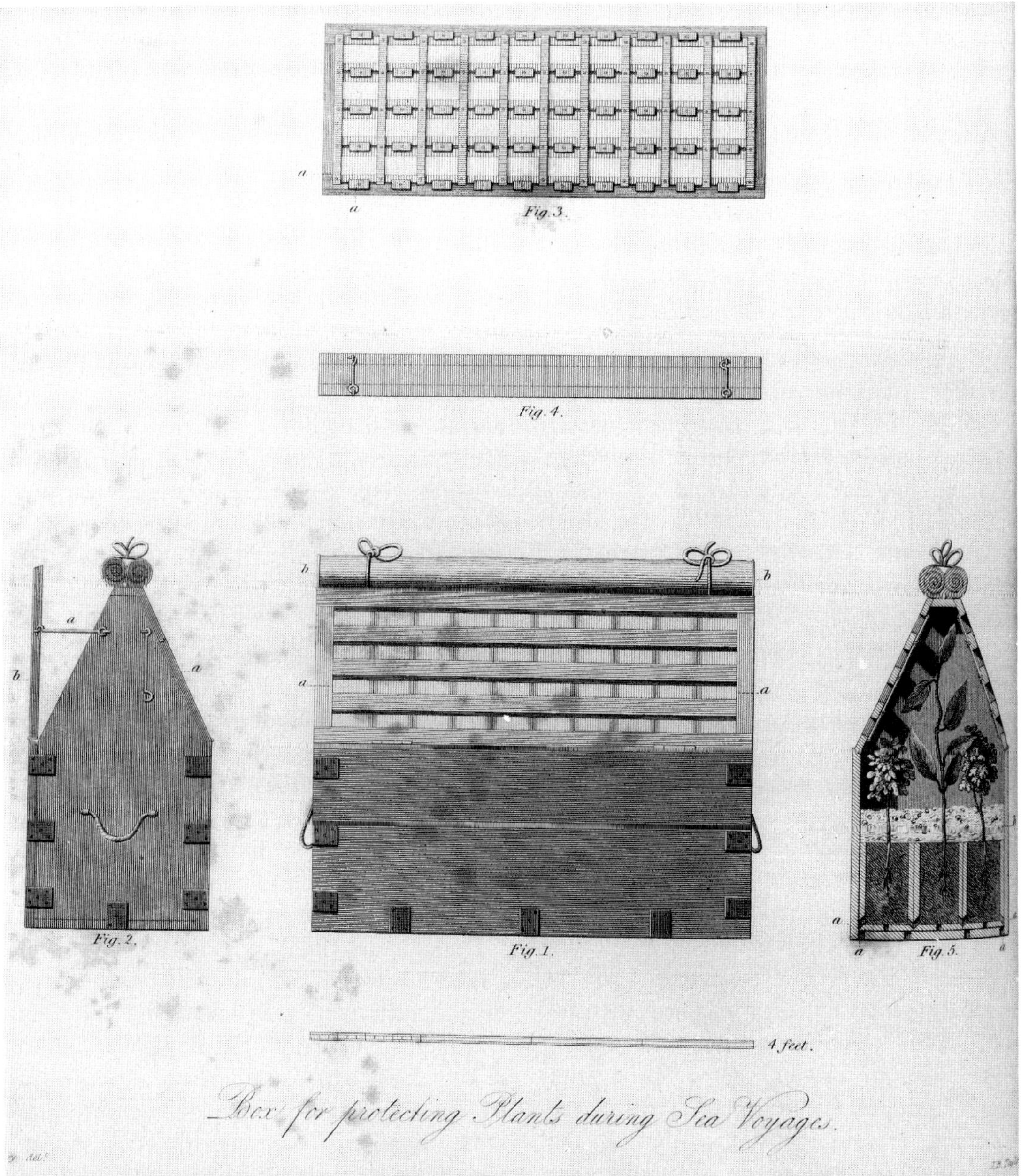

Figure 21.4 John Lindley's specification for a plant case stipulated square wooden pots for the plants, sloping sides which could be opened, and a tarpaulin which could be unrolled to protect plants from salt spray. (From *Transactions of the Horticultural Society of London*, Vol. 5, 1824, 192–200.)

seeds had been collected in the north Indian plains and were thus suitable only for cultivation in hothouses, recommended that the overland route should be restricted to hardy useful and ornamental trees and flowers; exotics of no particular importance could come to Britain by the longer Cape route[21]. Royle with characteristic efficiency had already drawn up a list of annuals and perennials that could now be introduced to India[22], preferably by the overland route. The success of the overland route was endorsed by Loddiges nursery which was pleased to tell Royle that nearly all the deodar seeds he had given them had germinated.

The advantage of getting seeds overland are most decided; we have had great quantities of Deodara sent by sea from time to time, and hardly ever got one to grow, whereas these quite surpassed all our expectations[23].

Plant cuttings, however, which travelled overland met with variable success; some slips of English fruit-trees sent in the winter of 1838–9 failed; the next consignment, made air-and water-tight with layers of brown paper and India-rubber cloth, was also dead on arrival at Saharanpur. An *Amherstia nobilis* died on being unduly delayed at Alexandria. Some mango trees packed in teak boxes, destined for Queen Victoria, also succumbed. Such losses convinced Royle that living plants fared much better by the Cape route; the difficulties they encountered on the overland route were 'much greater for the change of conveyances & the hurried transit over the desert'[24].

Although the overland route was used extensively, preserving the viability of the seeds for many species still remained a perplexing problem. The time of despatch was clearly crucial: experience had confirmed that no seeds should reach India before the appropriate time for sowing them. Those with tropical destinations needed to arrive by the beginning of the monsoons, that is the middle of June; October and November were the best months for the cultivation of plants in the northern plains; while March, or even April, was not too late for sowing them in Himalayan gardens. It was axiomatic that seeds should be absolutely ripe when collected and the practice of some nurseries of sending a mixture of old and new seeds was condemned. Every permutation in packaging had its advocate. Falconer favoured a double layer of India rubber; Jameson and the Agricultural and Horticultural Society of India used tin cases, hermetically sealed; and even as late as 1858 Jameson requested chestnuts encased in wax. Thwaites also employed sealed tin boxes, adding finely powdered charcoal to inhibit fungal growth; straightforward materials like canvas and paper also had their supporters. When Robert Fortune left for China in the spring of 1843 on a plant hunting expedition for the Horticultural Society of London, he was asked to test several containers for storing seeds: sealed bottles, tin-lined boxes, and canvas bags. He found on arrival in the East that the contents of some of the sealed bottles were moist and mouldy but that seeds stored in boxes and canvas appeared to be in good condition. He recommended canvas bags but conceded that it might be difficult 'to induce captains of ships or others to allow packages of this kind to be swinging about in cabins'[25]. Surprisingly, he also approved of the postal service for small quantities of seed, a method that William Carey had advocated as early as 1814[26]. From the 1840s it was not unusual for Kew Gardens to receive Indian seed by letter parcel, and cinchona seeds from South America were sometimes entrusted to the mail.

Dodging dangling seed bags was less of an inconvenience to ships' crews than large boxes cluttering decks, requiring watering and ventilation, and always a potential hazard in any storm. The plant boxes originally used by the East India Company were often no more than flour barrels with holes cut in their sides for ventilation, and with lath and tarpaulin covers which by the end of the eighteenth century had been superseded by glass tops. Large containers, or plant cabins as they were called, assumed the dimensions of small greenhouses and could weigh several tons. Thomas Manning, who was in charge of a consignment of plants, complained bitterly to Sir Joseph Banks of the crew's resentment:

the first mate does not approve of having a garden on the poop at all. He says it wracks the ship to pieces. The captain agrees in the same story and when the beams creak in the cuddy [i.e. the cabin] they all turn to me sometimes and damn the flower pots[27].

Wallich went out of his way to placate sailors and even to express his appreciation of their co-operation in a tangible manner.

I have made it a rule, on every occasion when a commander of a ship takes three or more boxes to present him with a box free of all expenses[28].

He also made sure that his plant boxes were never too heavy or too large for the crew to handle. They were glazed with stout glass, Chinese pearl oyster shell or thick talc. He suggested to Loddiges that their tightly sealed boxes could be improved by glazing them with very thick glass[29]. These glass-topped boxes had to be opened periodically for ventilation, a routine soon to be obviated by Dr Ward's newly invented 'closely-glazed cases'.

Nathaniel Ward (1791–1868), a physician in the East End of London, discovered accidentally in 1829 that some ferns and grasses had sprung up in the damp soil in a sealed bottle and subsequently continued to grow without fresh water or air for several years. In the presence of sunlight, evaporation and transpiration would fill any container with water vapour which then condensed on the glass and ran back into the soil. Provided there was sufficient light for photosynthesis and no extremes of temperature, this cycle would go on being repeated. A popular domestic use for his invention was the miniature conservatory, usually crammed with ferns, a decorative addition to many Victorian drawing-rooms. But Ward perceived a more practical application for his Wardian case, 'a ready mode of importing most plants, without risk, from the most

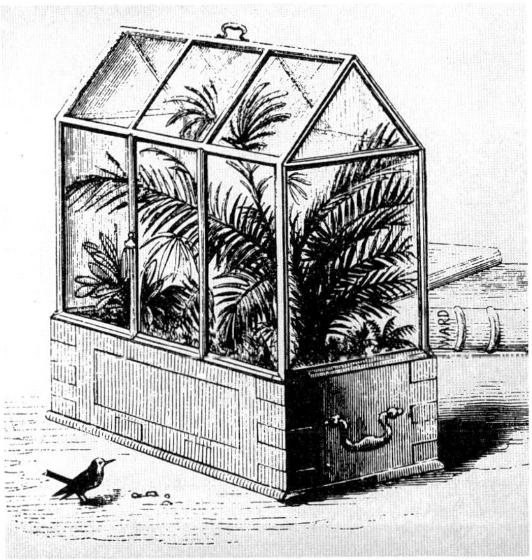

Figure 21.5 Wardian case. (From N. B. Ward (1852), *On the growth of plants in closely glazed cases*, (2nd edn), London.)

distant regions of the globe'[30] (Fig. 21.5). To test his theory he sent ferns, grasses, and some flowers in two Wardian cases to Sydney in June 1833. Stowed on the ship's poop, receiving only one watering during the entire four months' voyage, only a few ferns wilted, and a single primrose in full bloom excited home-sick settlers. The remarkably healthy state of the Australian plants brought back to England made a convert of Loddiges, which became the first nursery to adopt them extensively in their international trade. Before John Gibson went to India in 1835 as the Duke of Devonshire's collector, he paid a visit to Hackney to inspect Loddiges's Wardian cases, was duly impressed and subsequently used them. It was also in 1835 that Loddiges presented Wallich with his first Wardian case of plants. The Calcutta Botanic Garden forthwith switched to this latest method of plant transportation, and between 1836 and 1840 two-thirds of all the plant boxes used by Calcutta were Wardian cases. By 1842 when Royle was urging the Court of Directors also to use them, they had become more or less universally accepted. Robert Fortune was the first collector to use them intensively when he accompanied 18 Wardian cases placed on the poop of the ship that left Canton in December 1845, and triumphantly delivered his flourishing cargo to the Chiswick garden of the Horticultural Society of London.

When Fortune was engaged by the East India Company to send tea seeds to India, he first despatched them in canvas bags, in boxes of dry earth or in small packages. Their failure to germinate persuaded him to experiment with Wardian cases. In 1849 he sowed tea seeds in a case of mulberries and to his delight they had germinated by the time the ship docked at Calcutta. Encouraged by this result, he liberally added tea seeds to 16 Wardian cases before they were shipped to India. Once more they germinated freely in the soil in which the plants had been bedded. Fortune congratulated Ward.

> *We have done wonders with your cases in India, as well as in the country [i.e. China] . . . When I tell you that nearly twenty thousand tea-plants were taken in safety and in high health from Shanghae to the Himalayas, you will have some idea of our success.*

The now well-authenticated reputation of Wardian cases underwent another rigorous evaluation when Markham used 30 of them in the cinchona transfer. Each case, when planted with cinchonas in soil nine or ten inches deep, weighed just over three hundredweight. This cumbersome cargo reached England without mishap only to die in the intense heat of the Red Sea. Another batch of cinchonas also came by the overland route but passed through the Red Sea during the cool season and reached India in a reasonable state.

This traffic in Wardian cases revealed that they need not be hermetically sealed; McIvor at Ootacamund went so far as to recommend the glass covers always being open during fine weather. Kew Gardens which used Wardian cases right up to the Second World War ensured the circulation of fresh air by boring a small hole near the apex of each case (Fig. 21.6). Robert Wight blamed the death of many plants transported through the subcontinent in Wardian cases to excessive watering by over-zealous attendants. Falconer complained that the contents of numerous Wardian cases died during the slow journey by river and road from Calcutta to Saharanpur. Such failures, however, never undermined public confidence in Wardian cases. Sir William Hooker gave them a glowing testimonial.

> *. . . they have been the means, in the last fifteen years, of introducing more new and valuable plants to our gardens than were imported during the preceding century . . . I shall never forget the expression made use of by the late Mr Loddiges to me one day when speaking of your cases; 'whereas I used formerly to lose nineteen out of twenty of the plants I imported during the voyage, nineteen out of twenty is now the average of those that survive'*[31].

While Joseph Hooker was staying at the Calcutta Botanic Garden in March 1850, he witnessed the unloading of 391 frozen plants—apples, pears, peaches, carrots, gooseberries, and Venus' fly-trap (*Dionaea muscipula*)—from the hold of an American ice-ship.

> *More perfect success never attended an experiment: the plants were in vigorous bud, and the day after being released from their icy bonds, the leaves sprouted and unfolded, and they were packed in Ward's cases for immediate transport to the Himalaya mountains*[32].

An enterprising American had made his fortune in the 1820s by shipping blocks of ice to Calcutta; Bombay was included in these

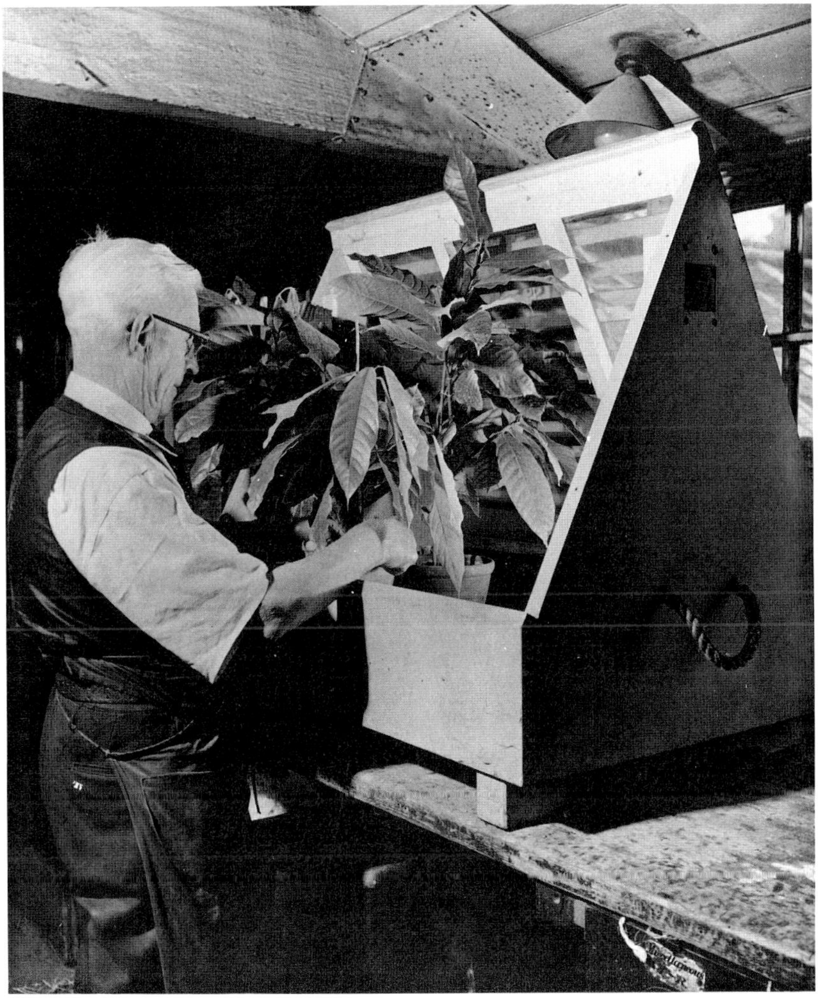

Figure 21.6 Wardian case being loaded with plants at the Royal Botanic Gardens, Kew.

regular deliveries from the 1830s. In Calcutta in 1840 Captain A. C. Lowe noticed that 'apples which are brought out in the iceships are now being hawked about the streets'. It was an obvious progression from shipping fruit to shipping the trees themselves. The Court of Directors expressed a keen interest when they learned about the successful attempt to transport a Wardian case of plants embedded in a block of ice.

The experiment of securing a winter temperature to these trees while on their way to their destination was ingenious and interesting, and we desire to be apprized of the result[33].

The plants, heavy with fat buds, were carried to Saharanpur by river steamer and bullock train and thence to the experimental garden at Mussoorie. Two dozen flourished there until an exceptionally heavy monsoon and a severe winter killed them. The experiment was repeated in 1856 with American fruit trees, packed this time in open boxes. They were established at Mussoorie and the conclusion reached by the Government of the North West Provinces was that

this method of transmitting fruit trees to distant localities could now be confidently recommended for general adoption for its success and cheapness[34].

Ice never became a serious contender as a regular means of protecting plants. In 1930 the Himalayan alpine, *Primula sonchifolia*, was conveyed through Burma in sections of bamboo stems with ice packing to the cold storage compartment of a homeward-bound steamer.

Air transport has now superseded the Wardian case after many years of sterling service. About 1934 Lionel de Rothschild had cuttings of *Erica pillansii* flown from South Africa for his garden at Exbury in Hampshire. Ludlow and Sherriff were the first plant collectors to make regular use of air freight. The aeroplane has since been the means of making available to the British gardeners plants that would not have survived any other form of transport.

NOTES

[1] P. Miller (1768), *Gardeners dictionary*. London
[2] Sir J. E. Smith (1821), *A selection of the correspondence of Linnaeus*, Vol. 2, London, 41.
[3] Ibid., Vol. 1, 1821, 136.
[4] J. C. Lettsom (1783), *Hortus Uptonensis*. London.
[5] Kyd to Governor-General, September 1786. B.M. (N.H.) D.T.C., Vol. 8, folios 63–70.
[6] Cannon (1970), Vol. 2, 771.
[7] Yonge to Banks, 20 August 1790. Kew Archives. Banks letters, Vol. 2, folio 8.
[8] *Horticultural Register and General Magazine*, **5**, 1836, 64–5, 98.
[9] Joseph Sabine to E.I.C., 20 December 1824. I.O.L.R. F/4/767, item 20773.
[10] Carey (1923), 393–4.
[11] Ibid, 395.
[12] Roxburgh (1814), vii.
[13] I.O.L.R. F/4/134, item 2427, folio 59. Reprinted in *Transactions Royal Society of Arts*, **27**, 1809, 236–8 and A. Beatson (1825) *Flora Sta. Helenica*, St Helena iii–v.
[14] *Transactions Royal Society of Arts*, **27**, 1809, 236.
[15] Wallich to Banks, 2 December 1819. B.L. Add. MSS 33982, folio 197.
[16] Wallich to Sabine, 16 July 1826. I.O.L.R. F/4/1761, folios 93–105.
[17] *Transactions of Horticultural Society of London*, (2nd series), **1**, 1835, 140–3.
[18] Instructions for packing living plants in foreign countries, especially within the tropics; and directions for their treatment during the voyage to Europe, *Transactions of Horticultural Society of London*, **5**, 1824, 192–200.
[19] Directions for collecting and preserving plants in foreign countries, *Journal of Asiatic Society of Bengal*, **1**, 1832, 113–15.
[20] *Gardener's Magazine*, **10**, 1834, 272.
[21] Lindley to E.I.C., 16 January 1839. I.O.L.R. L/F/2/38; J.F. Royle (1840), 429–39.
[22] Royle (1840), 439–51.
[23] Spry (1841), 173.
[24] Royle to E.I.C., 1 February 1847. I.O.L.R. L/F/2/106.
[25] *Journal of Horticultural Society of London*, **3**, 1848, 43.
[26] Roxburgh (1814), xi.
[27] Manning to Banks, 24 July 1806. Kew Archives. Banks letters, Vol. 2, folio 326.
[28] Wallich to Government of Bengal, 21 December 1840. Report on Calcutta Botanic Garden, 6.
[29] *Gardener's Magazine*, **12**, 1836, 209.
[30] *Gardener's Magazine*, **10**, 1834, 208.
[31] Hooker to Ward, 4 April 1851. N. B. Ward (1852), *On the growth of plants in loosely glazed cases*, London, 131–2.
[32] J. D. Hooker (1854), Vol. 2, 247–8.
[33] E.I.C. to Government of India, 25 August 1852. I.O.L.R. E/4/816, folio 1116.
[34] 11 March 1857. I.O.L.R. E/4/270, paras 106–9.

EPILOGUE

In January 1907 Sir Joseph Hooker complained to J. S. Gamble at Dehra Dun that 'Botany seems to be dead in India'[1]. With the disintegration of the Botanical Survey of India there was now no central authority to speak on behalf of botany, no organized scientific opinion to command the attention and respect of the Government of India. During the 40 years that remained of British rule, progress in the botanical sphere was intermittent, almost random; the spasmodic impetus behind the compilation of regional floras was dependent on the efforts of individuals. New research institutes concentrated on economic crops: for example, Agricultural Research Institute (1903), Imperial Forest Research Institute (1904), Sugarcane Breeding Institute (1912), Institute of Plant Industry (1924), Central Potato Research Institute (1935), and Central Rice Research Institute (1946). The Bose Research Institute, founded in 1917, made little progress with the botanical investigations of Chandra Bose. The isolation felt by individual botanists and their need to communicate fostered the formation of societies: the Indian Botanical Society in 1920 and the Botanical Society of Bengal in 1935; in 1939 Professor Birbal Sahni recruited a committee of palaeobotanists from which in 1946 at Lucknow emerged the Palaeobotanical Society; in 1941 New Delhi became the Headquarters of the Indian Society of Genetics and Plant Breeding.

A. C. Joshi's presidential address to the Indian Botanical Society in 1949, two years after the independence of his country, was a gloomy retrospective survey and a not very optimistic forecast for the immediate future[2]. The traditional neglect of the biological sciences persisted; the Government of India had proposed new laboratories for chemistry, physics, metallurgy, fuel, buildings, road, food, etc., but virtually ignored biology. Joshi sadly noted the decline of the Royal Botanic Garden at Calcutta during the previous 40 years and judged its herbarium 'very much out of date'. The way to professional respectability lay in schools, colleges, and universities with a more enlightened and purposeful teaching of botany. The growth of professional societies since 1947 was a tribute to the resolution and determination of scientists like Joshi. A joint session of the Indian Science Congress and the Pan-Indian Ocean Science Congress in 1951 spawned the International Society of Plant Morphologists; the Mycological Society in 1954 was followed by the Indian Society for Plant Physiology three years later. Most of these societies promoted periodicals, and foremost among them is the *Journal of Palynology* of the Palynological Society of India, founded in 1965. The Society for Economic and Taxonomic Botany launched its journal in 1980. In India's neighbours, the Botanical Society of Pakistan started the *Pakistan Journal of Botany* in 1969, and in 1972 the *Bangladesh Journal of Botany* became the protégé of the Bangladesh Botanical Society.

The first positive step towards the regeneration of official botanical activity in India was the resurrection of the Botanical Survey of India. After the retirement of C. C. Calder just before the outbreak of the Second World War, the directorship of the Botanical Survey had been allowed to lapse and only the services of the curator of the Industrial Section of the Indian Museum in Calcutta and a taxonomist at the Botanic Garden were retained. Following a plea for its reorganization in 1952, Father H. Santapau, Principal of St Xavier's College in Bombay and a competent botanist, was appointed temporary Director in 1954, to be followed in October 1955 by Dr J. C. Sen Gupta. With headquarters at Calcutta, four regions or circles were designated: Northern (Dehra Dun), Southern (Coimbatore), Eastern (Shillong), and Western (Poona). These regional centres were extended to include a Central Circle (Allahabad) in 1962, Andaman and Nicobar Circle and Arid Zone Circle (Jodphur) in 1972, and a Sikkim-Himalaya Circle in 1979. A Central Botanical Laboratory had a peripatetic existence in Calcutta, Lucknow, and Allahabad before joining the new five-storey building at the Botanic Garden. The remit of this laboratory, according to its first Director, was 'to conduct research which will cover a wide range of problems relating to the botany of India, and to supply information relating to the living plants of India for the utilization and exploitation of the botanical wealth of India for agriculture, horticulture, and forestry, trade, industry, and medicine'.

The Indian Botanic Garden, formerly the Royal Botanic Garden, was transferred to the Botanical Survey in 1963. The Garden's herbarium had already been adopted by the Botanical Survey in 1957 to form the nucleus of a national herbarium. Its collections had begun with the return of Nathaniel Wallich from England in 1832, were completely reorganized during William Griffith's brief tenure as Superintendent, and rehoused by Sir George King. Many collectors since the 1830s have conscientiously deposited specimens in the Calcutta herbarium which also houses a representative selection of the world's flora. One of its priceless possessions is the series of flower paintings begun by William Roxburgh in 1793, miraculously surviving the vicissitudes of the years and predatory insects. The combined resources of this herbarium and the regional herbaria in the Survey's Circles exceed 2000 000 specimens.

These herbaria perform a pivotal role in the Survey's floristic projects but equally indispensable for the compilation of floras are personnel with taxonomic expertise. There is something of a dearth of professional taxonomists in India because for years few universities offered courses at postgraduate level. 'Systematic Botany' wrote Father Santapau in 1962, 'seems to have fallen into evil days in our country, except perhaps for Mycology, which has very great and obvious economic importance'[3]. The Botanical Survey started its *Bulletin* in 1959 primarily to encourage taxonomic research. There are some signs of a revival of interest since the current concern to preserve the world's diminishing vegetation must involve taxonomic investigation and floristics.

A new national flora was a primary objective of the reconstituted Botanical Survey of India. Sir Joseph Hooker had never intended his *Flora of British India* to be comprehensive or infallible. It was, he told Gamble, 'past praying for in the matter of errors in detail, from bad identifications, bad specimens and bad examination'[4]. It was also uneven in its treatment of families. Since the publication of the last volume in 1897, many more taxa have been described, there have been changes in classification and nomenclature, and political boundaries have been adjusted. The regional floras of the early twentieth century made little if any use of post-Hooker herbarium material. As far back as 1930 Ethelbert Blatter had argued persuasively for a revision of Hooker's work, and 40 years later there was no opposition to the Botanical Survey's resolve to embark on a new Flora of India. The first of a projected long series of small fascicles on families or genera of vascular plants appeared in 1978. There is no predetermined sequence; parts are published as they are written but priority is given to plants of economic significance. The only framework is that of four series: flora of India, state flora analyses and check-lists, district floras, and special publications. Progress under the first category has been disappointing; the output of state and district floras has been more satisfactory—meriting special mention is the Carnatic Flora project with several substantial volumes on the flora of Tamilnadu by Father K. M. Matthew. The occasional publication of floras independent of the Botanical Survey programme still continues, some with foreign collaborators. An outstanding example is *An enumeration of the flowering plants of Nepal* (3 vols., London, 1978–82) by H. Hare, A. O. Chater, W. T. Stearn, and L. H. J. Williams summarizing the results of collecting in Nepal since the pioneer work of Buchanan and Wallich and providing modern nomenclature for some 6500 species. Three thorough reports on the Eastern Himalayas were published between 1966 and 1972, the result of several expeditions by botanists of the University of Tokyo.

The fourth series in the schedule of Survey publications, reserved for miscellaneous works, at the moment consists only of a few titles on grasses and ferns and bibliographies. The contribution that monographs can make has been emphasized by Blatter, Santapau, and Matthew. Blatter advocated short monographs on families and genera as a means of provisionally updating Hooker's *Flora of British India.* For Santapau they provide a firm foundation for floras. Father Matthew argues that monographs should precede the compilation of floras and regrets the abandonment of the monographic tradition established by the *Annals of the Royal Botanic Garden, Calcutta* from 1887 to 1939.

Although Pakistan has no Botanical Survey it has, since 1970, published more than 190 family parts in its national flora. Bangladesh which embarked on its national flora in 1972 had published 39 slim fascicles by 1987. The Royal Botanic Garden in Edinburgh undertook a *Flora of Bhutan* in 1975 with the first part appearing in 1983, the second in 1984, and the third in 1987. A co-operative effort on the part of the University of Peradeniya, the Department of Agriculture in Sri Lanka, and the Smithsonian Institution in Washington is revising Henry Trimen's *Handbook to the flora of Ceylon*; six volumes had been published by 1988.

Botanical gardens in India have proliferated, at least in name, since independence. The Indian Botanic Garden which owes its existence to the East India Company still retains its premier position. The Forest Research Institute at Dehra Dun maintains an arboretum of about 1000 species and 1200 cultivars. Although some research is pursued at the botanical gardens in the former hill stations at Darjeeling and Ootacamund, most residents and visitors treat them, quite understandably, as public parks. A small botanical garden adjoins the Tropical Research Institute at Trivandrum, the Eastern and Southern Circles of the Botanical Survey support a National Orchidarium, and at least 16 universities claim to have botanical gardens. Private initiative has converted an established garden at the Sarabhai Foundation in Ahmedabad into a systematic display of plant families and provides a simple explanatory guide to the collection for students and the general public. The secondment of an Indian liaison botanist to the Royal Botanic Gardens at Kew has for many years preserved the historical botanical links between the two countries, enabling a

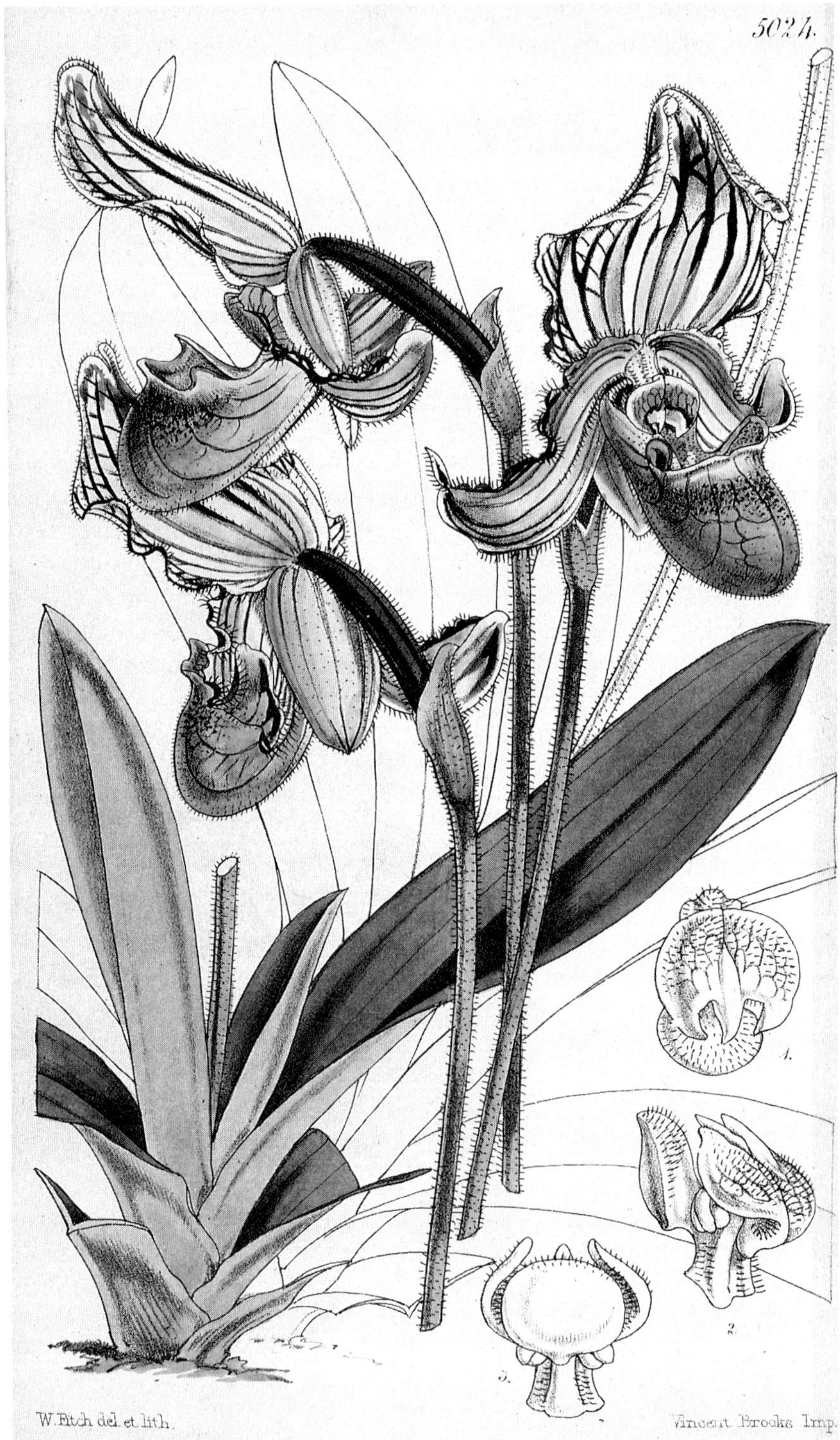

Figure 22.1 *Paphiopedilum fairrieanum.* (From *Curtis's Botanical Magazine*, 1857, plate 5024.) Drawn and lithographed by W. H. Fitch. Its existence is now threatened by the destruction of its habitat. Half a century after it was first introduced it had become so rare that in December 1904 the orchid nursery of Frederick Sander offered £1000 reward for its rediscovery. The orchid was found in Bhutan in 1905.

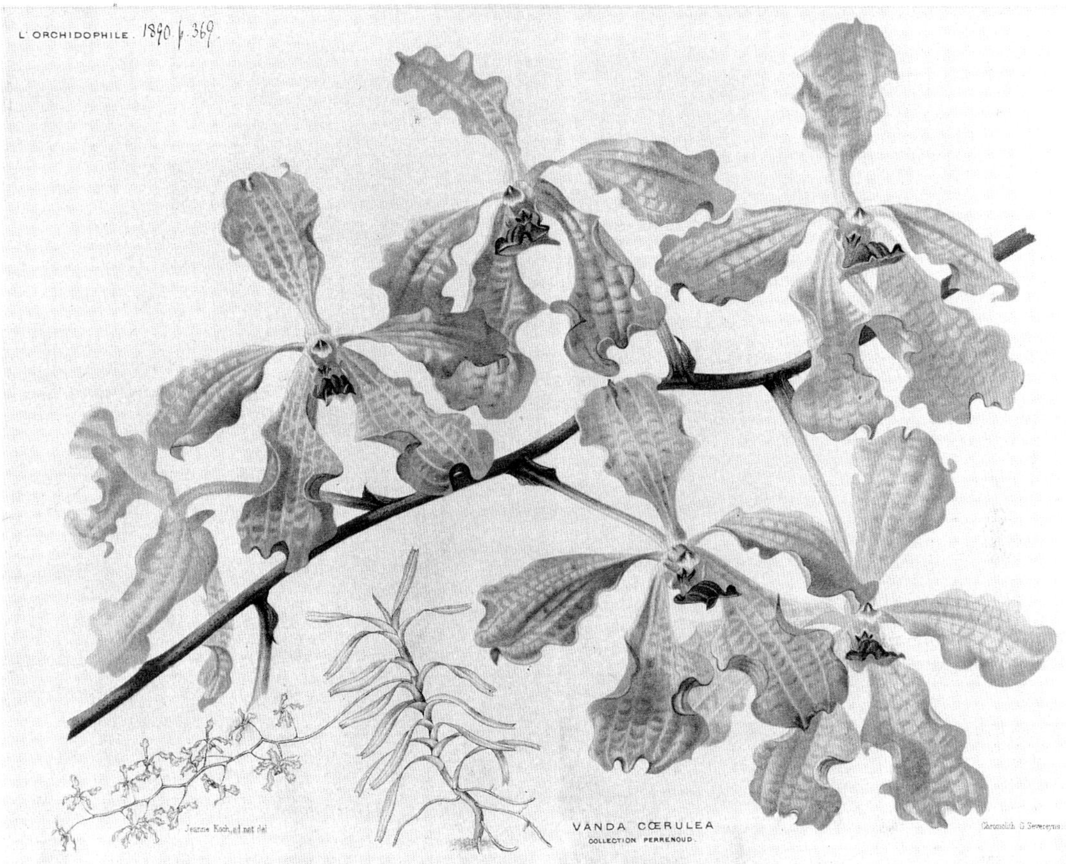

Figure 22.2 *Vanda coerulea.* (From *L'Orchidophile*, 1890, p. 369.) Drawn by Jeanne Koch. Chromolithograph. This orchid has now become a rarity through over-collecting.

succession of Indian botanists to extend their taxonomic expertise.

In January 1911, the office of Reporter on Economic Products, a post which had been held with such distinction by Sir George Watt, passed to the Botanical Survey. With it went the responsibility for the Indian Museum's Industrial Section which had its origins in the Bengal Economic Museum, founded in 1874. The Survey's Central Botanical Laboratory which now administers it has a commitment to the 'exploitation of the botanical wealth of India'. The programme of its Economic Botany Section includes the investigation of new vegetable drugs and a study of the use made of plants by India's tribal people. When shortages of raw materials during the Second World War highlighted the need for a new edition of Watt's *Dictionary of the economic products of India*, the Council for Scientific and Industrial Research convened an advisory board in 1942. The nation's raw materials were assessed in *The wealth of India*, published in 13 volumes between 1948 and 1976; the first volume of a new edition was published in 1985.

Much of India's future prosperity rests on the management and conservation of its natural resources. Climatic and topographical variations have endowed the subcontinent with one of the world's richest floras; every sort of habitat can be found—tropical rain forest, coniferous and broad-leaved forest, alpine meadow, dry scrubland, and

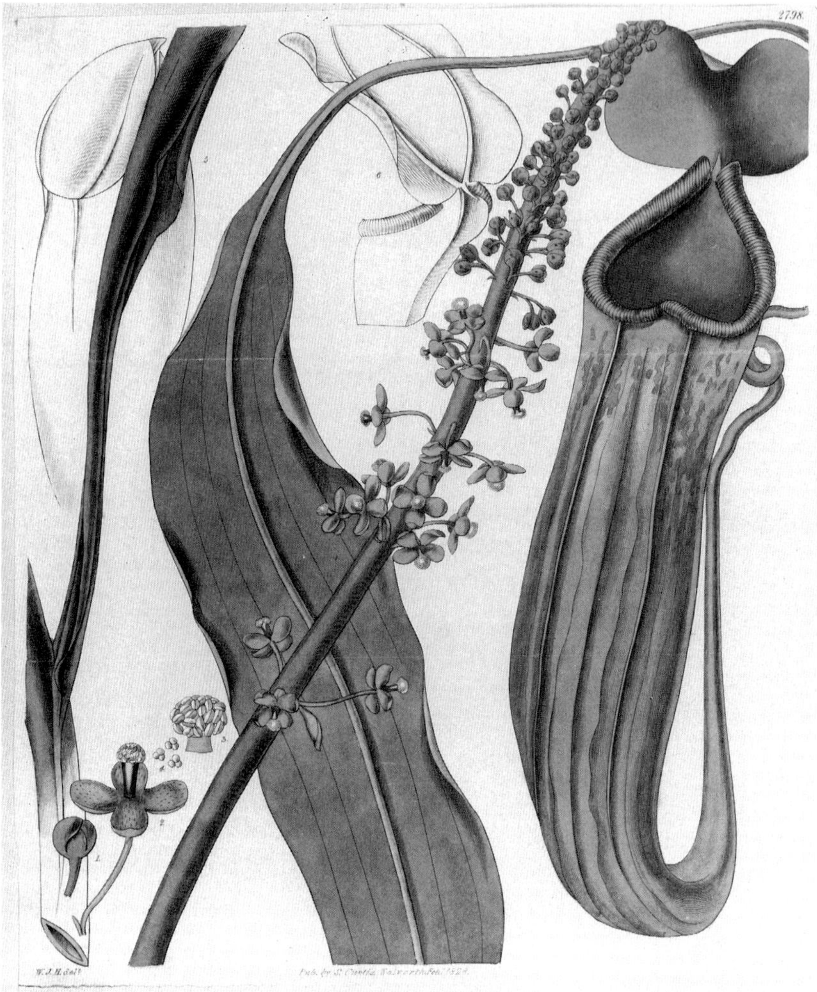

Figure 22.3 *Nepenthes khasiana.* (From *Curtis's Botanical Magazine*, 1828, plate 2798.) Hand-coloured engraving of a drawing by W. J. Hooker. This insectivorous plant is now threatened through over-collecting.

arid desert. But it is a flora under threat through population increase, urbanization, mining, logging, dams and hydro-electric power schemes, agricultural expansion, overgrazing, and now tourism (Fig. 22.1). Tribal villagers, especially in North East India, still cultivate their land by the cyclical burning of forests which, never recovering, degenerate into useless scrub. It has been estimated that about 14 per cent of India's virgin forest remains; in Bangladesh it is perhaps 9 per cent and in Pakistan as little as 6 per cent.

The destruction of natural habitats has been far more extensive and complete in India than perhaps anywhere else in the world. Of the various factors by which this has been brought about, deforestation must be considered as the most important. The forest has been systematically reduced and eradicated nearly over the whole of India[5].

Sir Joseph Hooker, who was privileged to explore Sikkim before the destruction of its forests, remembered the flanks of a mountain 'rose-coloured in spring' with *Magnolia campbellii*; J. S. Gamble a mere thirty years later reported it to be scarce. The orchid *Vanda coerulea* (Fig. 22.2) of which Hooker collected 'seven men's loads' is now an endangered species; *Cypripedium cordigerum* is no longer abundant at Gulmarg in Kashmir,

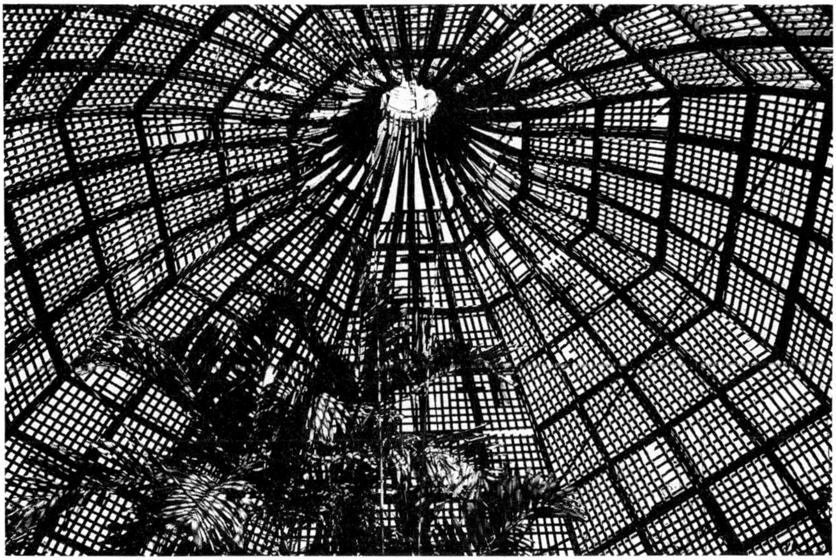

Figure 22.4 The roof of a splendid nineteenth-century conservatory in the Victoria Gardens in Bombay. It has been saved from demolition by the Bombay branch of the Indian Heritage Society and the generous support of local firms.

and three species of *Coelogyne* may now be extinct. Over-collecting is one of the reasons for a diminishing population of orchids; road-building in the Himalayas is destroying the habitat of rhododendrons, and the demands of pharmaceutical firms imperil the survival of *Rauvolfia serpentina* and *Dioscorea deltoidea*. It is not only deforestation and industrial development that are altering the ecosystem; the pattern and distribution of the Indian flora have been transformed by the introduction of exotics, particularly from Australia and tropical America, which have become naturalized, displacing indigenous species.

When the General Assembly of the International Union for the Conservation of Nature and Natural Resources met in New Delhi in 1969, the need for action to protect the Indian flora was unanimously recognized. The Botanical Survey of India which continued to monitor the situation, launched in 1980, with American financial aid, a Project on the Study, Survey, and Conservation of Endangered Species of the Flora. Within a few years more than a thousand threatened species were listed. The Government of India is a signatory to several international agreements, among them the Convention on International Trade in Endangered Species of Flora and Fauna and the Convention in Wetlands of International Importance. As a result of the India Wild Life (Protection) Act of 1972, more than 50 national parks and nearly 250 wildlife reserves have been created. The closure of the Valley of Flowers in the Nanda Devi National Park to mountaineers, trekkers, and grazing livestock in 1983 was further confirmation of the Government's commitment to conservation. A sanctuary for the rare pitcher plant, *Nepenthes khasiana*, has been established in Meghalaya State (Fig. 22.3).

Everywhere the explosion in the world's population is threatening delicate ecosystems. UNESCO pioneered the concept of biosphere reserves in 1973/4 dedicated to the preservation of biological diversity in plants, animals, and micro-organisms in natural and semi-natural ecosystems. The first national symposium on biosphere reserves held in India in September 1986 identified 13 prospective reserves and established the first one in the tropical and subtropical forests of the Nilgiris. India is especially rich in biological diversity with an estimated 65 000 animal and 45 000 plant species; some botanists have calculated that about 60 per cent of the country's vascular flora of 15 000 species is endemic.

Among the vital tasks of the subcontinent's botanical gardens are the protection of endangered species, the creation of gene banks, and the recruitment of public sympathy and support through educational programmes and publicity. There is a groundswell of public good will and signs of an emerging awareness of the catastrophic consequences of current despoliation. In 1988 campaigners on a 'Save the Western Ghats March' converged on Goa to denounce deforestation and to urge replanting with indigenous species of trees. The commercial exploitation of the forest resources in Kumaon and Garwhal Himalaya roused village women to protest against this devastation by symbolically embracing the trees. The leader of this Chipko Andolan movement, Sunderlal Bahaguna, justified their action with a religious text that compared a tree to ten sons. 'It gives ten valuable things: oxyen, water, energy, food, clothes, timber, medical herbs, fodder, flowers, and shade.' If such concern were shared by the population at large then the future of the subcontinent's still rich and diverse vegetation would be more secure.

NOTES

1 Huxley (1918), Vol. 2, 400.
2 *Journal of Indian Botanical Society*, v. **28**, 1949, 1–15.
3 *Science and Culture*, v. **28**, 1962, 45–8.
4 *Journal of Indian Botanical Society*, **9**, 1930,141-2.
5 M. S. Mani (ed.) (1974), 154.

BIBLIOGRAPHY

Anon. (1913). The Wallich herbarium. *Kew Blletin*, 255–63.

Acosta, C. (1578). *Tractado de las drogas y medicinas de las Indias orientales con sus plantas debuxadas al vivo, que las vio ocularmente*. Burgos.

Allan, M. (1967). *The Hookers of Kew, 1785–1911*. London.

Allen, C. (1986). *A glimpse of the burning plain: leaves from the Indian journals of Charlotte Canning*. London.

Anderson, J. (1788). *Letters to Sir Joseph Banks . . . on the subject of cochineal insects, discovered at Madras*. Madras.

Anderson, T. (1867). Notes on the foundation of the Royal Botanic Garden, Calcutta. *Journal of Agricultural and Horticultural Society of India*, **1**, 169–84.

Archer, M. (1955). Indian paintings for British naturalists. *Geographical Magazine*, (September) 220–30.

Archer, M. (1959). India and natural history: the rôle of the East India Company, 1785–1858. *History Today*, (November) 736–43.

Archer, M. (1962). *Natural history drawings in the India Office Library*. London.

Asiatic Society of Bengal (1885). *Centenary review of the Asiatic Society of Bengal; from 1784 to 1883*. Calcutta.

Babur (1922). *Babur-nama in English (memoirs of Babur)*. Translated from the original Turki text by A. S. Beveridge. London. 2 vols.

Bacon, T. (1837). *First impressions and studies from nature in Hindostan; embracing an outline of the voyage to Calcutta and five years' residence in Bengal and the Doab, from 1831 to 1836*. London. 2 vols.

Badger, G. P. (ed.) (1863). *Travels of Ludovico di Varthema in Egypt, Syria, Arabia Deserta and Arabia Felix, in Persia, India and Ethiopia*, AD *1503 to 1508*, (trans. J. W. Jones). London.

Bagchi, P. (1935). Victor Jacquemont in India. *Calcutta Review*, **56**, 51–60.

Ball, V. (1879–88). On the identification of the animals and plants of India which were known to early Greek authors. *Proceedings of the Royal Irish Academy*, (2nd series), **2**, 302–46.

Ball, V. (1889–90). A commentary on the Colloquies of Garcia De Orta, on the simples, drugs and medicinal substances of India. *Proceedings of the Royal Irish Academy*, **1**, 381–415, 647–85.

Banerji, S. C. (1980). *Flora and fauna in Sanskrit literature*. Calcutta.

Barbosa, D. (1866). *A description of the coasts of East Africa and Malabar in the beginning of the sixteenth century*, (trans. H. E. J. Stanley). London.

Basak, R. K. (1981). Robert Wight and his botanical studies in India. *Taxon*, **30**, 784–93.

Basak, R. K. (1983). *Botanical Survey of India—an account of its establishment, development and activities*. Calcutta.

Bastin, J. (1960). The changing balance of the early South East Asia pepper trade. *Papers on South East Asian Subjects*, No. 1. Kuala Lumpur.

Bastin, J. (1981). The letters of Sir Stamford Raffles to Nathaniel Wallich, 1819–1824. *Journal of Malayan Branch of the Royal Asiatic Society*, **54**, 1–73.

Bence-Jones, M. (1970). An English mansion in Bengal. Barrackpore: a relic of the Raj. *Country Life*, (February), 328–30.

Bence-Jones, M. (1973). *Palaces of the Raj: magnificence and misery of the Lord Sahibs*. London.

Bernier, F. (1934). *Travels in the Mogul Empire, AD 1656–1668*. London.

Beveridge, H. (1894). The Buchanan records. *Calcutta Review*, **197**, 1–17.

Beveridge, H. (ed.) (1909–14). *The Tuzuk-i-Jahangiri*, (trans. Alexander Rogers). London. 2 vols.

Bhattacharyya, P. K. (1982). Beginning of modern botany in India by Dutch in 16th–18th century (basic features and characteristics). *Indian Journal of History of Science*, **17**, 365–76.

Biswas, K. (1942). *The 150th anniversary volume of the Royal Botanic Garden, Calcutta*. Calcutta.

Biswas, K. (1950). *Original correspondence of Sir Joseph Banks relating to the foundation of the Royal Botanic Garden, Calcutta and the summary of the 150th anniversary volume of the Royal Botanic Garden, Calcutta*. Calcutta.

Bose, D. M., Sen, S. N., and Subbarayappa, B. V. (eds) (1971). *A concise history of science in India*. New Delhi.

Boulger, G. E. S. (1900). *History of Ceylon botany*. Vol. 5 of *Handbook to flora of Ceylon*, (H. Trimen and J. D. Hooker), pp. 367–81.

Boxer, C. R. (1963). *Two pioneers of tropical medicine: Garcia d'Orta and Nicolas Monardes*. London.

Boylan, P. J. (1977). *The Falconer papers, Forres*. Leicester.

Bretzl, H. (1903). *Botanische Forschungen des Alexanderzages*. Leipzig.

Britten, J. (1906). Hardwicke's botanical drawings. *Journal of Botany*, **44**, 235–41.

Britten, J. (1918). 'John' Roxburgh. *Journal of Botany*, **56**, 202–3.

Brockway, L. H. (1979). *Science and colonial expansion: the role of the British Royal Botanic Gardens*. London.

Brown, A. W. (1959). *Jacquemont*. Paris.

Buchanan, F. (1807). *A journey from Madras through the countries of Mysore, Canara and Malabar*. London. 3 vols.

Buchanan, F. (1826). Some notices concerning the plants of various parts of India and concerning the Sanscrita names of those regions. *Transactions Royal Society of Edinburgh*, **10**, 171–86.

Burkill, I. H. (1965). *Chapters on the history of botany in India*. Delhi.

Burrell, A. (1877). Indian tea cultivation: its origin, progress and prospects. *Journal of Society of Arts*, **25**, 199–215.

Candolle, A. de P. (1830). On the botany of India, and the facilities afforded for its investigation by the East India Company. *Edinburgh New Philosophical Journal*, 123–9.

Candolle, R. de and Radcliffe-Smith, A. (1981). Nathaniel Wallich, MD, PhD, FRS, FLS, FRGS, (1786–1854) and the Herbarium of the Honourable East India Company, and their relation to the de Candolles of Geneva and the Great Prodromus. *Botanical Journal of Linnean Society*, **83**, 325–48.

Cannon, G. (1970). *The letters of Sir William Jones*. Oxford. 2 vols.

Cannon, G. (1975). Sir William Jones, Sir Joseph Banks and the Royal Society. *Notes and Records of the Royal Society of London*, **29**, 205–29.

Cannon, G. (1985). Sir William Jones's founding and directing of the Asiatic Society. *India Office Library and Records Report 1984–85*, 11–28.

Carey, S. P. (1923). *William Carey*. London.

Carstensen, G. (1891). Landscape gardening in native states. *Journal of Bombay Natural History Society*, **6**, 72–83.

Carter, H. (1988). *Sir Joseph Banks, 1743–1820*. London.

Chakraverty, R. K. and Mukhopadhyay, D. P. (1983). Indian Botanic Garden, Howrah. *Bulletin of Botanical Survey of India*, **25**, 290–6.

Chatterjee, D. (1948). Early history of the Royal Botanic Garden, Calcutta. *Nature*, **161**, 362.

Chowdhury, K. A. (1969). Archaeological plant remains from pre- and proto-historic periods as a source of history of sciences. *Indian Journal of History of Science*, **4**, 5–14.

Cleghorn, H. (1861). *Forests and gardens of South India*. London.

Cleghorn, H. (1873). Obituary notice of Dr Robert Wight, FRS. *Transactions Botanical Society of Edinburgh*, **11**, 363–88.

Cleghorn, H. *et al.* (1851). Report of the committee appointed by the British Association to consider the probable effects in an oeconomical and physical point of view of the destruction of tropical forests. *British Association for Advancement of Science Report, 1851*, 78–102.

Coates, A. M. (1969). *The quest for plants*. London.

Cordiner, J. (1983). *A description of Ceylon*. London. Originally pub. 1807.

Council for Scientific and Industrial Research, India. (1948–76). *The wealth of India: raw materials*. Delhi. 13 vols.

Crawford, D. G. (1914). *A history of the Indian Medical Service, 1600–1913*. London, 2 vols.

Dalrymple, Alexander (1808). *Oriental repository, 1791–1808*. London. 2 vols.

Dames, M. L. (ed.) (1921). *The book of Duarte Barbosa. An account of the countries bordering on the Indian Ocean and their inhabitants . . . completed about the year AD 1518* (trans. M. L. Dames). London.

Dandy, J. E. (1958). *The Sloane herbarium: an annotated list of the horti sicci composing it; with biographical accounts of the principal contributors*. London.

De Wit, H. C.—D. (ed.) (1959). *Rumphius memorial volume*. Baarn.

Deb, D. B. (1977). The Indian Botanic Garden. *Bulletin of Botanical Survey of India* **19**, 263–8.

Desmond, R. (1977). *Dictionary of British and Irish botanists and horticulturists including plant collectors and botanical artists*. London.

Desmond, R. (1977). William Roxburgh's Plants of the coast of Coromandel, 1795–1820. *Hortus Aliquando*, (2), 22–41.

Desmond, R. (1982). *The India Museum, 1801–1879*. London.

Duran-Reynals, M. L. (1947). *The fever bark tree: the pageant of quinine*. London.

Dyson, K. K. (1978). *A various universe: a study of the journals and memoirs of British men and women in the Indian subcontinent, 1765–1856*. Delhi.

Falconer, H. (1854). On the quinine-yielding cinchonas and their introduction into India. *Journal of Agricultural and Horticultural Society of India*, **8**, 13–17.

Fazl, A. (1877). *Ain-i Akbari*, (trans. H. Blochman). London.

Firminger, T. A. C. (1863). *A manual of gardening for Bengal and Upper India*, (6th edn). Calcutta.

Fletcher, H. R. (1975). *A quest of flowers: the plant explorations of Frank Ludlow and George Sherriff told from their diaries and other occasional writings*. Edinburgh.

Floriju, P. J. (1987). Biographical notes about four plant collectors in Asia mentioned by N. C. Burman in his *Flora Indica* (1768). *Taxon*, **36**, 34–8.

Forbes, J. (1812–13). *Oriental memoirs: selected and abridged from a series of familiar letters written during seventeen years' residence in India*. London. 4 vols.

Foster, W. (1918). English commerce with India, 1608–1658. *Journal of Royal Society of Arts*, **66**, 361–84.

Foster, W. (1921). *Early travels in India, 1583–1619*. Oxford.

Fournier, M. (1987). Enterprise in botany: Van Reede and his Hortus Malabaricus. *Archives of Natural History* **14**, 123–58, 297–338.

Fraser, J. B. (1820). *Journal of a tour through part of the Snowy Range of the Himala Mountains and to the sources of the rivers Jumna and Ganges*. London.

Fryer, G. (1979). John Fryer, FRS and his scientific observations, made chiefly in India and Persia between 1672 and 1682. *Notes and Records of the Royal Society of London*, **33**, 175-206.

Fryer, J. (1909–15). *A new account of East India and Persia in eight letters being nine years' travels, 1672–1681*. London. 3 vols.

Gage, A. T. (1926). The Royal Botanic Garden, Calcutta. *Journal of Royal Horticultural Society*, **51**, 71–81.

Gardener, W. (1967). Six cinchona trees. *History Today*, **17**, 52–7.

Geoghegan, J. (1872). *Some account of silk in India, especially of the various attempts to encourage and extend sericulture in that country*. Calcutta.

Gowda, M. H. M. (1949). Early history of the botanic garden, Bangalore. *Science and Culture*, **15**, 181–3.

Gramiccia, G. (1988). *The life of Charles Ledger (1818–1905): alpacas and quinine*. Basingstoke.

Grey, E. (ed.) (1892). *Travels of Pietro della Valle in India* (From the old English translation of 1664, by G. Havers). London. 2 vols.

Grierson, A. J. C. and Long, D. G. (1983–7). *Flora of Bhutan including a record of plants from Sikkim*. Edinburgh.

Griffith, W. (1847–54). *Posthumous papers*. (arranged by J. McClelland). Calcutta. 6 vols.

Griffiths, P. (1969). *The history of the Indian tea industry*. London.

Grove, R. (1990). The origins of environmentation. *Nature*, **345**, 11–14.

Guerra, F. (1966). Drugs from the Indies and the political economy of the sixteenth century. *Analecta Medico-historica*, **1**, 29–54.

Gupta, R. K. (1966). Botanical explorations of Victor Jacquemont. *Indian Journal of History of Science*, **1**, 150–7.

Gupta, S. M. (1971). *Plant myths and traditions in India*. Leiden.

Harrison, S. (1978). Recent work on the herbarium of John Forbes Royle (1799–1858) at Merseyside County Museums. *Taxon*, **27**, 21–33.

Head, R. (1984). Divine flower power recorded: Sir William Jones (1746–94). *Country Life*, (September) 703–5.

Heber, R. (1828). *Narrative of a journey through the Upper Provinces of India from Calcutta to Bombay, 1824–1825*. London. 3 vols.

Hedge, I. C. and Lamond, J. M. (1990). Edinburgh's Indian botanical connections and collections. *Bulletin of the Botanical Survey of India*, **29**, 272–85.

Heniger, J. (1986) *Hendrik Adriaan van Reede tot Drakenstein (1636–1691) and* Hortus Malabaricus: *a contribution to the history of Dutch colonial botany*. Rotterdam.

Herodotus (1972). *The histories*, (trans. Aubrey de Sélincourt). London.

Heyne, B. (1814). *Tracts, historical and statistical on India; with journals of several tours through various parts of the Peninsula: also an account of Sumatra in a series of letters*. London.

Hobhouse, H. (1985). *Seeds of change: five plants that transformed mankind*. London.

Hodges, W. (1793). *Travels in India during the years 1780, 1781, 1782, and 1783*. London.

Hoffmeister, W. (1848). *Travels in Ceylon and continental India*. (from the German). Edinburgh.

Honig, P. and Verdoorn, F. (1945). *Science and scientists in the Netherlands Indies*. New York.

Hooker, J. D. (1849–51). *Rhododendrons of Sikkim-Himalaya*. London.

Hooker, J. D. (1854). *Himalayan journals*. London. 2 vols.

Hooker, J. D. (1855). *Illustrations of Himalayan plants*, (chiefly . . . made for the late J. F. Cathcart). London.

Hooker, J. D. and Thomson, T. (1855). Introductory essay. In *Flora Indica*, 1–289, London.

Hooker, W. J. (1831). Illustrations of Indian botany principally of the southern parts of the peninsula by Richard [sic] Wight. *Hooker's Botanical Miscellany*, **2**, 90–7.

House of Commons (1839). *Tea cultivation (India)*. Copy of papers received from India relating to the measures adopted for introducing the cultivation of the tea plant within the British possessions in India. London.

Howard, R. A. and Powell, D. A. (1965). The Indian Botanic Garden, Calcutta and the garden of the West Indies. *Bulletin of Botanical Survey of India*, **7**, 1–7.

Huegel, C. von (1845). *Travels in Kashmir and the Panjab* . . . with notes by Major T. B. Jervis. London.

Hunt Institute for Botanical Documentation (1980). *A selection of late 18th & early 19th century botanical paintings* . . . Pittsburgh.

Huxley, L. (1918). *Life and letters of Sir Joseph Dalton Hooker*. London, 2 vols.

Hyde, H. M. (1962). Dr George Govan and the Saharanpur Botanical Gardens. *Journal of Royal Central Asian Society*, **49**, 47–57.

Jackson, H. St J. (1891). Landscape gardening in native states. *Journal of Bombay Natural History Society*, **6**, 478–85.

Jacquemont, V. (1835). *Letters from India: describing a journey in the British dominions of India, Tibet, Lahore and Cashmeer, during the years 1828, 1829, 1830, 1831*. Undertaken by order of the French Government. London. 2 vols.

Jaffrey, A. T. (1855). *Hints to the amateur gardeners of Southern India*. Madras.

Jain, B. L. (1982). Development of life sciences in India in eighteenth–nineteenth century. *Indian journal of History of Science*, **17**, 114–31.

Jain, S. K. (1983). Botanical Survey of India: 1958–1983. *Bulletin of Botanical Survey of India*, **25**, 252–67.

Jones, W. (1790). The design of a treatise on the plants of India. *Asiatick Researches*, **2**, 345–52.

Jones, W. (1790). On the spikenard of the ancients. *Asiatick Researches*, **2**, 405–17.

Jones, W. (1795). Additional remarks on the spikenard of the ancients. *Asiatick Researches*, **4**, 109–18.

Jordanus, (1863). *The wonders of the East by Friar Jordanus*, (trans. Henry Yule). London.

Kapil, R. N. (1970). Biology in ancient and medieval India. *Indian Journal of History of Science*, **5**, 119–40.

Kapil, R. N. and Bhatnagar, A. K. (1976). Portuguese contributions to Indian botany. *Isis*, **67**, 449–52.

King, G. (1893). A short account of Colonel Kyd, the founder of the Royal Botanic Garden, Calcutta. *Annals of Royal Botanic Garden, Calcutta*, **4**, 1–11.

King, G. (1895). A brief memoir of William Roxburgh, author of the 'Flora Indica'. *Annals of Royal Botanic Garden, Calcutta*, **5**, 1–9.

King, G. (1899). A sketch of the history of Indian botany. *British Association for Advancement of Science, Transactions, Section K*, 904–19.

Kirtikar, K. R. (1905). Progress in natural history during the last century. *Journal of Bombay Branch of Royal Asiatic Society: memorial volume, 1905*, 353–71.

Knox, R. (1911). *An historical relation of the island Ceylon in the East-Indies*. London. Originally pub. 1681.

Kumar, D. (1980). Patterns of colonial science in India. *Indian Journal of History of Science*, **5**, 105–13.

Lach, D. F. (1965–77) *Asia in the making of Europe*. Vol. 1, *The century of discovery*. Vol. 2, *A century of wonder*: Book 3, *The scholarly disciplines*. Chicago.

Lamond, J. M. (1970). The Afghanistan collections of William Griffith. *Notes from Royal Botanic Garden Edinburgh*, **30**, 159–75.

'Landolicus' (1881). *The Indian amateur gardener*. Calcutta.

Lane Fox, R. (1973). *Alexander the Great*. London.

Larwood, H. J. C. (1958). Science in India before 1850. *British Journal of Educational Studies*, **7**, 36–49.

L'Escluse, C. de, (1567; 1574; 1579). *Aromatum et simplicium aliquot medicamentorum apud Indos nascentium historia*. Antwerp.

L'Escluse, C. de, (1582). *Aromatum et medicamentorum in orientali India nascentium liber*. Antwerp.

L'Escluse, C. de (1605). *Exoticorum libri decem, quibus animalium, plantarum, aromatum, aliorumque peregrinorum fructuum historiae describunter*. Antwerp.

Linschoten, J. H. van (1885). *The voyages of John Huyghen van Linschoten to the East Indies*. (From the old English translation of 1598.) London. 2 vols.

Long, D. G. (1979). The Bhutanese itineraries of William Griffith and R. E. Cooper. *Notes from Royal Botanic Garden, Edinburgh*, **37**, 355–68.

Lopez, R. S. (1943). European merchants in the medieval Indies: the evidence of commercial documents. *Journal of Economic History*, **3**, 164–84.

Lourteig, A. (1966). L'Herbier de Paul Hermann, base du Thesaurus Zeylanicus de Johan Burman. *Taxon*, **15**, 23–5.

Ludlow, F. (1972). Thomas J. Booth and the type localities of his Indian rhododendrons described by Nuttall. *Transactions Botanical Society of Edinburgh*, **41**, 351–63.

Mabberley, D. J. (1977). Francis Hamilton's Commentaries with particular reference to Meliaceae. *Taxon*, **26**, 523–40.

Mabberley, D. J. (1985). *Jupiter Botanicus: Robert Brown of the British Museum*. Braunschweig.

McClelland, J. (1845). Memoir of the late William Griffith. *Journal of Agricultural and Horticultural Society of India*, (part 1), 1–39.

McCrindle, J. W. (1877). *Ancient India as described by Megasthenes and Arrian*; being a translation of fragments of the *Indika* of Megasthenes collected by Dr Schwanbeck, and of the first part of the *Indika* of Arrian. London.

McCrindle, J. W. (1879). *The commerce and navigation of the Erythrean Sea*; being a translation of the Periplus Maris Erythraei by an anonymous writer. London.

McCrindle, J. W. (1882). *Ancient India as described by Ktesias the Knidian*; being a translation of the abridgement of his *Indika* by Photios and of the fragments of that work preserved in other writers. London.

McCrindle, J. W. (1893). *The invasion of India by Alexander the Great as described by Arrian, Q Curtius, Diodoros, Plutarch and Justin*. London.

McCrindle, J. W. (1897). *The Christian topography of Cosmas, an Egyptian monk*. London.

McCrindle, J. W. (1901). *Ancient India as described in classical literature*. London.

Maheshwari, P. (1960). Plants in ancient India. *Botanica*, **11**, 1–6.

Maheshwari, P. (1965). *Plants and human history*. Osmania, Hyderabad.

Maheshwari, P. and Kapil, R. N. (1958). A short history of botany in India. *Journal of University of Gauhati*, **9**, 1–34.

Major, R. H. (1857). *India in the fifteenth century*; being a collection of narratives of voyages to India in the century preceding the Portuguese discovery of the Cape of Good Hope from Latin, Persian, Russian and Italian sources. London.

Majunder, G. P. (1927). *Vanaspati: plants and plant-life as in Indian treatises and traditions*. Calcutta.

Malins, E. (1980). Indian influences on English houses and gardens at the beginning of the nineteenth century. *Garden History*, **8**, 46–66.

Mani, M. S. (ed.) (1974). *Ecology and biogeography in India*. The Hague.

Manilal, K. S. (ed.) (1980). *Botany and history of* Hortus Malabaricus. Rotterdam.

Manilal, K. S. (1984). *Hortus Malibaricus* and the ethnoiatrical knowledge of ancient Malabar. *Ancient Science of Life*, **4**, 96–9.

Markham, C. R. (1862). *Memorandum on Dr Anderson's report on the subject of the introduction of the chinchona into India*. London.

Markham, C. R. (1862). *Travels in Peru and India while superintending the collection of chichona plants and seeds in South America, and their introduction into India*. London.

Markham, C. R. (1876). The cultivation of caoutchouc-yielding trees in British India. *Journal of the Society of Arts*, **24**, 475–82.

Markham, C. R. (1880). *Peruvian bark. A popular account of the introduction of chinchona cultivation into British India, 1860–1880*. London.

Marr, J. R. (1972). An examination of some plant-names and identities in India. *Journal of Royal Asiatic Society*, 40–56.

Matthew, K. M. (1982). Botany and its technologies in peninsular India in the eighteenth and nineteenth centuries. *Indian Journal of History of Science*, **17**, 353–64.

Matthew, K. M. (1989). Plant taxonomy in India: present state and future tasks. *Journal of Swarmy Botanical Club*, **6**, 27–32.

Merrill, E. D. (1921). A review of the new species of plants proposed by N. L. Burman in his Flora Indica. *Philippine Journal of Science*, **19**, 329–88.

Michell, G. *et al.* (1982). *In the image of man: the Indian perception of the universe through 2 000 years of painting and sculpture*. London.

Miller, J. I. (1969). *The spice trade of the Roman empire, 29 BC to AD 641*. London.

Mitra, R. and Kapoor, L. D. (1976). Kamala: the national flower of India; its ancient history and uses in Indian medicine. *Indian Journal of History of Science*, **11**, 125–32.

Mitra, S. C. (1890). The pursuit of natural history among the natives of India. *Calcutta Review*, **91**, 159–76.

Mitra, S. C. (1891). A plea for the formation of a Linnean Society in Calcutta. *Calcutta Review*, **92**, 196–203, 408–10.

Mukerjee, S. K. (1959). Central National Herbarium. *Bulletin of Botanical Survey of India*, **1**, 30–2.

Mukerjee, S. K. and Nath, N. (1970). Horticultural science in ancient India. *Indian Journal of History of Science*, **5**, 183–6.

Mundy, P. (1914). *Travels in Asia, 1628–1634*. Vol. 2 of *The travels of Peter Mundy in Europe and Asia, 1608–1667*. London.

Nayer, M. P. and Das, A. R. (1983). William Roxburgh (1751–1815) & Nathaniel Wallich (1786–1854). *Bulletin of Botanical Survey of India*, **25**, 347–56.

Nicolson, D. H. *et al.* (1988). *An interpretation of Van Rheede's* Hortus Malabaricus. Königstein.

North, M. (1892). *Recollections of a happy life*. London. 2 vols.

Orta, G. da (1563). *Coloquios dos simples, e drogas he cousas mediçinais do India* . . . Goa.

Orta, G. da (1913). *Colloquies on the simples and drugs of India*, (trans. C. Markham). London.

Panikkar, K. M. (1929). *Malabar and the Portuguese*. Bombay.

Panikkar, K. M. (1931). *Malabar and the Dutch*. Bombay.

Pearson, L. (1960). *The lost histories of Alexander the Great*. New York.

Peiris, E. (1952). Paul Hermann, the father of Ceylon botany. *Journal of Ceylon Branch of Royal Asiatic Society*, (new series), **2**, 1–18.

Petch, T. (1920–1). The early history of botanic gardens in Ceylon with notes on the topography of Colombo. *Ceylon Antiquary*, **5**, 119–24; **7**, 63–73.

Petiver, J. (1698). An account of some Indian plants with their names, descriptions and vertues; communicated in a letter from Mr James Petiver . . . to Mr Samuel Brown, Surgeon at Fort St George. *Philosophical Transactions of Royal Society*, **20**, 313–35.

Petiver, J. (1700–3). An account of part of a collection of curious plants and drugs, lately given to the Royal Society by the East India Company. *Philosophical Transactions of Royal Society*, **22**, 1700–1, 579–94, 699–721, 843–58, 933–46, 1007–22; **23**, 1702–3, 1055–65, 1251–65, 1450–60.

Petiver, J. (1699). *Musei Petiveriani*. London.

Phillips, C. A. (1936). *Letters from India, 1829–1832: being a selection from the correspondence of Victor Jacquemont*, (trans. with an introduction by C. A. Phillips). London.

Pliny the Elder, (1958–62). *Natural history*, (trans. H. Rackham). London. 10 vols.

Prain, D. (1905). A sketch of the life of Francis Hamilton (once Buchanan), some time Superintendent in the Honourable Company's Botanic Garden at Calcutta. *Annals of Royal Botanic Garden, Calcutta*, **10** (part 2), i–lxxv.

Prain, D. (1919). 'John' Roxburgh. *Journal of Botany*, 28–34.

Radcliffe-Smith, A. and Parsons, T. J. (1970). *Microfiche of the Herbarium of the East India Company ('The Wallich Herbarium') at Kew*: prefactory notes, collectors, incorporated herbaria, cultivated material, lists, gazetteer . . . Zug.

Rajbhandri, K. R. (1976). History of botanical explorations in Nepal. *Journal of Bombay Natural History Society*, **73**, 468–81.

Randhawa, G. S. *et al.* (1971). *Famous gardens of India*. New Delhi.

Ravenstein, E. G. (ed.) (1898). *A journal of the first voyage of Vasco da Gama, 1497–1499*. London.

Reddy, D. V. S. (1974). Spread of knowledge of Indian herbs and drugs in Europe in XVI century. (The role of Garcia Da Orta, Cristoval Acosta and Charles Clusius and their publications). *Bulletin of Indian Institute of History of Medicine*, **4**, 102–15.

Rose, D. (1981). Dr Nathaniel Wallich. *The Garden*, **106**, 196–200.

Roxburgh, W. (1795–1820). *Plants of the coast of Coromandel; selected from drawings and descriptions presented to the Hon. Court of Directors of the East India Company*. London. 3 vols.

Roxburgh, W. (1814). *Hortus Bengalensis, or a catalogue of the plants growing in the Honourable East India Company's Botanic Garden at Calcutta*. Serampore.

Roxburgh, W. (1820–4). *Flora Indica*. Serampore, 2 vols.; 1832, 3 vols. (reprinted 1874); Cryptogams, *Calcutta Journal of Natural History*, **4**, 1844, 463–520.

Roy, M. (1970). Family relations of some plants in the Atharvaveda. *Indian Journal of History of Science*, **5**, 162–75.

Roy, M. (1978). Dyes in ancient and medieval India. *Indian Journal of History of Science*, **13**, 83–112.

Royle, J. F. (1837). *An essay on the antiquity of Hindoo medicine including an introductory lecture to the course of materia medica and therapeutics delivered at Kings College, London*. London.

Royle, J. F. (1840). *Essay on the productive resources of India*. London.

Royle, J. F. (1850). Report on the progress of the culture of the Chinese tea plant in the Himalayas from 1835 to 1847. *Journal of Agricultural and Horticultural Society of India*, **7**, 11–41.

Royle, J. F. (1855). *The fibrous plants of India fitted for cordage, clothing and paper; with an account of the cultivation and preparation of flax, hemp and their substitutes.* London.

Sangar, S. P. (1981). Intoxicants in Mughal India. *Indian Journal of History of Science*, **16**, 202–14.

Sangwan, S. (1983). Plant colonialism (1786–1857). *Proceedings Indian History Congress*, 414–24.

Sangwan, S. (1988). Indian response to European science and technology. *British Journal of History of Science*, **21**, 211–32.

Santapau, H. (1954). The revival of the Botanical Survey of India. *Science and Culture*, **19**, 377-81.

Santapau, H. (1958). The coconut, *Cocos nucifera* Linn: observations of the first English Jesuit priest in India. *Journal of Bombay Natural History Society*, **55**, 188–9.

Santapau, H. (1964). The Botanical Survey of India. *Science and Culture*, **30**, 2–11.

Santapau, H. (1965). The Indian Botanic Garden in the first 175 years. *Bulletin of Botanical Survey of India*, **7**, i–vii.

Santapau, H. (1966). The story of Indian tea. *Bulletin of Botanical Survey of India*, **8**, 103–7.

Sarton, G. (1937). Rumphius Plinius Indicus (1628–1702). *Isis*, **27**, 242–57.

Sastry, V. V. S. (1974). A brief account of the writers and books on Indian herbs, drugs and bazaar medicines in English and other European languages. *Bulletin of Indian Institute of History of Medicine*, **4**. 61–75.

Sastry, V. V. S. (1975). Hortus Indicus Malabaricus (Garden of Malabar of India). *Bulletin of Indian Institute of History of Medicine*, **5**, 143–9.

Sayers, D. (1988). Gardens of some Indian hill stations. *Hortus*, **7**, 63–72.

Sealy, J. R. (1956). The Roxburgh Flora Indica drawings at Kew. *Kew Bulletin*, 297–348.

Sealy, J. R. (1977). William Roxburgh's collection of paintings of Indian plants. *Endeavour*, 34, 84–9.

Seide, J. (1952). Nathaniel Wallich, surgeon and botanist (1786–1854). *Medicine Illustrated*, **6**, 91–2.

Seide, J. (1953). The relationship of Garcia da Orta's and Cristobal Acosta's botanical works. *Actes du Septième Congrès International d'Histoire des Sciences, Jérusalem*, 564–7.

Sen, S. (1949). *Indian travels of Thevenot and Careri.* New Delhi.

Sen, S. N. (1966). The character of the introduction of western science in India during the eighteenth and the nineteenth centuries. *Indian Journal of History of Science*, **1**, 112–22.

Sen Gupta, J. C. (1959). Botanical Survey of India; its past, present and future. *Bulletin of Botanical Survey of India*, **1**, 9–32.

Sharma, A. L. *et al.* (1972). Botany in the Vedas, part 2. *Indian Journal of History of Science*, **7**, 38–43.

Shukla, M. S. (1970). Arbori-horticulture in ancient India. *Indian Journal of History of Science*, **5**, 179–82.

Singh, G. and Joshi, P. D. (1982). Impact of European science and technology on the development of modern Ayurveda during 19th century. *Indian Journal of History of Science*, **17**, 313–25.

Singh, R. B. (1974). Cities and parks in ancient India. *Geographical Review of India*, **36**, 62–75.

Sinha, S. C. (1925). On the antiquity and the therapeutic uses of the Indian spikenard. *Journal of Bombay Natural History Society*, **30**, 777–87.

Skelton, R. (1972). A decorative motif in Mughal art. In *Aspects of Indian art*, (ed. P. Pal), pp. 147–52. Leiden.

Soares, A. X. (1921–3). Garcia d'Orta, a little known owner of Bombay. *Journal of Bombay Branch of Royal Asiatic Society*, **26**, 195–229.

Speede, G. T. Frederic S. B. (1842). *The Indian hand-book of gardening: or guide to the management of the kitchen, fruit and flower garden in India*, (2nd edn). Calcutta. Originally pub. 1840.

Speede, G. T. Frederic S. B. (1848–50). *The new Indian gardener and guide to the successful culture of the kitchen and fruit garden in India.* vol. 2: *Shrubbery and flower garden.* 1850. Calcutta.

Spruce, R. (1861). *Report on the expedition to procure seeds and plants of the* Cinchona succirubra *or red bark tree.* London.

Spry, H. H. (1841). *Suggestions received by the Agricultural and Horticultural Society of India for extending the cultivation and introduction of useful and ornamental plants with a view to the improvement of the agricultural and commercial resources of India.* Calcutta.

Srinivasan, K. S. (1961). William Carey; his activities and horticultural researches in India. *Bulletin of Botanical Survey of India*, **3**, 1–10.

Stafleu, F. A. (1971). *Linnaeus and the linnaeans: the spreading of their ideas in systematic botany, 1735–1789.* Utrecht.

Stansfield, H. (1953). The Royle herbarium. *North Western Naturalist*, (new series), **1**, 249–65.

Stansfield, H. (1954). Plant collecting in the Himalayas. *Bulletin of Liverpool Libraries, Museums and Arts Committee*, **3** (part 3), 5–36.

Stansfield, H. (1957). The missionary botanists of Tranquebar. *Bulletin of Liverpool Libraries, Museums and Arts Committee*, **6**, 19–42.

Stansfield, H. (1959). The Royle herbarium. *Pharmaceutical Journal*, (January), 74–75, 78.

Stearn, W. T. (1957). Botanical exploration to the time of Linnaeus. *Proceedings of Linnean Society of London 1956–7*, 173–96.

Stearn, W. T. (1976). Frank Ludlow (1885–1972) and the Ludlow–Sherriff expeditions to Bhutan and South-Eastern Tibet of 1933–1950. *Bulletin of British Museum (Natural History) Botany*, **5** (part 5).

Stearn, W. T. (1977). The earliest European acquaintance with tropical vegetation. *Gardens' Bulletin, Singapore*, **29**, 13–18.

Stearn, W. T. (1988). Carl Linnaeus's acquaintance with tropical plants. *Taxon*, **37**, 776–81.

Stearns, R. P. (1952). James Petiver: promoter of natural science. *Proceedings of American Antiquarian Society*, **62**, 243–365.

Stebbing, E. P. (1922–6). *The forests of India.* London. 3 vols.

Stewart, R. R. (1979). The first plant collectors in Kashmir and the Punjab. *Taxon*, **28**, 5–11.

Stewart, R. R. (1982). *History and exploration of plants in Pakistan and adjoining areas.* Islamabad.

Stewart, R. R. (1982). Missionaries and clergymen as botanists in India and Pakistan. *Taxon*, **31**, 57–64.

Strabo (1917–49). *The geography*; with an English translation by Horace Leonard Jones. London. 8 vols.

Styles, B. T. (1976). Sir William Jones' names of Indian plants. *Taxon*, **25**, 671–4.

Suresh, C. R. *et al.* (1983). A commentary on Rheede's aroids. *Taxon*, **32**, 126–32.

Swarmy, B. G. L. (1973–6). Sources for history of plant sciences in India. I: Epigraphy; II: The Rg Vedic soma plant; III: The maritime trade of ancient Tamils in plant products—a critique. *Indian Journal of History of Science*, **8**, 61–8; **2**, 11–30, 33–48.

Synge, P. M. (1950). Sir Joseph Hooker and the rhododendrons of Sikkim-Himalaya. *Rhododendron Society Yearbook 1950*, 38–51.

Tavernier, J.-B. (1925). *Travels in India*, (2nd edn. ed. W. Crooke), (trans. V. Ball). Originally pub. 1889. London.

Theophrastus, (1916). *Enquiry into plants and minor works on odours and weather signs*, (trans. Sir Arthur Hort). London. 2 vols.

Thomson, T. (1852). *Western Himalaya and Tibet: a narrative of a journey through the mountains of Northern India during the years 1847–8.* London.

Thomson, T. (1856). Notes on the herbarium of the Calcutta Botanic Garden, with especial reference to the completion of the Flora Indica. *Journal of Bengal Asiatic Society*, **25**, 405–18.

Trimen, H. (1887). Hermann's Ceylon herbarium and Linnaeus's Flora Zeylanica. *Journal of Linnean Society of London*, **14**, 129–55.

Turrill, W. B. (1953). *Pioneer plant geography: the phytogeographical researches of Sir Joseph Dalton Hooker.* The Hague.

Turrill, W. B. (1963). *Joseph Dalton Hooker: botanist, explorer and administrator.* London.

Valentia, Viscount, (1809) *Voyages and travels to India, the Red Sea, Abyssinia and Egypt in the years 1802, 1803, 1804, 1805 and 1806* London. 3 vols.

Van Reede tot Drakenstein, H. A. (1678–1703). *Hortus Malabaricus* [title varies]. Amsterdam. 12 vols.

Vigne, G. T. (1842). *Travels in Kashmir, Ladak, Iskardo, the countries adjoining the mountain-course of the Indus, and the Himalaya, north of the Panjab.* London.

Vishu-Mittre (1970). Biological concepts and agriculture in ancient India. *Indian journal of History of Science*, **5**, 144–61.

Voigt, J. O. (1845). *Hortus suburbanus Calcuttensis*. A catalogue of the plants which have been cultivated in the Hon. East India Company's Botanical Garden, Calcutta and in the Serampore Botanical Garden, generally known as Dr Carey's garden from the beginning of both establishments (1786 and 1800) to the end of August 1841. Calcutta.

Wallace, A. R. (ed.) (1908). *Notes of a botanist on the Amazon & Andes* . . . (Richard Spruce), (by A. R. Wallace). London. 2 vols.

Warmington, E. H. (1928). *The commerce between the Roman empire and India*. Cambridge.

Watt, G. (1889–96). *Dictionary of the economic products of India*. Calcutta. 7 vols.

Watt, G. (1907). Tea and the tea plant. *Journal of Royal Horticultural Society*, **32**, 64–96.

Watt, G. (1908). *The commercial products of India; being an abridgment of 'The dictionary of the economic products of India'* London.

Whiteway, R. S. (1899). *The rise of Portuguese power in India, 1497–1550*. London.

Wickham, H. (1908) *On the plantation, cultivation and curing of Para Indian rubber* (Hevea brasiliensis) *with an account of its introduction from the west to the eastern tropics*. London.

Wight, R. (1841). Recent botanical letters . . . addressed to G. A. W-Arnott. *Hooker's Journal of Botany*, **3**, 156–201.

Wight, R. (1861). Cotton cultivation in India. *Journal of Agricultural and Horticultural Society of India*, **12**, 1–26.

Wight, R. and Walker-Arnott, G. A. (1834). *Prodomus florae peninsulae Indiae Orientalis*: containing abridged descriptions of the plants found in the peninsula of British India. London.

Wijnands, D. O. (1983). *The botany of the Commelins: a taxonomical, nomenclatural and historical account of the plants depicted in the Moninckx Atlas and in the four books by Jan and Casper Commelin on the plants in the* Hortus Medicus Amstelodamensis. Rotterdam.

Wijnands, D. O. (1987). The *Hortus Medicus Amstelodamensis*: its role in shaping taxonomy and horticulture. *Kew Magazine*, **4**, 78–91.

Williams, D. (1962). Clements Robert Markham and the introduction of the cinchona tree into British India, 1861. *Geographical Journal*, **128**, 431–42.

Woodrow, G. M. (1910). *Gardening in India*, (6th edn). Bombay.

Wycherley, P. R. (1968). Introduction of *Hevea* to the Orient. *Planter, Kuala Lumpur*, **44**, 1–11, 127–37.

Yule, H. (1866). *Cathay and the way thither*. London. 3 vols.

Yule, H. (ed.) (1903). *The book of Ser Marco Polo, the Venetian, concerning the Kingdoms and marvels of the East*, (trans. Sir H. Yule). London. 2 vols.

Yule, H. and Burnell, A. C. (1903). *Hobson-Jobson*. A glossary of colloquial Anglo-Indian words and phrases, and of kindred terms, etymological, historical, geographical and discursive. London.

INDEX

Abelia triflora 301
Abelmoschus angulosus var. *grandiflorus* 168
Abies spectabilis 301
Abutilon auritum plate 7
Account of Kingdom of Nepal (Buchanan) 79
Account of Kingdom of Nepal (Kirkpatrick) 130
Account of manufacture of black tea . . . (Bruce) 238
Achabel garden 279
Acosta C. 16, 17, 33, 201, 212, 220
Adansonia digitata 85, 201
Aelium 258
Agave spp. 201, 204, 205, 275
Agra 111, 260–3
Agricultural and Horticultural Societies
  Bombay 281
  Calcutta 95, 96, 100, 101, 186, 187, 214–17, 240, 247, 252, 284, 318
  Lucknow 91
  Mysore 107
Ainslie W. 221
Aitchison J. E. T. 31
Aiton W. 294, 314
Akbar Emperor 27, 29, 145, 201, 259, 260, 262, 278
Alexander of Macedon 2, 3, 6, 56, 154
Allport T. plate 10
Alnus nepalensis 301
Alston A. H. G. 167
Alwis H. de 162, 163, 164, 165
Alwis W. de 165, 167, 168
Amarakosa 54
Amherst Lady 85, 87, 88, 182, 272, plate 14
Amherst Lord 235
Amherstia nobilis 86, 87, 90, 96, 99, 164, 182, 300–1, 318
Amomum subulatum 6
Amsterdam botanical garden 23, 37, 156, 157, 251, 288, 289, 291, 292
Anacardium occidentale 18, 201
Ananas comosus 18, 19, 201, 266
Anderson J. 40, 202, 203, 208–10
Anderson T. 31, 100–1, 195, 226, 228
Anemone vitifolia plate 13
Annals of Royal Botanic Gardens Calcutta 102, 152, 325
Annals of Royal Botanic Garden Peradeniya 167
Anneslea 184
Annona squamosa 201, 266
Anuradhapura garden 166
Araucaria cunninghamii 111
Areca catechu 14, 18, 32
Arenaria bryophylla 123
Argyreia nervosa 36
Arisaema griffithii 306
Arisaema nepenthoides 306
Arisaema speciosum plate 12
Aristobulus 2
Aristotle 2, 6
Arnold Sir A. 261
Aromatum et medicamentorum . . . (L'Escluse) 17
Aromatum et simplicium (L'Escluse) 17
artists, Indian
  employed by
    F. Buchanan 76, 78, 148, 149, 151
    J. F. Cathcart 143, 144, 148
    J. Hare 83
    B. Heyne 42, 148, 151
    Lady Impey 148, 149
    C. Martin 183
    Lady Monson 149
    A. Moon 162
    W. Roxburgh 47, 48, 49, 50, 82, 148, 149, 192
    N. Wallich 87, 89, 127, 128, 149, 151, 192
    R. Wight 116, 118, 149
  individual artists
    Alwis H. de 162, 163, 164, 165
    Alwis W. de 165, 167, 168
    Bhawani Das 149
    Gorachand 90, 133, 149, 150, 151, plates 8, 14
    Govindoo 118, 119, 121, 149
    Kasinaut 151
    Lachman Singh 82
    Manu Lal 149
    Ram Das 149
    Ramnaut 149, 151
    Rungia 90, 118
    Vishnupersaud 87, 90, 128, 149, 151
    Zain al-Din 149
  Mughal 26, 27, 144, 145, 147, 148, plate 1
  payment to 149, 151
Artocarpus heterophyllus 2, 9, 10, 14, 26, 160
Asiatic Society of Bengal 52–3, 185–6, 189–90, 220, 221, 316
Asiatick Researches 53, 71, 150, 151, 185, 186, 187
Asoka 258, 259
Assam 86, 91, 137, 138, 171, 236, 237, 252, plate 27

Assam Company 238, 240
Atharvaveda 25
Atkinson E. T. 188
Auckland Lord 95, 110, 111, 115, 217, 272, 274, 316
Augusta Princess 294, 295
Aungier G. 266

Babur Emperor 25–6, 27, 240, 259, 260
Babur-nama 25, 26, 145
Bacon T. 180, 263, 264
Badminton 292
Badulla garden 166, 167
Bailey F. M. 135
Baldaeus P. 155, 156
Ball V. 1, 2
balsams, *see Impatiens* spp.
bamboos, *see Bambusa* spp.
Bambusa spp. 2–3, 32, 33
banana, *see Musa* spp.
Bangalore 41, 43, 106
Bangladesh Journal of Botany 323
Banks Sir J. 23, 39, 40, 44, 56, 72, 159, 160, 161, 297, 314, 318
  Botanic Garden, Calcutta 58, 59, 60, 61, 62, 63, 83, 119, 281, 285
  East India Company 40, 43, 202, 204, 211, 233, 235, 294
  economic plants 54, 59, 202, 204, 205–6, 208, 209, 210, 211, 222, 233, 234, 235
  Plants of Coast of Coromandel 45, 46, 48, 49
Banks Lady 305, plate 11
banyan, *see Ficus benghalensis*
Barbosa D. 13
Barrackpore 66, 75, 76, 77, 272–3
Barron R. 278
Bartlet J. 211
Bateman J. 305, plate 13
Bazaar medicines (Waring) 221
bdellium, *see Commiphora mukul*
Beames J. 213–14, 281
Beaufort Duchess of 292
Beaumontia 100
Beddome R. H. 121, 149, 196
Bees Nursery, *see* Bulley A. K.
Begonia rex 307, plate 27
Belanger C. 121
Bellasis J. B. 274
Bengal 78, 141
Bengal dispensary (O'Shaughnessy) 221
Bengal pharmacopoeia (O'Shaughnessy) 221
Bennett J. W. 164
Bentham G. 88, 90, 247
Bentinck Lady 272
Bentinck Lord 43, 110, 189, 235, 236
Bentley R. 213
Berberis wallichiana 301
Bergenia ligulata 300, plate 9
Berkeley M. J. 165
Bernier F. 32, 35, 128, 207, 260, 263, 278, 279
Berry A. 67, 209
betel-nut palm, *see Areca catechu*
Bevere P. C. de 158
Beverningk H. 291
Bhagnagar 265
Bhawani Das 149
Bhutan 115, 130, 134, 135–7, 233, 234, 306, 325, 326
Biggs T. 180
Bird G. 163
Birdwood G. C. 245, 256
Biswas K. 103
Blackburn J. 293
Blackwood J. 212
Blane G. 54, 56
Blane W. 56
Blatter E. 325
Blinkworth R. 88, 93
Boehmeria nivea 205
Boerhaave H. 17, 288
Bogle G. 123, 135, 216, 233
Bolivia 222, 224, 225, 226, 247
Bombay 30, 121–2, 185, 186, 197, 216, 264, 266, 268, 274, 281, 284, 329
Bombay flora (Gibson & Dalzell) 121, 197
Bombay Natural History Society 186, 187
Booth T. J. 137–8, 303
Borassus flabellifer 9, 160
Borcht P. van der 287
Boscawen J. T. 301
Botanical Cabinet 298
Botanical Society of Edinburgh 188
Botanical Survey of India 102–3, 197, 323, 324–5, 329
Botany of Antarctic voyage (Hooker) 138
Bougainvillea glabra 111
Bradley R. 309
Brandis D. 197, 247, 248, 255, 256
Brazil 210, 211, 247, 248, 249, 251, 252
breadfruit, *see Pandanus leram*
Breyne J. 156
Brief directions . . . (Petiver) 310
Brief instructions . . . (Woodward) 309
Brindavan Gardens 281
British Museum 166, 192, 193
Brockway, L. H. 246
Broeger H. C. 273–4
Broome C. E. 165
Brown A. 35
Brown M. 183, 252
Brown R. 68, 89, 90, 95, 116, 124, 159
Browne S. 34, 35
Brownea qrandiceps 164, 167
Bruce C. A. 236, 237, 238, 240, 245
Bruce H. 88
Bruce R. 236
Buchanan F. 69, 89, 110, 116, 119, 148, 170, 204, 205, 220, 282, 299
  Bengal survey 78, 151, 169
  Lord Moira 78, 79

Mysore survey 73–5, 151, 169
Nepal 130, 131, 132, 133, 299, 306, plate 15
W. Roxburgh 69, 71, 72, 73, 74, 77
Lord Wellesley 72, 75, 76
Buddleia crispa 301
Bulkley E. 34, 35
Bulley A. K. 135, 138, 306
Bulmer W. 50
bungalow gardens 275–7
Burkill I. H. 133
Burlington Lord 292
Burma 71, 72, 86, 88, 91, 191, 253, 255
Burman J. 37, 157–8, 160, 290
Burman N. L. 24, 36, 37
Burnell J. 264, 265
Burt Mr 56
Bury E. 68
Bute Lord 294
Butea monosperma 27, 259, 275
Butler Mr 46
Bwye H. 277

Caesalpinia bonduc 221
Caesalpinia sappan 8, plate 4
Calcutta 52, 185–7, 189, 268, 271–2, 282
Calcutta Botanic Garden 137, 169, 184, 191, 193, 202, 215, 223, 228, 235, 239, 284, 285, 319, 323
T. Anderson 100
Sir J. Banks 44, 59, 60, 63, 64, 103–4
Botanical Survey of India 102–3
F. Buchanan 71, 74, 131
East India Company 58, 59, 63
H. Falconer 98, 100
T. Thomson 100
W. Griffith 93, 95 6, 98
G. King 101–2
R. Kyd 57, 58, 59, 60, 148, 210, 212, 233
W. Roxburgh 52, 61–6, 99, 252, 300
N. Wallich 81–8, 90–8, 124, 132, 149, 221, 300
Calcutta Gazette 268, 273
Calcutta Journal of Medicine 189
Calcutta Journal of Natural History 92, 96, 163, 164, 186, 187
Calcutta Medical College 189
Calcutta Review 216
Calder C. C. 103, 324
Caldwell Bishop 188
Cambessèdes J. 127
Camellia sinensis 44, 58, 86, 91–2, 93, 109, 111, 167, 202, 206, 217, 229, 231–40, 252, 311, 314, 320
Campbell G. 40
Cameron J. 107
Cameron W. 164
Campbell A. 139, 238
Candolle A. de 24, 88, 95, 116
Cannabis sativa 18, 25, 62, 164, 204–5, 315
Canning Lady 100, 181, 268–9, 272–3, plate 21
Canning Lord 224, 268
Cano J. S. del, 13
Cantor T. E. 191
Cape of Good Hope 19, 62–3, 81, 96, 202, 314
Capsicum annuum 201
Caraka 25
cardamom, *see Elettaria cardamomum*
Cardiocrinum giganteum 301, 302
Carey J. 314
Carey W. 66–7, 185, 300, 303, 314, 315, 318
Agricultural and Horticultural Society of India 215, 216
Catalogue of plants described by Dr Roxburgh 67
Roxburgh's *Flora Indica* 69, 85
Serampore garden 66, 93, 95, 273, 284
N. Wallich 82, 84, 88
Carica papaya 18, 201
Carruthers J. B. 167
Casearias J. 22
Casey T. 82
cashew nut, *see Anacardium occidentale*
cassava, *see Manihot esculenta*
Cassia fistula 2, 275
Cassiobury 291
Castilla elastica 247, 248, 250
Castro J. de 287
Catalogue of economic products . . . of Bombay (Birdwood) 256
Catalogue of ferns indigenous to Ceylon (Wall) 165
Catalogue of Indian medicinal plants . . . (Fleming) 220–1
Catalogue of indigenous and exotic plants . . . (Moon) 162, 164
Catalogue of plants growing in Bombay . . . (Graham) 197
Catalogue of plants of Kumaon (Strachey) 112
Catalogus plantarum Horti Medici Amstelodamensis (Commelin) 288, 289
Cathcart J. F. 143, 144, 148, 181, plate 20
Cedrus deodara 25, 111, 301–2, 318, plate 21
Century of Indian orchids (Hooker) 144
Ceropegia elegans 299
Ceylon 4, 13, 39, 40, 154, 184, 187, 194, 202, 288, 290, 291, 292, plate 30
Botanic Garden, 161–9
British occupation 159
cinchona 226, 228, 229
cinnamon 7, 9, 12, 19, 154, 155, 204
coffee 167, 252
Dutch occupation 18, 156–8
G. Gardner 163–4
W. Kerr 160–1
A. Moon 162
rubber 250
tea 167, 240
G. H. K. Thwaites 164–6
H. Trimen 166–7, 196, 325
J. C. Willis 167
Ceylon and its capabilities (Bennett) 164
Champion J. 163
Charlton Lieut. 236, 240, 247
Chatsworth 300, 303, 305

Chelsea Physic Garden 291, 292, 294, 305
chenar tree, *see Platanus Orientalis*
China 224, 231, 233, 234, 235, 236, 237, 239, 240
Chirston R. 222
Cinchona spp. 53, 99, 100, 198, 221–9
 Calcutta 224
 Ceylon 160, 166, 167, 202, 228, 229
 Nilgiris 100, 222, 223, 226, 227
 Saharanpur 107, 222
 Sikkim 227, 228
Cinnomomum spp. 2, 4, 6, 7, 8, 9, 12, 47, 57, 130, 154, 155, 160, 169, 203, 204, 288, 294, 300
Citrus aurantiifolia 201
Clarke C. B. 69, 101, 195, 196
Cleghorn H. F. C. vii, 98, 106, 107, 159, 255
Clematis montana 307, plate 14
Clifford G. 291
Clive Lord 41, 43, 73, 74, 268
clove, *see Syzygium aromaticum*
Clusius, *see* L'Escluse C de.
cochineal 44, 202, 208, 215, 217, 246
Cockerell Sir *C.* 296
cocoa, *see Theobroma cacao*
Coco-nut palm (Bennett) 164
coconut palm, *see Cocos nucifera*
Cocus nucifera 7, 9, 13, 14, 15, 18, 26, 29, 62, 155, 160
coffee 86, 202, 208, 215
 Calcutta 58, 62, 93, 251
 Ceylon 163, 165, 166, 228, 251, 252
 Samalkot 47, 203
Colebrooke T. H. 50, 53, 66, 68, 78, 79, 83, 149
Colesborne 305
Collett Sir H. 199
Collins J. 247, 248
Collinson P. 231, 293, 294
Coloquios dos simples . . . (Orta) 16, 17
Colombia 222, 224, 225, 246
Commelin J. 23, 156, 288, 289, 291
Commercial products of India (Royle) 217, 218
Commiphora mukul 6
Compton Bishop 292
Conolly H. 255
conservation in S. Asia vii, 163, 252–3, 255–6, 327–30
conservatories 287, 288, 289, 291, 292, 293, 294, 295, 296, 297, 299, 305
Conti N. de 9, 10
Cook J. 35
Cook M. 291
Cooke T. 103, 112–3, 197, 199
Cookson J. 182
Coonoor 281
Cooper J. 305
Cooper R. E. 134, 135, 170, 306
Corbyn Dr. 187
Cordiner J. 160
Cornwallis Lord 61, 76, 77
Coryphea taliera, plate 3
Corypha umbraculifera 9, 22, 155, 156
Cosmas 7, 154
Cossigny J. F. C. de 212
Costus, *see Saussurea costus*
Cotoneaster spp. 300, 301
cotton, *see Gossypium* spp.
Covilhã P. 12
Cowell J. 292
Cox Lieut. 58
Crawfurd J. 86
Crinum asiaticum 68
Crinum bracteatum 202
Crinum zeylanicum 292, 298
Cross R. 224, 225, 228, 229, 246, 250
Crotalaria juncea 205, 292
Crotalaria pulcherrima 298
Croton laccifer 156
Ctesias 1–2, 5
Cuneo M. do 246
Cunningham A. 129
Cunningham J. 231
Curtis W. 46, 49, 158, 298
Curtis's Botanical Magazine 182, 298, 299, 300, 301, 303, 306, 326, 328, plates 11, 12, 13, 25, 27
Curzon Lord 273, 277
custard-apple, *see Annona squamosa*
Cyananthus sherriffii 136
Cycas circinalis 96
Cyclopaedia (Rees) 132
Cymbidium aloifolium 305
Cynodon dactylon 275
Cypripedium cordigerum 328

Dactylopius spp. 208–11
Dalhousie Countess of 182
Dalhousie Lord 253, 255
Dalhousia 182
Dalzell N. A. 113, 121, 197
Daniell T. W. 174, 262, 296, 297, plate 32
Danish settlements 32, 39, 40, 41, 66, 67, 81, 149, 273
Danvers R. W. 282
Dapoori 113, 255, 271, 284
Darjeeling 139, 144, 224, 238
Darwin C. 142
date palm, *see Phoenix dactylifera*
Datura metel 18, 292
Davy J. 154
Daylesford 296
De la Condamine 221, 222, 246
De Leeuwenhorst 291
De Silva 132
Defabeck F. W. A. 281
deforestation 252–3, 255, 328–30
Delhi 263–4, 281, plate 31
Delile E. 127
Dellon G. 265
Delonix regia 164, 275
Delphinium cashmerianum 128
Dendrobium aphyllum 303

Dendrobium fimbriatum 303
Dendrocalamus strictus 2
Denison Sir W. 194
deodar, *see Cedrus deodara*
Description of Ceylon (Cordiner) 160
Description of the mangostan (Ellis) 310, 312
Desmodium gangeticum 293
Desmodium triquetrum 290
Devonshire Duke of 300, 303, 319
Dhwoj L. 134, 306
Diaz B. 12
Dictionary of economic products of India (Watt) 256, 327
Dillenius J. 292, 293
Dioscorea bulbifera 157
Dioscorea deltoidea 327
Dioscorides 6, 15, 56
Directions for taking care of growing plants at sea (Roxburgh) 315
Directions for bringing over seeds . . . (Ellis) 310, 313
Directions relative to preservation of growing plants . . . (Wallich) 315
Don D. 56, 132, 133
D'Oyly Sir C. 177
drugs, *see* medicinal plants
Drury H. 256
Dryander J. 46, 49
Du Bois C. 35, 294
Dublin, *see* Glasnevin botanical garden
Dufferin Lady 272, 283–4
Dufferin Lord 277
durian, *see Durio zibethinus*
Durio zibethinus 9, 18
Dutch settlements 18–24, 30, 72, 154–5, 159, 264–5
Duthie J. F. 102, 103, 111–2, 133, 196, 199

East India Company (British) 29, 30, 49, 50, 72–3, 76, 188, 231
  Sir J. Banks 40, 44, 45, 46, 47, 48, 60, 64, 202
  Calcutta Botanic Garden 57, 58, 59, 63, 64, 91
  Hooker and Thomson's *Flora Indica* 142
  India Museum 190–3
Ecuador 222, 224, 225, 246
Eden E. F. 179, 272, 281
Edgeworth M. P. 98
Edgeworthia gardneri 132
Edinburgh Botanic Garden 299, 325
Edinburgh Philosophical Journal 188
Edwards G. 159
Elettaria cardamomum 6, 7, 58, 74, 120, 167
Ellenborough Lord 272
Ellis J. 39, 231, 233, 298, 311, 312, 313
Eltham 292, 293
Elwes H. J. 133, 305, 306, plate 28
English botany (Smith & Sowerby) 116, 118
Ensete glaucum 148
Enumeratio plantarum Zeylaniae (Thwaites) 165, 196
Enumeration of flowering plants of Nepal (Hare *et al*) 135, 325
Eratosthenes 4
Erythrina stricta plate 5
Erythrina variegata 275
Essay on productive resources of India (Royle) 217, 256
Eucalyptus spp. 107
Eupatorium zeylanicum 158
Euphorbia antiquorum 292
Evelyn J. 292
Exotic botany (Smith) 75, 132, 303, 304
Exotic flora (Hooker) 303
Exoticorum libri decem (L'Escluse) 17

Fagel G. 291
Falconer H. 139, 142, 192, 253, 261, 263, 318, 320
  Calcutta Botanic Garden 98–100, 138, 223
  Saharanpur Botanic Garden 110–1, 124, 191, 236, 237, 238, 260
Farrer R. 138
Farris Mr 248
Fazl A. 27
Fedrici C. de 13–4
Ferguson W. 163–4
Ferns of British India (Beddome) 121
Ferns of Southern India (Beddome) 121, 149
fibres 204–5
Fibrous plants of India (Royle) 256
Ficus benghalensis 3, 6, 31, 63, 85, 99, 101, 103, 104, 174, 176, 259, 288
Ficus elastica 247, 248
Ficus religiosa 1, 25, 293
Ficus retusa 111
Fidler L. L. 121
Finch W. 30, 259
Finlayson G. 89
Firminger T. 282, 284
First steps in botany (Watt) 189
Fischer C. E. C. 199
Fitch R. 29, 260
Fitch W. H. 99, 143, 144, 180, 326, plates 13, 17, 19, 22, 25, 27
Fitzhugh T. 239
flame of the forest, *see Butea monosperma*
flax *see Linum usitatissimum*
Fleming J. 50, 53, 60, 63, 68, 78, 83, 130, 220, 221
Flora of British India (Hooker) 69, 138, 167, 195–6, 197, 199, 324, 325
Flora Graeca (Sibthorp & Smith) 241
Flora Indica (Burman) 36, 37
Flora Indica (Hooker & Thomson) 138, 142, 194, 195
Flora Indica (Roxburgh) 68–9, 74, 85, 86, 118, 183, 196
Flora Londinensis (Curtis) 46, 49
Flora of Presidency of Bombay (Cooke) 199
Flora of Presidency of Madras (Gamble) 199
Flora Simlensis (Collett) 199

Flora sylvatica for Southern India (Beddome) 121, 149
Flora of Upper Gangetic Plain (Duthie) 112, 199
Flora Zeylanica (Linnaeus) 156, 158
Flowers drawn and painted after nature . . . (Cookson) 182
Forbes J. 183, 273, 294–5
Forest flora of Bombay Presidency (Talbot) 197
Forest flora of North-West and Central India (Stewart) 197
Forestry 85, 86, 121, 189, 197, 253–6
Forests of Nepal (Stainton) 134
Forrest E. 175
Fortune R. 224, 238–9, 318, 320
Fothergill J. 233, 293, 311
Foxe J. 35
Fraser J. B. 67, 175–7, plate 23
French settlements 32, 72, 119, 120, 121, 266
Fryer J. 31, 32, 260, 264, 265
Fuchsia spp. 93
Fulham Palace 292

Gage A. T. 103, 197
Galen 6, 15
Gama V. de 12
Gamble J. S. 189, 199, 322, 324, 328
Ganeshkhind 112, 113, 271
Garcinia mangostana 63, 93, 314
Garcinia morella 159
Gardeners dictionary (Miller) 294
Gardens
  European in India 264–85
  Hindu 258–64, 278–9, 282
  in Europe 287–307
  Mughal 260–4
Gardner E. 85, 88, 132, 236, 300, plate 10
Gardner G. 163–4, 187
Garwhal 123, 124, 127, 129, 130, 237, 330
Gentiana depressa 306
Gentiana ornata 306
Gentiana pedicellata 119
Geranium spp. 300
Gerard A. J. & P. 125, 127
Gerard J. 287
Gibson A. 98, 113, 121, 197, 255
Gibson J. 291, 303, 319
Gifford Mr 35
Gilder Dr 98
Gillan H. 211
Gillespie Lieut. 58
ginger, *see Zinger officinale*
Gladstone W. E. 245, 246
Glasnevin botanical garden 129, 301, 302
Gleanings in science 186, 187
Gloriosa superba 166, 275, 291, 298, plate 6
Goa 12, 264
Goedkint A. J. 15, 22
Goldingham J. 268
Goldring W. 282, 283
Gomez W. 88
Gondalpara 265
Good P. 62
Gorachand 90, 133, 149, 150, 151, plates 8, 14
Gordon G. J. 236, 237, 239, 252
Gordon J. 298
Gossypium spp. 1, 2, 4, 9, 31, 74, 128, 175, 208, 215, 217, 233
  Sir J. Banks 44, 54, 202, 206–7
  Ceylon 164
  experimental farms 117, 207
Govan G. 88, 107, 109, 127, 169, 170, 222, 235, plate 8
Govindoo 118, 119, 121, 149
Graham J. 197
Graham M. 65, 148
Graham R. 88, 163, 182
Grant G. 188
Grant Duff M. 273
Gray A. 196
Great Exhibition 1851, 218
Greece 2, 3, 5
Green J. 202
Grevillea robusta 101, 107
Griffith W. 31, 114–15, 118, 124, 138, 169, 187, 191, 192, 197, 306
  Calcutta Botanic Garden 93, 94, 95, 96, 97, 98, 99, 100, 324
  Nepal 135
  Roxburgh's *Flora Indica* 69
  N. Wallich 90, 91–2, 95, 237
Grisebach A. H. R. 194
Grosier Abbé 234
Guatemala 211, 212
Gulmarg 125, 278, 328
Günther A. 158

Hakgala garden 166, 228
Hales S. 307
Hamilton F., *see* Buchanan F.
Hamilton W. 68
Hampton Court 291–2
Handbook to ferns of British India (Beddome) 121
Handbook to flora of Ceylon (Trimen) 167, 196–7, 325
Hardinge Lord 264
Hardwicke T. 76, 148, 180, 192, 304
  Calcutta Botanic Garden 67
  Ceylon 161
  W. Roxburgh 65
  N. Wallich 82, 84, 88, 132
  in western Himalayas 71, 123, 127, 303
Hare J. 50, 79, 82, 83
Hartecamp 291
Harvey W. H. 165–6, 192
Hasskarl J. C. 223, 225, 226
Hastings Lady 272
Hastings Lord 271, 272
Hastings W. 72, 188, 190, 273, 296
  Asiatic Society of Bengal 52
  Calcutta Botanic Garden 58

interest in economic plants 123, 135, 216, 233, 253
opium 240, 241, 242
Hatton Sir C. 291
Havell R. 175
Hawal Bagh 111
Haworth A. H. 90
Hay T. 306
Heber Bishop 85, 174, 215, 264, 268, 272
Hecateus 1
Hedychium flavum 299
Hedychium gardnerianum 300, plate 10
Helfer J. W. 191, 253
Hemileia vastatrix 165
hemp, *see Cannabis*; *Crotalaria*
Heneratgoda garden 166, 250
Herball (Gerard) 287
Herbarium Amboinense (Rumphius) 23, 37, 45, 48, 80, 84
Herbert J. D. 169, 186
Herbert W. 298, 299
Hermann P. 20, 156–7, 158, 167, 288, 291, 292, 293
Herodotus 1, 5, 6
Hevea brasiliensis 202, 247, 248, 249, 250, 251
Heyne B. 31, 67
Botanist Madras Presidency 41, 43, 74, 89, 106, 169, 171, 203, 216
Indian artists 42, 148, 151, 184
Hibiscus rosa-sinensis 26
Highcliffe 294
Hill J. 24
Himalayan journals (Hooker) 141, 142–3
Himalayas 123–53, 301–7
Hindous (Solvyns) 175
Hints on gardening in India (Woodrow) 275, 284
Hippocrates 5
Histoire naturelles des quinquinas (Weddell) 222
Historia plantarum (Ray) 23, 35
Hodges W. 173, 260, 262, 282
Hodgson B. 139, 180, 191, plate 17
Hodgsonia heteroclita plate 17
Hoffmeister W. 97, 129, 164
Hooker Sir J. D. 180, 181, 198, 273, plates 17–19
balsams 197, 306–7
botanical organization in India 102, 323
Calcutta Botanic Garden 93, 96, 99
Ceylon 164, 165, 167, 196–7, 252
cinchona 229
J. F. Duthie 112
Flora of British India 194–6, 199, 324
Flora Indica (with Thomson) 142, 194
Himalayan journals 142
Illustrations of Himalayan plants 143, 144
Imperial gazetteer 197
India Museum 192, 193
Indian artists 144, 148
Khasia Hills 141, 301
Nepal 133, 139, 170
opinions on floral affinities 126, 177, 179
Rhododendrons of Sikkim–Himalaya 144
rubber 248, 250
Sikkim 137, 139, 141, 169, 171, 174, 303, 305, 306, 320, 328, plates 18, 25
T. Thomson 100
R. Wight 117, 118, 141
Hooker W. 183
Hooker Sir W. J. 142, 227, 239, 316, 320
East India Company 89, 285
Ceylon 160–1, 164, 165
Cinchona 229
Colonial floras 193–4
Curtis's Botanical Magazine 141, 182, 299, 303, 328, plate 27
G. Gardner 163
W. Griffith 192
J. Nimmo 121
T. Thomson 100
N. Wallich 90, 93, 96, 97, plate 9
R. Wight 116, 117
Hope J. 31, 47, 71
Horsfield T. 190–1
Horti academi Lugdano-Batavi catalogus (Hermann) 288
Horticultural Society of London 223, 239, 274, 299, 301, 306, 312, 314, 315, 316, 317, 318, 320, plate 26
Hortus Bengalensis (Roxburgh) 67, 95, 202
Hortus Botanicus Vindobonensis (Jacquin) plate 2
Hortus Cliffortianus (Linnaeus) 291
Hortus Elthamensis (Dillenius) 292, 293
Hortus Kewensis (Aiton) 294
Hortus Malabaricus (Van Reede) 8, 10, 15, 20–4, 31, 32, 33, 34, 35, 37, 45, 48, 49, 74, 79, 80, 84, 116, 156, 158, 205, 220, 288, 291
Hortus Suburbanus Calcuttensis (Voigt) 95, 284
Hortus Zeylanicus (Trimen) 167
hothouses, *see* conservatories
Hough H. 107
Hove A. 46, 169, 205, 206
Howard J. E. 226, 227, plate 22
Huegel K. A. F. von 119, 125, 169, 279
Hughes T. 61
Humayun Emperor 144, 145, 240, 259
Humboldt A. von 129, 222
Hume Sir A. and Lady A. 68, 77, 202, 298, 299
Hunter W. 63, 86
Hunzikar J. 180
Hutton Mrs 182
Hyne G. 43, 84
Hynmers Mr 35
Hypericum oblongifolium 301

Ibn Batuta 154
ice for transporting plants 320–2
Ichnocarpus frutescens 292
Icones Plantarum Indiae Orientalis (Beddome) 121, 149
Icones Plantarum Indiae Orientalis (Wight) 118, 149, 163, 164, 165, 197

Illustrations of the botany . . . of Himalayan Mountains (Royle) 55, 125, 127–9, 149, 217, 222, 235
Illustrations of Himalayan plants (Hooker) 143, 144, plates 17, 19
Illustrations of Indian botany (Wight) 118
Illustrations of nueva quinologie of Pavon (Howard) plate 23
Impatiens glandulifera 306–7
Impatiens spp. 197, 294, 306
Imperial gazetteer of India 197
Impey Lady M. 148, 149
India Museum London 43, 89, 97, 99, 129–30, 142, 190–3, 218, plate 12
India Office 193, 194, 212, 224, 229, 248, 249
India Review 187
Indian Agriculturist 216
Indian Association for Advancement of Natural Science 186
Indian Association for Cultivation of Science 189
Indian Forester 216
Indian handbook of gardening (Speede) 284
Indian Museum Calcutta 53, 190
Indian Tea Gazette 216
Indian trees (Brandis) 256
Indigo, *see Indigofera*
Indigofera spp. 30, 44, 47, 62, 175, 202, 212–14, 233, 246
Ipomoea batatas 201, 216
Ipomoea pestigridis 36
Ipomoea quamoclit 181
I'timad-ud-Daula's tomb 263
Itinerario (Linschoten) 17, 18, 154

Jack W. 95
jack-fruit, *see Artocarpus heterophyllus*
Jacquemont V. 84–5, 93, 109, 110, 124, 125–6, 142, 169, 170
Jacquin N. J. von plate 2
Jahangir Emperor 27, 30, 31, 145, 202, 214, 240, 259, 260, 278, 279, plate 1
Jamaica botanical garden 202
Jameson W. 98, 111, 238, 239, 318
Jardin des Plantes Paris 93, 125, 142, 222
Jasminum officinale 294, 298
Java 223, 226, 227, 231, 247, 251, 252
Jenkins F. 236
John of Montecorvino 9
John C. S. 39, 40, 41, 61, 81
Johnston *Sir* A. 160
Johnstone J. 277
Jones Sir W. 52–7, 313
  Asiatic Society of Bengal 52–3, 186, 188, 202, 220
  Sir J. Banks 54, 56, 206
  Indian artist 149
  J. G. Koenig 40, 159
  C. Linnaeus 53–4
  W. Roxburgh 47
  P. Russell 45
  Spikenard 54–6, 130
  Van Reede's *Hortus Malabaricus* 23
Jonville E. de 159–60, 190
Jordanus of Séverac 9
Joshi A. C. 323
Journal of Asiatic Society 186–7
Journal of Palynology 323
Journal of tour through . . . Himala Mountains (Fraser) 175–76
Junghuhn F. W. 223, 226
Jussieu 75, 95, 110, 116

Kaempfer E. 239
Kaine Mr 261
Kalutara 161
Kama Sutra (Vatsyayana) 258
Kane J. 91
Kanjilal U. 189
Kashmir
  F. Bernier 128–9, 278, 279
  J. F. Duthie 111
  H. Falconer 99, 110
  T. Hardwicke 71, 123, 127
  V. Jacquemont 126
  Jahangir 27, 145, 278
  W. Moorcroft 124, 281
  J. F. Royle 126–7
  G. T. Vigne 125, 278
  E. Winterbottom 129
Kasinaut 151
Kerr J. 245
Kerr W. 67–8, 160, 161, 162, 169
Kew Gardens vi 59, 92, 142, 186, 210, 285, 294, 295, 314, 318, 325
  Sir J. Banks 44
  Calcutta Botanic Garden 62, 93, 299
  Ceylon 160–1, 163, 166
  cinchona 226, 229
  C. B. Clarke 195
  Sir H. Collett 199
  R. Cross 299
  J. F. Duthie 112
  W. H. Fitch 144
  W. Griffith 115
  J. D. Hooker 141
  A. Hove 46, 206
  India Museum 192, 193
  Kew gardeners 106
  Lal Bagh garden 106, 107
  W. G. McIvor 277
  E. Madden 301
  Museum at Kew 191
  Orchids 303, 305
  rubber 248, 249, 250, 251
  R. Spruce 225
  W. T. Thisetton-Dyer 121
  Wardian cases 320, 321
  R. Wight 117
Khasia Hills 141, 195, 216, 223, 300, 301, 303, 305
King Sir G. 31, 101–2, 248, 250, 281, 324

Kingdon Ward F. 123, 135, 138, 169, 171, 305, 306
Kinnaur, *see* Kunawar
Kirkpatrick W. 130
Kirtikar K. P. 188
Klein J. G. 41, 89
Kleinwort Sir C. 297
Knight J. 163
Knowlton J. 292, 298
Knox Capt. 130
Knox R. 155–6, 160
Knypersley 305, plate 13
Koenig C. 64, 159, 220
Koenig J. G. 31, 39–40, 43, 44, 45, 46, 47, 64, 116, 294
Kunawar 110, 126
Kyd R. 61, 85, 202, 236, 294, 312–13, 314
  Cinnamon 203
  Coffee 251
  formation of Calcutta Botanic Garden 57–60, 281
  Indian artists 148, 149
  indigo 212
  opuntia 209, 210
  tea 233–4
  teak 253
  tobacco 214

Ladakh 125, 129
Lal Bagh 106–7, 281
Lambert A. B. 68, 132
Lamorran Rectory Truro 301, 302
Landolphia spp. 247
Lansdowne Lord 277
Latter Col. 236
Laurus 2
Law J. S. 121
Lawrence L. 300
Lawson M. A. 107
Lear E. 176, 261, 277, 281
Lear J. G. 163
Ledger C. 223, 226–7, 229
Leiden botanical garden 17, 22, 156, 157, 223, 288, 292
Leschenault de la Tour L. 88, 119, 120
L'Escluse C. de 14, 16, 17, 18, 33, 287, 288
Lettsom J. C. 233, 239, 311, 313
Leycester W. 85, 215
Liliacées (Redouté) plate 6
Lilium martagon 145, 146, 147
Lilium nepalense plate 28
Lilium sherriffiae 137
Lilium Wallichianum var. *neilgherrense* 301
Lindley J. 88, 90, 92, 111, 116, 131, 133, 192, 217, 301, 305, 316, 317
Linnaean Society of Calcutta 186
Linnaeus C. 37–8, 56, 222, 294, 309
  Flora Zeycanica 156, 158
  influence in India 53–4, 110, 116
  tea 231, 233, 239, 311
  use of his books in India 40
  Van Reede's *Hortus Malabaricus* 23, 291
Linnean Society of London 80, 89, 97, 187, 191, 192
Linschoten J. H. van 17, 18, 29, 154, 156, 201
Linum usitatissimum 204
lithography 88, 89–90, 117–8, 127, 151–2, 183
Liverpool botanical garden 68, 93, 215, 300, 303, plate 10
Livingstone D. 96
Lloyd botanical garden 102
Lobb T. 301, 305
Loddiges nursery 93, 239, 298, 300, 303, 305, 314, 318, 319, 320, plate 15
Lodi Gardens 281
Londesborough 292
London G. 291, 292
Loten J. G. 158–9
lotus, *see Nelumbo nucifera*
Loudon J. C. 272, 298, plates 14, 15
Low H. plate 28
Lowe A. C. 216, 321
Lowndes D. G. 134
Lucknow 282
Ludlow F. 135–7, 169, 305, 306, 322
Ludwig Baron 96
Luffa acutangula plate 2
Lush C. 113
Lushington S. 277
Lutyens Sir E. 264, 284, plate 31
Lysaght P. plate 30
Lytton Lord 272, 277

Macartney Lord 211, 234
Macaulay T. B. 188, 189, 213, 274, 277, plate 30
McClelland J.
  Calcutta Journal of Natural History 186, 187
  W. Griffith 92, 98, 99, 100, 115
  report on Burmese forests 253, 255
  Search for tea in Assam 91, 237
  N. Wallich 92, 96
MacDowall Major-General 160
McIvor W. 107, 226, 227, 277–8, 320
Mackenzie C. 41, 73, 78, 106
Mackenzie D. 49, 72
Mackintosh C. 247
Macmillan H. F. 167, 284
McNab W. 299
MacPherson J. 58
Macrae J. 162, 169
Madden E. 129, 170, 301, 302
Madras 34, 35, 39, 40, 216, 267, 268, 269–70, 274
Madras Journal of Literature and Science 187
Magnolia campbellii 301, 328
mahogany, *see Swietenia mahogoni*
maize, *see Zea mays*
Makurpuri Palace 282, 283
Malabar 173, 203, 251, 252, 291, plate 6
  C. Belanger 121
  S. Browne 34, 35
  F. Buchanan 74, 79

forestry 253, 255
V. Jacquemont 126
J. Ray 35
R. Wight 116
Malaya, *see* Penang, Singapore
Mangifera indica 17, 18, 19, 26, 32, 63, 175, 292, 309, 318
mango, *see Mangifera indica*
mangosteen, *see Garcinia mangostana*
Manihot esculenta 201
Manihot glaziovii 247, 250
Manning T. 318
Mansur 27, 145, plate 1
Manu Lal 149
Manual of gardening for Bengal... (Firminger) 282, 284
Manucci N. 32
Maries C. 282
Markham C. R. 207, 224–6, 228, 229, 246, 247, 250, 320, plate 22
Martel M. 225
Martin C. 183, plate 5
Martin R. M. 78
Martius K. F. P. von 88, 192
Martyr P. 246
Masson F. 40
Masters J. W. 91
Materia medica of Hindoostan (Ainslie) 221
Matthew K. M. 325
Matthew of St Joseph 20, 22
Mattioli P. 7, 287
Maxwell E. 126
Mayo Lady 272
Meconopsis grandis 137
Meconopsis napaulensis 141, 305–6
Meconopsis regia 134
Meconopsis simplicifolia 137, 306, plate 18
Meconopsis superba 137
medicinal plants
C. Acosta 16, 200
W. Ainslee 221
F. Buchanan 200
Calcutta Botanic Garden 83, 84, 95, 221
Ceylon 156, 157
cinchona 221–9
Nepal 132
G. da Orta 16, 200
J. Fleming 200
P. Hermann 157
B. Heyne 43
Sir W. Jones 23, 53, 54, 200
J. G. Koenig 40, 200
W. O'Shanghessy 221
W. Roxburgh 50, 220, 221
J. F. Royle 109, 220, 222, 223, 224
J. Russell 45
Reede tot Drakenstein 20, 24, 156
E. J. Waring 221
Medicinal plants (Bentley & Trimen) 213
Medinilla Walkeri 166
Megasthenes 3, 4, 154
Melchet Park 296
Melville L. 301
Memoirs of Wernerian Natural History Society 187
menagerie Barrackpore 75–7
Menonville T. de 208
Merremia vitifolia 182
Metcalfe Sir C. 90
Metroxylon spp. 18, 57, 58, 59, 63, 85, 293
Metz F. 119
Mewse B. 35
Mexico 208, 209, 246
Michael J. 255
Michelia champaca 27, 161, 258, 275
Michelia doltsopa 133
Mill Hill 293
Miller P. 292, 293, 294, 298, 309
Minto Lord 77, 272, 277, 282
Mitchell W. S. 43
Mitra D. B. 214
Mitra S. C. 186, 188
Mocca 251
Moira Lord 78, 79, 82
Moluccas 63
Monandrian plants (Roscoe) 300, plate 10
Money J. W. B. 227
Monograph of genus Lilium (Elwes) plate 28
Monson Lady A. 149
Montgomerie W. 86
Moon A. 162, 163, 164, 169, 252
Moorcroft W. 124–5, 126, 169, 281
Morus spp. 203, 207, 208
Mucuna pruriens 42
Mudd W. 271
Mukherjee J. 189
mulberry, *see Morus*
Müller J. 247
Mundy P. 13, 259, 260, 261, 262, 264, 266, 267
Munro Sir T. 203
Munro W. 98
Murray J. 294
Musa spp. 2, 6, 33, 175, 205
Musaeum Zeylanicum (Hermann) 157, 158
Musei Petiveriani (Petiver) 34–5
museums 189–90, 256; *see also* India Museum
Mussoorie 109, 110, 111, 112, 321
Myristica fragrans 18, 29, 31, 58, 59, 63, 86, 184, 203, 204

Naini Tal 112, 280
Nardostachys grandiflora 6, 54–6, 130
Natural history of the tea-tree (Lettsom) 233, 239, 313
nature printing 179–80
Nearchos 4
Nederlantze Hesperides (Commelin) 291
Neilson Capt. 211
Nelumbo nucifera 25
Neolamarckia cadamba 258
Nepal 306, 325

F. Buchanan 75, 76, 130–2, 133
D. Don 133
garden at Khatmandu 282
W. Hoffmeister 129
J. D. Hooker 133, 139
W. Kirkpatrick 130
O. Polunin 134
J. D. A. Stainton 135
tea 235
N. Wallich 85, 88, 132–3, 300
Nepenthes khasiana 328, 329
Nerium indicum 26, 288
Nerium tinctorium 212
New W. 106
New account of East-India . . . (Fryer) 31, 32
New Indian gardener (Speede) 284
Nicol G. 48, 49
Nicotiana spp. 62, 164, 202, 214, 215
Nilgiri Hills 118–19, 182, 329
cinchona 222, 223, 226, 227
coffee 252
Ootacamund gardens 107, 277, 278, 279, 281
potatoes 216
tea 238
W. McIvor 107, 226, 277
R. Wight 116
Nimmo J. 121, 197
Normansell H. T. 163
Norris Sir W. 165
North F. 159, 160, 184, 190
North M. 149, 166, 178, 181, 261, plate 24
Notes on . . . principal indigenous tonics of India (Waring) 221
nutmeg, *see* Myristica fragrans
Nuttall T. 137, 138, 303
Nyctanthes arbor-tristis 287
Nymphaea alba 127

Oakes Mr 273
Ocinum sanctum 283
oleander, *see* Nerium indicum
On the cultivation of tea . . . (Royle) 217
On the culture . . . of cotton (Royle) 256
On the growth of plants . . . (Ward) 319
Ondaatje W. C. 163
Onesicritus 2
Ootacamund *see* Nilgiri Hills
opium, *see Papaver somniferum*
Opuntia spp. 47, 202, 203, 208–11, 265, 275
orchids 93, 117, 131, 139, 144, 301, 303, 305, 328–9
Orford Hall 293
Oriental memoirs (Forbes) 183
Oriental scenery (Daniell) 296
Orta G. da 14, 15–16, 17, 33, 56, 212, 220, 264
Osborne Sir J. 296
O'Shaughnessy W. 221
Oud en nieuw Oost – Indien (Valentjin) 156
overland route to India 316–18
Ovington J. 31

Padbrugge R. 156
painters, Indian *see* artists
Pakistan 325
Pakistan Journal of Botany 323
palmyra palm, *see Borassus flabellifer*
Paludanus B. 18
Pandanus leram 47, 58, 62, 203
Pandanus tectorius 2, 26
Pandanus thwaitesii 156
Pantling Mr 152
Papaver somniferum 18, 240–4, 246
papaya, *see Carica Papaya*
Paphiopedilum fairrieanum 326
Pappe C. W. L. 96
Paradisus Batavus (Hermann) 157
Parel 268–9
Parkinson S. 159
Parks F. 181
Parry R. 149
Paxton J. 300, 305
Pearse Colonel 58
Peel Sir L. 273
Pemberton R. B. 135
Penang 63, 86, 204, 251
Pennant T. 159
pepper, see *Piper*
Peradeniya botanical garden 149, 162–9
Periplus of the Erythraean Sea 5, 56, 212
Peru 208, 212, 222, 223, 225, 226, 246, 247
Petiver J. 33–5, 156, 231, 232, 309, 310
Petre Lord 293, 298
Pharmacopoeia of India (Waring) 221
Phoenix dactylifera 19, 26, 57, 58, 59
photography 180
Phytographia (Plukenet) 35, 45
Picea smithiana 301, plate 8
picturesque in India 173–80
Picturesque tour along River Ganges . . . (Forrest) 175
Pigot G. 268
pineapple, *see Ananus comosus*
Pinetum Britannicum (Ravenscroft) plate 21
Pinjore 124
Pinus wallichiana 300
Piper spp
Ceylon 288
V. de Gama 12
R. Fitch 29
Malabar 19, 58, 74
P. Mundy 13
Pliny the Elder 6
Rome 7
W. Roxburgh 47, 202–3
L. di Varthema 14
Pistacia vera 2
Pitt T. 267, 270
Plantae Asiaticae rariores (Wallich) 87, 89–90, 115, 117, 149, 152, plates 8, 14
Plants of Coast of Coromandel (Roxburgh) 45–50, 68, 89, 148, 205, 254, plates 3, 4

Platanus orientalis 145, 278–9, plate 1
Pliny the Elder 5–7, 15, 56, 212
Plukenet L. 35, 45, 231
Poa annua 177
Polo M. 8–9, 14, 212, 259
Polunin O. 134
Polygonum spp. 307
Pondicherry 32, 119, 120, 121, 266
Portuguese settlements 12–18, 72, 154, 155, 201, 264
potato, *see Solanum tuberosum*
Potter G. 82, 85, 86, 88, 91
Potus Theae (Linnaeus) 233
Praecursores (Hooker & Thomson) 195
Prain D. 31, 103, 197
Prangos pabularia 110, 124
prickly pear, *see Opuntia*
Primula alta 171
Primula denticulata 306, plate 26
Primula eburnea 134, 135
Primula kingii 137
Primula prolifera 150
Primula sikkimensis 137, 141, plate 25
Primula sonchifolia 322
Primula spp. 306, 307
Prinsep J. 186, 187
Pritchett G. J. 225, 226, 228
Prodromus florae Nepalensis (Don) 56, 133
Prodromus florae Peninsulae Indiae Orientalis (Wight & Walker-Arnott) 116–17
Punjab plants (Stewart) 197

Raffles Sir S. 86, 88, 151, 190
Ram Das 149
Rambagh 260, 261
Ramnaut 150, 151
Randal Mr 35
Rauvolfia serpentina 329
Ravenscroft E. plate 21
Ray J. 23, 33, 35, 37
Redouté P.-J. plates 6, 15
Reede tot Drakenstein *see* Van Reede
Report on Caoutchouc of Commerce (Collins) 247
Repton H. 297
Rhea, *see Boehmeria nivea*
Rhododendrons 138, 139, 141, 154, 303–4
Rhododendron arboreum 71, 132, 148, 303, 304, 315
Rhododendron arboreum sub sp. *nilagiricum* 154
Rhododendron ciliatum 303
Rhododendron dalhousiae 303
Rhododendron formosum 303
Rhododendron fulgens 303
Rhododendron grande 137
Rhododendron falconeri 303
Rhododendron hodgsonii 144
Rhododendron imperator 171
Rhododendron maddenii 301, 303
Rhododendron nuttallii 138, 303
Rhododendron tanastylum 138
Rhododendron thomsonii 198
Rhododendrons of Sikkim-Himalaya (Hooker) 137, 144, 303
Rhyn W. ten 22
Ridley H. N. 251
Rig Veda 25, 258
Ripon Lady 273
Roberts E. 180, 181, 268
Roe Sir T. 259, 260
Rollisson nursery 93
Rome 5–7, 12
Rosa clinophylla plate 15
Rosa macrophylla plate 16
Roscoe W. 300, plate 10
Roses 260, 275, plates 15, 16
Ross R. 171
Roth A. W. 43, 191
Rottler J. P. 40–1, 89, 110, 159
Rough notes of a journey through the wilderness . . . (Wickham) 249, 250
Roxburgh B. 69
Roxburgh James 69
Roxburgh John 50, 61, 81, 82
W. Roxburgh 31, 43, 53, 79, 84, 99, 110, 116, 132, 161, 208, 212, 214, 294, plate 15
  F. Buchanan 71, 72, 73, 74, 77, 78
  Calcutta Botanic Garden 41, 52, 61–6, 234, 281, 298, 300
  coffee 252
  Flora Indica 67, 68–9, 118, 183, 196
  Hortus Bengalensis 95
  Indian artists 149, 192, 193, 324
  J. G. Koenig 40
  R. Kyd 59, 61
  medicinal plants 220, 221
  native fibres 205
  Plants of Coast of Coromandel 47–50, 68, 89, 90, 117, 148, plates 3, 4
  rubber 247
  Samalkot 47, 106, 202–3, 209, 252
  P. Sonnerat 120
  teak 253, 254
  transportation of plants 314, 315
  N. Wallich 81, 96, 97, 151
Roxburgh W. junior 63, 64, 65, 77
Royal Asiatic Society London vi 185, 187, 193, 216, 217
Royal Botanic Gardens Kew *see* Kew Gardens
Royal Society of Arts vi 205, 240
Royal Society of London vi 34, 46, 53, 54, 77, 159, 187
Royle J. F. 56
  Illustrations of botany of . . . Himalayan mountains 55, 118, 125, 127, 222, 235
  India House London 98, 191, 192, 256
  medicinal plants 109, 220, 222, 223, 224
  Royal Asiatic Society 216–17
  rubber 247
  Saharanpur Botanic Garden 88, 109–10, 126, 205, 220
  tea 235, 237, 239

transportation of plants 316, 317, 318, 319
Roystonea regia 101
rubber 246–51
Rubus wardii plate 20
Rumphius G.E. 23, 36–7, 90
Runjia 90, 118
Ruskin J. 181
Russell P. 43, 53, 89, 202
Sir W. Jones 56
J. G. Koenig 39, 40
Naturalist Madras Presidency 44–5
Roxburgh's *Plants of Coast of Coromandel* 45–9

Sabine J. 315
Saccharum officinarum 4, 175, 191, 202, 203, 208, 214, 215
Sadras 265
sago palm, *see Metroxylon*
Saharanpur Botanic Garden 108, 149, 282, 303
J. F. Duthie 102, 103, 111, 199
H. Falconer 110–1, 260
G. Govan 107–9, 222, 235
W. Jameson 111
G. King 101
J. F. Royle 88, 98, 109–10, 111, 126, 205, 220, 222, 235
St Helena 210, 268, 314
St Vincent botanical garden 54, 60, 162, 202, 312, 313
Salt H. plate 30
Samalkot 41, 47, 61, 106, 116, 120, 202, 203, 204, 252
Sambach R. 35
sandalwood, *see Santalum album*
Sander F. 326
Sansevieria zeylanica 204
Santalum album 18, 58, 74, 253
Santapau M. 324, 325
Saraca declinata 164
Saunders R. 135
Saussurea costus 6, 99
Schlagintweit brothers 129–30, 133, 169
Schlich W. 255
Schmid B. 118–19, 182
Scott D. 236
screw-pine, *see Pandanus tectorius*
Scully J. 133
Select views of India (Hodges) 173
Selection of Hexandrian plants (Bury) 68
Selection of rare and curious fruits . . . (Bennett) 164
Sen Gupta J. C. 324
Serampore 66, 67, 81, 149, 273
Sertum orchidaceum (Lindley) 131
Sezincote 296–7, plate 32
Shah Jahan Emperor 145, 260, 263
Sharma N. K. 134, 306
Sheh Singh 202
Shepherd J. 300
Sherard J. 292, 293, 298
Sherard W. 157, 291, 293
Sherriff G. 135–7, 169, 305, 306, 322
Sherwill W. S. 242, 243
Shorea robusta 123
Shuter J. 43, 116
Sibthorp J. 241
Sikandra 260, 261, 262, 263
Sikkim 130, 133, 135, 137, 138–41, 195, 227, 228, 248, 303
silk 207–8, 231, 240
Simla 125, 126, 176, 199, 216, 277
Sinclair Sir J. 202
Sinclair J. V. E. 182
Singapore 86, 251
Singh B. 132
Singh Ranjit 125, 126, 129
Sircar M. L. 189
sisal, *see Agave*
Sloane Sir H. 209, 231
Smith C. 61–2, 63, 86, 184, 204, 211
Smith H. 179, 180
Smith Sir J. E. plate 8
Sir J. Banks 40
F. Buchanan 71, 75, 76, 77, 78, 79, 80, 132
Flora Graeca 241
T. Hardwicke 303, 304
C. Linnaeus 54, 116
W. Roxburgh 48, 64
N. Wallich 300
Smith M. R. 67, 247
Snelling L. 134, 136, plate 28
Societies 185–6; *see also* Agricultural and Horticulties Societies; Asiatic Society of Bengal
Solander D. 45
Solanum tuberosum 41, 113, 131, 135, 201, 215–16, 277
Solvyns F.B. 174–5
Soma 25
Sonnerat P. 120
Sorbus cashmiriana 301
Sowerby J. 46
Soymida febrifuga 221
Specimens of nature printing . . . (Smith) 179, 180
Speede G. T. F. 96, 284
spices 5–6, 7, 8, 12, 13, 14, 19, 29, 86; *see also individual spices*
Spicilegium Neilgherrense (Wight) 118, 119
spikenard, *see Nardostachys grandiflora*
Spiti Valley 125
Splinter M. 22
Spofforth 298
Spring Grove 211, 294
Sproat R. 211
Spruce R. 224, 225, 226, 228, 229, 246, 247
Sri Lanka *see* Ceylon
Stainton J. D. A. 134
Stanmore Hill 294
Stavorinus J.S. 260
Steam ships 217, 316
Stearn W. T. 6, 137

Stellaria decumbens 123
Stemona tuberosa 47, 298
Stevens T. 29
Stewart J. L. 197
Stoves, *see* conservatories
Strabo 2, 4, 5
Strachey H. 129
Strachey Sir R. 112, 129, 191
Strychnos colubrina 155
sugar, *see Saccharum officinarum*
Sullivan J. 277, 278
sunn hemp, *see Crotalaria juncea*
Surat 30, 31, 35, 259, 260, 266, 267
Swainson W. 97
Swartz O. P. 40
sweet potato, *see Ipomoea batatas*
Swietenia mahogoni 202, 252
Swinden T. 66
Syen A. 22, 23, 156, 157, 288
Sykes W. H. 121, 134, 191, 193
Symes M. 71, 72, 148
Systematic list of trees . . . of Bombay (Talbot) 197
Syzygium aromaticum 6, 12, 29, 58, 59, 63, 86, 184, 203, 204
Syzygium jambos 16

Taj Mahal 260–2
Talbot W. A. 197
Talipot palm, *see Corypha umbraculifera*
tanner's bark 292
Taprobane *see* Ceylon
Tamarindus indica 201
Tavernier J. 32, 259
tea, *see Camellia sinensis*
teak, *see Tectona grandis*
Tectona grandis 58, 62, 63, 85, 99, 252, 253, 254, 255
Tennant *Sir* J. E. 163
Tennant W. 148
Tentamen florae Napalensis illustratae (Wallich) 86, 117, 128, 132, 133, 149, 151, 302
Terminalia catappa 101
Terry E. 30–1, 201, 215, 251, 259, 260
Thalictrum foliosum 221
Theobroma cacao 167, 202, 252
Theophrastus 2–3, 5, 6, 56
Thesaurus Zeylanicus (Burman) 37, 84, 158, 160, 290
Thespesia populnea plate 24
Thévenot J. de 259, 260
Thiselton-Dyer Sir W. 102, 121, 249, 251, 252
Thomas G.S. 297
Thomson T. 31, 98, 117, 118
  Calcutta Botanic Garden 181
  Flora of British India 194, 195
  Flora Indica (with Hooker) 142
  J. D. Hooker 139, 141, 142, 170, 194, 195
  Ladakh 129, 138
  Saharanpur Botanic Garden 100
Thorndon Hall 293
Thunberg C. P. 159
Thunbergia laurifolia 111
Thwaites G. H. K. 164–6, 174, 196, 250, 306, 318
Tibet 126, 129, 133, 135, 139, 141, 233, 234
Tilman H. W. 134
tobacco, *see Nicotiana*
Torrens H. 187
Tractado de las drogas . . . (Acosta) 16, 17
Traité de la culture du Nopal (Menonville) 208
Tranquebar 32, 39, 40, 41
Transactions of Medical and Physical Society 187
transportation of plants 309–22
Trapa bispinosa 30
Trees of Madras Presidency (Beddome) 121
Trimen H. 165, 166–7, 196, 213, 250, 251, 325
Tropical planting and gardening (Macmillan) 284
Tughlaq 259
Turner S. 135
Twenty-four views in St Helena (Salt) plate 30
Twining T. 148
Typhonium divaricatum 294
Tytler J. 236

Ulloa A. de 222
Upton Park 293
Useful plants of India (Drury) 256

Valentia Lord 63, 148, 159, 183, 184, 201
Valentijn F. 156, 265
Valle P. della 32–3
Vallet P. 145
Van Reede tot Drakenstein 8, 10, 15, 19–24, 32, 33, 35, 90, 156, 220, 288, 309
Vanda coerulea 303, 327, 328
Vanda tessellata plate 11
Vandes Comtesse 298
Varthema L. di 14
Vatsyayana 258
Venezuela 222, 247
Veitch J. 301
Vere J. 298, 305
Vetch Lieut. 247
Vicary N. 98
Victoria amazonica 101
Victoria and Albert Museum London 193
Views in the Himala Mountains (Fraser) 175, plate 23
Views in India (Barron) 278
Views in India (White) 176, 278
Vigne G.T. 125
Viridarium Orientale 20
Vishnupersaud 87, 90, 128, 149, 151
Voigt J.O. 93, 95, 284
Voyage to the islands Medera Barbados . . . (Sloane) 209

Wager Sir C. 303
Waldemar Prince 129, 164, 176
Waldo D. 35
Walker A. 162
Walker G.W. 159, 163

Walker-Arnott G.A. 116
Wall G. 165
Wallich N. 31, 163, 170, 182, 183, 282
Asiatic Society of Bengal 53, 185, 189
F. Buchanan 80
Burma 86, 253
Calcutta Botanic Garden 78, 81–98, 106, 124, 221, 281, 284, 324
W. Carey 82
coffee 252
East India Company 88–9, 90
Flora Indica 69, 85
W. Griffith 91, 93, 95, 96, 237
J. Hare 82, 83
J. D. Hooker 142
India Museum 89
Indian artists 83, 90, 149–51, 192, plates 7–9, 14, 26
J. McClelland 91, 96, 237
Malaya 86
plant collections 56, 88, 110, 127, 141, 191, 298, 300, 303, 305, 306, plate 10
Sir S. Raffles 86
J. P. Rottler 41
W. Roxburgh 81, 96, 97
tea 91, 235–7, 240
transportation of plants 315, 316, 319
J. Wight 96, 97
Wanderings of a pilgrim in search of the picturesque (Parks) 181
Ward N. 91, 309, 319
wardian case 319–20
Warming E. J. 221
Warner R. 293
Watson Capt. 253
Watson J. F. 218
Watson J. G. 88, 163, 169
Watt G. 189, 201, 245, 256, 306, 327
Watts J. 291, 292
Waugh Major 106
Wealth of India 327
Weddell H. A. 222, 223, 224, 225, 226
Weir J. 224, 225, 229
Wellesley Lord 63, 72–3, 79, 183, 215, 271
Barrackpore 75, 77, 272
F. Buchanan 75, 76
Indian drawings 50, 149
Wentworth 305, 315
Westland A. B. 262
White G. F. 176, 281
Whitehill C. 251
Wickham H. A. 249, 250
Wight R. 98, 115–18, 221, 320
Agri-Horticultural Society of Madras 107, 216
R.H. Beddome 121
Govindoo 118, 121, 149
Madras Journal of Literature and Science 187
Naturalist Madras Presidency 43, 89, 116
plants presented to Kew 141
publications 116–18, 163, 164, 165, 197
N. Wallich 96, 97, 237
Wild flowers in southern and western India (Brown) 183
Wilder R. 207
Wilkins Sir C. 190
William of Orange (later William III) 288, 291
Williams L. H. J. 134
Williamson Mr 113
Willingdon Lady 281
Willis J. C. 167, 169
Winchester Bishop 300
Winterbottom J.E. 129
Wissett R. 205
Wollaston A. F. R. 133, 306
Wood Sir C. 189, 194
Woodfordia 2
Woodrow G. M. 271, 275, 281, 284
Woodward J. 309
Wormleybury 298, 299
Wrightia tinctoria 212
Wycherley P. R. 249

Yonge *Sir* G. 202, 313, 314
Young C. W. 252

Zain al-Din 149
Zea mays 202
Zingiber officinale 8, 9, 12, 18, 29, 164